Günter Hofstetter
Herbert A. Mang

**Computational Mechanics
of
Reinforced Concrete Structures**

Fundamentals and Advances in the Engineering Sciences

Grundlagen und Fortschritte der Ingenieurwissenschaften

Prof. Dr.-Ing. Dr.-Ing. E. h. *Wilfried B. Krätzig*, Ruhr-Universität Bochum
Prof. em. Dr.-Ing. Dr.-Ing. E. h. *Theodor Lehmann*[†] Ruhr-Universität Bochum
Prof. Dr.-Ing. Dr.-Ing. E. h. *Oskar Mahrenholtz*, TU Hamburg-Harburg
Prof. Dr. *Peter Hagedorn* TH Darmstadt

Konvektiver Impuls-, Wärme- und Stoffaustausch
von Michael Jischa

Einführung in Theorie und Praxis der Zeitreihen- und Modalanalyse
von Hans G. Natke

Mechanik der Flächentragwerke
von Yavuz Basar und Wilfried B. Krätzig

Introductory Orbit Dynamics
von Fred P. J. Rimrott

Festigkeitsanalyse dynamisch beanspruchter Offshore-Konstruktionen
von Karl-Heinz Hapel

Abgelöste Strömungen
von Alfred Leder

Strömungsmechanik
von Klaus Gersten und Heinz Herwig

Konzepte der Bruchmechanik
von Reinhold Kienzler

Dünnwandige Stab- und Stabschalentragwerke
von Johann Altenbach, Wolfgang Kissing
und Holm Altenbach

Thermodynamik der Strahlung
von Stephan Kabelac

Berechnung von Phasengleichgewichten
von Ralf Dohrn

Simulation von Kraftfahrzeugen
von Georg Rill

Computational Mechanics of Reinforced Concrete Structures
von Günter Hofstetter und Herbert A. Mang

Günter Hofstetter
Herbert A. Mang

Computational Mechanics of Reinforced Concrete Structures

CIP-Titelaufnahme der Deutschen Bibliothek

Vieweg is a subsidiary company of Bertelsmann Professional Information.

All rights reserved
© Friedr. Vieweg & Sohn Verlagsgesellschaft mbH, Braunschweig/Wiesbaden, 1995

No part of this publication may be reproduced, stored in a retrieval system or transmitted, mechanical, photocopying or otherwise, without prior permission of the copyright holder.

Printed and bound by W. Langelüddecke, Braunschweig
Printed on acid-free paper
Printed in Germany

ISBN 3-528-06390-4

Preface

Almost three decades have passed since publication of the pioneering work by Ngo and Scordelis [Ngo (1967)] on finite element analyses of reinforced concrete beams. This period of time is characterized by extensive research activities in the fields of constitutive modelling of plain and reinforced concrete and of the development of powerful algorithms for nonlinear finite element analyses of reinforced concrete structures. These research endeavors have resulted in an impressive increase of the scientific level of material modelling of plain and reinforced concrete and of the efficiency of software for nonlinear finite element analyses of reinforced concrete structures.

The impact of these research activities on engineering practice, however, is less impressive. This situation was addressed by Scordelis, in 1985, in a state-of-the-art report [Scordelis (1985)]: "While extensive research on finite element analysis of reinforced concrete structures has been conducted over the past eighteen years, its implementation in the design of actual structures has been disappointing. Present designs are still generally based on determining internal forces and moments by elastic analyses in which the reinforced concrete or prestressed concrete system is assumed to be uncracked, homogeneous, isotropic and linearly elastic."

In spite of the growing awareness of the inadequacy of these assumptions and the increasing acceptance of nonlinear finite element analysis of reinforced concrete structures during the last decade, the dominance of traditional methods of analysis of such structures has remained. From the standpoint of engineering practice, this is understandable. The motivation for replacing conventional methods of analysis of reinforced concrete structures by mechanically more sophisticated and considerably more complicated analysis methods will remain small as long as the expected increase in quality of the design is not commensurate to the increase in mechanical sophistication and computational complexity. Nevertheless, nonlinear finite element analyses of reinforced concrete structures have gained momentum during the last decade. At present, this mode of analysis is mainly used as a tool for the assessment of the structural safety of non-standard reinforced concrete structures.

The importance of the mentioned class of finite element analyses as a powerful research tool in the field of computational mechanics of reinforced concrete structures is undisputed. The two constituent parts of this scientific field are constitutive modelling of reinforced concrete and nonlinear structural analysis of reinforced concrete structures. In the opinion of the writers these two constituent parts are drifting apart. The conclusions of a survey paper on "Nonlinear finite element analysis of reinforced and prestressed concrete structures" [Mang (1991)] contain a remark on this observation.

This paper was presented at the Second International Conference on Computer Aided Analysis and Design of Concrete Structures. One of the main goals of this scientific conference was to bridge the gap between constitutive modelling and structural analysis. Although this goal was not reached, there was a consensus of opinion that computational mechanics of reinforced concrete structures rests on both constitutive modelling and structural analysis. Constitutive modelling without application of the developed material models to structural analysis runs the risk of becoming an end in itself. Structural analysis based on obsolete, mechanically inconsistent material models may yield useless results.

The writers feel committed to the concept of a synthesis of constitutive modelling of reinforced concrete and of structural analysis of reinforced concrete structures. This commitment has led to the title of the book: *Computational Mechanics of Reinforced Concrete Structures*. It is tacitly understood that the term "computational mechanics" includes constitutive modelling, with special emphasis on algorithmic aspects, and structural analysis.

The book consists of four chapters. Their titles are:

1. *Materials, Physical Phenomena, Experiments*

2. *Mathematical Models*

3. *The Finite Element Method for Reinforced Concrete and Prestressed Concrete Structures*

4. *Application to Engineering Problems*

The experiments described in chapter 1 provide the basis for the formulation of mathematical models for the simulation of the material behavior. These models are presented in chapter 2. In chapter 4 they are applied to the analysis of reinforced and prestressed concrete structures by means of finite element methods described in chapter 3. The numerical analyses reported in chapter 4 have been carried out at the Institute for Strength of Materials of the University of Technology of Vienna during the last fifteen years.

There is a large body of literature relevant to the topic of the book. It is growing at a great pace. Hence, a selection of subtopics to be treated in the book was necessary. This selection was based on two criteria. The first criterion was the possibility of a consistent and well-balanced treatment of a subtopic. In this context, the terms "consistent" and "well-balanced" refer to the four chapters of the book. The purpose of this criterion was to counteract the aforementioned drifting apart of the two constituent parts of continuum mechanics of reinforced concrete. The second criterion was the pertinent research experience of the writers. These two criteria have resulted in the restriction to static, deterministic mathematical models.

Basic courses on mathematics, mechanics, material sciences, reinforced concrete and finite element methods, representing standard parts of civil engineering curricula at institutions of higher learning, provide the necessary grounding to benefit from the book.

The writers are indebted to the former doctoral students H. Floegl and H. Walter and to the former doctoral students and present colleagues J. Eberhardsteiner and G. Meschke for many helpful discussions and for the permit to use material from their dissertations.

Special thanks go to O. Graf for the drawing of the figures and to Ms. E. Koglbauer for the cooperation in producing the LaTeX-file for the camera-ready manuscript.

The writers also wish to thank the publishers, Vieweg-Verlag, Braunschweig/Wiesbaden, and the editors of the book series "Fundamentals and Advances in the Engineering Sciences", W.B. Krätzig, O. Mahrenholtz and P. Hagedorn, for the good cooperation and the patience in connection with the delayed completion of the manuscript. Last, but not least, thanks are due to the late Th. Lehmann, formerly co-editor of the book series, for his part in the realization of the book project.

Vienna, February 1995

G. HOFSTETTER and H. A. MANG

Contents

1 Materials, Physical Phenomena, Experiments   **1**
- 1.1 Introduction 1
- 1.2 Concrete 2
  - 1.2.1 Introduction 2
  - 1.2.2 Uniaxial Behavior 3
  - 1.2.3 Multiaxial Behavior 10
  - 1.2.4 Time-dependent Behavior of Concrete 19
- 1.3 Reinforcing and Prestressing Steel 27
  - 1.3.1 Material Properties of Reinforcing Steel 27
  - 1.3.2 Material Properties of Prestressing Steel 29
  - 1.3.3 Cyclic Loading 30
  - 1.3.4 Effect of Low and High Temperatures 31
  - 1.3.5 Relaxation 31
- 1.4 Reinforced Concrete 32
  - 1.4.1 Bond 32
  - 1.4.2 Aggregate Interlock 38
  - 1.4.3 Dowel Action 40
  - 1.4.4 Experiments on Reinforced Concrete Specimens 41

2 Mathematical Models   **46**
- 2.1 Introduction 46
- 2.2 Failure Criteria 49
  - 2.2.1 Introduction 49
  - 2.2.2 Ultimate Strength Curve for Biaxial Stress States 50
  - 2.2.3 Ultimate Strength Surface for Triaxial Stress States 51
- 2.3 Time-Independent Material Models for Concrete 54
  - 2.3.1 Linear-Elastic Fracture Models 54
  - 2.3.2 Nonlinear-Elastic Fracture Models 55
  - 2.3.3 Models Based on the Theory of Plasticity 79
  - 2.3.4 Plastic Fracturing Theory 110
  - 2.3.5 Endochronic Theory of Plasticity 111
- 2.4 Material Models for the Time-Dependent Behavior of Concrete . . . 112
  - 2.4.1 Introduction 112
  - 2.4.2 Material Models for Shrinkage of Concrete 114
  - 2.4.3 Models for Creep of Concrete 117
  - 2.4.4 Material Models for Concrete Aging 128
  - 2.4.5 Discussion of Creep and Shrinkage Laws 130
- 2.5 Material Models for Reinforcing and Prestressing Steel 131
  - 2.5.1 Time-Independent Material Models 131
  - 2.5.2 Time-Dependent Material Model for Prestressing Steel . . . 133

	2.6	Models for Consideration of the Interface Behavior	134
		2.6.1 Models for Consideration of Aggregate Interlock	134
		2.6.2 Models for Consideration of Bond	136

3 The Finite Element Method for RC and PC Structures — 139

- 3.1 Review of the Finite Element Method — 139
  - 3.1.1 Introduction — 139
  - 3.1.2 Principle of Virtual Displacements — 139
  - 3.1.3 Linearization of the Mathematical Formulation of the Principle of Virtual Displacements — 142
  - 3.1.4 Finite Element Discretization — 146
  - 3.1.5 Newton Iteration — 148
  - 3.1.6 Alternatives to the Newton Iteration — 152
- 3.2 Update of the Stresses — 156
  - 3.2.1 Introduction — 156
  - 3.2.2 Nonlinear Elastic Constitutive Models — 157
  - 3.2.3 Hypoelastic Constitutive Models — 158
  - 3.2.4 Elastic-Plastic Constitutive Models — 160
  - 3.2.5 Detection of Material Failure and Modelling of Post-Failure Material Behavior — 170
  - 3.2.6 Implementation of Material Models — 197
- 3.3 Formulations for Time-Dependent Problems — 199
  - 3.3.1 Introduction — 199
  - 3.3.2 Algorithms for the Computation of Creep Strains — 202
  - 3.3.3 Algorithm for Consideration of Relaxation of the Prestressing Steel — 208
- 3.4 Finite Element Modelling of Reinforced and Prestressed Concrete — 209
  - 3.4.1 Introduction — 209
  - 3.4.2 Finite Elements for Concrete — 211
  - 3.4.3 Representation of the Reinforcing Steel — 238
  - 3.4.4 Representation of Tendons — 243
  - 3.4.5 Models for Consideration of the Interface Behavior — 258

4 Application to Engineering Problems — 267

- 4.1 Introduction — 267
- 4.2 Nonlinear FE-Analyses of 3D Reinforced Concrete Structures — 268
  - 4.2.1 Short Reinforced Concrete Cantilever — 268
  - 4.2.2 Reinforced Concrete Cylinder Subjected to a Concentrated Compressive Force — 276
- 4.3 Nonlinear FE-Analyses of RC Panels, Slabs and Shells — 282
  - 4.3.1 Reinforced Concrete Cooling Tower — 283
- 4.4 Nonlinear FE-Analyses of PC Panels, Slabs and Shells — 290
  - 4.4.1 Prestressed Concrete Slab — 291
  - 4.4.2 Prestressed Concrete Shell with Edge Beams — 299
  - 4.4.3 Prestressed Concrete Secondary Containment Structure — 307

A	**Notation**	**321**
	A.1 Displacements	321
	A.2 Strains	321
	A.3 Stresses	322
	A.4 Moduli	323
	A.5 Miscellaneous	324
B	**Bibliography**	**327**
C	**Index**	**355**

1 Materials, Physical Phenomena, Experiments

1.1 Introduction

In general, the mechanical behavior of reinforced concrete (RC) and prestressed concrete (PC) structures is very complex. It is characterized by material and, to less extent, geometric nonlinearity. Depending on the load intensity, the degree of nonlinearity may be significant. In order to gain insight into the physical behavior of the individual material components (concrete, reinforcing steel, prestressing steel) and of the composite materials (reinforced concrete, prestressed concrete) it is useful to begin with a survey of experimental investigations.

Much effort has been put into experimental investigations of the material behavior of concrete with special emphasis on the modes of material failure. Such investigations are often rather complicated because of the strong dependence of the behavior of concrete on the type of loading. Concrete under tension behaves completely different than under compression. The behavior of concrete under uniaxial compression differs significantly from the one under biaxial or triaxial compression. If tensile stresses are predominant, failure of concrete will be caused by cracking. In this case concrete behaves like a brittle material. If compressive stresses are dominating, concrete will fail by crushing. In this case concrete behaves like a ductile material. In general, neither tension nor compression is prevailing. Therefore, the behavior of concrete under mixed states of stress must be investigated.

Load-controlled experiments are characterized by a well-defined, homogeneous stress field. Thus, the stresses are known and the strains are measured. In practice, there is always some unwanted interaction between the specimen and the testing device. Especially in the vicinity of the interfaces of the specimen and the platens of the testing device this may have a considerable influence on the experimental results. It is important to reduce and to control the deviations from a homogeneous stress field. Load-controlled experiments allow studying the material behavior up to the ultimate strength. Deformation-controlled experiments also permit investigation of the behavior of the material under strains which are larger than the ones corresponding to the ultimate strength.

In addition to consideration of the short-term behavior of concrete, the time-dependent characteristics of concrete must be taken into account. Experimental results show time-dependent deformations of concrete under sustained loading. This effect is termed as creep. Another reason for time-dependent strains in concrete is the drying process. The respective phenomenon is called shrinkage. The effect of aging contributes to the complexity of concrete behavior. It may have a significant influence not only on the compressive strength but also on creep deformations.

In contrast to concrete, the mechanical behavior of reinforcing and prestressing steel is characterized by uniaxial states of stress. Compared to concrete, the mechanical properties of steel do not vary strongly. Steel is a ductile material. It is able to sustain large plastic deformations under tension as well as under compression. A characteristic feature of steel is the hysteretic behavior under cyclic loading.

After introductory comments on the component materials concrete and steel a few remarks on the composite material reinforced concrete will be made. An important aspect of the behavior of reinforced concrete is the behavior at the interface between concrete and reinforcing steel. It is governed by bond and dowel action. Bond enables the transmission of forces from concrete to steel. Dowel action contributes to the transmission of shear forces across cracks intersected by reinforcing bars.

The experiments described in this chapter serve as the basis for the formulation of mathematical models, which will be presented in chapter 2. Implementation of these models into a finite element code provides an analysis tool for the purpose of more realistic predictions of the complex behavior of reinforced and prestressed concrete structures. Nevertheless, the availability of such a tool does not imply that conventional modes of analysis no longer have an adequate place in the design of such structures.

1.2 Concrete

1.2.1 Introduction

Concrete consists of hardened cement paste in which aggregates are embedded. Hence, it is a heterogeneous material.

Material models can be formulated at the microscopic, the mesoscopic and the macroscopic level. The modelling of concrete at the microscopic level is beyond the scope of this book. A survey concerning this mode of modelling of concrete can be found in [Wittmann (1977), Wittmann (1985)].

Material models at the mesoscopic level help to gain insight into the origin and nature of the non-linear behavior of concrete. At this level, hardened cement paste, representing the matrix of the material, can be described as a viscoelastic material, consisting of a load-bearing skeleton and of pores with a wide range of pore sizes. The pore size distribution and the porosity mainly depend on the water/cement ratio and the degree of hydration [Wittmann (1985)]. With increasing porosity higher stress concentrations occur in the load-bearing skeleton. Compared with the matrix, the aggregates usually have a higher strength. Their mechanical behavior is approximately linear-elastic. Hence, in general, time-dependent deformations are caused by the matrix. Both the stiffness of the aggregates and the porosity of the hardened cement paste have an influence on the internal stress distribution. Because of the different properties of the basic constituents of concrete, cracks are formed at matrix-aggregate interfaces and within the matrix in consequence of drying.

1.2 Concrete

The material models used in structural engineering represent macroscopic material models. They are characterized by consideration of heterogeneities and structural defects in an average sense. From this point of view concrete is regarded as a nonlinearly behaving homogeneous material.

1.2.2 Uniaxial Behavior

Compressive Behavior

Stress-strain curves for concrete under uniaxial compression are commonly obtained from tests with cylindrical, cubic or prismatic specimens at an age of 28 days. Typical stress-strain curves and the corresponding relationship between stress and volumetric strain ε_V^C are shown in Fig.1.1 [Hognestad (1955), Kupfer (1969)].

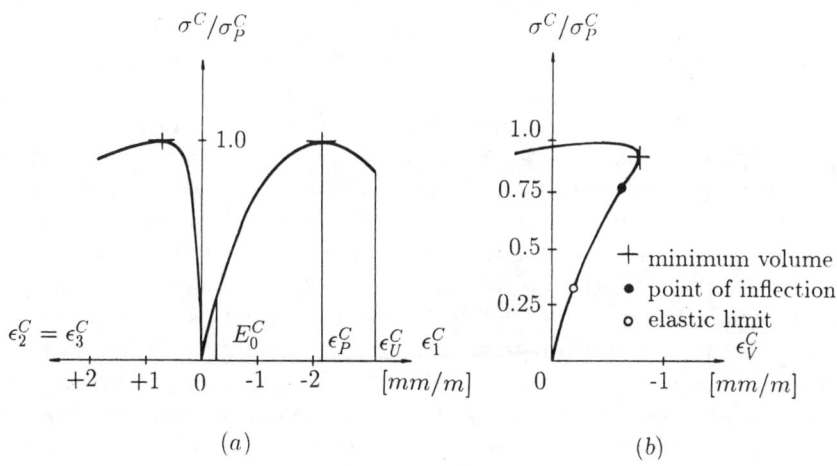

Figure 1.1: Concrete under uniaxial compression: (a) typical stress-strain curves, (b) stress - volumetric strain diagram

The peak stress or uniaxial compressive strength σ_P^C, the corresponding strain ε_P^C, the initial tangent modulus E_0^C and the ultimate strain at failure ε_U^C can be regarded as characteristic values of the uniaxial stress-strain curve in compression [Carreira (1985)]. The values of σ_P^C are commonly within the range of -20 to -75 MPa, with -20 to -25 MPa for low-strength concrete, -35 to -40 MPa for normal-strength concrete and -60 to -75 MPa for high-strength concrete, respectively. Values for E_0^C are ranging from $20\,000$ MPa to $40\,000$ MPa. The strain at σ_P^C is within the range of 0.2 to 0.4 %.

The value of σ_P^C depends on the shape of the specimen. This effect can be explained by frictional restraint at the interfaces between the specimen and the platens of the testing device. It is more pronounced for a cube than for cylinders or prisms with a ratio of height to diameter of commonly 2.5. The latter are employed in order to minimize the mentioned restraint effects [Schickert (1980), Kotsovos (1984a), Bakht (1989)]. In addition, for specimens of the same shape a decreasing uniaxial compressive strength is observed with increasing size [Bonzel (1959), Schickert (1980)]. This is a typical size effect.

The "true" compressive strength, obtained after elimination of the influence of the testing method and of the geometric form of the specimen on σ_P^C, has been extracted from experiments with different testing conditions [Schickert (1980)].

The stress-strain curve and the stress - volumetric strain curve in Fig.1.1 can be subdivided into several regions, characterized by different material behavior [Slate (1963), Hsu (1963), Shah (1966), Shah (1968), Wang (1978), Schickert (1980), Carrasquillo (1981), Ziegeldorf (1983), Smadi (1989)]. Before application of the load to the concrete, randomly orientated cracks can be observed mainly at the interfaces between coarse aggregates and the matrix. They are primarily caused by tensile stresses resulting from volume changes in consequence of hydration and shrinkage of the matrix [Hsu (1963), Smadi (1989)]. Depending on the strength of concrete, up to 30 - 70 % of σ_P^C there is only a small increase in pre-existing bond cracks (first stage of microcracking). Hence, the stress-strain diagram is approximately linear. Because of the composite structure of concrete, even the strain field generated by a uniaxial compressive test is characterized by local strain concentrations in the vicinity of existing flaws. The latter are sources of load-induced cracking. Beyond the elastic limit bond cracks are beginning to propagate and the width of existing cracks is increasing (second stage of microcracking). Therefore, the stress-strain curve is becoming nonlinear. The rate of volume decrease is increasing up to the point of inflection on the σ^C-ε_V^C diagram. Within this region the cracks are stable up to the so-called discontinuity stress [Su (1988)]. For normal-strength concrete the discontinuity stress occurs at approximately 70 to 75 % of the ultimate strength and for high-strength concrete at approximately 85 to 90 %. Upon further loading, matrix cracks are increasing significantly. They are joining with bond cracks forming continuous cracks (third stage of microcracking). Their primary orientation is parallel to the direction of the stress. For high-strength and light-weight concrete also cracks passing through the aggregates are observed. At approximately 88 % to 95 % of the ultimate strength, the minimum of the volumetric strain is observed.

In the case of load-controlled experiments, continuous cracking accompanied by increasing volumetric strain finally leads to failure of the specimen at σ_P^C. Fig.1.2 contains illustrations of crack patterns, obtained from x-ray examination, for various levels of uniaxial loading [Liu (1972)].

In the case of displacement-controlled experiments beyond the uniaxial ultimate strength a gradual decrease in load carrying-capacity, at increasing strain, can be observed. Eventually, failure because of crushing occurs at the ultimate strain ε_U^C. The descending part of the stress-strain curve is called strain-softening branch. Its shape strongly depends on the testing device and on the frictional restraint between the platens of the testing device and the specimen. For these reasons it could be argued that strain softening is not a material phenomenon but merely the result of the behavior of the testing device [Kotsovos (1984b)]. In order to refute this

1.2 Concrete

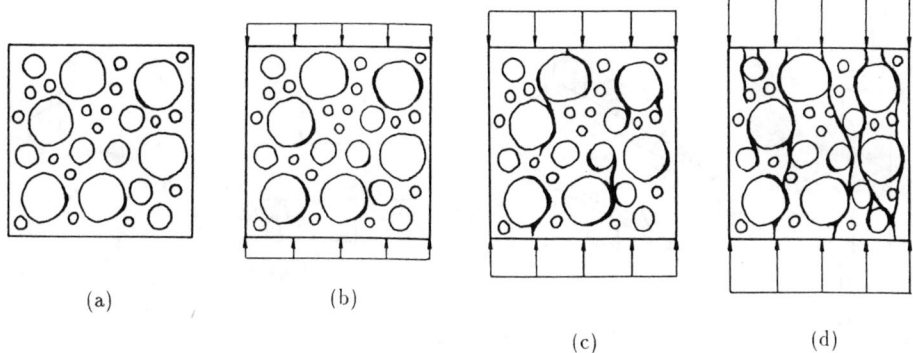

Figure 1.2: Crack patterns for various degrees of uniaxial loading

argument, a testing method characterized by almost complete elimination of the end-restraints was developed [Shah (1987)]. Using this method, a pronounced strain-softening region was observed. It is characterized by uniformly distributed matrix cracking. The primary direction of the cracks is the one of the applied load. In contrast to testing procedures with pronounced frictional end restraints, no inclined cracks indicating shear-type cracking were found. Nevertheless, the shape of the stress-strain curve in the strain-softening region depends on the gage length and the size of the specimen.

At all stages the total amount of matrix cracking is considerably less than that of bond cracking. Hence, bond is the weakest link of concrete [Hsu (1963), Shah (1966)]. Since bond cracks are primarily formed at the interfaces between coarse aggregates and the matrix, σ_P^C is also influenced by the size and the proportion of such aggregates.

Compared with normal-strength concrete, high-strength concrete is characterized by a smaller amount of bond cracking, less inelastic deformation and an initiation of the second and third stage of microcracking at higher stress levels [Smadi (1989)].

Usually, standard material tests are conducted quasi-statically with a strain rate of the order of 10^{-6} to 10^{-5} per second. Experiments simulating dynamic loading conditions with strain rates up to 0.2 per second [Rasch (1962), Dilger (1984), Soroushian (1986)] show an increase in strength, with increasing strain rate, of up to 35%.

Tensile Behavior

Although concrete is primarily designed to withstand compression, its tensile properties are important for a complete description of the material behavior. The uniaxial tensile strength σ_T^C can be determined either by direct or indirect methods (Fig.1.3). The former are characterized by applying the tensile force to the specimen (Fig.1.3(a)). Hence, σ_T^C is obtained directly from the test [Heilmann (1969), Gopalaratnam (1985), Carreira (1986), Reinhardt (1986), Zhen (1987), Ansari (1987),

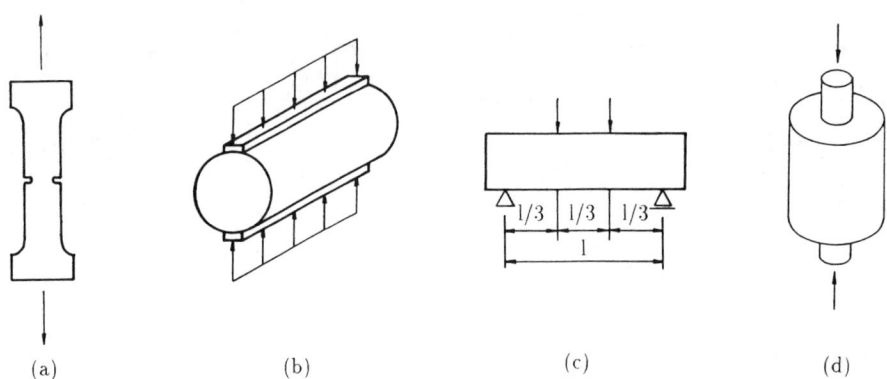

Figure 1.3: Direct and indirect tension tests: (a) direct tension test, (b) bending test, (c) splitting cylinder test, (d) double punch test

Nianxiang (1989)]. The latter are based on stress fields generated by tests other than direct tension tests. From the state of stress at failure the tensile strength can be calculated by means of the theory of elasticity or by limit analysis. Popular indirect tension tests are the splitting cylinder test (Fig.1.3(b)) [Bonzel (1964)] and the bending test (Fig.1.3(c)) [Bonzel (1963), Raphael (1984)]. In the splitting cylinder test a compressive load along two opposite generators of a cylindrical specimen is applied. It can be shown analytically [Timoshenko (1943)] that this type of loading yields an almost uniform distribution of the tensile stresses in the vertical plane of symmetry. In the bending test, pure bending at the midsection of a simple supported beam of span l is achieved by applying two point loads at $l/3$ and $2l/3$, respectively. An alternative to the splitting cylinder test is the double punch test [Chen (1980), Bortolotti (1988), Marti (1989)], in which a compressive load is applied along the axis of a cylindrical specimen by means of one steel punch each on the top and the bottom surface of the specimen (Fig.1.3(d)). In the double punch test uniformly distributed stresses are obtained for all diametrical planes. According to [ACI-Committee 224], the flexural tensile strength, also denoted as modulus of rupture, is 40 to 80 % higher than the splitting tensile strength.

It is easier to perform indirect tests than direct tests. Indirect tests yield more reliable results. Remarkably, up to now no standard direct tension test exists [ACI-Committee 224]. The main drawbacks of direct tests are their sensitivity with respect to eccentric application of the tensile force and to shrinkage. If specimens are not stored under wet conditions until testing, shrinkage will cause partial cracking. Hence, the tensile strength of field concrete cores is usually lower than that of laboratory cores [Raphael (1984)]. Shrinkage cracks occur at the surface of the specimen. Consequently, these surface cracks have a greater influence on the result from a direct test than on the one from a splitting cylinder test, where surface cracks are likely in the compressive region. However, indirect tension tests only permit determination of σ_T^C. Stress-strain curves can only be obtained from direct tension

1.2 Concrete

tests.

A load-controlled direct tension test yields a linear stress-strain curve terminated by brittle failure [Carreira (1986)]. A displacement-controlled test results in a σ-ε diagram deviating from linearity before σ_T^C is reached. The post-peak part of the diagram characterizes the strain-softening region. A typical stress-elongation curve for concrete subjected to uniaxial tension is shown in Fig.1.4(a).

Up to about 60 % of σ_T^C the stress-elongation relationship is linear [Reinhardt (1986), Zhen (1987)]. Pre-existing bond cracks are distributed randomly throughout the specimen. Beyond the elastic limit bond cracks are extending. The stress-strain relationship becomes nonlinear. Shortly before the peak stress is reached, additional microcracks accumulate in a weak cross-section of the specimen, forming a microcrack band (Fig.1.4(b)) [Ansari (1987), Cedolin (1987), Duda (1991)]. It is characterized by a strain which is considerably larger than the strains in the remaining part of the specimen. When the peak stress is reached, the width of the microcrack band decreases. In the post-peak region the strains are increasing rapidly within the microcrack band, forming a single crack. In the remaining part of the specimen, however, the strains are decreasing. In spite of aggregate debonding, because of cogging of aggregates the specimen is still able to carry a small amount of tensile forces across the critical section [Cedolin (1987)]. The described mechanism of force transfer in the last part of the softening region is only relevant for monotonic loading.

Softening is restricted to a narrow crack band. In the remaining part of the specimen the strains are decreasing, indicating unloading. Hence, softening is a discrete phenomenon [Reinhardt (1986)]. Specimens of different lengths yield almost identical stress-displacement diagrams (Fig.1.4(c)). However, no unique stress-strain relationship is existing in the post-peak region (Fig.1.4(c)) [Gopalaratnam (1985), Ansari (1987)]. The descending branch of the stress-strain diagram is becoming increasingly steeper with increasing length ℓ of the specimen because the localized displacements at the crack are smeared over a larger length of the specimen, resulting in smaller strains (Fig. 1.4(c)) [Van Mier (1984)]. In addition, the shape of the softening branch is influenced by the stiffness of the testing machine and the length and position of the strain gages [Zhen-hai (1987), Ansari (1987)]. Finally, the crack extends across the entire cross-section of the specimen, which is separated into two parts. The separation primarily occurs at the interfaces between coarse aggregates and the matrix or within the matrix. For light-weight concrete or high-strength concrete, cracks propagating through aggregates are also observed. A rough estimate of the unaxial tensile strength and the modulus of rupture is 10% and 15%, respectively, of the uniaxial compressive strength. A more detailed investigation yields decreasing values of $\sigma_T^C/|\sigma_P^C|$ with increasing compressive strength [Raphael (1984)]. The axial strain, corresponding to the peak stress, is of the order of 0.01 % to 0.02 %.

The increase in strength with increasing strain rate is more pronounced in tension than it is in compression. For dynamic loading at high strain rates an increase of more than 50 % has been reported [Suaris (1983), Raphael (1984), Mlakar (1985)].

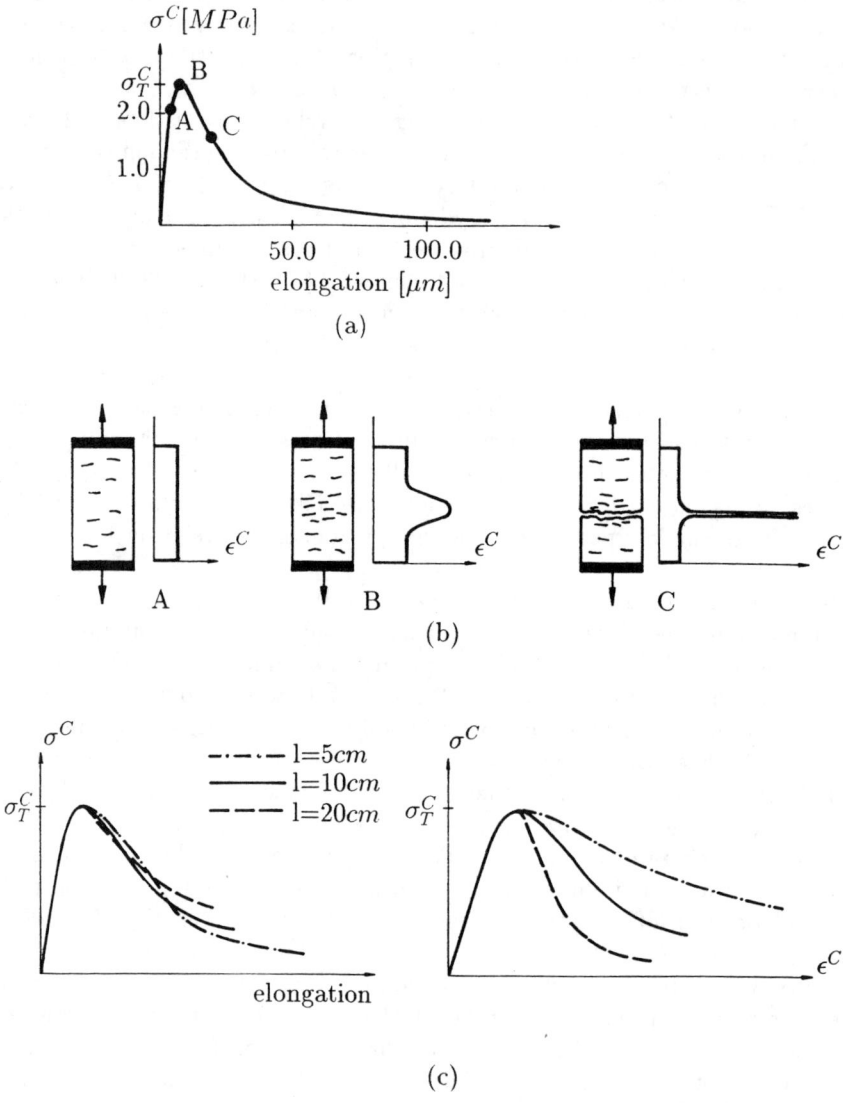

Figure 1.4: Behavior of concrete under uniaxial tension: (a) stress-elongation diagram, (b) strain localization, (c) stress-elongation diagrams for different lengths of specimens compared with the corresponding stress-strain diagrams

1.2 Concrete

Cyclic Loading

So far only a monotonically increasing load or deformation has been considered. In this subsection the behavior of concrete under cyclic or repeated compressive or tensile loading will be discussed. Typically, cyclic loading may either consist of a large number of load cycles at a relatively low stress level (high-cycle, low-amplitude loading) or of a relatively small number of cycles at a relatively high stress level (low-cycle, high-amplitude loading). The former is relevant, e.g., for machine foundations, and the latter, e.g., for earthquakes.

Experiments [Suaris (1987)] show that with increasing number of load cycles the continued growth of microcracks leads to a reduction of the strength of concrete. The stress-strain relationship of concrete under cyclic loading is characterized by an envelope defined as the limiting curve of all stress-strain curves regardless of the load pattern (Fig.1.5) [Sinha (1964), Karsan (1969), Reinhardt (1986)].

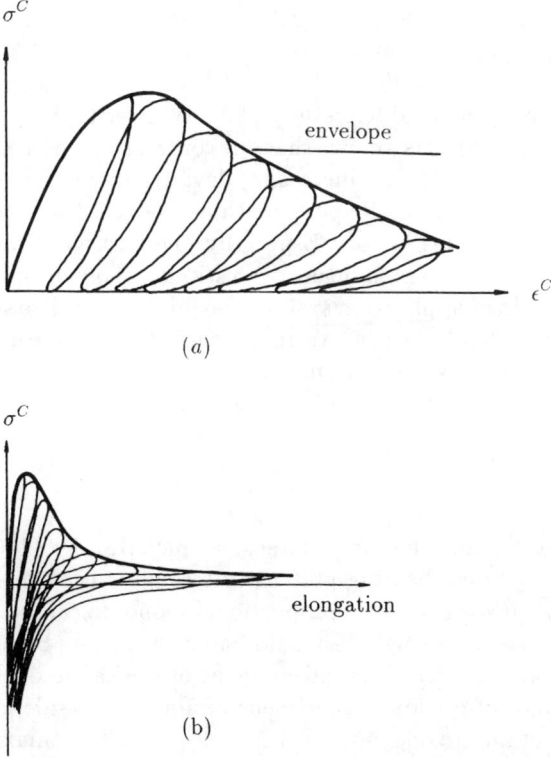

Figure 1.5: Stress-strain curves for (a) cyclic compressive loading and (b) cyclic tensile loading

The envelope is unique, i.e., a specimen can be loaded until the envelope curve is reached, regardless of the strain accumulated prior to the current load cycle. This curve is identical with the stress-strain diagram for monotonically increasing strain. Failure occurs when the envelope is reached.

The fatigue strength, defined as the maximum stress level sustained by a specimen on the basis of two million load cycles without failure, is about 45 to 55 % of the uniaxial compressive strength [Nelson (1988)]. In these experiments the minimum stress within one load cycle was chosen as 10% of the maximum stress.

1.2.3 Multiaxial Behavior

Numerous biaxial and triaxial tests of concrete have been performed within the last decades [Bellamy (1961), Iyengar (1965), Fumagalli (1965), Kupfer (1969), Mills (1970), Liu (1972), Kupfer (1973), Andenaes (1977), Schickert (1977), Tasuji (1978), Gerstle (1980), Scavuzzo (1983), Mier (1984), Winkler (1984), Buyukozturk (1984), Wang (1987), Yin (1989)]. A comparison of different experiments, e.g., of the ones reported in [Kupfer (1969), Andenaes (1977), Gerstle (1980), Wang (1987), Yin (1989)]), shows a relatively large scatter of the results. The differences are primarily caused by the different types of testing machines and the different techniques of applying the load [Gerstle (1980), Winkler (1984)]. A comparison by Winkler [Winkler (1984)] has shown that, in contrast to one-part machines (Fig.1.6(a)) which are characterized by increasing deviations from symmetric loading with increasing deformation, multi-part machines (Fig.1.6(b)) allow symmetric loading throughout the loading history. Especially, one-part machines with rigid steel platens (Fig.1.6(c)) result in a significant overestimation of the strength of concrete. With respect to the load application system, flexible systems like steel brushes (Fig.1.6(d)) are superior to rigid platens. An improvement of the steel brushes are steel bristle packets supported by an elastomer cushion, allowing a better approximation of the desired uniform stress distribution [Winkler (1984)].

Biaxial Behavior

Instead of using solid bearing platens, Kupfer [Kupfer (1969), Kupfer (1973)] employed flexible brush bearing platens to test plate-type specimens with the dimensions of 20 × 20 × 5 cm under proportional monotonically increasing biaxial loading ($\sigma_1/\sigma_2 = $ const. for a single test). He found that using brush bearing platens yields a uniaxial strength, which is independent of the shape of the specimens and equal to the strength of prismatic specimens of a height to side length ratio of 4.

The type of failure depends on the ratio σ_1/σ_2. For uniaxial and biaxial compression as well as for combined tension and compression with the tensile stress smaller than about 1/15 of the compressive stress, microcracks parallel to the free surface of the specimens are formed. The increase of the ultimate strength under biaxial compression of the same intensity in both directions is approximately 16 % of the

1.2 Concrete

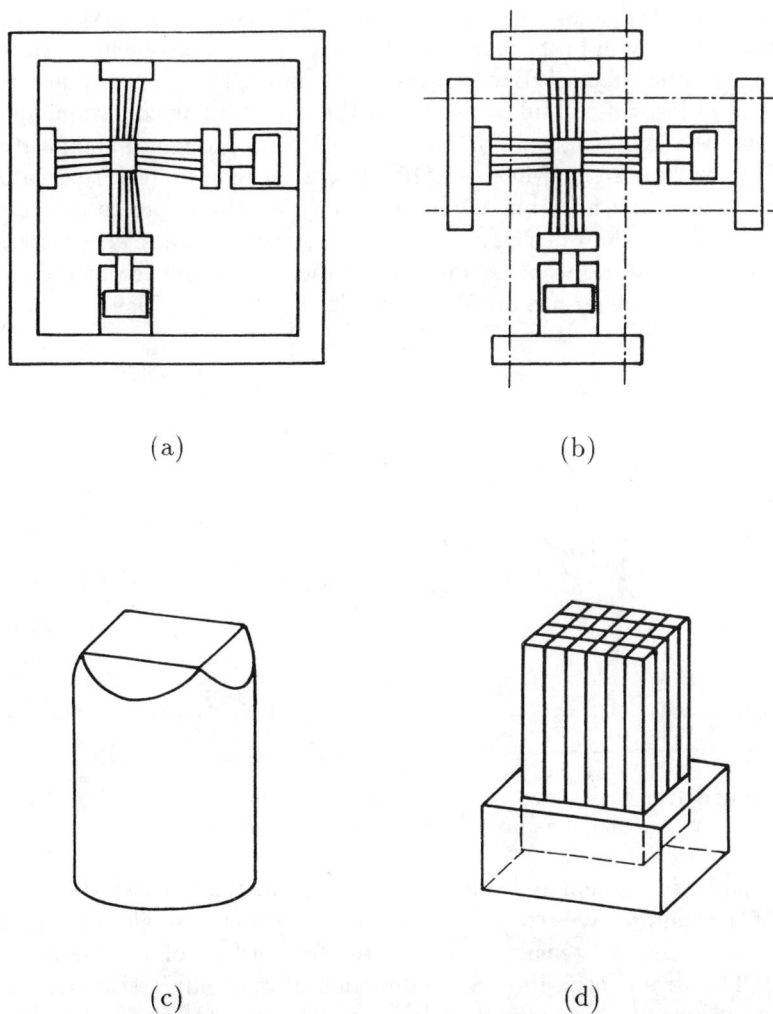

Figure 1.6: Different testing devices and load application systems: (a) one-part machine, (b) multi-part machine, (c) rigid bearing platens, (d) steel brushes

uniaxial compressive strength σ_P^C. This result corresponds very well with experiments conducted by Andenaes [Andenaes (1977)] on cubic specimens subjected to fluid pressure. Andenaes found an increase of strength under biaxial compression of the same intensity in both directions of 13% over the uniaxial strength. In order to demonstrate the effect of frictional restraints on the specimen, Andenaes also used steel platens instead of fluid pressure. For the same loading condition, an increase of 28 % over the uniaxial strength was observed. The maximum biaxial strength was found to be about 25 % [Andenaes (1977)] to 27 % [Kupfer (1969)] greater than the uniaxial strength. Under biaxial compression the strain corresponding to the larger (in absolute value) principal compressive stress increases with increasing stress at failure. Stress-strain curves for a few ratios σ_1/σ_2 and corresponding stress - volumetric strain diagrams are shown in Fig. 1.7 [Kupfer (1969)].

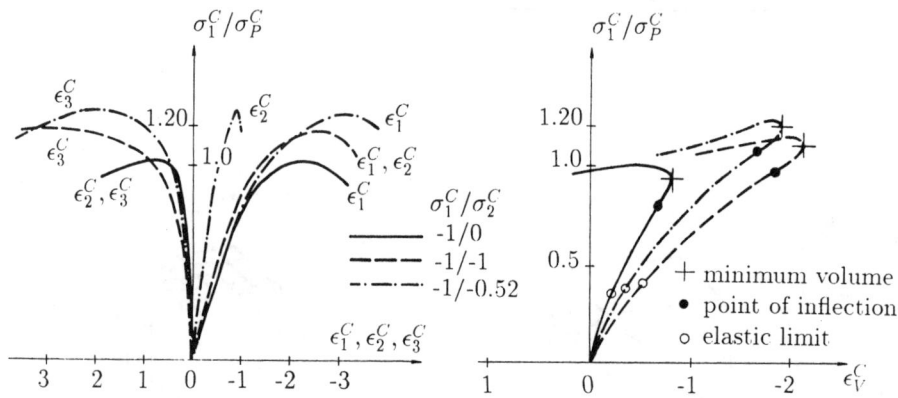

Figure 1.7: Stress-strain curves and corresponding stress - volumetric strain diagrams for different stress ratios σ_1^C/σ_2^C

For combined tension and compression, with the tensile stress larger than about 1/15 of the compressive stress, and for biaxial tension, a single crack perpendicular to both the principal tensile stress and the free surface of the specimen occurs at failure. The strains at failure in the direction of the compressive stress decrease in absolute value with increasing tensile stress. The strength of concrete under biaxial tension is almost equal to the uniaxial tensile strength σ_T^C.

The curves corresponding to the elastic limit, the point of inflection of the stress - volumetric strain diagram, the extremum of the volumetric strain and the ultimate load are shown in Fig.1.8 [Kupfer (1969)] for the entire range of biaxial states of stress.

The elastic range of the stress-strain diagram extends up to approximately 35 % of σ_P^C. Beyond the elastic limit the rate of volume decrease increases until a point of inflection on the stress-strain diagram is reached at approximately 80 - 90% of σ_P^C. This point can be viewed as the onset of major microcracking. The minimum volume is observed at about 95% of σ_P^C. Further loading results in an increase of

1.2 Concrete

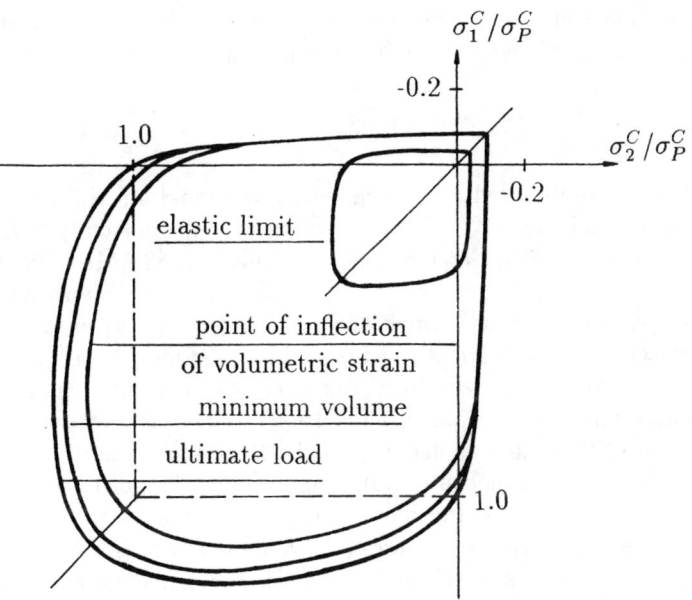

Figure 1.8: Typical loading curves for biaxial stress states

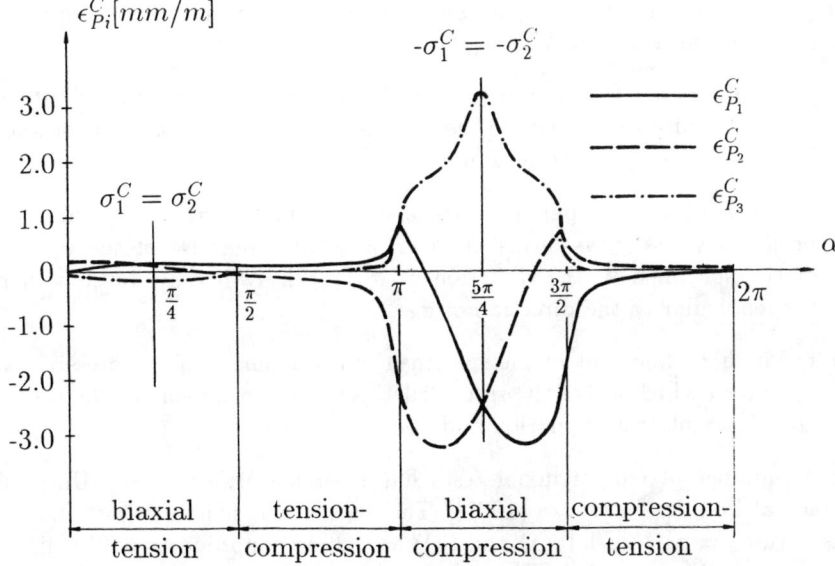

Figure 1.9: Strains at ultimate strength for biaxial stress states

volume of the specimen and, finally, in its failure. The outermost curve in Fig.(1.8) represents the ultimate strength envelope. The strains at the ultimate strength are plotted in Fig.1.9 as functions of the angle $\alpha = \arctan \sigma_1/\sigma_2$.

Triaxial Behavior

Triaxial tests can be subdivided into so-called conventional triaxial tests with cylindrical specimens [Bellamy (1961), Buyukozturk (1984)] and truly triaxial tests [Mills (1970), Schickert (1977), Gerstle (1980), Scavuzzo (1983), Mier (1984), Winkler (1984), Wang (1987)]. The former are restricted to load combinations with two equal principal stresses ($\sigma_2 = \sigma_3$) which are generated by fluid pressure acting on the curved surface of the cylindrical specimen. Thus, defining σ_1 as the stress acting parallel to the axis of the cylindrical specimen, conventional triaxial tests are restricted to stress paths in the plane defined by the σ_1-axis and the hydrostatic axis ($\sigma_1 = \sigma_2 = \sigma_3$). This plane is denoted as the Rendulic plane. In truly triaxial tests the three principal stresses can be varied independently. Such tests are usually conducted with cubic specimens. In order to achieve the desired uniform stress distribution in the specimen, friction between the loading surfaces and the specimen has to be reduced to a minimum [Winkler (1984)]. For conventional triaxial tests this problem is not so severe because frictional restraints only occur at the top and bottom surface of the cylindrical specimens. By choosing the same ratio of height to side length as for uniaxial tests, the influence of frictional restraints becomes relatively small.

Three different failure modes can be observed for triaxial compression with $\sigma_1 > \sigma_2 > \sigma_3$ (in absolute value) [Wang (1987)]:

a) If σ_2 and σ_3 are less than $0.1\sigma_1$, cracks will occur in planes perpendicular to the directions of σ_2 and σ_3. These cracks cause the separation of the specimen into individual concrete columns.

b) If the smallest principal stress σ_3 is smaller than $0.1\sigma_1$ but the intermediate principal stress σ_2 is larger than $0.15\sigma_1$, then, because of the confinement of the specimen in the direction of σ_2, cracks will only develop in planes perpendicular to the direction of σ_3.

c) For high confinement of the specimen in consequence of compressive stresses σ_2 and σ_3 which are of the same order as σ_1 the specimen is able to sustain a large amount of deformation and volume change.

A great number of truly triaxial tests has been performed at the University of Colorado at Boulder [Scavuzzo (1983)]. These tests include nonproportional loading paths with $\sigma_2 = \sigma_3$ (Fig.1.10(a) and (b)) as well as nonproportional loading paths under general three-dimensional stress states such as the starlike loading path of Fig.1.10(c). It consists of hydrostatic loading up to a certain stress level, followed by different stress paths in a deviatoric plane, including unloading and reloading.

1.2 Concrete

Deviatoric planes are defined as planes of constant hydrostatic stress. Hence, these planes are perpendicular to the hydrostatic axis.

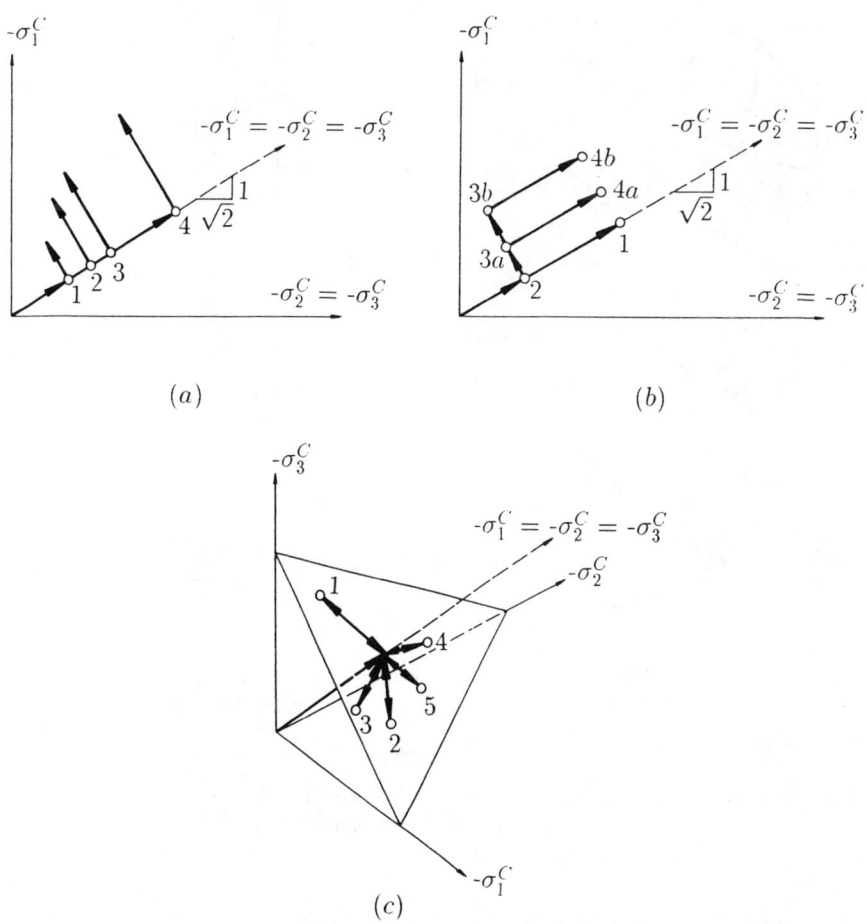

Figure 1.10: Selected stress paths for triaxial tests: (a) and (b) nonproportional loading paths with $\sigma_2^C = \sigma_3^C$, (c) nonproportional loading path under general three-dimensional stress states

By analogy to the curve defining failure for biaxial loading (Fig.1.8), for three-dimensional stress states failure is defined by the failure surface in the principal stress space (Fig.1.11(a)). The intersection of the failure surface with a deviatoric plane yields a failure curve for a particular hydrostatic stress level (Fig.1.11(b)). The intersection of the failure surface with the Rendulic plane yields two curves, called the compressive and the tensile meridian, respectively. These two curves are shown in Fig.1.11(c) [Ahmad (1982)], where $\|\mathbf{s}\|$ denotes the norm of the deviatoric stress tensor $\mathbf{s}$.

It follows from Fig.1.11(b) and Fig.1.11(c) that the distance of a point on the

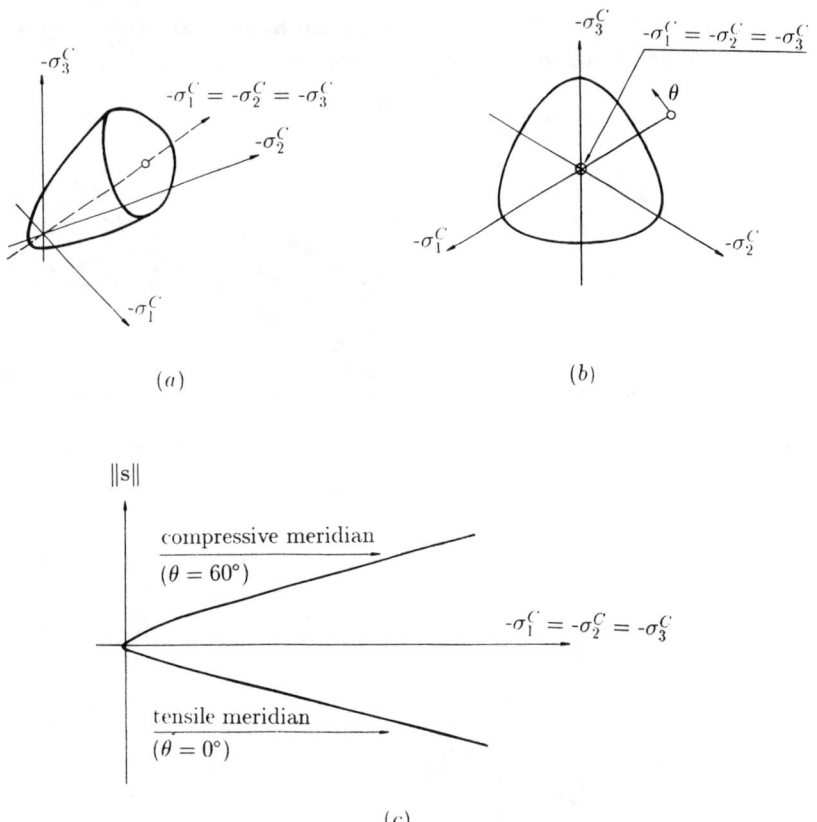

Figure 1.11: Failure surface for triaxial loading: (a) in the principal stress space, (b) corresponding failure curve in a deviatoric plane, (c) corresponding compressive and tensile meridian

failure surface from the hydrostatic axis depends on the hydrostatic stress and on the angle θ denoted as angle of similarity or Lode angle. Failure points on the compressive meridian ($\theta = 60°$) are obtained by superimposing a compressive stress $\Delta\sigma_1^C$ on a hydrostatic stress state. Failure points on the tensile meridian ($\theta = 0°$) are obtained by superimposing a tensile stress $\Delta\sigma_1^C$ on a hydrostatic stress state. The uniaxial compressive strength and the biaxial strength for equal tensile stresses represent special points on the compressive meridian. The uniaxial tensile strength and the biaxial strength for equal compressive stresses represent special points on the tensile meridian [Chen (1982)]. With increasing confinement, the strength of concrete is increasing. A transition from strain-softening to strain-hardening up to failure takes place. Fig.1.12 shows plots of the axial stress σ_1^C versus the axial strain and the radial strain for triaxial compression tests on cylindrical specimens at different levels of confining pressure $\sigma_r^C = \sigma_2^C = \sigma_3^C$ [Smith (1989), Lanig (1991a)].

1.2 Concrete

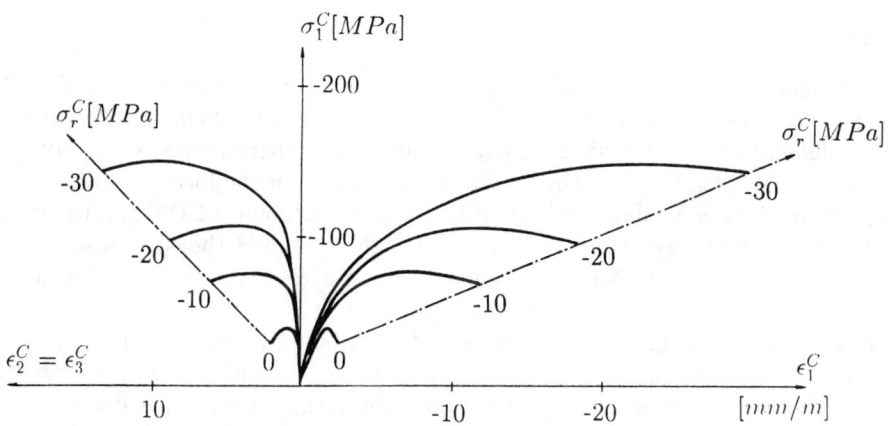

Figure 1.12: Stress-strain diagrams at different levels of confining pressure

Bažant [Bažant (1986)] conducted confined compression tests with cylindrical specimens. The axial stress was increased up to 2000 (!) MPa. In contrast to conventional triaxial tests where the lateral normal stresses are kept constant, in the confined compression tests almost uniaxial strain conditions were achieved because of the extremely large stiffness of the pressure vessel. A typical stress-strain diagram obtained under these conditions is characterized by the initial decrease of the slope of the diagram, followed by a stiffening response (Fig.1.13). The initial softening is attributed to the collapse of pores, whereas the subsequent stiffening is explained by the closure of pores.

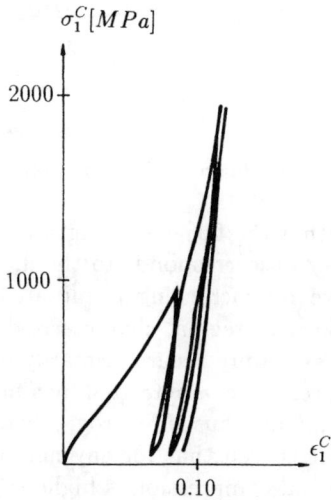

Figure 1.13: Stress-strain diagram for an approximately uniaxial state of strain

Cyclic Loading

The fatigue strength of biaxially loaded plate-type specimens for a range of $1 - 2 \cdot 10^6$ load cycles was investigated by Su [Su (1988)]. These experiments can be viewed as a supplement to the biaxial tests with monotonically increasing stresses, conducted by Kupfer [Kupfer (1969)]. The results are presented in the form of Wöhler curves, also denoted as S-N diagrams. They relate the maximum of the applied stress S to the number of load cycles N at which fatigue failure of the specimen occurs. In [Su (1988)] the rate of loading was chosen as one cycle per second. The ratio of the minimum to the maximum stress was equal to 0.05.

In contrast to earlier investigations [Tepfers (1979)], in [Su (1988)] a considerable difference of the slope of the S-N curve for low-cycle and high-cycle loading was found. From 1 to approximately 10^3 cycles the fatigue strength is decreasing at a considerably higher rate than for a number of cycles beyond 10^4 (Fig.1.14).

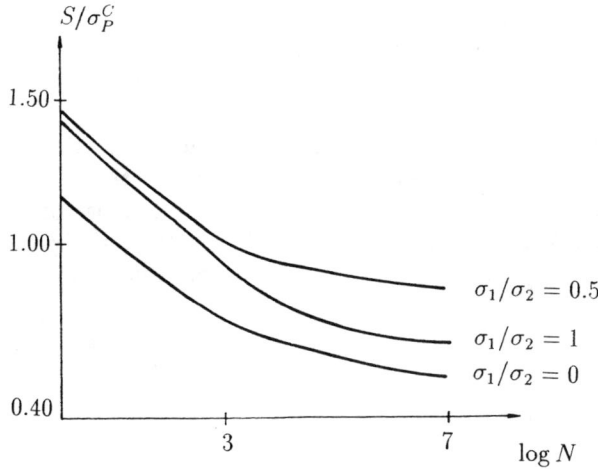

Figure 1.14: S-N curves for biaxially loaded specimens

It is interesting to note that the fatigue strength at the border between the low-cycle and the high-cycle region corresponds to the discontinuity stress, i.e., to the stress at which progressive microcracking is initiated. Hence, it is concluded in [Su (1988)] that in the low-cycle region, characterized by a fatigue strength larger than the discontinuity stress, failure occurs primarily because of progressive matrix cracking. In the high-cycle region, characterized by a fatigue strength below the discontinuity stress, failure is mainly caused by progressive bond cracking. With respect to the stress ratio σ_1/σ_2, it is seen that for any number of load cycles the fatigue strength of concrete in biaxial compression is higher than in uniaxial compression. For $N = 2 \cdot 10^6$, the fatigue strength ranges from 56% of σ_P^C ($\sigma_1/\sigma_2 = 0$) up to 85% of σ_P^C ($\sigma_1/\sigma_2 = 0.2$ and $\sigma_1/\sigma_2=0.5$). The increase in fatigue strength for biaxial loading in comparison with uniaxial loading can be explained by the confinement of

1.2 Concrete

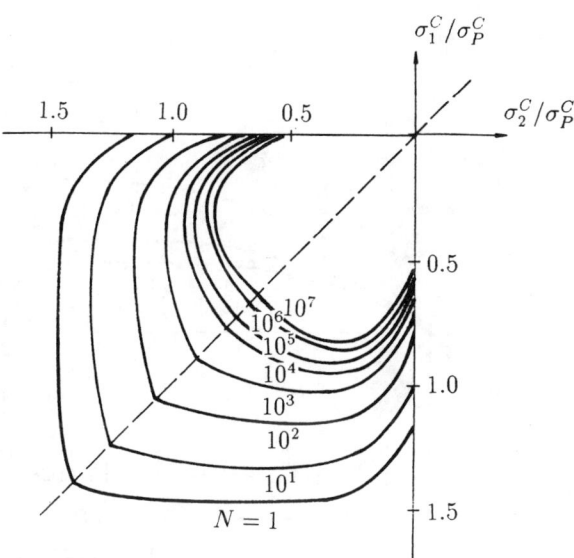

Figure 1.15: Fatigue strength envelopes for biaxially loaded specimens

the specimen in the lateral direction, yielding an increase of the discontinuity stress. Analogous to the curve representing the ultimate strength of concrete for monotonically increasing biaxial loading (Fig.1.8), curves defining the fatigue strength under biaxial loading are plotted in Fig.1.15. The strength for N = 1 is larger than the one reported by Kupfer [Kupfer (1969)] and Tasuji [Tasuji (1978)]. For uniaxial loading the strength for N = 1 is 17 % higher than the uniaxial compressive strength σ_P^C. This can be explained by the considerably higher loading rate used in the tests performed by Su [Su (1988)].

In general, up to 0.10 N the strain increases rapidly, followed by a uniform increase up to 0.85 N and, subsequently, by a rapid increase until failure.

1.2.4 Time-dependent Behavior of Concrete

Aging

Because of progressive hydration of concrete, the ultimate compressive and tensile strength and the modulus of elasticity increase with increasing age of concrete. The effect of aging of concrete on the uniaxial compressive strength and on the modulus of elasticity is illustrated in Fig.1.16 [Rüsch (1983)], where t denotes the age of concrete, measured in days.

The curves shown in Fig.1.16 hold for a constant temperature of approximately 20°C. At higher temperatures, hydration is accelerated. Therefore, aging is accelerated. At lower temperatures, hydration is decelerated. Therefore, aging is decelerated.

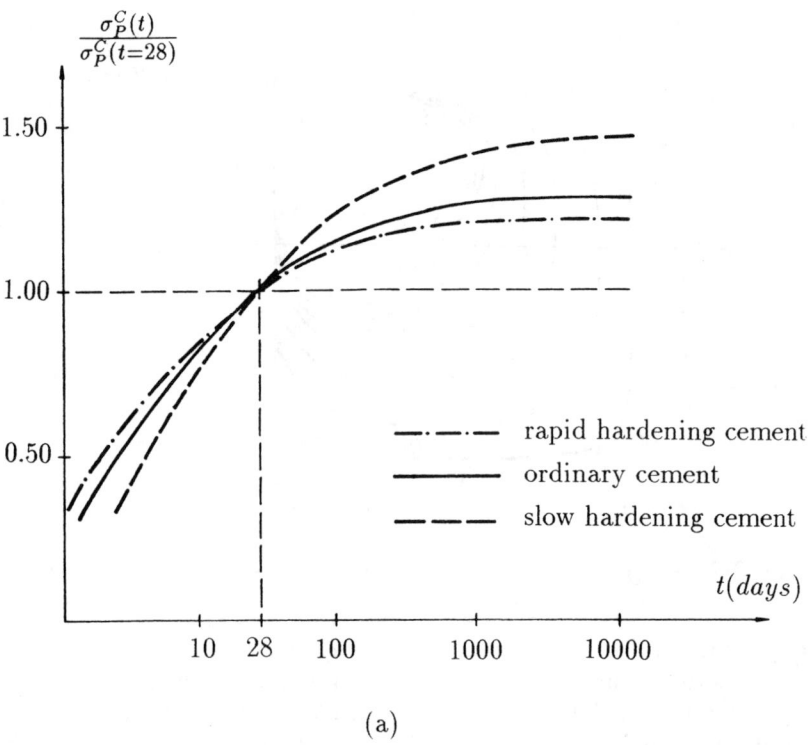

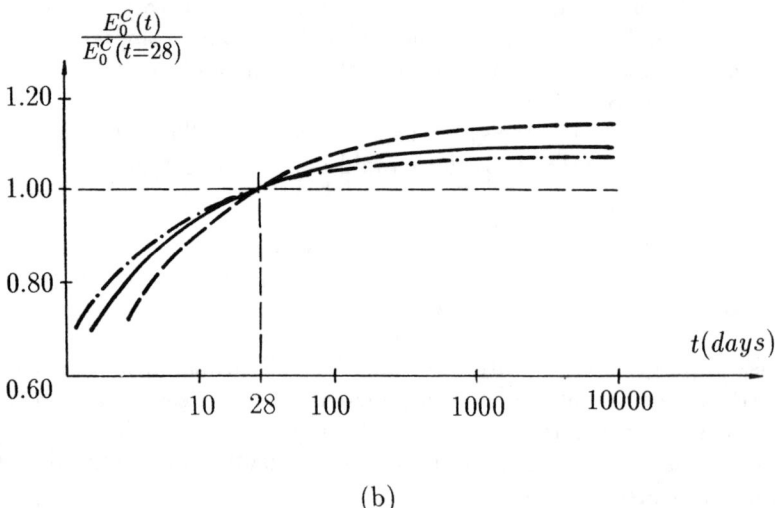

Figure 1.16: Effect of aging of concrete on the material properties: (a) uniaxial compressive strength, (b) modulus of elasticity

In order to obtain aging curves for different temperatures, the age of concrete has to be substituted by the effective age which depends on the degree of hydration [Rüsch (1983)].

Shrinkage

The time-dependent volume decrease of concrete at constant relative environmental humidity and temperature without the action of applied external loads is denoted as shrinkage. This phenomenon is caused by chemical, thermal and hygral processes [Acker (1989)]. The volume decrease in consequence of hydration is termed as chemical shrinkage. Hydration of concrete is an exothermal process. Hence, during hydration the temperature of concrete is greater than the temperature of the surrounding environment. With ceasing hydration, the natural cooling of concrete to the level of the temperature of the surrounding environment results in thermal shrinkage. The most important contribution to shrinkage comes from the drying process of concrete. Initially, there is a surplus of water in the concrete. This surplus is not consumed for hydration. The reduction of the water content of concrete to the moisture content of the surrounding environment, resulting from diffusion of capillary and adsorbed water, is called drying shrinkage. Obviously, the time-dependent reduction of the moisture content in a specimen is beginning at the surface. In Fig.1.17(a) the moisture distribution throughout the depth d of a plate-type specimen, at a relative humidity of the surrounding environment of $h_r = 50\%$, is plotted for different points of time $t = t_i$, $i = 1, 2, 3$, where $t_3 > t_2 > t_1$.

With increasing time, the reduction of moisture proceeds to the interior of the specimen. This process is very slow. Therefore, shrinkage of the relatively dry neighborhood of the surfaces of the specimen is hindered by the relatively moist inner part.

In contrast to the observed restrained shrinkage, unrestrained shrinkage would only be possible for infinitely thin sheets of concrete (Fig.1.17(b)) [Wittmann (1985)]. In this case, drying shrinkage would be free of restraint effects. Such unrestrained shrinkage can be viewed as a material property, primarily depending on the type of cement, the water/cement ratio, the temperature and the degree of hydration at the beginning of drying. An approximately linear relationship may be assumed between unrestrained shrinkage and the relative humidity [Wittmann (1985)]. In contrast to infinitely thin sheets of concrete, in a specimen of finite thickness the rate of water diffusion to the surface depends on both the size and the shape of the specimen. Hence, these geometric properties of the specimen influence the rate of shrinkage [Bryant (1987)]. The drying shrinkage of three specimens of different sizes and shapes is shown in Fig.1.17(d) [Almudaiheem (1987)]. At each point of time the moisture content is varying throughout the depth of the specimen. Because of the resulting restraint, stresses are produced (Fig.1.17(c)). If they exceed the tensile strength, cracks will be formed in the relatively dry vicinity of the surface of the specimen. Because of these cracks the surface of the specimen is increased. Consequently, the loss of water will be accelerated. Thus, even the tensile strength has an influence on drying shrinkage. According to [Rüsch (1983), Hansen (1987), Smadi (1987)], the ultimate shrinkage of concrete primarily depends on the relative humidity and the temperature, the water/cement ratio, the drying shrinkage of the cement paste, the aggregate content, and the ratio of Young's moduli of the aggregates and the hydrated cement [Almudaiheem (1987)]. Because of the almost linear-elastic behavior of the aggregates, shrinkage of the cement paste is restrained. Hence, the amount of shrinkage decreases with increasing aggregate content.

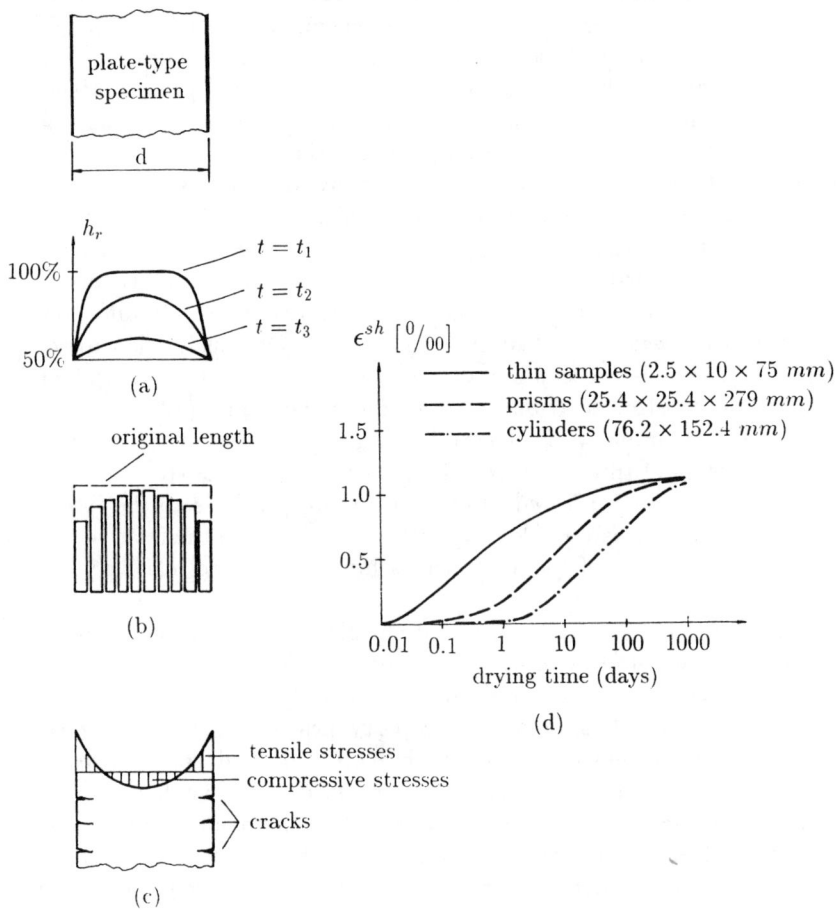

Figure 1.17: Restrained and unrestrained shrinkage of concrete: (a) moisture distribution through the depth of a plate-type specimen; (b) unrestrained shrinkage for infinite thin sheets of concrete; (c) stresses produced by restrained shrinkage; (d) drying shrinkage of specimens of different sizes and shapes

1.2 Concrete

Creep

The time-dependent deformation of concrete in consequence of the action of sustained loads can be subdivided into short-time creep and long-time creep. The mechanism of the former is explained as a stress-induced redistribution of capillary water within the hardened cement paste, whereas the latter is primarily caused by displacements of particles of the hardened cement paste [Wittmann (1985)]. According to [Trost (1991)], one day after loading, up to 20-25% of the total creep deformation may already be present.

In a model explaining the mechanism of long-time creep [Wittmann (1985)] the particles of the cement paste are considered to be restrained in their position by coupling forces. In order to move a particle to a new position, a certain amount of energy, termed as activation energy of the creep process, is required. Without the action of applied loads there would be no preferred direction of motion of the particles and, hence, no observable deformation. However, because of the stresses produced by the external loads, a preferred direction of motion of the particles is generated, yielding a time-dependent deformation. Since the hydration of the cement paste proceeds with increasing time, the positions of the particles are becoming increasingly more stable. Therefore, the amount of creep deformation depends on the age of concrete at loading. Provided the sustained stress level is sufficiently low, the rate of creep decreases with increasing age of concrete. Additional important factors which have an influence on the amount of creep deformations are the temperature and the relative humidity. Creep deformations are increasing at higher temperatures. Under identical loading conditions, at 70^oC the creep deformations are up to 3.5 times larger than at room temperature. Beyond 70^oC the creep deformations are decreasing. At 95^oC they are only about 1.7 times of the creep deformations at room temperature [Nasser (1965)]. Only for values larger than 50 % the relative humidity has a pronounced influence on creep. For such values of relative humidity the creep deformations are increasing rapidly with increasing relative humidity. This effect can be explained by the reduction of the amount of energy required to move the particles into new positions because of the adsorption of water films leading to a destabilization of the microstructure of concrete [Wittmann (1985)]. Thus, compared with the wetter inner regions of a drying specimen, the dryer outer regions exhibit less creep. As drying of the specimen is proceeding, the creep rate is decreasing.

Uniaxial creep tests on low-, medium- and high-strength concretes for stress levels from 40 to 80 % of the 28-day uniaxial compressive strength are described in [Smadi (1985), Smadi (1987)]. The creep strains, measured after a loading period of 60 days, are plotted in Fig.1.18(a) as a function of the ratio of the applied stress over the 28-day compressive strength, termed as stress intensity s.

The value of s at which the concrete specimens eventually fail under sustained loading was found to be 75% of σ_P^C for low- and medium-strength concrete and 80% for high-strength concrete, respectively. These results correspond to values reported for low- and medium-strength concrete in [Rüsch (1960)]. Fig.1.18(a) also shows that a linear relationship between the applied stress and the creep strain holds up to a

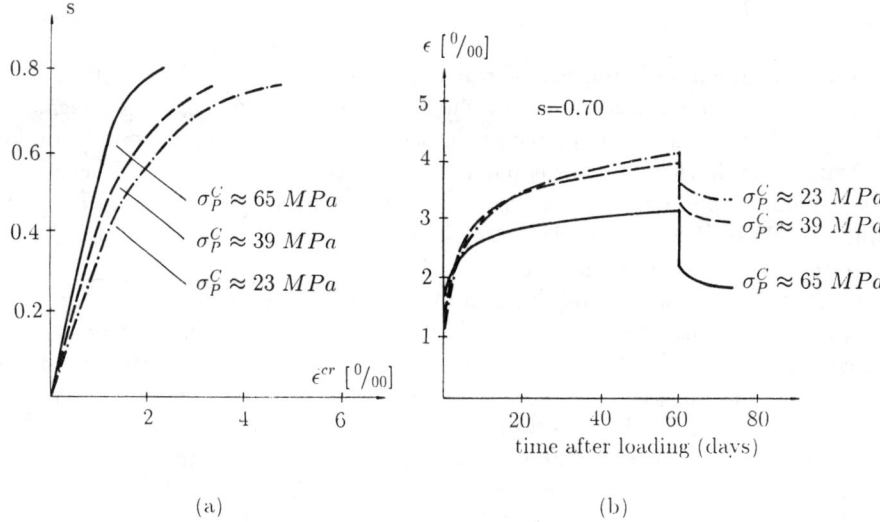

Figure 1.18: (a) Creep strains after a loading period of 60 days and (b) creep recovery after unloading for different concretes

stress intensity of approximately 45 % for low- and medium-strength concrete and up to 65% for high-strength concrete, respectively. Within this region the increase of interfacial bond cracks, caused by the sustained stress, is almost negligible in comparison to the increase of such cracks, resulting from drying shrinkage. For a certain stress intensity the measured creep strains are decreasing with increasing σ_P^C. Hence, analogous to the short-time behavior, low-and medium-strength concrete exhibits a more ductile time-dependent behavior than high-strength concrete.

Upon further loading, the deviation of the stress-strain diagrams from linearity is increasing. The reason for this is the significant increase of interfacial bond cracks [Smadi (1989)]. However, no significant cracking can be noticed.

For loading beyond the ultimate long-term strength, bond cracks, matrix cracks and combined cracks increase rapidly until failure occurs. By analogy to the mechanism of failure under short-time loading, failure under sustained loads is caused by continuous microcracking [Smadi (1989)].

In the tests reported in [Smadi (1985), Smadi (1987)] the sustained loads were removed after 60 days and creep recovery was measured. It follows from Fig.1.18(b) that most of the creep recovery takes place during the first few days after unloading. The creep recovery is less than the initial strain. The larger σ_P^C, the greater the creep recovery for a certain stress intensity. In Fig.1.18(b) the stress intensity is 70%. For lower stress intensities the differences between low-, medium- and high-strength concretes are less significant.

Uniaxial creep tests on specimens with sustained loads acting for 11 and 25 years, respectively, are reported in [Rüsch (1968), Probst (1978), Lanig (1991b)]. In Fig.1.19 the longitudinal strains of prismatic specimens under uniaxial sustained

1.2 Concrete

compressive loading are plotted. The stress intensities $s = \sigma^C/\sigma^C_{P,56}$ vary from 47% to almost 100% of the uniaxial cylindrical strength of $\sigma^C_{P,56} = -35.6$ MPa at the age of 56 days. The latter corresponds to the age at loading.

The specimens with s larger than 70% eventually failed. The specimen with $s = 66\%$, however, so far did not fail. The specimens characterized by $s = 57\%$ and 47%, respectively, were unloaded after 11 years.

Creep experiments for determination of the ultimate strength and fatigue strength under sustained biaxial loading on plate-type specimens, as have been used by Kupfer [Kupfer (1969)] for short-time tests, are described in [Linse (1976a)]. The tests did not show a significant difference of the behavior of concrete under biaxial sustained loading and its behavior under uniaxial sustained loading. The ultimate strength under sustained biaxial loads was found to be 88% of the respective short-time ultimate strength σ^C_P. The latter was within the range of -22 MPa to -25 MPa. Deformations at failure are two to five times larger than deformations at failure in a short-time test.

The values of the fatigue strength for biaxial loading were found to be within the range of 65 to 80% of σ^C_P [Linse (1976a)]. This is different from the range of values of 45 to 55% of σ^C_P, reported in [Nelson (1988)]. Different ratios of maximum over minimum stress within one load cycle seem to be the reason for the difference.

Creep tests on sealed cylindrical specimens under triaxial sustained compressive loading are described in [Lanig (1991a)]. In these tests strains were measured over a period of 90 days of acting loads. The ratio of the radial stress σ^C_r over the longitudinal stress σ^C_1 was varied between 0 to 1, i.e., the experiments cover the range from uniaxial to hydrostatic loading. In addition, the stress intensity s, defined as the ratio of σ^C_1 over σ^C_P, was varied from 30 (Fig.1.20) to 210%. Stress intensities larger than 100% are only possible in case of the existence of radial pressure.

Figure 1.20 contains two sets of parameter lines. One of the two parameters is the ratio of the radial over the longitudial stress, σ^C_r/σ^C_1. The other parameter is the time of sustained loading. The ordinate is the total strain. The curve corresponding to 1 minute of acting loads may be viewed as the ε^C-σ^C_r/σ^C_1 diagram for short-time loading. Expectedly, for a certain point of time the longitudinal strains are decreasing with increasing radial pressure. Accordingly, the lateral extension for sustained uniaxial loading is decreasing with increasing radial pressure. For higher values of the radial pressure a change from lateral extension to contraction occurs.

Simultaneous Creep and Shrinkage

In order to identify the contributions of creep and of drying shrinkage to the total time-dependent deformation of a loaded specimen, experiments are usually conducted simultaneously on a sealed loaded specimen and on a drying unloaded specimen. Remarkably, the time-dependent deformation of a drying loaded specimen is greater than the sum of the time-dependent deformations of the sealed loaded specimen and the drying unloaded specimen. This can either be attributed to the special mechanism of drying creep or to load-induced shrinkage. In [Wittmann (1985)] the

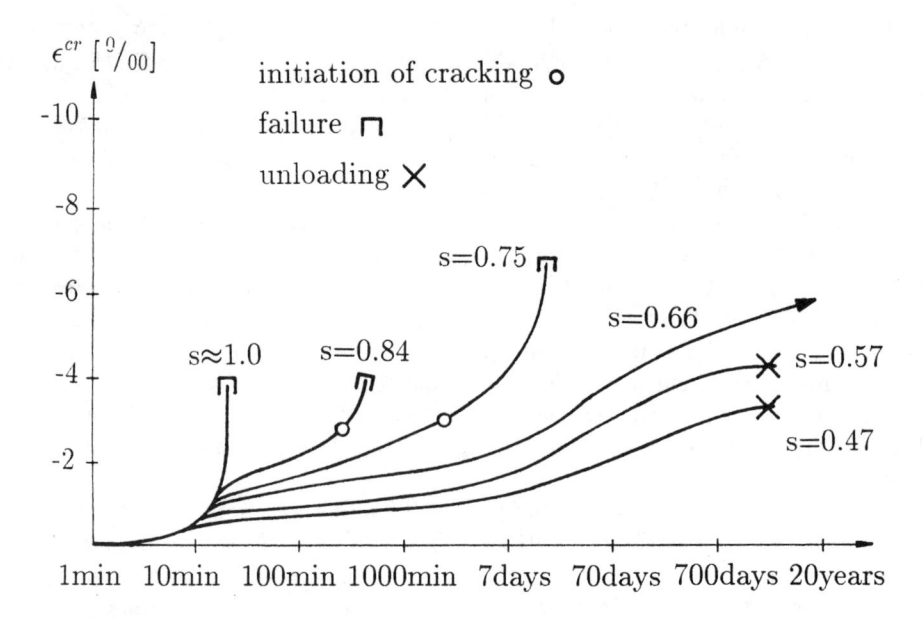

Figure 1.19: Creep tests with different levels of compressive loading

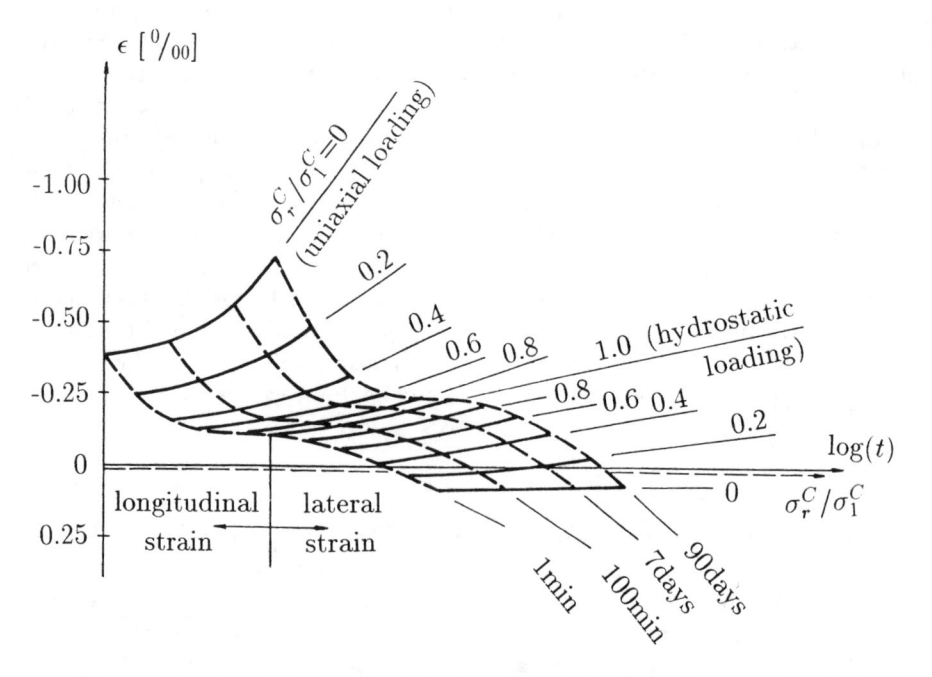

Figure 1.20: Time-dependent strains of cylindrical specimens under triaxial sustained compressive loading for $s = 0.30$.

1.3 Reinforcing and Prestressing Steel

rate of shrinkage is shown to depend on the stress in a specimen produced by an external load. Compressive stresses resulting from external loads lead to a higher rate of shrinkage. Tensile stresses yield a lower rate of shrinkage (Fig.1.21).

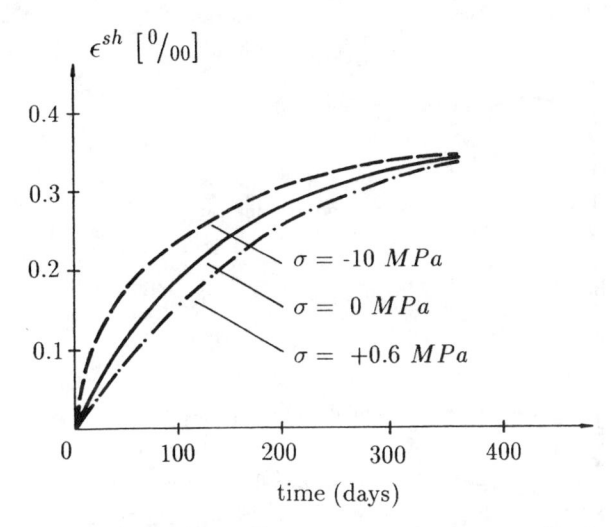

Figure 1.21: The dependence of shrinkage on stresses resulting from external loads

Nevertheless, the final value of shrinkage is hardly affected by external stresses. Because of the dependence of shrinkage on the external loads, the usual separation of the total time-dependent deformation into creep and shrinkage deformation, respectively, is criticized in [Wittmann (1985)]. Instead of this separation rather the total time-dependent deformation should be viewed as resulting from the total stresses, the relative humidity, the temperature and the degree of hydration.

1.3 Reinforcing and Prestressing Steel

1.3.1 Material Properties of Reinforcing Steel

The material behavior of reinforcing steel is linear-elastic almost up to the yield stress σ_Y^S. With regards to as-rolled steel, the linear-elastic material domain is followed by a pronounced yield plateau, a strain-hardening region up to the ultimate tensile strength σ_T^S and, finally, by a strain-softening region until failure occurs at the ultimate strain ε_U^S (Fig.1.22). The decrease of stress with the strain increasing beyond the strain corresponding to σ_T^S will only be noticed if the stress is referred to the original area of the cross-section of the specimen. If, however, the stress is referred to the actual area, then the stress will increase until failure. The yield stress and

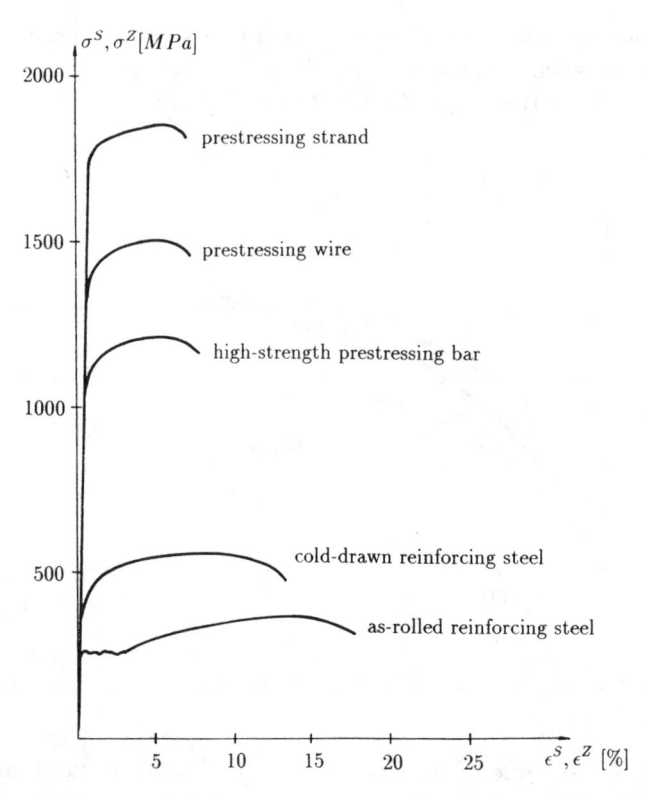

Figure 1.22: Stress-strain curves for reinforcing and prestressing steel

the ultimate strength of cold-drawn steel are greater than the corresponding stresses of as-rolled steel. Cold-drawn steel does not have a definite yield stress. Hence, for this type of steel the yield stress must be defined somewhat artificially. According to [ACI-Committee 439], the yield stress for cold-drawn steel is defined as the stress at a total strain of 0.0035, whereas according to the German code DIN 1045, the yield stress is defined as the stress at 0.002 permanent strain. As-rolled steel is characterized by an almost identical stress-strain curve in tension and compression. For cold-drawn steel, however, the absolute value of the yield stress in compression is smaller than the yield stress in tension. The reason for this phenomenon, which is known as the Bauschinger effect, is the drawing process. In general, the modulus of elasticity is about 200 000 MPa. Typical values of the yield stress and the tensile strength for as-rolled steel are within the range of 220 to 420 MPa and 340 to 500 MPa, respectively. For cold-drawn steel σ_Y^S varies between 420 and 520 MPa and σ_T^S between 500 to 690 MPa [Leonhardt (1980), Naaman (1982)], respectively. The ultimate strain at failure, measured over a length of ten times the diameter of the specimen, is within the range of 8 to 16%.

Plain bars are usually made of steel with low yield strength. Especially for re-

1.3 Reinforcing and Prestressing Steel

inforcing steel of higher quality it is important to improve the bond between steel and concrete. The bond properties of deformed bars with lugs or ribs rolled onto the surface of the bars are better than the ones of plain bars. Therefore, deformed bars are generally preferred to plain bars.

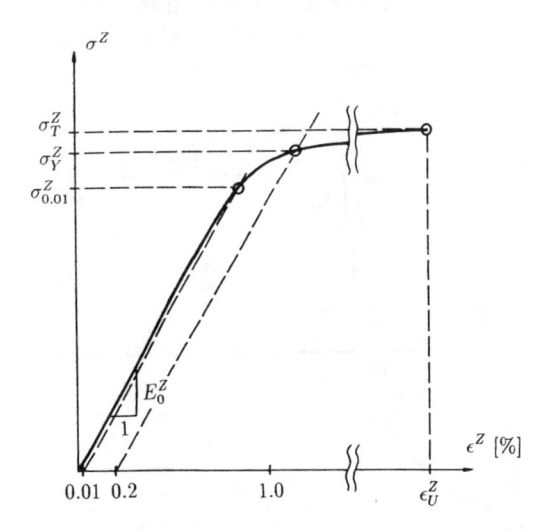

1.3.2 Material Properties of Prestressing Steel

Desirable properties of prestressing steel are linear-elastic material behavior up to a relatively large stress, a large tensile strength permitting a large elongation, sufficient ductility, good bond properties, a low amount of loss of stress because of relaxation, a large fatigue strength and a high degree of resistance to corrosion. The material properties depend on the type of the production process. Common procedures to meet the quality requirements of prestressing steel are the use of alloy steel, heat treatment, cold-drawing and stress-relieving.

Figure 1.23: Material properties of prestressing steel

Specified material properties (Fig.1.23) are the modulus of elasticity E_0^Z (170 000 - 200 000 MPa), the proportional limit $\sigma_{0.01}^Z$ (750 - 1 400 MPa), the yield strength σ_Y^Z (850 - 1 600 MPa), the tensile strength σ_T^Z (1 000 - 1 900 MPa) and the strain at failure ε_U^Z (6 - 7%). $\sigma_{0.01}^Z$ is defined as the stress at 0.01% permanent strain. It is regarded as the stress at which the stress-strain curve is beginning to deviate from linearity. The yield strength for prestressing steel without a definite yield plateau is defined as the stress at 0.1% or 0.2% permanent strain. It is denoted as $\sigma_{0.1}^Z$ or $\sigma_{0.2}^Z$, respectively. ε_U^Z is determined from a specimen with a ratio of length to diameter equal to ten [Leonhardt (1980), Naaman (1982)].

A comparison of stress-strain diagrams for reinforcing and prestressing steel (Fig. 1.22) shows that the substantially higher tensile strength of prestressing steel goes along with a significant reduction of the strain at failure. Hence, the ability to sustain large inelastic deformations is reduced.

Prestressing steel is available in the form of bars, wires and strands. The latter consist of several wires twisted together. Their bond properties are better than the ones of single wires.

1.3.3 Cyclic Loading

The behavior of reinforcing and prestressing steel under cyclic loading (Fig.1.24) is characterized by an unloading path almost parallel to the linear-elastic part of the stress-strain diagram. At initial loading, a well-defined yield plateau may exist. This property disappears in the subsequent load cycles. The aforementioned Bauschinger effect consists of an increase in tensile strength after repeated load cycles, accompanied by approximately the same amount of decrease in compressive strength. The cyclic behavior shown in Fig.1.24 is not found in the reinforcement of concrete structures. The stresses in the reinforcing steel, caused by cyclic loading, are usually limited to the tensile region and to small compressive stresses.

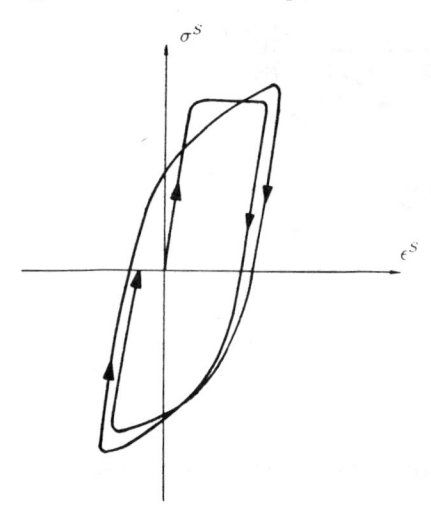

Figure 1.24: Behavior of reinforcing and prestressing steel under cyclic loading

The fatigue strength of reinforcing and prestressing steel is less than the ultimate short-time strength. It depends primarily on the average stress and the difference between the minimum and the maximum stress within the load cycles. The fatigue strength can be determined from the so called Smith-diagram. In this diagram the minimum and the maximum stress within the load cycles which can be sustained 2 million times without failure are plotted over the mean stress within one cycle, σ_m (Fig.1.25).

1.3 Reinforcing and Prestressing Steel

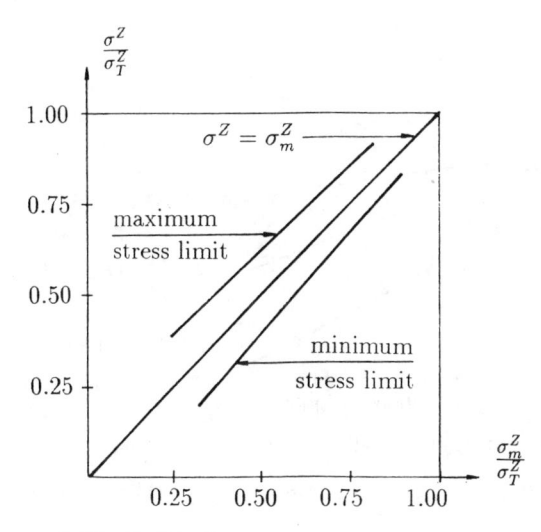

Figure 1.25: Smith-diagram

1.3.4 Effect of Low and High Temperatures

For temperatures exceeding 300 to 350°C, both the yield strength and the tensile strength are decreasing rapidly. For reinforcing or prestressing steel which has sustained temperatures above 400°C and has been cooled down slowly, the behavior of cold-drawn and as-rolled steel is different. As-rolled steel is regaining its original material properties. This is not the case with cold-drawn steel. With decreasing temperatures, both the yield strength and the tensile strength are increasing. After an abrupt decrease of temperature, steel may lose its ductile behavior.

1.3.5 Relaxation

Relaxation is defined as the time-dependent loss of stress in a material under constant strain. From this definition and from the one for creep it follows that the basic mechanisms of relaxation and creep are the same. Since creep and shrinkage of concrete as well as changes of the applied external loads induce changes of lengths of the tendons, the theoretical conditions of relaxation are not satisfied completely. However, relaxation tests take account of the major factors affecting stress relaxation. These factors are the initial stress level, the type of steel and the temperature [Magura (1964)]. For a stress below one half of the tensile strength, the loss of stress because of relaxation is negligibly small. For much larger stresses such losses of stress can be larger than 10% of the initial stress (Fig.1.26) [Leonhardt (1980)].

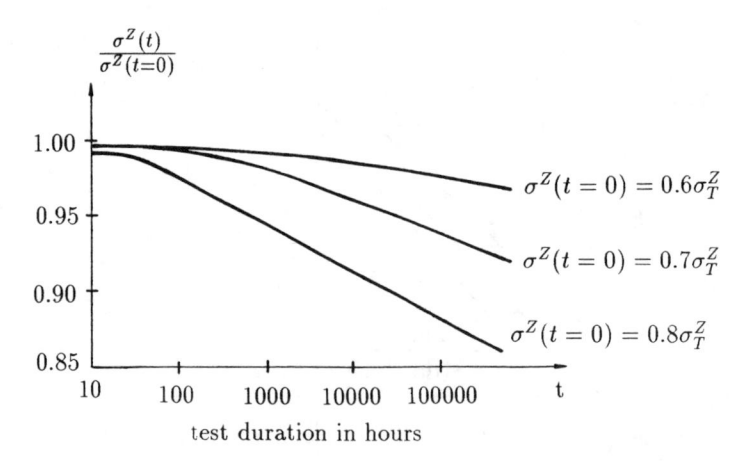

Figure 1.26: Typical relaxation losses for a prestressing steel at a temperature of 20°C

1.4 Reinforced Concrete

The mechanical behavior of the composite material reinforced concrete differs considerably from the one of its two basic constituents. Therefore, the interaction between steel and concrete needs to be investigated. The mechanism of the stress transfer between the reinforcement and the surrounding concrete through bond and dowel action is of interest. Bond is characterized by the transfer of longitudinal forces, whereas dowel action contributes to the transfer of shear forces. In a reinforced concrete specimen subjected to uniform tension, bond between steel and concrete results in a tensile stress in the reinforcement which is larger than average at cracks and lower than average midway between neighboring cracks. The tensile stresses in the concrete are zero at cracks and larger than average midway between neighboring cracks.

Another mechanism contributing to the load-carrying capacity of cracked concrete is aggregate interlock. It has to do with the transfer of shear stresses across rough cracks. Aggregate interlock is also present in plain concrete. However, usually the crack plane is intersected by reinforcing bars contributing to the resistance against relative displacements in this plane. Therefore, aggregate interlock is described in this subsection.

1.4.1 Bond

The load-carrying capacity of reinforced concrete strongly depends on bond between concrete and steel. A characteristic quantity for mathematical models of bond is the bond stress τ^B. It is defined as the shear force per unit area of the surface of the reinforcing bar, acting at the interface of concrete and steel, parallel to the direction

1.4 Reinforced Concrete

of the bar. Bond stresses are present whenever stresses are transferred from concrete to steel or vice versa, as is the case, e.g., in anchorage zones and in the vicinity of cracks. The bond stresses can be defined as the rate of change of the stress in the reinforcing bar [Ferguson (1966)].

For plain bars, bond is caused by chemical adhesion between the concrete paste and the surface of the reinforcing bar and by sliding friction at the interface of concrete and steel. Adhesion is destroyed even if the slip is small. Within regions of bond slip, stresses between the reinforcement and the concrete can only be transferred by friction. The transferability of stresses by friction strongly depends on the roughness of the steel surface and on the pressure acting perpendicular to the interface of concrete and steel. In the case of high-strength reinforcement, such as cold-drawn reinforcing bars, the described actions of bond may not be sufficient to prevent bond failure. Hence, especially for this type of reinforcement, deformed bars are employed (Fig.1.27(a)). In this case bond is strongly influenced by the bearing capacity and the shear strength of the concrete sections between the lugs of the bars. The stresses are transferred from the steel bars to the concrete at the front surfaces of the lugs. Shortly after the appearance of primary cracks, cone-shaped secondary cracks are formed in the concrete at the lugs (Fig.1.27(b)) [Goto (1971)]. The conditions at the lugs differ totally from those between the lugs. Consequently, for deformed bars the term "bond stress distribution" is to be understood in an average sense.

As long as the concrete sections between the lugs do not fail, slip between concrete and steel caused by local consolidation or crushing of concrete is small. Bond failure is either caused by shearing of the concrete sections in front of the lugs of the reinforcing bar or by splitting of concrete adjacent to the bar. Splitting of concrete is a much more sudden type of failure than shearing of concrete sections. For cyclic loading, bond deterioration is of special interest. It is caused by progressive crushing of the concrete in front of the lugs and by the continuous growth of microcracks [Pochanart (1989)].

If the tensile stresses in a reinforced concrete specimen subjected to tension in the direction of the reinforcing bar (Fig.1.27(a)) are below the tensile strength of concrete, then perfect bond between concrete and steel may be assumed. In this case the strains in the reinforcing bar and the adjacent concrete are compatible. Consequently, in both materials the tensile stresses are uniformly distributed along the entire length of the specimen with exception of the end regions where the stresses are transferred from the steel bar to the concrete through bond. If the external force is increased such that the tensile strength of concrete is reached and cracking occurs (Fig.1.27(b)), then the tensile stresses in the concrete at the cracks will be reduced to zero whereas the intact concrete between the cracks will still carry tensile stresses.

Global equilibrium requires a sudden increase of the tensile stresses in the reinforcing bars at the cracks. Fig.1.27(c) refers to the situation after this increase. The transfer of tensile stresses between the concrete and the steel is achieved through bond. Especially in the vicinity of cracks high bond stresses are acting (Fig.1.27(e)). They are accompanied by relative displacements between the steel and the concrete in the direction of the bars (bond slip). Fig.1.27 shows that the uniform longitudinal force to which the specimen is subjected produces nonuniform distributions of the stresses in the steel and the concrete. At the cracks, the concrete stresses are zero whereas the steel stresses attain a maximum. Midway between the cracks, the steel stresses attain a minimum whereas the concrete stresses reach a maximum.

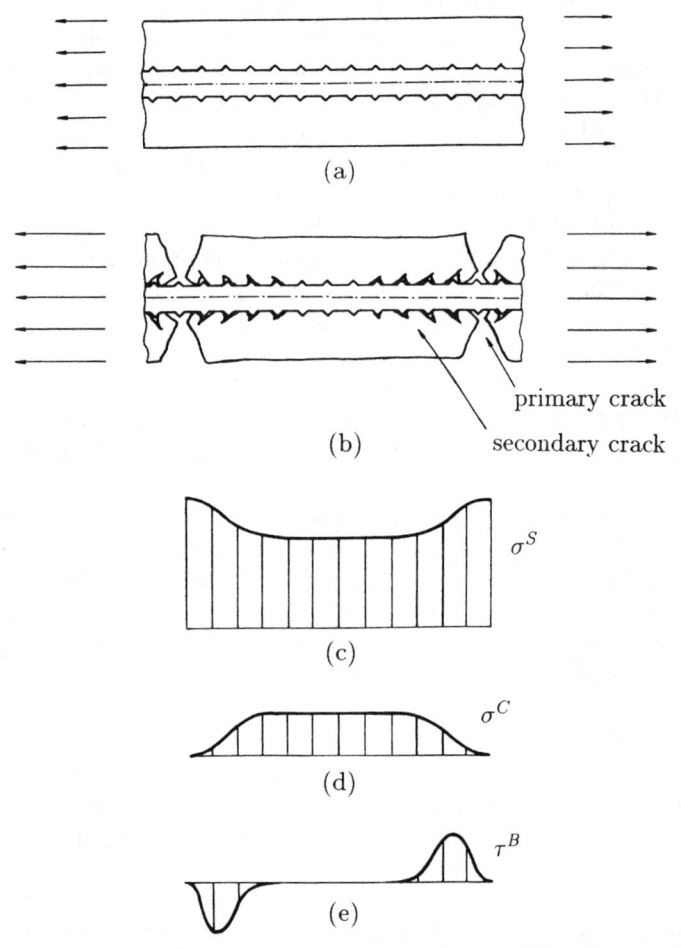

Figure 1.27: Part of a reinforced concrete specimen subjected to tension: (a) uncracked specimen, (b) cracked specimen, (c) distribution of steel stresses, (d) distribution of concrete stresses, (e) distribution of bond stresses

Because of the residual tensile load-carrying capacity of the cracked concrete adjacent to a reinforcing bar, the response of the cracked reinforced specimen under tensile loading is stiffer than the one of an identical steel bar which is not embedded in cracked concrete. This effect is known as tension stiffening (Fig.1.28(a)) [Shima (1990)].

1.4 Reinforced Concrete

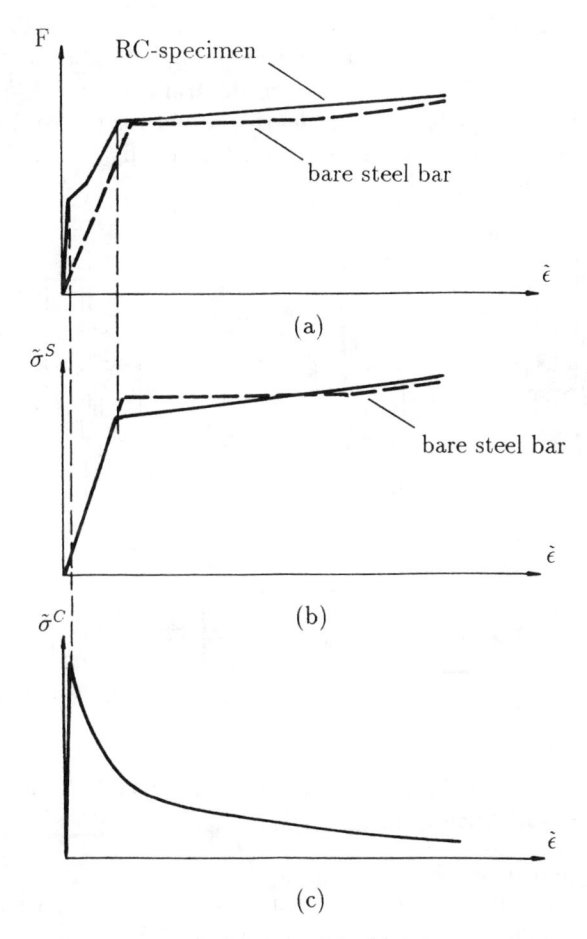

Figure 1.28: Tension stiffening effect: (a) total force - average strain relationship, (b) average steel stress and (c) average concrete stress versus average strain

If, after the onset of cracking, the longitudinal external force F is increased further, new cracks will be formed between the existing cracks. Hence, the intact regions of concrete between the cracks are shortened. Consequently, as can be seen in Fig.1.28(c), the tensile stresses, transferred from the steel to the concrete through bond, are decreasing with increasing external forces.

The diagram of the average steel stress $\tilde{\sigma}^S$ versus the average strain $\tilde{\varepsilon}$ of the specimen, shown in Fig.1.28(b), lies slightly below the stress-strain curve of a bare steel bar. This is a consequence of the fact that yielding of the reinforcement starts at the cracks where the steel strains attain maximum values which are considerably larger than the average strain of the specimen. Hence, yielding of the steel is already initiated before the average strain attains the elastic limit.

The diagram of the average tensile stress $\tilde{\sigma}^C$ in the concrete versus the average strain $\tilde{\varepsilon}$ of the specimen (Fig.1.28(c)) is characterized by a similar shape as the respective diagram for plain concrete, shown in Fig.1.4(c). However, there is an important difference. The ultimate average tensile strain at which the ability to carry tensile stresses is lost completely is considerably larger for the concrete component of reinforced concrete than for plain concrete. Commonly, the difference is about one order of magnitude.

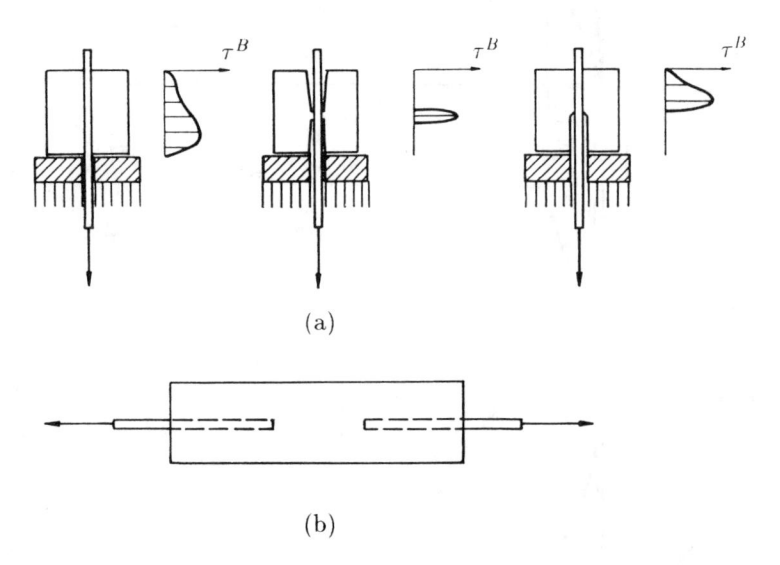

Figure 1.29: Tests for the determination of bond strength: (a) pull-out tests, (b) tension test

The ultimate bond strength τ_P^B can either be determined from pull-out tests or from tension tests (Fig.1.29). In both cases the reinforcing bars are loaded progressively up to failure. The bond stress varies considerably along the embedded bar. Therefore, experiments with long embedment lengths only allow determination of an average value of the bond strength. Consequently, in order to obtain the peak bond stress, anchorage tests with very short embedment lengths have been conducted [Rehm (1961)]. In pull-out tests the surrounding concrete is in compression (Fig.1.29(a)). This is not the case in tension tests (Fig.1.29(b)). The conditions in such tests are close to the conditions of the tensile zone of cracked concrete. The results of tension tests conducted with specimens of different ages and, hence, different strengths, with both plain and deformed bars, are reported in [Chapman (1987)].

The load-slip relationship for a specimen with a plain bar is shown in Fig.1.30(a) [Chapman (1987)]. The load-slip curve is linear up to a peak value corresponding to a critical value of slip. For larger slips, adhesive bond is destroyed, resulting in a sharp decrease of the load with increasing slip. Within the descending region of the load-slip curve, pull-out of the bar is mainly prevented by friction. Remarkably, the compressive strength of concrete has almost no influence on the bond strength of plain bars. Hence, it can be concluded that bond resulting from chemical adhesion

1.4 Reinforced Concrete

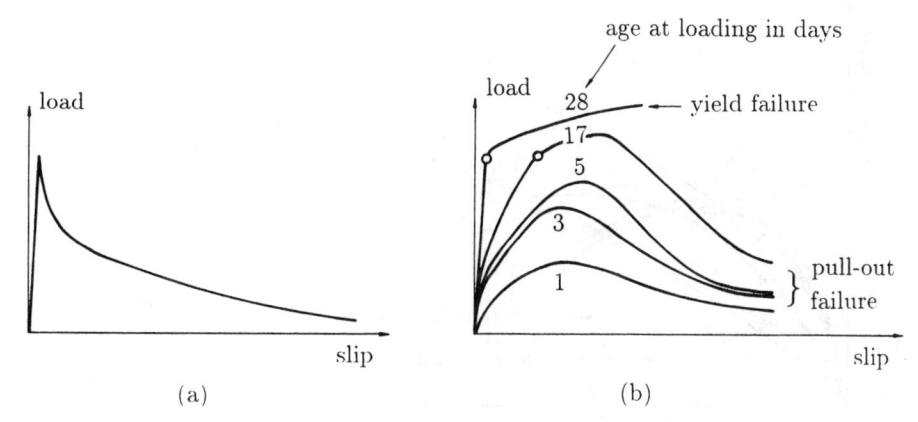

Figure 1.30: Load-slip relationships for (a) plain bars and (b) deformed bars

is almost independent of the compressive strength of concrete.

Tests conducted with deformed bars, however, have shown that the bond strength increases with increasing age at loading (Fig.1.30(b)) [Chapman (1987)]. The reason for the different behavior of plain bars and deformed bars is the mechanical interlock of the deformed bars. It depends on the compressive strength of concrete. Different types of failure were observed. They are a function of the age at loading and, hence, of the compressive strength, and of the embedment length of the deformed bar. For short embedment lengths and for a concrete with a small compressive strength, pull-out of the bar is the dominant failure mechanism. For long embedment lengths and a concrete with a large compressive strength, the failure mode is yielding of the bar followed by excessive slip. For situations between these two extreme cases, namely, for short embedment lengths and a small compressive strength, on the one hand, and for long embedment lengths and a large compressive strength, on the other hand, also failure in consequence of splitting of concrete may occur.

In the experiments shown in Fig.1.30 slip was measured as the displacement of the bar relative to the end-surface of the concrete specimen. What is needed is the relationship between the local bond stress and the local slip at the concrete-steel interface along the bar. Determination of this relationship is a difficult task [Jiang (1984), Tassios (1984), Lahnert (1986)]. Usually, the bond stress is only a relatively small fraction of the steel stress. Consequently, a small error of the measured steel strain may cause a large error of the local bond stress. Hence, the results of such experiments show a large scatter (Fig.1.31) [Lahnert (1986)]. The influence of the size of the specimen and the bar in pull-out tests on the type of bond failure was reported by [Bažant (1988)]. For smaller specimens with smaller bar sizes, failure because of shearing of the bar against the adjacent concrete was dominating. For larger specimens with larger bar sizes, a tendency to a more brittle type of failure was observed.

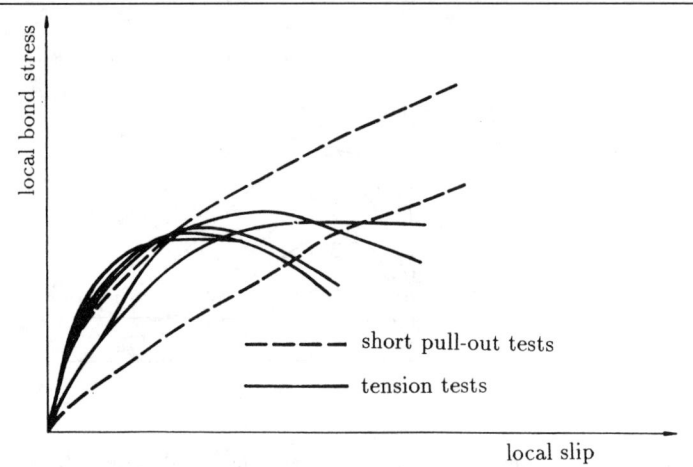

Figure 1.31: Local bond stress - bond slip relationships

1.4.2 Aggregate Interlock

A substantial amount of shear forces can be transmitted across cracks by aggregate interlock. In general, cracking occurs at the aggregate-matrix interfaces, resulting in rough cracks with a crack width considerably smaller than the size of most aggregates. According to [Walraven (1981)], the particles with a diameter of at least twice the crack width contribute to aggregate interlock. The roughness of a crack is composed of a global and a local roughness [Fardis (1979)]. The former is determined by the location and the size of the protruding aggregates. The latter is governed by small asperities and small protruding particles. It is present even at smooth concrete interfaces.

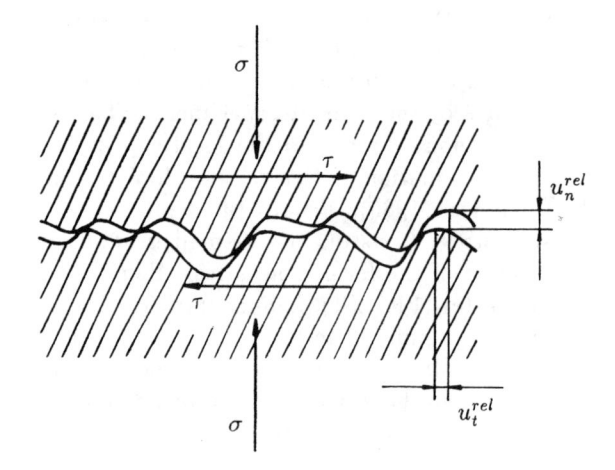

Figure 1.32: Relative displacements at crack planes

1.4 Reinforced Concrete

Because of relative displacements u_t^{rel} parallel to the crack surface, also denoted as shear displacements or shear slip, contact zones on the surfaces of the aggregates, protruding from opposite sides of a crack, are formed (Fig.1.32). The formation of these zones is accompanied by large contact stresses, causing irreversible deformations of the matrix. Because of the protruding aggregates in the case of rough crack faces, shear displacements are accompanied by an increase of the crack width u_n^{rel}. This increase is denoted as crack dilatancy. For the transfer of substantial shear forces it is essential that the increase of the crack opening is resisted by compressive forces acting perpendicular to the crack. Such forces can result from external loading, from dead load of the concrete, or from reinforcing bars intersecting the crack. If such bars are present, the increase of the crack opening will activate additional tensile stresses in the reinforcing bars [Bažant (1980)]. They are balanced by compressive stresses in the concrete, acting perpendicular to the crack faces. Hence, the main parameters for aggregate interlock are the shear displacement, the crack width, the shear stress and the normal stress at the crack surfaces [Yoshikawa (1989)]. Apart from the reinforcement, the most important factors for the shear resistance of crack faces are the friction between the aggregates and the matrix, the strength of concrete and, for relatively large crack widths, also the size of the aggregates [Walraven (1981)].

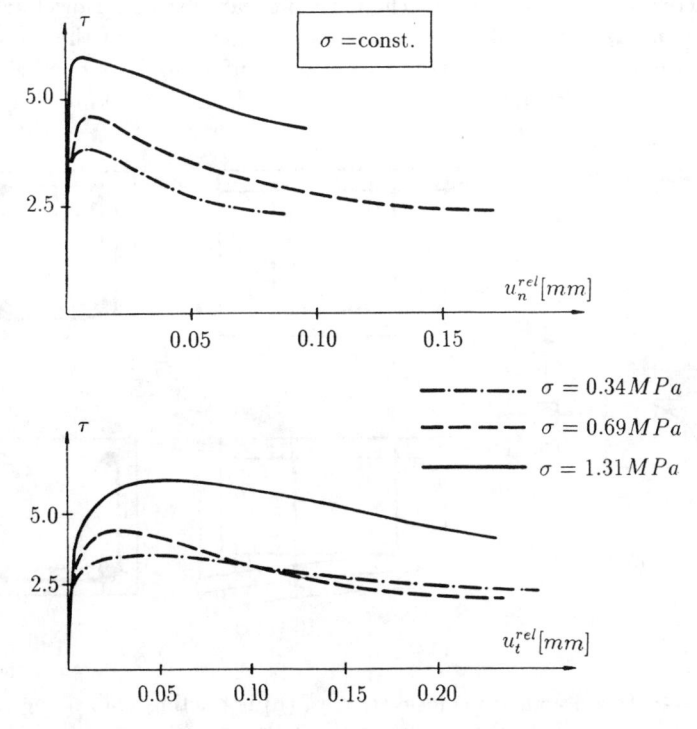

Figure 1.33: Results of direct shear tests at constant normal stress

Simple experimental investigations of aggregate interlock are conducted by means of direct shear tests. Either the compressive stress perpendicular to the crack plane or the crack width is kept constant [Daschner (1982)]. A more realistic test set-up allows for the variation of the normal compressive stress with increasing shear displacement and crack width. Obviously, it is by far more difficult to perform such tests. In order to reduce the complexity of these tests, the normal stresses and the shear stresses were decoupled in the tests reported by [Divakar (1987)]. Typical relationships between the shear stress, on the one hand, and the crack width and the shear displacement, on the other hand, at constant normal stress, are shown in Fig.1.33 for a concrete characterized by $\sigma_P^C = 35$ MPa [Divakar (1987)]. For cyclic loading a considerable difference between the results for the first cycle and the results for subsequent cycles is observed.

1.4.3 Dowel Action

In addition to aggregate interlock the reinforcement contributes to the transfer of shear forces accross cracks. These reinforcing bars act like dowels resisting the shear displacements at the crack faces. The effect of aggregate interlock decreases with increasing crack width. Therefore, for relatively large crack openings dowel action may become the dominating mechanism of shear transfer. Dowel action is a local phenomenon characterized by local bending and shearing of the bars and by strongly local compressive stresses exerted by the reinforcing bars on the adjacent concrete. There are also tensile stresses which in turn lead to cracking (Fig.1.34(a)) [CEB (1991)].

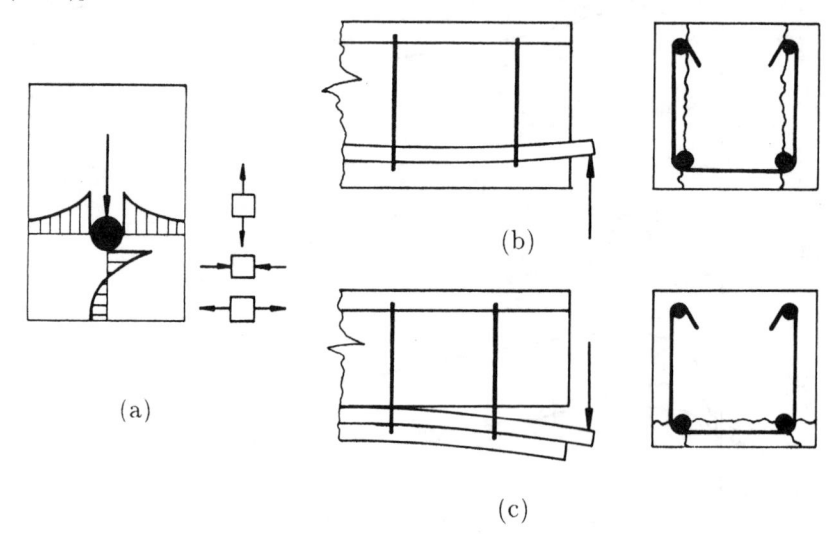

Figure 1.34: Dowel action: (a) local stresses, (b) bar acting against the concrete core, (c) bar acting against the concrete cover

1.4 Reinforced Concrete

Dowel action is different depending on whether the bar is acting against the concrete core or the concrete cover. In the first case the reinforcing bar is acting like a beam on elastic foundation. Failure usually occurs by splitting of the concrete under the bar in the plane formed by the dowel load and the dowel bar (Fig.1.34(b)) [Soroushian (1986b), Soroushian (1988)]. The ultimate resistance depends on the bearing strength of concrete, i.e., on the compressive strength of the confined concrete adjacent to the bar. It can be as large as three times the compressive strength of concrete [Soroushian (1987b)]. In the second case the reinforcing bar may be viewed as a bar on elastic foundation until cracks are formed in the concrete cover, separating the cover from the core. As the bar tends to slip longitudinally through the concrete, it acts like a wedge. This action contributes to the splitting of concrete. In the region of splitting cracks the bar is only supported by the stirrups. Failure may either occur as soon as the cover splits or well after concrete splitting. In the latter case yielding of the bar or of the stirrup closest to the crack faces is the reason for the failure [Soroushian (1987a)] (Fig.1.34(c)).

Tests [Vintzeleou (1987)] have shown that doubling of a small concrete cover may yield a substantial increase of dowel strength. An increase of the concrete cover from 20 to 40 mm, e.g., results in a 60% increase of dowel strength. Additional factors which have an influence on the dowel strength are the arrangement of the stirrups and the diameters of the reinforcing bars.

1.4.4 Experiments on Reinforced Concrete Specimens

Experiments on thirty plate-type concrete specimens with the dimensions of 890 × 890 × 70 mm were conducted at the University of Toronto. The specimens were reinforced with two layers of welded wire meshes consisting of plain bars. An extensive report is contained in [Vecchio (1982)]. A short summary of the experimental program together with a proposal for an analytical material model for reinforced concrete can be found in [Vecchio (1986)]. The specimens are characterized by different strengths of concrete and different yield strengths of the reinforcement. Moreover, the percentage of the reinforcing steel as well as the ratio of the amount of transverse to longitudinal reinforcement were varied. The specimens were subjected to various combinations of in-plane loading including pure shear, uniaxial compression and various combinations of shear and biaxial states of normal stresses. Reversed cyclic shear loading was also considered. The respective forces were applied by means of shear keys, each of which was connected with two hydraulic jacks (Fig.1.35). Most of the tests were carried out for proportional loading. However, also a few non-proportional loading histories were taken into account.

Four tests were selected for an international analytical competition to predict the behavior of reinforced concrete under different plane stress conditions. These tests were chosen such that the ultimate load carrying capacity was not controlled by yielding of the reinforcement. Hence, the quality of the constitutive models for both concrete and interface behavior played an important role in this competition. A summary of the results is contained in [Collins (1985)].

This experimental study was supplemented by a series of thirty panel tests [Bhide (1987)]. In this series of tests most of the specimens were only reinforced in one direction. The specimens were subjected to uniaxial tension, pure shear and to various combinations of uniaxial tension and shear, respectively. In a few cases a compressive loading was added to such a load combination in the transverse direction. Moreover, three additional tests on reinforced concrete elements subjected to reversed cyclic shear are documented in [Stevens (1987)].

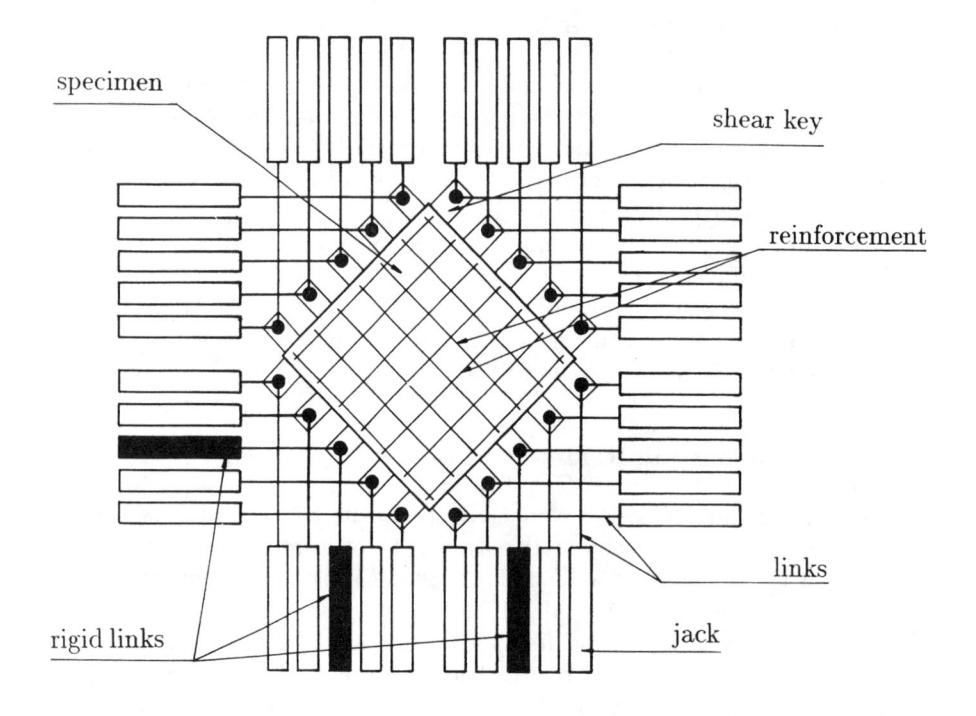

Figure 1.35: Test-setup for the experiments on reinforced concrete specimens, conducted by Vecchio and Collins

Although the first experiments of this series were conducted more than ten years ago, the test-setup is still considered as the best test setup yet conceived [Schnobrich (1991)]. The test results are a valuable contribution to a better understanding of the mechanical behavior of reinforced concrete. All of the described phenomena which govern the interface behavior of reinforced concrete, such as bond, aggregate interlock and dowel action, contribute to the load-carrying capacity of the test specimens. A point of criticism of the tests is the use of reinforcing meshes consisting of plain bars. Deformed bars would have better represented the conditions usually encountered in practice. Because of different mechanisms of bond for plain bars and deformed bars (see subsection 1.4.1), some of the experimental results, such as the ones for crack spacings and crack widths, cannot be directly applied to concrete

1.4 Reinforced Concrete

which is reinforced with deformed bars.

The most essential conclusions drawn from the experimental investigations are the following:

- Depending on the type of loading, for commonly used percentages of reinforcement distributed equally in the longitudinal and the transverse direction, three different types of failure of the test panels were observed. At high biaxial tensile stresses failure occurs in consequence of yielding of the reinforcement at cracks, whereas at high biaxial compressive stresses failure is governed by the compressive strength of concrete. If there is compression in one direction and tension in the other one, failure of concrete will occur at a principal compressive stress which may be considerably lower than the uniaxial compressive strength. This observation agrees well with the experimentally obtained failure envelope for biaxial stress states of plain concrete, obtained by Kupfer [Kupfer (1969), Kupfer (1973)].

- For a very small percentage of reinforcement, the ultimate load is reached at the initiation of concrete cracking. For a larger amount of reinforcement, failure occurs because of yielding of the reinforcing steel. For a very large amount of reinforcement, the ultimate load is governed by failure of concrete.

- Failure involving yielding of the reinforcement is preceded by ductile behavior of the panel, whereas failure of concrete without yielding of the reinforcement is preceded by brittle behavior.

- Initial cracks are perpendicular to the direction of the principal tensile stress obtained on the basis of linear elasticity. Additional cracks resulting from continued loading of specimens with different percentages of reinforcement in the longitudinal and the transverse direction, however, enclose an acute angle with the direction of the stronger reinforcement. This is the case especially after yielding of the weaker reinforcement. If concrete fails before yielding of the stronger reinforcement, then a single crack will occur. It is parallel to the direction of the stronger reinforcement. This crack extends across the full width of the panel.

This type of mechanical behavior is shown in Fig.1.36. The percentage of the reinforcement in the longitudinal direction is twice the one in the transverse direction. The specimen denoted as PV20 in [Vecchio (1982)] was subjected to monotonically increasing pure shear. Initial cracking was observed at a shear stress of $\tau_{xy} = 2.21$ MPa. The angle enclosed by the cracks and the longitudinal direction is approximately $45°$. Further loading resulted in an increase of the crack widths. At $\tau_{xy} = 4.14$ MPa, the weaker transverse reinforcement yielded. Thereafter, a pronounced rotation of the crack direction took place. The failure plane was almost parallel to the longitudinal direction. At failure, the compressive concrete stress was only about 40% of the uniaxial cylindrical strength.

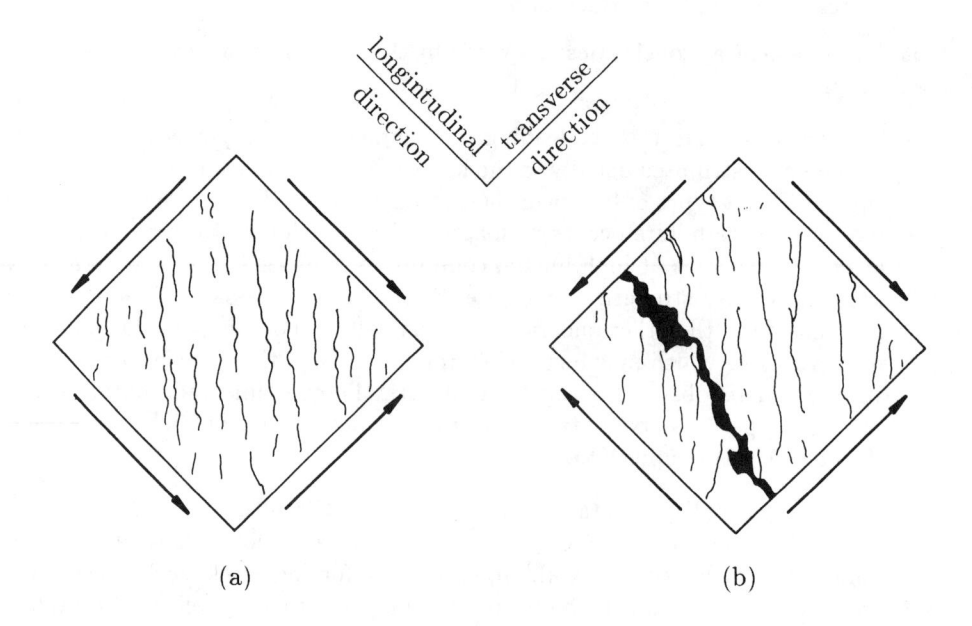

Figure 1.36: Panel PV20 (a) shortly before failure and (b) after failure

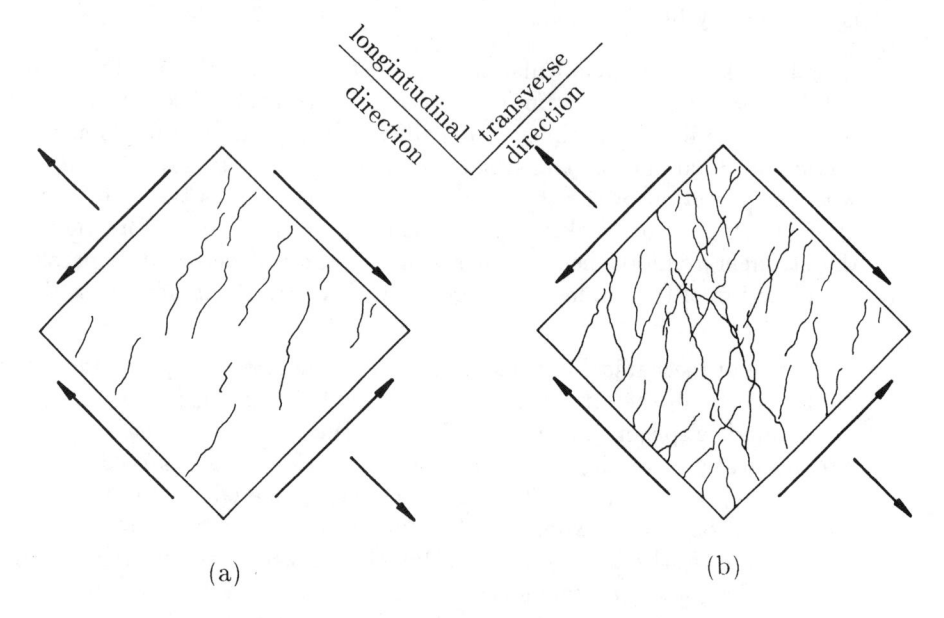

Figure 1.37: Panel PB21 (a) at about two thirds of the ultimate load and (b) at the ultimate load

1.4 Reinforced Concrete

A similar behavior is shown in Fig.1.37. It refers to a concrete panel with reinforcement only in the longitudinal direction. In [Bhide (1987)] this panel was denoted as PB21. The panel was subjected to uniaxial tension in the longitudinal direction and to shear loading with a constant ratio of 3.08 : 1. The primary cracks are approximately perpendicular to the direction of the principal tensile stress computed on the basis of linear elasticity. However, upon further loading the directions of newly formed cracks change gradually, approaching the direction of the reinforcement. Failure is characterized by a rapid increase of the widths of a few cracks.

- For shear loading combined with biaxial tension, cracking is initiated at a somewhat lower load level than for pure shear loading, Also the load-carrying capacity is somewhat lower. However, if shear loading is accompanied by compressive stresses, then the load level at the initiation of cracking is increased substantially. The same is true for the ultimate load.

- Loading histories including reversed cyclic loading or preloading in tension result in a decrease of the load-carrying capacity.

- The cracking strength of the concrete panels is lower than the tensile strength obtained from the split cylinder test or the double punch test. In the experiments [Bhide (1987)] the observed tensile strength of the panels was only 63% and 71% of the tensile strength from the split cylinder test and the double punch test, respectively.

- Because of the load-carrying capacity of the uncracked concrete between two adjacent cracks, the concrete is able to carry tensile forces between the cracks. The tensile forces are transmitted across cracks by bond slip, aggregate interlock and dowel action. If cracking occurs perpendicular to the reinforcement, then the tensile stresses in the cracked concrete will be smaller than for cracks, which are inclined to the direction of the reinforcement. The increase of tensile load-carrying capacity in the latter case is attributed to the action of aggregate interlock and dowel action.

Some of the findings are intuitively clear. Some of them have been corroborated in previous experiments on reinforced concrete beams. However, the experiments conducted at the University of Toronto are distinguished by the large variety of panels tested with a single test-setup. Hence, it is not only possible to compare the effects of changes in loading, concrete strength, yield strength of the reinforcing steel, layout and amount of the reinforcement, etc., on the behavior of reinforced concrete panels but also to quantify the individual effects.

Similar experiments on reinforced concrete panels were conducted by Mehlhorn and Kollegger [Mehlhorn (1990)].

2 Mathematical Models

2.1 Introduction

Till now there is no constitutive law for concrete, which is generally accepted. Nevertheless, a large number of material models based on different theories has been published within the last decades. Most of these constitutive models are characterized by acceptable deviations of numerical results from experimental data for a certain class of loading-paths. However, for more general loading histories these deviations may be unacceptable. The numerical modelling of the mechanical behavior of concrete is complicated by the fact that, depending on the state of stress, the material behavior of concrete ranges from brittle to ductile and failure of concrete in tension is different from failure in compression. Compressive failure of concrete is characterized by crushing and tensile failure by cracking.

Material models for concrete and reinforcing or prestressing steel consist of two main components, namely, of constitutive relations for the "intact" material and of failure criteria. Usually, constitutive models for concrete are not able to reproduce all of the experimentally observed material phenomena. The required complexity of a constitutive model strongly depends on the type of problem to be solved. Hence, even simple constitutive models can be applied successfully within the framework of finite element analyses, provided the limitations of these models are not exceeded in these analyses.

The constitutive relations can be classified according to the underlying theory. Frequently, they are subdivided into models based on the theory of elasticity [Saleeb (1982)] and models within the framework of the theory of plasticity [Chen (1982), Lubliner (1990)]. For three-dimensional stress-states failure criteria are defined in terms of ultimate-strength surfaces, and for states of plane stress in terms of ultimate-strength curves. Usually, different types of failure are treated separately. However, more recently a unified treatment of cracking and crushing was proposed [Lubliner (1989), Feenstra (1993a)]. It offers computational advantages.

Elastic materials are characterized by the fundamental assumption that the strains only depend on the current state of stress and on the temperature. For a given temperature there exists a one-to-one relationship between the state of stress and the state of strain. Such an assumption excludes time-dependent behavior such as creep under constant load or stress relaxation under constant strain.

However, almost all real materials exhibit some rate-dependence under cyclic loading and time-dependence under long-time static loading. Hence, for such materials the strains also depend on the history of the stress. Materials which exhibit rate-dependent or time-dependent behavior but have a well-defined natural state to

2.1 Introduction

which they eventually return when all the external actions are removed, are called viscoelastic materials. The special case of a linear dependence of the strains on the history of the stress is denoted as linear viscoelasticity.

Elastic models can also be classified on the basis of their mathematical formulation by means of total (or secant) or rate constitutive relations. The former are either Cauchy-elastic or hyperelastic models, the latter include the so-called hypoelastic models. For Cauchy-elastic materials the current stress σ is a function of the current strain ε and the temperature T, i.e.,

$$\sigma = \mathbf{f}(\varepsilon, T) \ . \tag{2.1}$$

A shortcoming of (2.1) is the possible violation of the laws of thermodynamics because energy may be generated for certain load cycles [Chen (1982)].

If, for a given temperature, the stresses of an elastic material are obtained from a strain-energy density function W as

$$\sigma = \frac{\partial W}{\partial \varepsilon} \ , \tag{2.2}$$

then no energy will be generated in any loading cycle. Hence, the laws of thermodynamics are satisfied. Such models are referred to as Green-elastic or hyperelastic. It follows from

$$C_{ijkl} = \frac{\partial^2 W}{\partial \varepsilon_{ij} \partial \varepsilon_{kl}} = \frac{\partial^2 W}{\partial \varepsilon_{kl} \partial \varepsilon_{ij}} = C_{klij} \tag{2.3}$$

that for hyperelastic materials the tensor of the material moduli $\mathbf{C}$ is symmetric with respect to the subscripts ij and kl. $\mathbf{C}$ is also known as the Hessian matrix of the potential function W. It can be shown that uniqueness and stable material behavior require W to be convex, yielding a positive definite matrix $\mathbf{C}$ [Chen (1982)]. Positive definiteness of $\mathbf{C}$ also insures a unique inverse of the constitutive relations. Material instability occurs, when $\mathbf{C}$ becomes singular.

Cauchy-elastic and hyperelastic material models are characterized by total stress-strain relations. Consequently, the state of strain is uniquely determined from the current state of stress. Hence, it does not depend on the loading history. It follows that no distinction between loading and unloading is made. Therefore, constitutive relations based on Cauchy-elasticity or hyperelasticity are only suitable for monotonic, proportional loading paths. Models belonging to this class are mere extensions of constitutive models based on linear elasticity. The constant material parameters are replaced by secant moduli [Cedolin (1977b)] or tangent moduli [Gerstle (1981a), Gerstle (1981b)], expressed in terms of stress or strain invariants. More sophisticated material models of this class include the coupling between the deviatoric stress and volumetric strain and vice versa [Kotsovos (1977), Kotsovos (1978), Kotsovos (1979)]. A different approach to constitutive models within this category are the equivalent-uniaxial models. They are characterized by considering concrete as a stress-induced orthotropic material [Liu (1972b), Ahmad (1986)]. For each of the principal axes of orthotropy the stress-strain response is described separatel· by means of equivalent-uniaxial constitutive relations.

In contrast to Cauchy-elastic and hyperelastic materials, hypoelastic constitutive relations are given in rate form [Truesdell (1955)], i.e.,

$$\dot{\sigma} = \mathbf{f}(\dot{\varepsilon}, \varepsilon, \sigma, T) \; . \tag{2.4}$$

For this class of models the stress rate may be a function of the strain rate, the total strain, the total stress and the temperature. The term "hypoelastic" denotes the fact that the response of such a material is regarded as elastic only on the rate level. Hence, the dependence of the stress-strain response on the loading history can be taken into account. Since the stresses are not derived from a potential function, the laws of thermodynamics may also be violated for hypoelastic models.

A frequently used special case of (2.4) is

$$\dot{\sigma} = \mathbf{C}(\sigma)\dot{\varepsilon} \quad \text{or} \quad \dot{\sigma} = \mathbf{C}(\varepsilon)\dot{\varepsilon} \; . \tag{2.5}$$

According to the degree of the dependence of $\mathbf{C}$ on σ or ε, hypoelastic models can be subdivided into zero-order models, first-order models, etc. A zero-order model is characterized by a tangent material stiffness matrix $\mathbf{C}$ with coefficients which do not depend on σ or ε. Hence, this special case represents a linear Cauchy-elastic material. In a first-order hypoelastic model the coefficients of $\mathbf{C}$ depend linearly on σ or ε. Typical representatives of this class of constitutive models are rate formulations of the equivalent-uniaxial type [Darwin (1977), Elwi (1979)] and invariant-based hypoelastic models [Stankowski (1985)].

Usually, elastic behavior is restricted to a certain range of stress and temperature. For loading beyond this range, parts of the deformations will remain even if the material is unloaded completely. Hence, depending on the load or temperature history of the material, inelastic deformations will be present even if all external actions are removed. The theory of plasticity allows determination of inelastic deformations.

Within the framework of this theory the dependence of the stresses on the history of the material can be described by means of an array of additional variables $\boldsymbol{\xi}$, denoted as internal variables. This gives

$$\sigma = \mathbf{f}(\varepsilon, T, \boldsymbol{\xi}) \; . \tag{2.6}$$

The internal variables serve as a means to describe hardening or degradation of the material. In the latter case they are referred to as plastic damage variables [Lubliner (1989)]. Since the internal variables describe irreversible material behavior, their evolution is governed by rate equations. Material models on the basis of the theory of plasticity are usually characterized by formulating the yield surface in the stress space. However, also strain-space plasticity models for concrete have been developed [Mizuno (1992), Pekau (1992)]. A brief summary of the theoretical foundations of the theory of plasticity, with special emphasis on stress-space plasticity, will be given at the beginning of subsection 2.3.3.

By analogy to the classification of elastic models as total and as rate-type constitutive relations, also elastic-plastic models can be subdivided into these two categories. Constitutive relations for models on the basis of the flow theory of plasticity are given in rate form. On the other hand, material models based on the deformation theory of plasticity are characterized by total constitutive relations. It is assumed that the total plastic strain can be determined from the stress. However, this theory is contradictory for loading paths consisting of unloading from the yield surface followed by loading to a different state of stress on the yield surface. Although the plastic strains should not change for such an unloading and reloading path, they do change on the basis of the deformation theory of plasticity.

Developments ranging beyond the scope of the classical theory of plasticity are the plastic-fracturing theory [Bažant (1979)] and the endochronic theory of plasticity [Bažant (1978b), Bažant (1980c)].

All of the aforementioned material models describe the constitutive behavior on the macroscopic level in a phenomenological manner. E.g., the internal variables, governing the material behavior, have no direct physical meaning. They are not related to the microstructural behavior. A different approach makes use of the mechanics of the microstructure in order to formulate the constitutive relations on the macrostructural level [Bažant (1988b), Bažant (1990)].

A large number of constitutive models has been developed during the past two decades. Summaries of elasticity-based models can be found in [Saleeb (1982), Eberhardsteiner (1991)], summaries of plasticity-based models are contained in [Chen (1982), Meschke (1991)]. One important reason for the differences between the numerically predicted behavior of concrete structures and the experimentally observed structural behavior is the application of inadequate material models. Hence, it is necessary to assess the range of applicability of these models.

In the following, a few selected material models, which have been used in finite element analyses, will be described in detail. An important common feature of the different constitutive models is an appropriate failure criterion. It allows to decide whether or not material failure will occur. Moreover, the failure criterion allows to distinguish between cracking and crushing of concrete. For this reason, the most widely used ultimate strength curve for biaxial stress states and a relatively sophisticated ultimate strength surface for triaxial stress states will be presented. In the following section the superscript C, standing for "concrete" is omitted, because there is no need for a distinction between different materials.

2.2 Failure Criteria

2.2.1 Introduction

Frequently, there is a difference between failure as occurring in a load-controlled experiment and failure as observed in a displacement-controlled experiment. In a load-controlled experiment, failure is associated with the maximum load which can be sustained by the specimen. The stress-strain relationship only consists of an ascending branch. The stresses at failure, σ_{Pi}, are peak values. The corresponding strains are denoted as ε_{Pi}. In a displacement-controlled experiment, failure is associated with the maximum displacement which can be sustained by the specimen. The strains at maximum displacement, denoted as ultimate strains ε_{Ui}, are often much larger than the strains ε_{Pi} at peak stress. The stresses at σ_{Ui}, however, are smaller than the peak stresses σ_{Pi}. The region beyond the strains at peak stress is characterized by descending stresses in the stress-strain diagram. This region is known as the softening region. Softening behavior can be seen, e.g., in Fig.1.1 and in Fig.1.4. In some cases, as in conventional triaxial compression tests with sufficiently high levels of confinement in the lateral direction, no softening region is observed.

For biaxial and triaxial stress states, failure is defined by failure curves in the stress plane and failure surfaces in the stress space, respectively. Both are also denoted as failure envelopes. Frequently, the terms failure curve, failure surface and failure envelope are associated with the maximum load. However, more precisely, they should be denoted as ultimate strength curves and ultimate strength surfaces. They should be distinguished from failure envelopes associated with the ultimate strain. For the construction of such failure envelopes till now there are almost no experimental data available.

2.2.2 Ultimate Strength Curve for Biaxial Stress States

The experimentally obtained ultimate strength envelope of concrete subjected to biaxial stress states, shown in Fig.1.8, is approximated analytically as [Kupfer (1973)]:

$$\left(\frac{\sigma_{P1}}{\sigma_P} + \frac{\sigma_{P2}}{\sigma_P}\right)^2 - \frac{\sigma_{P2}}{\sigma_P} - 3.65\frac{\sigma_{P1}}{\sigma_P} = 0, \qquad 1 \geq \alpha_2 \geq 0,$$

$$\frac{\sigma_{P1}}{\sigma_T} - \left(1 - \frac{\sigma_{P2}}{\sigma_P}\right)^{1/3} = 0, \qquad 0 \geq \alpha_2 \geq -\infty, \qquad (2.7)$$

$$\sigma_{P1} - \sigma_T = 0, \qquad \infty \geq \alpha_2 \geq 1.$$

In (2.7) $\sigma_{Pi}, i = 1, 2$, are the ultimate strengths under biaxial loading, expressed by the uniaxial compressive (prism) strength σ_P and the tensile strength σ_T; α_2 is a ratio given as $\alpha_2 = \sigma_1/\sigma_2$.

Failure because of crushing of concrete occurs for biaxial compression and for tension-compression loading provided the tensile stress is smaller than 1/15 of the absolute value of the compressive stress. This type of failure is characterized by the complete loss of the ability to carry forces. As shown in Figs.1.1 and 1.7, crushing of concrete under uniaxial and biaxial loading occurs at the ultimate strain ε_{Ui} which is larger than the strain ε_{Pi} corresponding to the ultimate strength. The stress at failure is smaller than σ_{Pi}.

2.2 Failure Criteria

A different type of failure is cracking. It is observed for tension-compression loading provided the tensile stress is larger than 1/15 of the absolute value of the compressive stress and for biaxial tension. Cracking will occur in the direction perpendicular to the principal tensile stress. If there is only one crack at a material point, then concrete will still carry forces parallel to the crack as long as $\sigma_P < \sigma_2 < \sigma_T$, where σ_2 is the stress parallel to the crack. σ_2 follows from the stress-strain relationship for a uniaxial state of stress.

As can be seen in Fig.1.8, in the tension-compression region ($\rightarrow \alpha_2 < 0$) both the tensile and the compressive strength of concrete are considerably lower than the respective uniaxial strengths. This effect is accounted for in (2.7$_2$).

Because of cracking the tensile stress is reduced to zero and, consequently, α_2 changes from an initially negative value to zero. Taking into account only the actual value for α_2, the degradation of the compressive strength in the tension-compression region will be cancelled erroneously after cracking [Crisfield (1989)]. The same problem is also encountered for three-dimensional stress states.

2.2.3 Ultimate Strength Surface for Triaxial Stress States

In order to determine the peak stresses σ_{Pi}, $i = 1, 2, 3$, for any ratio of the principal stresses σ_i, an ultimate strength surface for triaxial stress states is needed. Fig.1.11 shows an experimentally obtained ultimate strength surface of concrete. It is seen that the meridians of this surface are curved. The ultimate strength curves in the deviatoric planes change from a rather triangular-type of shape at small hydrostatic pressure to a rather circular shape at larger hydrostatic pressure. Extensive reviews of different types of ultimate strength criteria for concrete are given in [Chen (1982), Saleeb (1982)]. These reviews start with simple one-parameter criteria such as the maximum tensile stress criterion of Rankine for brittle fracture of concrete and the pressure-independent von Mises yield criterion and Tresca yield criterion for ductile materials. Then, two-parameter models such as the pressure-dependent Mohr-Coulomb criterion and Drucker-Prager criterion are discussed. The reviews close with more sophisticated four- and five-parameter models [Ottosen (1977), Willam (1975)]. In contrast to the two-parameter models, which are characterized by straight meridians and circular ultimate strength curves in the deviatoric planes, the four- and five-parameter models are distinguished by curved meridians and noncircular ultimate strength curves in the deviatoric planes. The five-parameter model, proposed by Willam and Warnke [Willam (1975)], has shown good agreement of the computed values of the ultimate strength with experimental data. This is true for different ratios of the principal stresses for a broad range of concrete strengths.

The experimentally obtained ultimate strength surface in Fig.1.11 is plotted in the principal stress space. For a given stress point in the principal stress space the component in the direction of the hydrostatic axis ξ is equal to $I_1/\sqrt{3}$. The hydrostatic axis is characterized by $\sigma_1 = \sigma_2 = \sigma_3$.

$$I_1 = \boldsymbol{\sigma} : \mathbf{1} \tag{2.8}$$

is the first invariant of the stress tensor. In (2.8) the symbol : denotes a tensor contraction ($\boldsymbol{\sigma} : \mathbf{1} = \sigma_{ij}\delta_{ij}$); $\mathbf{1}$ and δ_{ij} are the second-order unit tensor and the Kronecker delta, respectively. The distance ρ_u of a stress point on the ultimate strength surface from the hydrostatic axis is a function of I_1 and of the angle of similarity θ following from

$$\cos 3\theta = \frac{3\sqrt{3}}{2} \frac{J_3}{\sqrt{J_2^3}} . \tag{2.9}$$

J_2 and J_3 are the second and third invariant of the deviatoric stress tensor

$$\mathbf{s} = \boldsymbol{\sigma} - \sigma_0 \mathbf{1} . \tag{2.10}$$

According to (2.10), $\mathbf{s}$ is obtained by subtracting the hydrostatic stress

$$\sigma_0 = \frac{1}{3} I_1 \tag{2.11}$$

from the total stress. Consequently, the first invariant of the deviatoric stress tensor, J_1, is equal to zero. The norm of $\mathbf{s}$, denoted as ρ, is given as

$$\rho = \|\mathbf{s}\| = \sqrt{\mathbf{s} : \mathbf{s}} . \tag{2.12}$$

J_2 can be expressed in terms of $\mathbf{s}$ as follows:

$$J_2 = \frac{1}{2} \mathbf{s} : \mathbf{s} . \tag{2.13}$$

Combining (2.12) and (2.13) yields

$$\rho = \sqrt{2 J_2} . \tag{2.14}$$

The dependence of the ultimate strength surface on the hydrostatic stress can be formulated in terms of σ_0, I_1 or ξ. The distance of a stress point from the hydrostatic axis can be formulated in terms of ρ or J_2. In what follows, the variables σ_0 and ρ are preferred because they are directly related to the decomposition of the stress tensor into the hydrostatic and the deviatoric stress. For the derivation of (2.9) and for further geometric interpretations in the principal stress space reference to [Chen (1982), Saleeb (1982)] is made.

Because of threefold symmetry of the ultimate strength curve in an arbitrary deviatoric plane (Fig.1.11) it is sufficient to formulate the analytical expression for ρ_u for the region $0° \leq \theta \leq 60°$. The respective part of the ultimate strength curve is chosen as an ellipse, resulting in

$$\rho_u(\sigma_0, \theta) = \frac{2\rho_C(\rho_C^2 - \rho_T^2)\cos\theta + \rho_C(2\rho_T - \rho_C)\sqrt{4(\rho_C^2 - \rho_T^2)\cos^2\theta + 5\rho_T^2 - 4\rho_T\rho_C}}{4(\rho_C^2 - \rho_T^2)\cos^2\theta + (\rho_C - 2\rho_T)^2} , \tag{2.15}$$

2.2 Failure Criteria

where $\rho_C = \rho_C(\sigma_0)$ and $\rho_T = \rho_T(\sigma_0)$ are the values of ρ_u on the compressive and the tensile meridian, respectively [Willam (1975)].

The tensile and the compressive meridian of the ultimate strength surface are approximated as quadratic parabolae, i.e.,

$$\rho_T = a_0 + a_1\sigma_0 + a_2\sigma_0^2 , \qquad \theta = 0° ,$$
$$\rho_C = b_0 + b_1\sigma_0 + b_2\sigma_0^2 , \qquad \theta = 60° . \qquad (2.16)$$

Constraining the two meridian curves to intersect the hydrostatic axis at the same point, the six parameters $a_i, b_i, i = 0, 1, 2$, in (2.16) are reduced to five independent constants. For determination of these parameters, use of the uniaxial tensile and compressive strength, the strength for biaxial compresssion of the same intensity in both directions and of two additional stress points is made. These two points are characterized by a large hydrostatic pressure. One of them is located on the compressive meridian and the other one on the tensile meridian.

The values for the ratio ρ_C/ρ_T are restricted to

$$1 \leq \frac{\rho_C}{\rho_T} \leq 2 . \qquad (2.17)$$

For $\rho_C/\rho_T = 1$, the ellipse degenerates to a circle, yielding an ultimate strength surface with circular ultimate strength curves in the deviatoric planes. Hence, in this case the ultimate strength is independent of θ or, equivalently, independent of J_3. For $\rho_C/\rho_T = 2$, the elliptic curve degenerates to a straight line, resulting in triangular shapes of the ultimate strength curves in the deviatoric planes. Hence, in the limiting case the smoothness of the ultimate strength surface is lost.

Apart from purely tensile loading and nearly hydrostatic loading it is difficult to define the mode of material failure of concrete precisely. In [Han (1987)] different modes of failure are listed: cracking, crushing and a mixed type of failure. This specification of failure modes was originally proposed in [Hsieh (1982)].

Cracking is assumed to occur if

$$\sigma_1 > 0 , \quad \sigma_1 \geq \sigma_2 \geq \sigma_3 , \qquad (2.18)$$

holds for a stress point on the ultimate strength surface. (2.18) can be rewritten in terms of I_1 and J_2 as [Chen (1982)]

$$\sqrt{J_2} \cos\theta + \frac{I_1}{2\sqrt{3}} > 0 , \qquad |\theta| \leq 60°. \qquad (2.19)$$

Crushing is assumed to occur if

$$\varepsilon_1 < 0 , \quad \varepsilon_1 \geq \varepsilon_2 \geq \varepsilon_3 , \qquad (2.20)$$

implying $\varepsilon_2 < 0$ and $\varepsilon_3 < 0$. Using Hooke's law, (2.20) is rewritten as

$$\sigma_1 - \nu(\sigma_2 + \sigma_3) < 0 , \qquad \sigma_1 > \sigma_2 > \sigma_3 . \qquad (2.21)$$

At failure, the assumption of linear-elastic material behavior is not valid. Therefore, (2.21) can only be viewed as a means to obtain an approximate formula for the determination of the failure mode. Reformulation of (2.21) in terms of the stress invariants yields

$$\sqrt{J_2}\cos\theta + \frac{1-2\nu}{(1+\nu)}\frac{I_1}{2\sqrt{3}} < 0, \qquad |\theta| \leq 60°. \tag{2.22}$$

Defining the crushing coefficient α as

$$\alpha = \frac{-I_1}{2\sqrt{3}\sqrt{J_2}\cos\theta}, \tag{2.23}$$

if follows from (2.19) and (2.22) that

$$\begin{aligned} \alpha &< 1 & &\Rightarrow \text{cracking}, \\ 1 \leq \alpha &\leq \frac{1+\nu}{1-2\nu} & &\Rightarrow \text{mixed failure mode}, \\ \alpha &> \frac{1+\nu}{1-2\nu} & &\Rightarrow \text{crushing}. \end{aligned} \tag{2.24}$$

Crushing is accompanied by the complete loss of strength and stiffness. On the contrary, if a single crack is formed at a material point, then, for the three-dimensional case, the material at this point will be considered as transversely isotropic. Analogous to the biaxial case, additional cracks may be formed at a material point. For the mixed mode of failure, a multiaxial softening law was proposed [Han (1987)]. The fundamentals of this concept were reported in [Han (1987)]. However, specific assumptions for its application to concrete are missing. In [Meschke (1991)] the mixed failure mode was treated simply on the basis of ideally-plastic material behavior.

2.3 Time-Independent Material Models for Concrete

2.3.1 Linear-Elastic Fracture Models

Early applications of the finite element method to ultimate load analysis of concrete structures were based on linear-elastic fracture models [Ngo (1967)]. Hence, out of the two main causes for the nonlinearity of the material behavior, namely, the nonlinearity of the stress-strain relations for the intact concrete and cracking and crushing of concrete, only cracking and crushing was taken into account. Such material models may prove to be sufficiently accurate for structures where cracking is the most important source of the nonlinearity of the structural response. This is the case especially for beams, panels and thin shells.

These constitutive models represent the first attempts to consider the nonlinear behavior of concrete structures within the framework of finite element analyses.

2.3.2 Nonlinear-Elastic Fracture Models

Nonlinear Uniaxial Constitutive Relationship

A widely used expression for the numerical representation of the stress-strain curve for concrete subjected to uniaxial compression is an equation of the form [Kabaila (1964)]

$$\sigma = \frac{A + B\varepsilon}{1 + C\varepsilon + D\varepsilon^2} \ . \tag{2.25}$$

The constants A, B, C and D are determined from the following conditions (see Fig.(1.1)):

$$\begin{aligned} \text{for} \quad \varepsilon = 0 \ , & \quad \sigma = 0 \ , \\ & \quad \frac{d\sigma}{d\varepsilon} = E_0 \ , \\ \text{for} \quad \varepsilon = \varepsilon_P \ , & \quad \sigma = \sigma_P \ , \\ & \quad \frac{d\sigma}{d\varepsilon} = 0 \ , \end{aligned} \tag{2.26}$$

where E_0 is the initial tangent modulus of concrete and ε_P is the strain at the ultimate concrete strength σ_P. It follows that

$$\sigma = \frac{E_0 \varepsilon}{1 + \left(\dfrac{E_0 \varepsilon_P}{\sigma_P} - 2\right) \dfrac{\varepsilon}{\varepsilon_P} + \left(\dfrac{\varepsilon}{\varepsilon_P}\right)^2} \ . \tag{2.27}$$

A slightly different expression for σ is reported in [Bathe (1989)]. In order to obtain this expression, also the ultimate compressive strain and the corresponding stress at crushing were taken into account. It was noted, however, that it may be difficult to obtain accurate values for this point on the stress-strain curve.

For uniaxial tension, the stress-strain curve may be considered as linear up to the tensile strength σ_T.

Equivalent-Uniaxial Models for Biaxial Stress States

As was observed in experiments with biaxial and triaxial stress states, for monotonic proportional loading, the individual σ_i-ε_i curves in the principal directions are qualitatively similar. This has led to the idea of describing the biaxial or triaxial constitutive behavior of concrete by means of equivalent-uniaxial constitutive relations for each one of the principal directions. These equivalent-uniaxial stress-strain curves also serve as the basis for determination of the tangent stiffness moduli.

The constitutive equations for a homogeneous, isotropic, linear-elastic material under plane-stress conditions are given as

$$\left\{\begin{array}{c}\sigma_x\\ \sigma_y\\ \tau_{xy}\end{array}\right\} = \frac{E}{1-\nu^2}\begin{bmatrix}1 & \nu & 0\\ \nu & 1 & 0\\ 0 & 0 & \frac{1-\nu}{2}\end{bmatrix}\left\{\begin{array}{c}\varepsilon_x\\ \varepsilon_y\\ \gamma_{xy}\end{array}\right\}, \qquad (2.28)$$

where E and ν are the modulus of elasticity and Poisson's ratio, respectively, and σ_x (ε_x) and σ_y (ε_y) are the normal stress (strain) in the x- and y-direction, respectively, of a two-dimensional Cartesian coordinate system and τ_{xy} (γ_{xy}) is the shear stress (engineering shear strain). With the help of the ratio α_i of the principal stress σ_j orthogonal to the principal stress σ_i in the direction considered, i.e.,

$$\alpha_i = \frac{\sigma_j}{\sigma_i}, \quad i,j = 1,2, \quad i \neq j, \qquad (2.29)$$

(2.28) can be rewritten for principal directions as

$$\sigma_i = \frac{E\varepsilon_i}{1-\nu\alpha_i}. \qquad (2.30)$$

Hence, the material stiffness for biaxial loading of an isotropic linear-elastic material differs from the one for uniaxial loading by the factor $1/(1-\nu\alpha_i)$. The reason for this difference is the Poisson effect. It follows that multiplication of (2.25) (with ε_i instead of ε) by the factor $1/(1-\nu\alpha_i)$ results in the constitutive equation for an isotropic nonlinear-elastic material subjected to a plane state of stress. Consequently, this equation is given as

$$\sigma_i = \frac{A + B\varepsilon_i}{(1-\nu\alpha_i)(1+C\varepsilon_i + D\varepsilon_i^2)}, \quad i = 1,2, \qquad (2.31)$$

where σ_i and ε_i are the principal stresses and the principal strains, respectively.

As follows from Fig.1.7, the shape of the stress-strain curve and the peak stress strongly depend on the value of α_i. According to [Cedolin (1976)], the whole range of values for the ratio $\alpha_2 = \sigma_1/\sigma_2$ can be subdivided into a subdomain extending from biaxial compression to tension-compression ($1 \geq \alpha_2 > -0.325$), characterized by nonlinear material behavior, and a subdomain, extending from tension-compression to biaxial tension ($-0.325 \geq \alpha_2 \geq -\infty$ and $\infty \geq \alpha_2 \geq 1$), characterized by linear material behavior. The nonlinear region can be subdivided into a subregion of biaxial compression ($1 \geq \alpha_2 \geq 0$), for which the stress-strain curve for a given value of α_2 is characterized by a horizontal tangent at the peak stress, a subregion in the tension-compression domain, for which the condition of a horizontal tangent of the stress-strain curve at peak stress no longer holds ($-0.1 \geq \alpha_2 \geq -0.325$) and a subregion, characterized by a continuous transition between these two domains ($0 > \alpha_2 > -0.1$).

Introducing boundary conditions analogous to (2.26) and replacing E_0 in (2.26$_2$) by $E_0/(1-\nu\alpha_i)$, the constitutive relations for biaxial states of compressive stress (i.e., for the quadrant in Fig.1.8, characterized by negative values of σ_1 and σ_2, or, equivalently, by $1 \geq \alpha_2 \geq 0$) are obtained as [Liu (1972b)]

2.3 Time-Independent Material Models for Concrete

$$\sigma_i = \frac{E_0 \varepsilon_i}{(1-\nu\alpha_i)\left[1+\left(\dfrac{E_0 \varepsilon_{Pi}}{\sigma_{Pi}(1-\nu\alpha_i)}-2\right)\dfrac{\varepsilon_i}{\varepsilon_{Pi}}+\left(\dfrac{\varepsilon_i}{\varepsilon_{Pi}}\right)^2\right]}, \quad 1 \geq \alpha_2 \geq 0, \quad (2.32)$$

where σ_{Pi} and ε_{Pi} denote the peak stress and the corresponding strain for a certain value of α_i. In (2.32) the Poisson effect and the confinement of microcracking in the presence of biaxial compressive stresses are included. The former is considered by the factor $1/(1-\nu\alpha_i)$. The latter is accounted for by σ_{Pi} and ε_{Pi}, depending on α_i.

The condition (2.26$_4$) is not valid in the biaxial tension-compression region. Hence, for this region the condition (2.26$_4$) and the term $D\varepsilon^2$ in (2.31) must be omitted [Cedolin (1977)], yielding

$$\sigma_i = \frac{E_0 \varepsilon_i}{(1-\nu\alpha_i)\left[1+\left(\dfrac{E_0 \varepsilon_{Pi}}{\sigma_{Pi}(1-\nu\alpha_i)}-1\right)\dfrac{\varepsilon_i}{\varepsilon_{Pi}}\right]}, \quad -0.1 \geq \alpha_2 > -0.325 . \quad (2.33)$$

For the remaining part of the nonlinear region, characterized by $0 > \alpha_2 > -0.1$, an expression which guarantees a continuous transition between (2.32) and (2.33) is formulated as follows:

$$\sigma_i = \frac{E_0 \varepsilon_i}{(1-\nu\alpha_i)\left[1+\left(\dfrac{E_0 \varepsilon_{Pi}}{\sigma_{Pi}(1-\nu\alpha_i)}-1\right)\dfrac{\varepsilon_i}{\varepsilon_{Pi}}-(1+10\alpha_2)\left(1-\dfrac{\varepsilon_i}{\varepsilon_{Pi}}\right)\left(\dfrac{\varepsilon_i}{\varepsilon_{Pi}}\right)\right]}.$$
$$0 > \alpha_2 > -0.1 . \quad (2.34)$$

The constitutive equations for the entire range of values of α_2, including both the linear and the nonlinear material region, can be combined to one equivalent-uniaxial constitutive equation [Floegl (1981)]:

$$\sigma_i = \frac{E_0 \varepsilon_i}{(1-\nu\alpha_i)\left[1+g_1(\alpha_2)\left(\dfrac{E_0 \varepsilon_{Pi}}{\sigma_{Pi}(1-\nu\alpha_i)}-1\right)\dfrac{\varepsilon_i}{\varepsilon_{Pi}}-g_2(\alpha_2)\left(1-\dfrac{\varepsilon_i}{\varepsilon_{Pi}}\right)\dfrac{\varepsilon_i}{\varepsilon_{Pi}}\right]},$$
$$(2.35)$$

where the functions $g_1(\alpha_2)$ and $g_2(\alpha_2)$ are defined as

$$g_1(\alpha_2) = \begin{cases} 1 & \text{for } 1 \geq \alpha_2 > -0.325 \\ 0 & \text{otherwise} \end{cases}, \quad (2.36)$$

$$g_2(\alpha_2) = \begin{cases} 1 & \text{for } 1 \geq \alpha_2 \geq 0 \\ 1+10\alpha_2 & \text{for } 0 \geq \alpha_2 \geq -0.1 \\ 0 & \text{otherwise.} \end{cases} \quad (2.37)$$

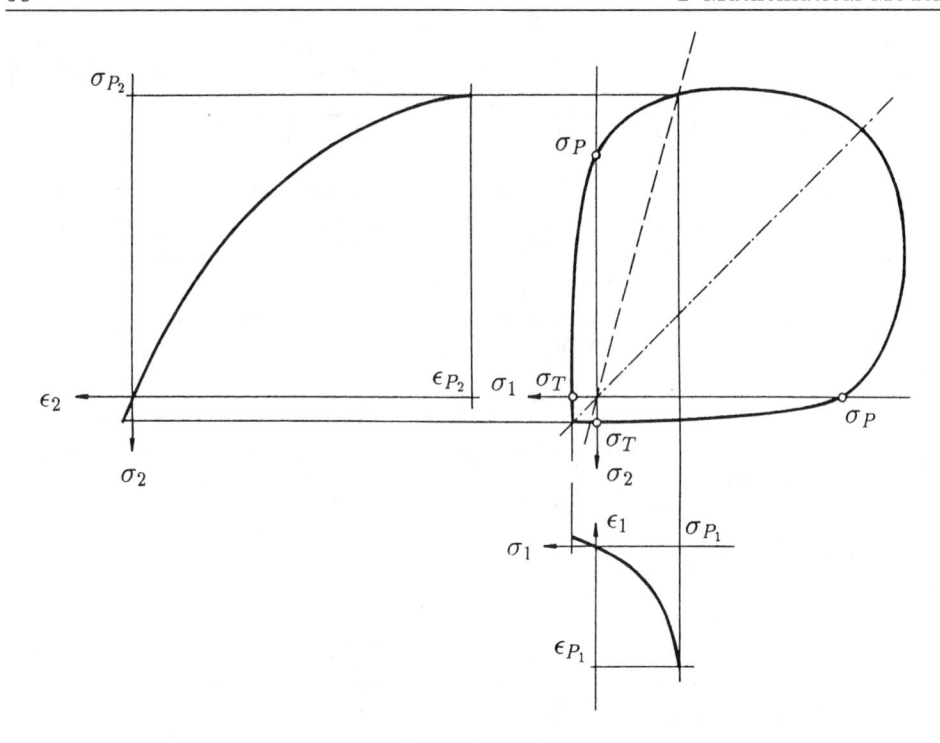

Figure 2.1: Ultimate strength envelope and equivalent-uniaxial stress-strain curves

(2.35) is a total (or secant) stress-strain relation. The stress is uniquely determined from the current strain. Hence, the stress-strain response does not depend on the loading history and no difference between loading and unloading is made. The ultimate strengths σ_{Pi}, $i = 1, 2$, are computed from the ultimate strength envelope which is based on the biaxial experiments conducted by Kupfer [Kupfer (1973)] and described in subsection 2.2.2. Elimination of σ_{P_1} in (2.7) by means of $\alpha_2 = \sigma_{P_1}/\sigma_{P_2}$ yields σ_{P2} as a function of σ_P and/or σ_T and of α_2. The peak stress in the perpendicular direction is $\sigma_{P1} = \alpha_2 \sigma_{P2}$. For the strain ε_{Pi} at the ultimate strength σ_{Pi} piecewise linear approximations of the experimental results [Kupfer (1973)] shown in Fig.1.9 are used. Fig.2.1 contains the ultimate strength envelope according to (2.7) and the equivalent-uniaxial stress-strain curves for a particular value of α_2 [Walter (1988)]. Fig.2.2 allows a comparison of the computed stress-strain curves for two different stress ratios with the corresponding experimentally obtained curves [Floegl (1981)].

The data used for the formulation of equivalent-uniaxial stress-strain relations were obtained from biaxial tests. Consequently, these relations cannot be applied to triaxial states of stress. In particular, the described constitutive model for biaxial states of stress does not account for the pronounced influence of the hydrostatic pressure on the mechanical behavior of concrete subjected to triaxial stress states.

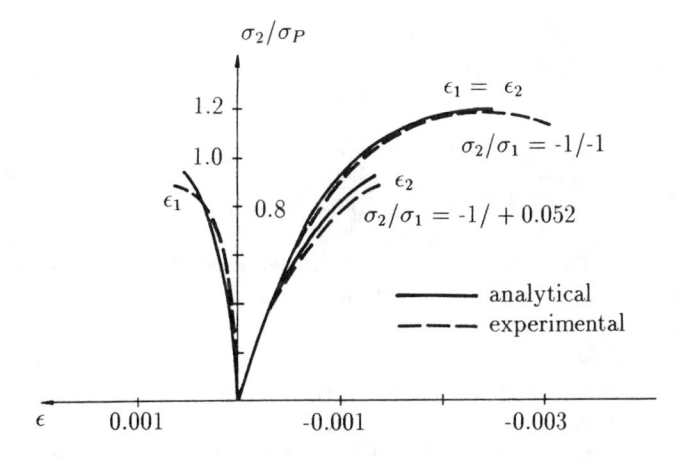

2.3 Time-Independent Material Models for Concrete

Figure 2.2: Computed and experimentally obtained stress-strain curves

Computation of the principal stresses for given principal strains according to (2.35) is relatively complicated because the ratio α_i is not known *a priori*. Moreover, the ultimate strength σ_{Pi} and the corresponding strain ε_{Pi} depend on α_i. Hence, the principal stresses must be determined iteratively.

In general, the material properties of concrete in the two principal directions change during loading. Assuming, e.g., proportional tension-compression loading, the tangent modulus in the direction of the tensile stress can be considered as constant up to failure. The tangent modulus in the direction of the compressive stress, however, decreases with increasing loading. Hence, concrete behaves like an orthotropic material. For a linear-elastic orthotropic material under plane stress conditions the constitutive equations are given as

$$\left\{ \begin{array}{c} \sigma_{11} \\ \sigma_{22} \\ \tau_{12} \end{array} \right\} = \frac{1}{1-\nu_1\nu_2} \left[\begin{array}{ccc} E_1 & \nu_2 E_1 & 0 \\ \nu_1 E_2 & E_2 & 0 \\ 0 & 0 & (1-\nu_1\nu_2)G \end{array} \right] \left\{ \begin{array}{c} \varepsilon_{11} \\ \varepsilon_{22} \\ \gamma_{12} \end{array} \right\}, \qquad (2.38)$$

where E_1, ν_1 and E_2, ν_2 are the moduli of elasticity and Poisson's ratios in the directions of the principal axes of orthotropy and G denotes the shear modulus with respect to these axes. Hence, in (2.38) the stresses and the strains are referred to the principal axes of orthotropy. Because of the symmetry requirement for hyperelastic materials (see equation (2.3)), the condition

$$\nu_2 E_1 = \nu_1 E_2 \qquad (2.39)$$

must hold. As follows from the tangent material stiffness matrix in (2.38), for an orthotropic material coupling between normal strains and shear stresses, on the one hand, and between shear strains and normal stresses, on the other hand, is not taken into account.

For a nonlinear orthotropic material the constitutive equations for plane stress conditions, referred to principal axes of orthotropy, can be formulated in rate form as

$$\left\{ \begin{array}{c} \dot{\sigma}_{11} \\ \dot{\sigma}_{22} \\ \dot{\tau}_{12} \end{array} \right\} = \frac{1}{1-\nu_1\nu_2} \left[\begin{array}{ccc} E_{T1} & \nu_2 E_{T1} & 0 \\ \nu_1 E_{T2} & E_{T2} & 0 \\ 0 & 0 & (1-\nu_1\nu_2)G_T \end{array} \right] \left\{ \begin{array}{c} \dot{\varepsilon}_{11} \\ \dot{\varepsilon}_{22} \\ \dot{\gamma}_{12} \end{array} \right\}, \quad (2.40)$$

where E_{T1} and E_{T2} are the tangent moduli in the directions of the principal axes of orthotropy and G_T denotes the tangent shear modulus with respect to these axes.

The orthotropic behavior of concrete is stress-induced, i.e., in its initial stress-free state concrete is an isotropic material. The principal material axes of stress-induced orthotropic materials are not known *a priori*. These directions depend on the state of stress [Bažant (1983)].

A basic assumption for stress-induced orthotropy is that the principal axes of orthotropy coincide with the principal directions of stress. Hence, making use of (2.39), the rate-type constitutive equations for consideration of the nonlinear material behavior of concrete, formulated in the principal directions of stress, are obtained from (2.40) as

$$\left\{ \begin{array}{c} \dot{\sigma}_1 \\ \dot{\sigma}_2 \end{array} \right\} = \frac{E_{T1}}{\frac{E_{T1}}{E_{T2}} - \nu^2} \left[\begin{array}{cc} E_{T1}/E_{T2} & \nu \\ \nu & 1 \end{array} \right] \left\{ \begin{array}{c} \dot{\varepsilon}_1 \\ \dot{\varepsilon}_2 \end{array} \right\}. \quad (2.41)$$

In (2.41) $\nu = \nu_1$ is assumed to be equal to the initial value of Poisson's ratio, as obtained from a uniaxial compression test. Alternatively, an "equivalent" Poisson's ratio, defined as $\nu = \sqrt{\nu_1\nu_2}$, could be introduced into (2.38) [Darwin (1976)]. Since Poisson's ratio is assumed to be constant, the increase in volume, occurring shortly before the peak stress is reached, cannot be taken into account in this constitutive model. In order to obtain the uniaxial tangent moduli, the fictitious uniaxial stress-strain relationship

$$\sigma_i^{(f)} = \sigma_i(1 - \nu\alpha_i) \quad (2.42)$$

was defined in [Liu (1972b)]. Substitution of (2.35) into (2.42) shows that the Poisson effect is not considered in the expression for $\sigma_i^{(f)}$. Differentiation of this expression with respect to ε_i yields

$$\frac{d\sigma_i^{(f)}}{d\varepsilon_i} = \frac{E_0\left[1 - \left(\frac{\varepsilon_i}{\varepsilon_{Pi}}\right)^2\right]}{(1-\nu\alpha_i)\left\{1 + g_1(\alpha_2)\left[\frac{E_0\varepsilon_{Pi}}{\sigma_{Pi}(1-\nu\alpha_i)} - 1\right]\frac{\varepsilon_i}{\varepsilon_{Pi}} - g_2(\alpha_2)\left(1 - \frac{\varepsilon_i}{\varepsilon_{Pi}}\right)\frac{\varepsilon_i}{\varepsilon_{Pi}}\right\}^2}.$$

$$(2.43)$$

2.3 Time-Independent Material Models for Concrete

In [Liu (1972b)] $d\sigma_i^{(f)}/d\varepsilon_i$ is interpreted as the uniaxial tangent modulus E_{Ti} of (2.41).

Because of formulation of the constitutive equations in the principal directions, the shearing behavior is not addressed. In a nonlinear analysis these directions usually change as the applied load is changed. Hence, in general, the tensor of incremental stresses obtained for a certain load increment within the framework of incremental finite element analysis, also contains shear stresses referred to the principal directions of the preceding load increment. Because of lack of experimentally obtained information, the definition of the shear modulus G_T in [Liu (1972b)] is based on the assumption of the invariance of $1/G_T$ with respect to the rotation of the coordinate axes. For such a rotation the components $D_{T_{ijkl}}$ of the tangential compliance tensor $\mathbf{D}_T$ transform as a fourth-order tensor. Hence,

$$D_{T_{pqrs}} = n_{pi} n_{qj} n_{rk} n_{sl} D_{T_{ijkl}} , \qquad (2.44)$$

where

$$n_{ij} = \cos(x_i', x_j) \qquad (2.45)$$

denotes the direction cosine of the angle enclosed by the rotated coordinate axis x_i' and the original axis x_j. Thus, $1/G_T = 2D_{T_{1212}}$ cannot change if the coordinate axes x_1 and x_2 rotate about the x_3-axis perpendicular to the plane in which the stresses are acting. It follows from (2.44), making use of (2.39), that this invariance condition is satisfied for

$$G_T = \frac{E_{T1} E_{T2}}{E_{T1} + E_{T2} + 2\nu E_{T2}} \qquad (2.46)$$

where E_{Ti}, $i = 1, 2$, are the uniaxial tangent moduli according to (2.43). In [Darwin (1976)] the shear modulus is taken as $G_T = 0.25(E_{T1} + E_{T2} - 2\nu\sqrt{E_{T1}E_{T2}})$. This expression was derived by requiring G_T to be independent of any particular direction.

After reaching the ultimate strength, concrete either fails because of crushing or cracking. The type of failure is determined by the ratio of the principal stresses (see subsection 2.2.2).

Within the framework of finite element analyses, stresses and strains and the tangent stiffness tensor must be transformed from the principal directions of orthotropy to the directions of the axes of the chosen global system of reference. The tangent stiffness tensor is transformed according to (2.44). The stresses and strains are transformed according to the transformation law of second-order tensors, i.e.,

$$\sigma_{kl} = n_{ki} n_{lj} \sigma_{ij} , \qquad \varepsilon_{kl} = n_{ki} n_{lj} \varepsilon_{ij} . \qquad (2.47)$$

The described total constitutive model was recast by Walter in rate form to account for nonproportional loading as well as unloading and reloading [Walter (1988)]. Darwin reported on an orthotropic constitutive law in rate form [Darwin (1976), Darwin (1977)]. This material law is also based on equivalent-uniaxial stress-strain relations. It was extended to allow for consideration of general biaxial cyclic loading. The description of the material behavior under cyclic loading is based on experimental work by Karsan and Jirsa [Karsan (1969)].

Equivalent-Uniaxial Models for Triaxial Stress States

A three-dimensional hypoelastic orthotropic material model for concrete within the framework of the equivalent-uniaxial stress-strain concept was developed by Elwi [Elwi (1979)]. Assuming the principal axes of orthotropy to coincide with the principal stress directions, the constitutive equations can be written in rate form as

$$\left\{\begin{array}{c}\dot{\varepsilon}_1\\ \dot{\varepsilon}_2\\ \dot{\varepsilon}_3\end{array}\right\} = \left[\begin{array}{ccc} E_{T1}^{-1} & -\nu_{12}E_{T2}^{-1} & -\nu_{13}E_{T3}^{-1} \\ -\nu_{21}E_{T1}^{-1} & E_{T2}^{-1} & -\nu_{23}E_{T3}^{-1} \\ -\nu_{31}E_{T1}^{-1} & -\nu_{32}E_{T2}^{-1} & E_{T3}^{-1} \end{array}\right] \left\{\begin{array}{c}\dot{\sigma}_1\\ \dot{\sigma}_2\\ \dot{\sigma}_3\end{array}\right\}. \qquad (2.48)$$

Symmetry of the tangent compliance matrix requires

$$\nu_{12}E_{T2} = \nu_{21}E_{T1}, \quad \nu_{13}E_{T3} = \nu_{31}E_{T1}, \quad \nu_{23}E_{T3} = \nu_{32}E_{T2}. \qquad (2.49)$$

Making use of (2.49) yields

$$\left\{\begin{array}{c}\dot{\varepsilon}_1\\ \dot{\varepsilon}_2\\ \dot{\varepsilon}_3\end{array}\right\} = \left[\begin{array}{ccc} E_{T1}^{-1} & -\mu_{12}(E_{T1}E_{T2})^{-1/2} & -\mu_{13}(E_{T1}E_{T3})^{-1/2} \\ & E_{T2}^{-1} & -\mu_{23}(E_{T2}E_{T3})^{-1/2} \\ \text{sym.} & & E_{T3}^{-1} \end{array}\right] \left\{\begin{array}{c}\dot{\sigma}_1\\ \dot{\sigma}_2\\ \dot{\sigma}_3\end{array}\right\}, (2.50)$$

where

$$\mu_{12} = \sqrt{\nu_{12}\nu_{21}}, \quad \mu_{13} = \sqrt{\nu_{13}\nu_{31}}, \quad \mu_{23} = \sqrt{\nu_{23}\nu_{32}}. \qquad (2.51)$$

Solution of (2.50) for the stress rates yields the rate form of the stress-strain relations:

$$\dot{\sigma} = \mathbf{C}_T \dot{\varepsilon}. \qquad (2.52)$$

The tangent stiffness matrix $\mathbf{C}_T$ is given as

$$\mathbf{C}_T = \phi \left[\begin{array}{ccc} (1-\mu_{23}^2)E_{T1} & (\mu_{13}\mu_{23}+\mu_{12})\sqrt{E_{T1}E_{T2}} & (\mu_{12}\mu_{23}+\mu_{13})\sqrt{E_{T1}E_{T3}} \\ & (1-\mu_{13}^2)E_{T2} & (\mu_{12}\mu_{13}+\mu_{23})\sqrt{E_{T2}E_{T3}} \\ \text{sym.} & & (1-\mu_{12}^2)E_{T3} \end{array}\right] \qquad (2.53)$$

with

$$\phi = \frac{1}{1-\mu_{12}^2-\mu_{23}^2-\mu_{13}^2-2\mu_{12}\mu_{13}\mu_{23}}. \qquad (2.54)$$

Specialization of (2.53) and (2.54) for plane stress conditions yields the tangent stiffness matrix for the rate form of the orthotropic constitutive relations for biaxial stress states, proposed in [Darwin (1976), Darwin (1977)].

The constitutive relations (2.52) with $\mathbf{C}_T$ according to (2.53) can be recast on the basis of fictitious equivalent-uniaxial strains ε_i^{eq}. Such a strain is defined as the strain which would be obtained if a given principal stress σ_i represented a uniaxial state of stress rather than an ingredient of a multiaxial state of stress. Hence, the rates of the fictitious equivalent-uniaxial strains are given as

2.3 Time-Independent Material Models for Concrete

$$\dot{\varepsilon}_i^{eq} = \frac{\dot{\sigma}_i}{E_i} \quad \rightarrow \quad d\varepsilon_i^{eq} = \frac{d\sigma_i}{E_i}\ . \tag{2.55}$$

The total fictitious equivalent-uniaxial strains are obtained as

$$\varepsilon_i^{eq} = \int \frac{d\sigma_i}{E_i}\ . \tag{2.56}$$

The σ_i-ε_i^{eq} diagrams permit determination of the tangent stiffness moduli. It follows from (2.50) that $\dot{\varepsilon}_i^{eq}$ can be interpreted as the strain rate in the direction, indicated by the subscript i, which would occur if the material was subjected to the uniaxial stress rate $\dot{\sigma}_i$. Hence, except for the uniaxial case, the equivalent-uniaxial strains ε_i^{eq} are fictitious quantities. In order to identify these quantities, (2.52) is formally rewritten as

$$\left\{\begin{array}{c}\dot{\sigma}_1\\ \dot{\sigma}_2\\ \dot{\sigma}_3\end{array}\right\} = \left[\begin{array}{ccc}E_{T1}B_{11} & E_{T1}B_{12} & E_{T1}B_{13}\\ E_{T2}B_{21} & E_{T2}B_{22} & E_{T2}B_{23}\\ E_{T3}B_{31} & E_{T3}B_{32} & E_{T3}B_{33}\end{array}\right]\left\{\begin{array}{c}\dot{\varepsilon}_1\\ \dot{\varepsilon}_2\\ \dot{\varepsilon}_3\end{array}\right\}\ . \tag{2.57}$$

The coefficients B_{ij} are obtained by comparison of the matrix coefficients in (2.57) with those in (2.53). It follows from (2.57) that

$$\dot{\sigma}_i = E_{Ti}(B_{i1}\dot{\varepsilon}_1 + B_{i2}\dot{\varepsilon}_2 + B_{i3}\dot{\varepsilon}_3) = E_{Ti}\dot{\varepsilon}_i^{eq}\ . \tag{2.58}$$

In (2.58) the expression in parenthesis can be regarded as the rate of the equivalent-uniaxial strain,

$$\dot{\varepsilon}_i^{eq} = B_{i1}\dot{\varepsilon}_1 + B_{i2}\dot{\varepsilon}_2 + B_{i3}\dot{\varepsilon}_3\ . \tag{2.59}$$

This interpretation yields constitutive relations which are formally analogous to the relation for a uniaxial state of stress. (2.53) is referred to the principal axes of orthotropy. These axes are assumed to coincide with the principal stress axes. The coefficients B_{ij} are obtained by comparing respective coefficients of the tangent stiffness matrices in (2.53) and (2.57). Hence, the equivalent-uniaxial strains are also defined in the principal stress directions. It follows from (2.56) that the definition of the equivalent-uniaxial strains becomes questionable when the loading history causes the principal stress axes to rotate. In this case ε_i^{eq} is the result of the integration over a stress differential $d\sigma_i$ which does not maintain its direction.

The relationship between the principal stresses and the fictitious equivalent-uniaxial strains is based on a generalization of (2.25), yielding

$$\sigma_i = \frac{E_0\varepsilon_i^{eq}}{1 + \left(R + \frac{E_0\varepsilon_{Pi}^{eq}}{\sigma_{Pi}} - 2\right)\left(\frac{\varepsilon_i^{eq}}{\varepsilon_{Pi}^{eq}}\right) - (2R-1)\left(\frac{\varepsilon_i^{eq}}{\varepsilon_{Pi}^{eq}}\right)^2 + R\left(\frac{\varepsilon_i^{eq}}{\varepsilon_{Pi}^{eq}}\right)^3}\ , \tag{2.60}$$

where

$$R = \frac{E_0\varepsilon_{Pi}^{eq}}{\sigma_{Pi}}\cdot\frac{\frac{\sigma_{Pi}}{\sigma_{Si}} - 1}{\left(\frac{\varepsilon_{Si}^{eq}}{\varepsilon_{Pi}^{eq}} - 1\right)^2} - \frac{\varepsilon_{Pi}^{eq}}{\varepsilon_{Si}^{eq}}\ . \tag{2.61}$$

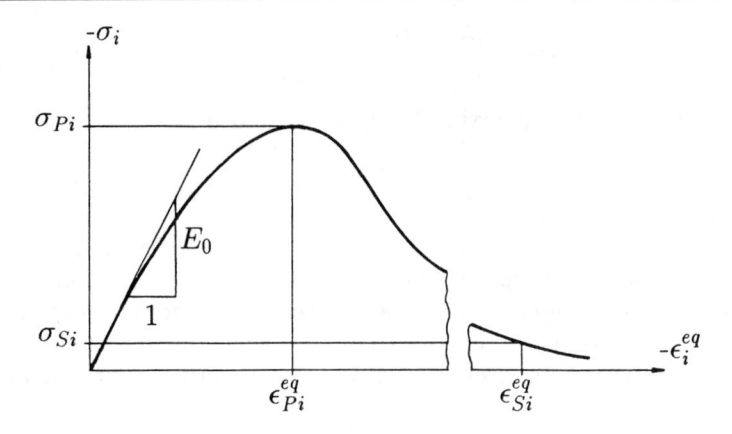

Figure 2.3: Stress - equivalent uniaxial strain curve

In (2.60) and (2.61) σ_{Pi} is the peak stress in the principal direction defined by the subscript i, ε_{Pi}^{eq} is the corresponding fictitious equivalent-uniaxial strain and σ_{Si} and ε_{Si}^{eq} are the stress and the corresponding fictitious equivalent-uniaxial strain for an arbitrary point on the stress - equivalent strain curve in the post-peak region. Fig.2.3 illustrates the stress - equivalent strain diagram according to (2.60). The parameters chosen for the definition of this curve are shown in the figure.

The tangent moduli E_{Ti} are obtained from (2.60) as

$$E_{Ti} = \frac{d\sigma_i}{d\varepsilon_i^{eq}} = \frac{E_0 \left[1 + (2R-1)\left(\frac{\varepsilon_i^{eq}}{\varepsilon_{Pi}^{eq}}\right)^2 - 2R\left(\frac{\varepsilon_i^{eq}}{\varepsilon_{Pi}^{eq}}\right)^3 \right]}{\left[1 + \left(R + \frac{E_0 \varepsilon_{Pi}^{eq}}{\sigma_{Pi}} - 2\right)\left(\frac{\varepsilon_i^{eq}}{\varepsilon_{Pi}^{eq}}\right) - (2R-1)\left(\frac{\varepsilon_i^{eq}}{\varepsilon_{Pi}^{eq}}\right)^2 + R\left(\frac{\varepsilon_i^{eq}}{\varepsilon_{Pi}^{eq}}\right)^3 \right]^2}. \tag{2.62}$$

An empirical expression for Poisson's ratio is determined from data based on uniaxial compression [Kupfer (1969)]. On the basis of this expression for uniaxial compression, three independent Poisson's ratios are defined as follows:

$$\nu_i = \nu_0 \left[1 + 1.3763 \left(\frac{\varepsilon_i^{eq}}{\varepsilon_{Pi}^{eq}}\right) - 5.36 \left(\frac{\varepsilon_i^{eq}}{\varepsilon_{Pi}^{eq}}\right)^2 + 8.586 \left(\frac{\varepsilon_i^{eq}}{\varepsilon_{Pi}^{eq}}\right)^3 \right], i = 1, 2, 3. \tag{2.63}$$

In (2.63) ν_0 denotes the initial value of Poisson's ratio as obtained from a standard uniaxial compression test. (2.51) can now be recast in terms of ν_i as

$$\nu_{12} = \sqrt{\nu_1 \nu_2}, \qquad \nu_{23} = \sqrt{\nu_2 \nu_3}, \qquad \nu_{13} = \sqrt{\nu_1 \nu_3}. \tag{2.64}$$

In order to ensure that ϕ in (2.53) is non-negative, values for ν_i according to (2.63) which exceed 0.5 are not admitted. The limiting value of Poisson's ratio of 0.5 corresponds to a vanishing rate of volume change.

2.3 Time-Independent Material Models for Concrete

The peak stresses or ultimate strengths σ_{Pi} for any ratio of the principal stresses are determined by means of the five-parameter ultimate strength surface proposed by Willam and Warnke [Willam (1975)] (see subsection 2.2.3). For the purpose of defining the fictitious equivalent-uniaxial strains ε_{Pi}^{eq}, corresponding to the peak stresses σ_{Pi}, a suitable surface in the equivalent-uniaxial strain space was assumed [Elwi (1979)]. This surface has the same shape as the five-parameter ultimate strength surface. However, experimental results are only available from the uniaxial tension test and the uniaxial compression test. The fictitious equivalent-uniaxial strains corresponding to the three remaining points for definition of the ultimate strength surface (see subsection 2.2.3) cannot be obtained directly from experiments, because, for multiaxial loading, the equivalent-uniaxial strains are fictitious quantities. As a remedy it was proposed in [Elwi (1979)] to determine these fictitious equivalent-uniaxial strains by trial and error. An alternative is determination of the fictitious equivalent-uniaxial strains according to an empirical σ_{Pi}-ε_{Pi}^{eq} relationship obtained from triaxial experiments as [Buyukozturk (1985)]

$$\frac{\sigma_{Pi}}{\sigma_P} = 0.764 + 0.211 \frac{\varepsilon_{Pi}^{eq}}{\varepsilon_P} + 0.025 \left(\frac{\varepsilon_{Pi}^{eq}}{\varepsilon_P}\right)^2 . \tag{2.65}$$

The constitutive model is completed by defining an arbitrary point on the σ_i-ε_i^{eq} diagram in the softening region. The experimentally obtained shape of the softening branch strongly depends on the type of the testing device. Hence, instead of taking test data, σ_{Si} and ε_{Si}^{eq} are chosen as

$$\sigma_{Si} = 0.25 \sigma_{Pi} , \qquad \varepsilon_{Si}^{eq} = 4\varepsilon_{Pi}^{eq} . \tag{2.66}$$

Stress - equivalent-strain curves are key items for this constitutive model. Difficulties to obtain realistic values for the parameters defining these corresponding relationships are serious drawbacks of this material model.

In the form in which this constitutive model was proposed in [Elwi (1979)] it does not contain a criterion allowing to distinguish between loading and unloading. For uniaxial loading and proportional multiaxial loading the distinction between virgin loading, unloading and reloading is obvious. However, this is not the case for nonproportional multiaxial loading. In [Kotsovos (1978)] a criterion in terms of octahedral stresses was proposed, permitting this distinction. In [Stankowski (1985)] the respective criterion was formulated in terms of the principal stresses. For the latter formulation a considerably better agreement of experimental and numerical results was obtained. Therefore, it was proposed in [Eberhardsteiner (1991)] to include the principal stress criterion for distinction between loading and unloading in the three-dimensional hypoelastic constitutive model. According to this criterion, virgin loading causing inelastic deformations will occur, if at least one of the principal stresses exceeds its previously attained maximum value. Otherwise unloading or reloading takes place, which can be described approximately by means of the initial values of the elastic material parameters, E_0 and ν_0. As an example, the constitutive relations in rate form for unloading in the principal stress direction defined by the subscript 1 and loading in the other two principal directions are given as

$$\left\{\begin{array}{c}\dot{\varepsilon}_1\\ \dot{\varepsilon}_2\\ \dot{\varepsilon}_3\end{array}\right\} = \left[\begin{array}{ccc} E_0^{-1} & -\nu_0 E_0^{-1} & -\nu_0 E_0^{-1}\\ & E_{T2}^{-1} & -\sqrt{(\nu_2\nu_3)/(E_{T2}E_{T3})}\\ \text{sym.} & & E_{T3}^{-1}\end{array}\right]\left\{\begin{array}{c}\dot{\sigma}_1\\ \dot{\sigma}_2\\ \dot{\sigma}_3\end{array}\right\}. \qquad (2.67)$$

In [Elwi (1979)] the constitutive model was verified by comparing computed stress-strain curves with curves obtained from some of the biaxial experiments conducted by Kupfer et al. [Kupfer (1969)] and from a few of the triaxial tests carried out by Schickert and Winkler [Schickert (1977)]. An extensive numerical investigation of the model is contained in [Eberhardsteiner (1991)]. Numerical results were not only compared to the aforementioned experimentally obtained results but also to test results reported in [Kotsovos (1978)] and [Scavuzzo (1983)]. Apart from one-dimensional and two-dimensional loading, for which good correspondence of numerical results with experimental data has been obtained, this comparison also contains several conventional triaxial tests (Figs.1.10(a) and (b)) and truly triaxial tests (Fig.1.10(c)), both with nonproportional loading paths. The loading paths in the conventional triaxial tests are restricted to the Rendulic plane in the principal stress space, characterized by $\sigma_2 = \sigma_3$. These tests include hydrostatic loading, hydrostatic loading followed by loading paths in different deviatoric planes (Fig.1.10(a)) and hydrostatic loading followed by partial unloading and deviatoric reloading, continued by loading parallel to the hydrostatic axis (Fig.1.10(b)). The loading paths in the truly triaxial tests are nonproportional loading paths resulting from general triaxial loading conditions (Fig.1.10(c)). Figs.2.4 - 2.6 permit a comparison of experimentally determined material responses with computed ones for the loading paths shown in Fig.1.10. Table 2.1 contains the values of σ_0 and τ_0 for selected stress points on the loading paths.

In Figs.2.4 - 2.6, σ_0 is the octahedral normal stress (which is equal to the hydrostatic stress),

$$\varepsilon_0 = \frac{I'_1}{3} \qquad (2.68)$$

is the octahedral normal strain (which is equal to one third of the volumetric strain ε_V) and

$$\tau_0 = \sqrt{\frac{2}{3}J_2}, \quad \gamma_0 = 2\sqrt{\frac{2}{3}J'_2} \qquad (2.69)$$

are the octahedral shear stress and the octahedral engineering shear strain, respectively. I'_1 is the first invariant of the strain tensor. J'_2 ist the second invariant of the deviatoric strain tensor. In the following, the evaluation of the constitutive model is summarized according to the results presented in [Elwi (1979)] and [Eberhardsteiner (1991)].

For hydrostatic loading, followed by loading in a deviatoric plane, Fig.2.4 shows very good agreement of the experimentally obtained τ_0/σ_P-γ_0 diagram with the computed one. The τ_0/σ_P-ε_0 diagram demonstrates the ability of the model to account for the increase of the octahedral normal strain ε_0 shortly before failure.

2.3 Time-Independent Material Models for Concrete

loading paths of Fig.1.10(a)		loading paths of Fig.1.10(b)			loading path of Fig.1.10(c)		
stress point	σ_0 [MPa]	stress point	σ_0 [MPa]	τ_0 [MPa]	stress point	σ_0 [MPa]	τ_0 [MPa]
1	-15.5	1	-55.1	0.0	1	-55.1	27.6
2	-22.8	2	-27.6	0.0	2	-55.1	10.3
3	-30.8	3a	-27.6	6.9	3	-55.1	22.5
4	-53.6	4a	-55.1	6.9	4	-55.1	3.4
		3b	-27.6	20.7	5	-55.1	17.2
		4b	-55.1	20.7			

Table 2.1: Values σ_0 and τ_0 for selected stress points on the loading paths

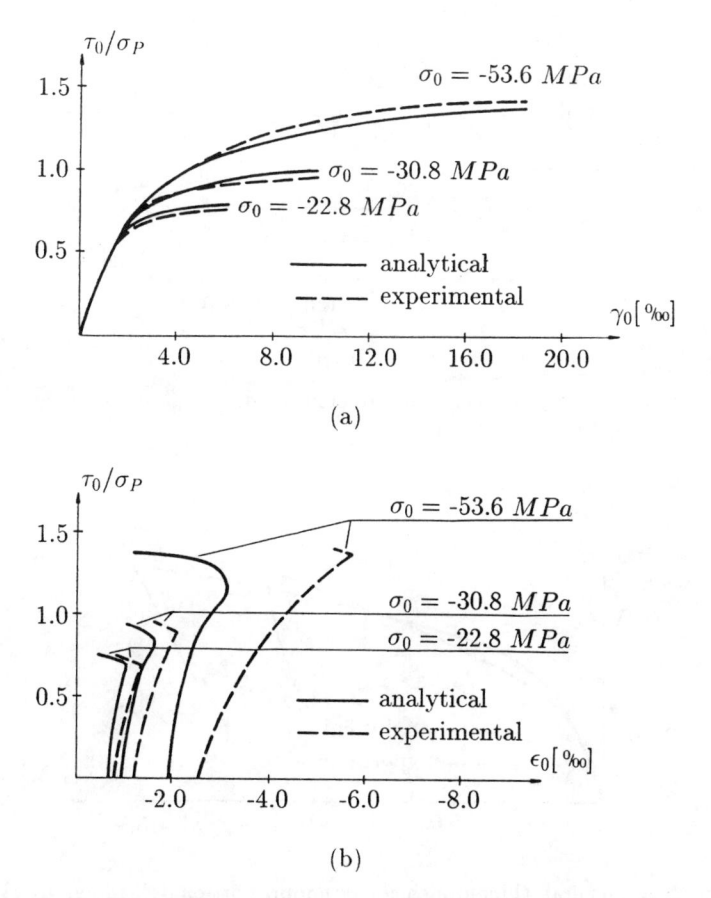

Figure 2.4: Deviatoric axisymmetric loading from different levels of hydrostatic loading: (a) τ_0/σ_P-γ_0 diagram, (b) τ_0/σ_P-ϵ_0 diagram

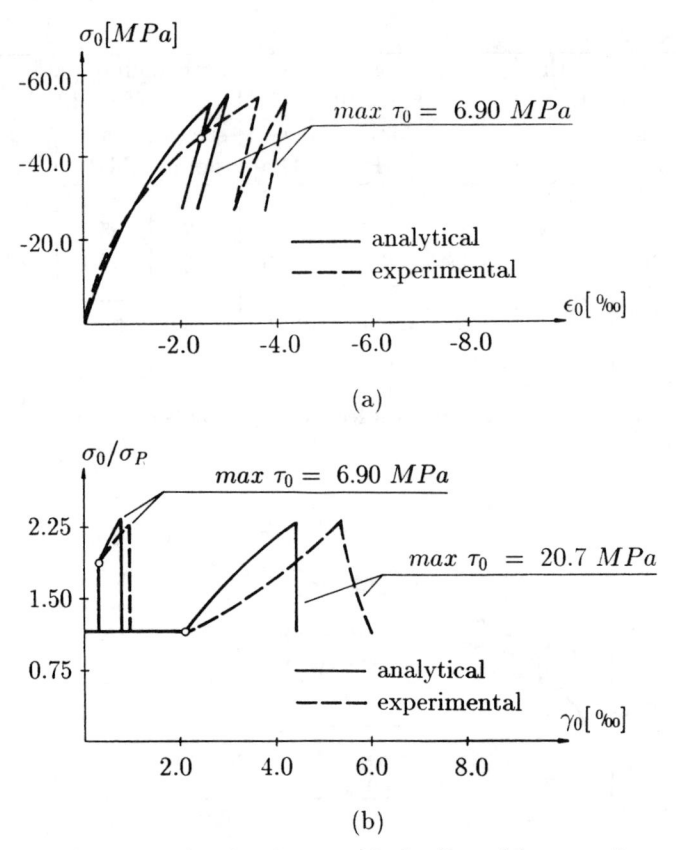

Figure 2.5: Nonproportional axisymmetric loading: (a) σ_0-ε_0 diagram, (b) σ_0/σ_P-γ_0 diagram

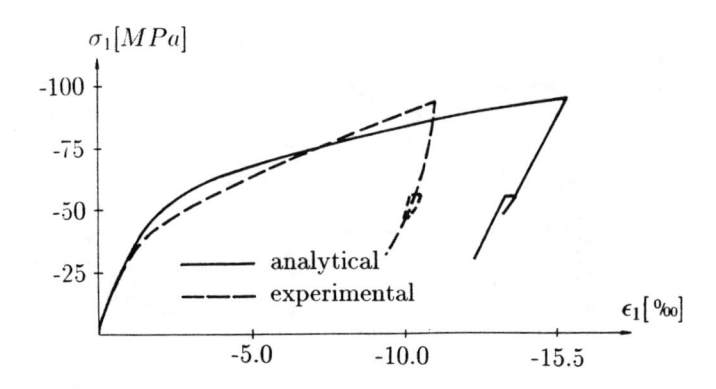

Figure 2.6: General three-dimensional nonproportional loading: σ_1-ε_1 diagram

However, the initial parts of the τ_0/σ_P-ε_0 curves in Fig.2.4, referring to initial hydrostatic loading, show considerable differences between the observed and the predicted values of the octahedral normal strain under hydrostatic loading. The main reason for this is lack of knowledge of the form of the surface in the equivalent-uniaxial strain space defining the fictitious equivalent-uniaxial strains ε_{Pi}^{eq}, which correspond to the peak stresses σ_{Pi}. In general, the available experimental data do not permit determination of all values for ε_{Pi}^{eq} which are needed. Hence, the missing values must be estimated by means of (2.65). However, the quality of constitutive modelling strongly depends on the quality of these values.

Analogous to the situation illustrated in Fig.2.4, it is shown in Fig.2.5 that the analytically obtained values for ε_0 for the loading path (a) in Fig.1.10(b) are significantly smaller than the corresponding experimental values. However, the shapes of experimentally obtained and corresponding computed σ_0-ε_0 curves are similar. The criterion distinguishing between loading and unloading also plays an important role. The points on the σ_0-ε_0 diagram and σ_0/σ_P-γ_0 diagram, for which virgin loading in the principal direction 1 is beginning after unloading and reloading has taken place, are marked by circles.

Fig.2.6 shows that even for nonproportinal, truly triaxial loading paths, including unloading and reloading along a different stress path, the predicted material response is satisfactory, except for large compressive stresses σ_1, for which the analytically obtained strains are too large.

Concrete failure is not treated in [Elwi (1979)]. The failure modes can be determined on the basis of the ultimate strength surface and the failure criterion described in subsection 2.2.3.

Apart from the aforementioned shortcomings of the material model, the analytical description of the material response is quite good. Even for relatively complicated loading paths, the analytically obtained results, in general, agree reasonably well with the corresponding test results. For loading paths containing unloading and reloading, the predictive capabilites of the model depend strongly on the criterion employed to distinguish between loading and unloading.

In [Ahmad (1986)] a nonlinear orthotropic constitutive law, characterized by total σ-ε relations, is presented to model both the pre-peak and the post-peak behavior of concrete. Similar to the orthotropic model described previously, stress-strain relations are specified for the principal axes of orthotropy which are assumed to coincide with the principal stress directions. However, because of the secant formulation there is no difference between loading and unloading. The three secant material moduli and the three Poisson's ratios are the material parameters. At any stage of the loading they are expressed as functions of the respective values at maximum strength.

Comments on Equivalent-Uniaxial Models

The main advantage of nonlinear-elastic orthotropic models based on equivalent-uniaxial constitutive relations is their simplicity. No sophisticated theoretical concepts are necessary to model the material behavior. It is approximated by curve fitting based on experimentally obtained data. For the biaxial constitutive model the ultimate strength envelope and the corresponding strain envelope are formulated as functions of the uniaxial ultimate strength and the corresponding strain resulting from a conventional uniaxial test. Thus, the constitutive behavior of concrete under plane stress can be described solely by means of material parameters obtained from conventional uniaxial tests. Hence, from the engineering point of view, especially the biaxial model is quite attractive.

Compared with this material model, the formulation of the constitutive model for triaxial stress states is more complicated. Instead of an ultimate strength curve and a corresponding strain curve for biaxial states of stress, an ultimate strength surface in the principal stress space and a corresponding strain surface in the equivalent-uniaxial strain space must be defined. However, in general, not all of the experimental data needed for the specification of these surfaces are available. The use of empirical formulae such as (2.65) to obtain estimates of the missing data seems to be the reason for the differences between the computed and the measured material behavior.

If equivalent-uniaxial models are applied within the framework of finite element analyses of concrete structures, it is necessary to be aware of the severe limitations of this class of constitutive models. A serious objection to such orthotropic models is that, in general, they do not satisfy the form-invariance condition for initially isotropic materials [Bažant (1983)]. As mentioned previously, concrete is an initially isotropic material with stress-induced orthotropic material behavior. Hence, the directions of the principal axes of orthotropy are not fixed and not known *a priori*. They depend on the directions of the principal stresses. Form invariance for initially isotropic materials implies that, regardless of the coordinate system used as the material reference frame, for a given loading path the same stress-strain response must be predicted by the constitutive model. It is shown in [Bažant (1983)] that form invariance for this class of models is only guaranteed if the directions of the principal stresses coincide with the directions of the principal strains. Moreover, if the directions of the principal stresses are rotating during the loading sequence, this rotation must coincide with the one of the principal strains. Only in this case an unambiguous definition of the principal axes of orthotropy is possible. However, even then the rotation of the axes of orthotropy is questionable for physical reasons. The decrease of the tangent stiffness with increasing loading is caused by material defects such as microcracks, loss of bond at the aggregate-mortar interfaces, etc., which are the results of the loading history up to the present state. Rotation of the axes of orthotropy implies the rotation of such material defects relative to the material. Obviously, this is impossible.

Strictly speaking, orthotropic constitutive models should only be applied in case of loading histories with no rotation of the principal stresses and strains relative to the material. However, for a given loading history in a real-life engineering problem it is unlikely that the directions of the principal stresses will not rotate.

In spite of their fundamental mechanical shortcomings, equivalent-uniaxial mate-

2.3 Time-Independent Material Models for Concrete

rial models have been used sucessfully for ultimate load analyses of reinforced and prestressed concrete structures within the framework of the finite element method.

Invariant-Based Biaxial Constitutive Model

A constitutive formulation for biaxial stress states, belonging to the category of variable moduli models, is presented in [Gerstle (1981a)]. In order to account for the nonlinearity of the material behavior, the constant material parameters in the constitutive relations for an isotropic, linear-elastic material are replacced by the respective tangent material quantities. The mentioned equations for linear elasticity can be formulated in terms of the octahedral stresses and the octahedral strains as

$$\sigma_0 = 3K_0\varepsilon_0 \, , \qquad \tau_0 = G_0\gamma_0 \, , \qquad (2.70)$$

where K_0 and G_0 denote the bulk modulus and the shear modulus, respectively. The tangent material quantities corresponding to K_0 and G_0 are the tangent bulk modulus K_T and the tangent shear modulus G_T. K_T is a function of the hydrostatic stress. G_T is a function of the octahedral shear stress. Thus, (2.70) is replaced by

$$\dot\sigma_0 = 3K_T(\sigma_0)\dot\varepsilon_0 \, , \qquad \dot\tau_0 = G_T(\tau_0)\dot\gamma_0 \, . \qquad (2.71)$$

Since the volumetric and the deviatoric response are strictly separated, (2.71) represents an uncoupled formulation. Experimentally observed coupling effects between volumetric (deviatoric) strains and deviatoric (volumetric) stresses, are ignored.

The general shape of a τ_0-γ_0 diagram (Fig.2.4) is approximated as

$$\tau_0 = \tau_{P0}\left(1 - \exp^{-G_0\gamma_0/\tau_{P0}}\right) , \qquad (2.72)$$

where τ_{P0} denotes the octahedral shear strength. Differentiation of (2.72) with respect to γ_0 yields the tangent shear modulus:

$$G_T = \frac{d\tau_0}{d\gamma_0} = G_0 \exp^{-G_0\gamma_0/\tau_{P0}} . \qquad (2.73)$$

Elimination of γ_0 in (2.73) by means of (2.72) gives

$$G_T = G_0\left(1 - \frac{\tau_0}{\tau_{P0}}\right) . \qquad (2.74)$$

Hence, the assumption of an exponential τ_0-γ_0 relation yields a tangent shear modulus, which is linearly decreasing from $G_T = G_0$ at $\tau_0 = 0$ to $G_T = 0$ at $\tau_0 = \tau_{P0}$.

Similarly, the tangent bulk modulus can be described as a linearly decreasing function

$$K_T = K_0\left(1 - C_K\frac{\sigma_0}{\sigma_{P0}}\right) . \qquad (2.75)$$

C_K is a constant which, according to [Gerstle (1981a)], can be chosen as 2/3, yielding the lower bound of the tangent bulk modulus as $K_T = K_0/3$; σ_{P0} is the hydrostatic stress at failure in a biaxial test. Substitution of $\alpha_2 = \sigma_1/\sigma_2$ into (2.11) and (2.69$_1$), respectively, and use of (2.13) yields

$$\sigma_{P0} = \frac{1}{3}(1+\alpha_2)\sigma_{P2} ,$$

$$\tau_{P0} = \frac{\sqrt{2}}{3}\sqrt{1-\alpha_2+\alpha_2^2}\,\sigma_{P2} ,$$

(2.76)

where $\sigma_{P2} = \sigma_{P2}(\alpha_2)$ denotes the major compressive principal stress on the ultimate strength envelope for biaxial stress states.

This simple formulation is restricted to monotonic loading. Hence, no criterion for loading and unloading is required. Neither volume dilatation, occurring shortly before the ultimate strength is reached, nor softening is taken into account. However, within the limitations of this constitutive model, good agreement between the measured and the computed stress-strain response was found.

Invariant-Based Three-Dimensional Hypoelastic Constitutive Model

The hypoelastic constitutive model proposed by Stankowski and Gerstle [Stankowski (1985)] is a typical representative of an empirical material model. Its formulation is based on extensive experimental tests conducted at the University of Colorado at Boulder [Scavuzzo (1983)]. The concrete used in this experimental investigation has a uniaxial compressive strength of 25.0 MPa. The initial values of the bulk modulus and the shear modulus are 14 500 MPa and 11 700 MPa, respectively.

The constitutive equations are given in rate form. They represent relations between the octahedral stress rates and strain rates:

$$\left\{ \begin{array}{c} \dot{\varepsilon}_0 \\ \dot{\gamma}_0 \end{array} \right\} = \left[\begin{array}{cc} (3K_T)^{-1} & H_T^{-1} \\ Y_T^{-1} & (2G_T)^{-1} \end{array} \right] \left\{ \begin{array}{c} \dot{\sigma}_0 \\ \dot{\tau}_0 \end{array} \right\}. \quad (2.77)$$

In (2.77) H_T and Y_T are tangent coupling moduli. They allow consideration of the coupling between the deviatoric stress and the volumetric strain and between the hydrostatic stress and the deviatoric strain. The dependence of ε_0 on τ_0 and of γ_0 on σ_0 can be observed experimentally (Figs. 2.4 and 2.5).

It follows from (2.11) and (2.69$_1$) together with (2.13) that

$$\dot{\sigma}_0 = \frac{1}{3}\dot{I}_1 , \qquad \dot{\tau}_0 = \frac{1}{3\tau_0}\mathbf{s}:\dot{\mathbf{s}} , \quad (2.78)$$

and from (2.68) and (2.69$_2$) (noting that J_2' is defined by replacing J_2 in (2.13) by J_2' and $\mathbf{s}$ by the deviatoric strain tensor $\mathbf{e}$) that

$$\dot{\varepsilon}_0 = \frac{1}{3}\dot{I}_1' , \qquad \dot{\gamma}_0 = \frac{2}{3\gamma_0}\mathbf{e}:\dot{\mathbf{e}} , \quad (2.79)$$

2.3 Time-Independent Material Models for Concrete

where
$$\mathbf{e} = \boldsymbol{\varepsilon} - \varepsilon_0 \mathbf{1} . \tag{2.80}$$

Hence, the deviatoric octahedral stress and strain rates do not only depend on the rates of the deviatoric stress and the deviatoric strain tensor, respectively, but also on the current values of these tensors.

For given principal stress rates $\dot{\sigma}_i, i = 1,2,3$, the corresponding octahedral stress rates $\dot{\sigma}_0$ and $\dot{\tau}_0$ are uniquely determined. However, only for the special case of axisymmetry, for which $\dot{\varepsilon}_2 = \dot{\varepsilon}_3$, the two equations (2.79) suffice for determination of the strain rates $\dot{\varepsilon}_i, i = 1,2,3$. For general three-dimensional problems a third equation is obtained by assuming the strain rates to be proportional to the stress rates, i.e.,

$$\frac{\dot{e}_2}{\dot{e}_1} = \frac{\dot{s}_2}{\dot{s}_1} = \frac{\dot{\sigma}_2 - \dot{\sigma}_0}{\dot{\sigma}_1 - \dot{\sigma}_0} = \text{const.} . \tag{2.81}$$

Combining (2.79) and (2.81), the principal strain rates can be formulated in terms of $\dot{\varepsilon}_0$, γ_0 and $\dot{\gamma}_0$ [Stankowski (1985)]. Experimental results have shown, however, that the hypothesis (2.81) is not generally valid.

The material moduli in (2.77) were obtained by data fitting of extensive experimental data. The value for the tangent bulk modulus obtained from the empirical formula

$$K_T = K_0(1 + C_K \sigma_0) \geq C'_K K_0 \tag{2.82}$$

decreases linearly with increasing hydrostatic pressure (σ_0 is negative) until the lower bound $C'_K K_0$ is reached. For the concrete used in the tests [Scavuzzo (1983)], $C_K = 0.025$ MPa^{-1} and $C'_K = 0.14$.

For determination of the tangent shear modulus the following empirical formula is used:

$$G_T = G_H \left(1 - \frac{\tau_0}{\tau_{P0}}\right), \tag{2.83}$$

where
$$G_H = G_0(1 + C_G \sigma_0) \geq C'_G G_0 . \tag{2.84}$$

For a given hydrostatic stress σ_0, G_T is linearly decreasing with increasing τ_0 such that $G_T = 0$ if $\tau_0 = \tau_{P0}$. τ_{P0} is the octahedral shear strength with respect to the deviatoric plane defined by σ_0 and the angle of similarity θ for the current stress state. τ_{P0} can be computed, e.g., on the basis of the five-parameter ultimate strength surface, proposed by Willam and Warnke [Willam (1975)]. For a given value of τ_0, G_T is decreasing linearly with increasing hydrostatic pressure until the lower bound $C'_G G_0 (1 - \tau_0/\tau_{P0})$ is attained. For the concrete used in the tests [Scavuzzo (1983)], $C_G = 0.021$ MPa^{-1} and $C'_G = 0.47$. It is noted that the relations (2.82) and (2.83) for the tangent bulk and the tangent shear modulus are similar to the respective expressions (2.75) and (2.74) for the corresponding biaxial model [Gerstle (1981a)].

The coupling moduli H_T and Y_T are given by empirical formulae as

$$H_T = 1\,000 \left(10 + \frac{43.5}{\sigma_0 - 10}\right) \text{MPa} \quad \text{for} \quad \sigma_0 < -10\,\text{MPa}, \tag{2.85}$$

$$Y_T = \frac{4 \cdot 10^9}{\tau_0^2} \text{MPa}. \tag{2.86}$$

Because of the dependence of K_T, G_T, H_T and Y_T on the octahedral stresses (see (2.82) - (2.86)) constitutive models of the described type are referred to as variable moduli models. In the considered variable moduli model the stress-strain response in the softening region is not taken into account.

Formulation of the constitutive equations in terms of octahedral stresses and strains suggests using a criterion for loading and unloading which is based on octahedral stresses. However, experimental results have indicated that the criterion for loading and unloading should be based on the principal stresses. In the context of the three-dimensional equivalent-uniaxial hypoelastic model described in the previous subsection such a principal stress criterion was introduced. According to this criterion, virgin loading causing inelastic deformations will occur, if at least one of the principal stresses exceeds its previously attained maximum value. Otherwise unloading or reloading will take place. Unloading and reloading can be described approximately by means of the initial values of the elastic material parameters, K_0 and G_0, and by infinitely large values of the coupling moduli. For some loading paths, the mentioned criterion for loading and unloading leads to considerable differences between computed and experimental results [Eberhardsteiner (1991)]. This is the case, e.g., for hydrostatic loading up to a certain level of hydrostatic pressure, followed by unloading of σ_1 while keeping σ_2 and σ_3 constant. In this case the loading/unloading criterion based on principal stresses signals unloading. However, when proceeding on this loading path, the tensile meridian of the ultimate strength surface is hit. Since τ_0 increases during such "unloading" from $\tau_0 = 0$ up to the octahedral shear strength τ_{P0}, deviatoric loading takes place. Thus, in such a case the octahedral shear stress should also be taken into account in the criterion for loading and unloading.

In [Stankowski (1985)] similarities between the empirical constitutive model described above and certain elements of the theory of plasticity were addressed. For this purpose the loading/unloading criterion was recast in terms of a loading surface in the principal stress space. This surface consists of three planes which are perpendicular to the axes of the principal stress space. The locations of these three planes are determined by the previously attained maximum values for each one of the three principal stresses. If, e.g., the stress σ_1 exceeds its previously attained maximum value, then the plane perpendicular to the σ_1-axis will be translated with increasing σ_1 such that σ_1 will remain on this plane. If, on the other hand, unloading occurs, then the loading surface will remain unchanged. Hence, a moving loading surface defines virgin loading. It also follows from the experimental investigation that virgin loading in one of the principal directions is associated with an increasing inelastic deformation in the direction of the increasing principal stress. Thus, in a space of plastic strains, associated with the principal stress space, the plastic strain

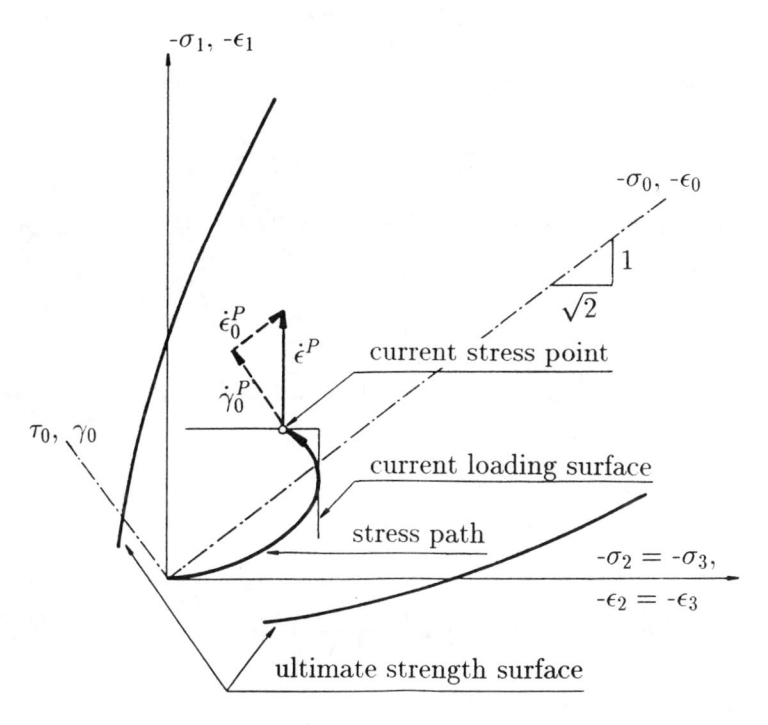

2.3 Time-Independent Material Models for Concrete

Figure 2.7: Ultimate strength surface, loading surface and plastic strain vector in the Rendulic plane

rate $\dot{\varepsilon}^p$ can be represented as a vector. Its direction is parallel to the σ_1-axis. $\dot{\varepsilon}^p$ can be decomposed into the component parallel and the component perpendicular to the hydrostatic axis. The former yields the rate of the plastic volumetric strain, $\dot{\varepsilon}_0^p$, and the latter the rate of the plastic octahedral engineering shear strain, $\dot{\gamma}_0^p$. Fig.2.7 illustrates this interpretation of the plastic strain rate for axisymmetric conditions. It enables determination of the coupling moduli as functions of the bulk and the shear modulus. Hence, the empirical formulae (2.85) and (2.86) are no longer needed. In the following, determination of the coupling moduli is explained for the case of virgin loading in the principal direction 1, with the vector of the plastic strain rate pointing in this direction. Hence, decomposition of $\dot{\varepsilon}^p$ parallel and perpendicular to the hydrostatic axis yields (Fig.2.7)

$$\frac{\dot{\varepsilon}_0^p}{\dot{\gamma}_0^p} = \frac{1}{\sqrt{2}} \ . \tag{2.87}$$

Additive decomposition of $\dot{\varepsilon}_0$ and $\dot{\gamma}_0$ into the elastic and the plastic part yields

$$\dot{\varepsilon}_0 = \dot{\varepsilon}_0^e + \dot{\varepsilon}_0^p = \frac{\dot{\sigma}_0}{3K_0} + \dot{\varepsilon}_0^p,$$

$$\dot{\gamma}_0 = \dot{\gamma}_0^e + \dot{\gamma}_0^p = \frac{\dot{\tau}_0}{2G_0} + \dot{\gamma}_0^p . \tag{2.88}$$

For purely deviatoric loading, characterized by $\dot{\sigma}_0 = 0$, (2.77) results in

$$\dot{\varepsilon}_0 = \frac{\dot{\tau}_0}{H_T} , \qquad \dot{\gamma}_0 = \frac{\dot{\tau}_0}{2G_T} . \tag{2.89}$$

Substituting (2.89) into (2.88) and inserting the result into (2.87) gives

$$H_T = \frac{2\sqrt{2}G_T}{1 - \dfrac{G_T}{G_0}} . \tag{2.90}$$

For purely hydrostatic loading, characterized by $\dot{\tau}_0 = 0$, (2.77) is reduced to

$$\dot{\varepsilon}_0 = \frac{\dot{\sigma}_0}{3K_T} , \qquad \dot{\gamma}_0 = \frac{\dot{\sigma}_0}{Y_T} . \tag{2.91}$$

Substituting (2.91) into (2.88) and inserting the result into (2.87) yields

$$Y_T = \frac{3K_T}{\sqrt{2}\left(1 - \dfrac{K_T}{K_0}\right)} . \tag{2.92}$$

Replacing (2.87) by

$$\frac{\dot{\varepsilon}_0^p}{\dot{\gamma}_0^p} = \sqrt{2} \tag{2.93}$$

allows determination of the coupling moduli for virgin loading in the directions associated with σ_2 and σ_3. If σ_1 as well as σ_2 and σ_3 exceed their previously attained maximum values, then the current stress point will be located at the corner of the loading surface shown in Fig.2.7. Hence, in this case, the direction of $\dot{\varepsilon}^p$ is not unique. As a remedy for this nonuniqueness, in [Stankowski (1985)] the strain rates are assumed to be proportional to the stress rates (see(2.81)).

The introduction of a loading surface consisting of three planes which are perpendicular to the directions of the principal stresses, will lead to a deficiency of the constitutive model if the stress point approaches the ultimate strength surface. In this case the loading surface should attain the shape of the ultimate strength surface. In this way the transition from volume compaction to volume expansion shortly before failure could be modelled. However, this requirement is not met by the loading surface. In order to overcome this deficiency, in (2.87) and (2.93) the constant value of the ratio $\dot{\varepsilon}_0^p/\dot{\gamma}_0^p$ is replaced by a function in terms of τ_0/τ_{P0}. The ratio τ_0/τ_{P0} serves as a measure of the distance of the stress point from the ultimate strength surface. The respective function is chosen such that for $\tau_0/\tau_{P0} = 0$ the vector of plastic strain rates is parallel to the direction of virgin loading. With increasing values of τ_0/τ_{P0}, this vector is assumed to rotate. For $\tau_0/\tau_{P0} = 1$, it is perpendicular to the ultimate strength surface.

2.3 Time-Independent Material Models for Concrete

For practical applications, the constitutive equations in rate form are approximated by incrementally linear constitutive equations containing the incremental quantities $\Delta\varepsilon_0$, $\Delta\gamma_0$, $\Delta\sigma_0$ and $\Delta\tau_0$.

Extensive numerical investigations of the model can be found in [Stankowski (1983), Eberhardsteiner (1991)]. In order to be able to compare the potential of this model to the one of the equivalent-uniaxial model described in the previous subsection, Figs.2.8 - 2.10 were prepared. They contain experimentally obtained as well as computed stress-strain diagrams for the same stress paths as the ones shown in Figs.2.4 - 2.6. Values of σ_0 and τ_0 for selected stress points on the respective stress paths, plotted in Fig.1.10 are listed in Table 2.1. A close examination of Figs.2.4 - 2.6 and of Figs.2.8 - 2.10 shows that there are no significant differences between the predictive capabilities of the two constitutive models.

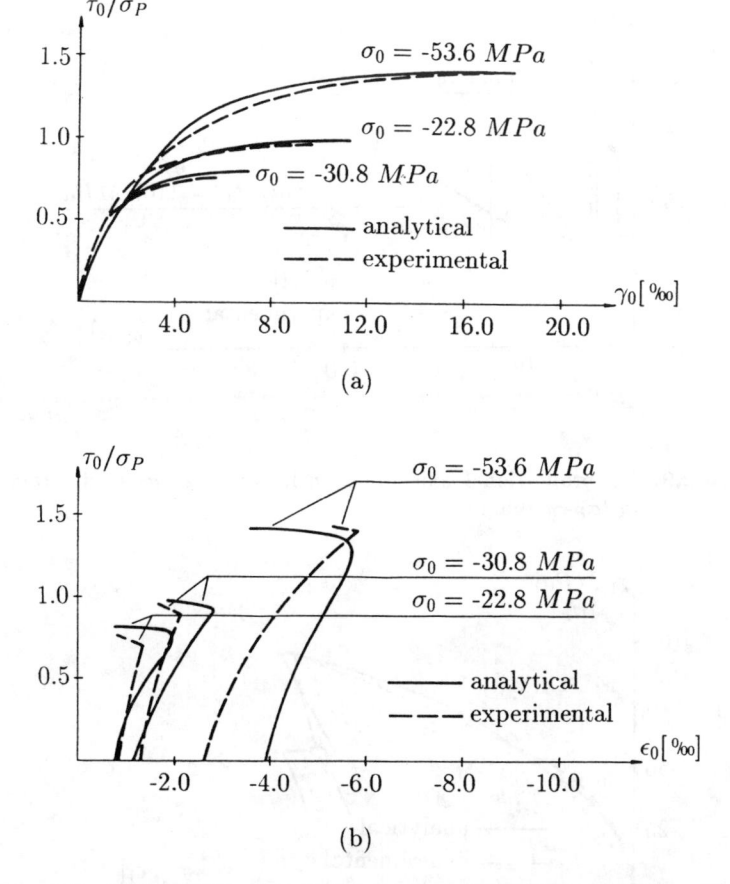

Figure 2.8: Deviatoric axisymmetric loading from different levels of hydrostatic loading: (a) τ_0/σ_P-γ_0 diagram, (b) τ_0/σ_P-ε_0 diagram

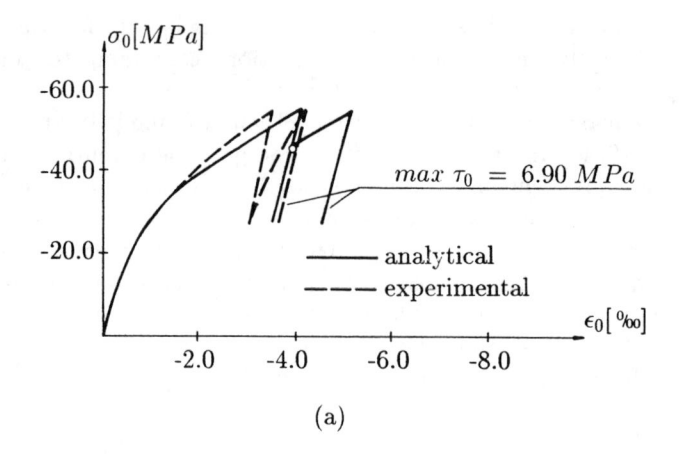

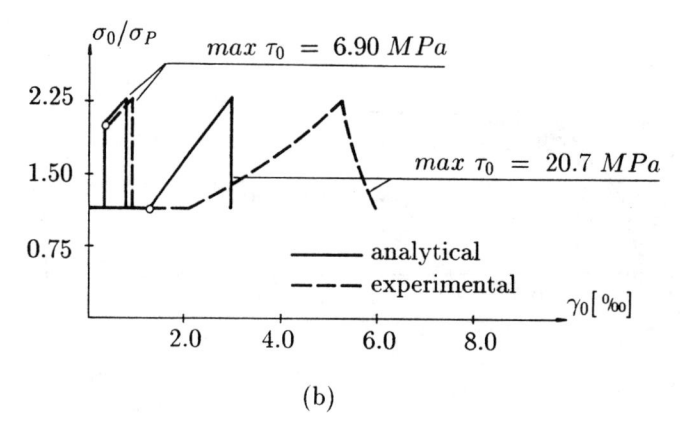

Figure 2.9: Nonproportional axisymmetric loading: (a) σ_0-ϵ_0 diagram, (b) σ_0/σ_P-γ_0 diagram

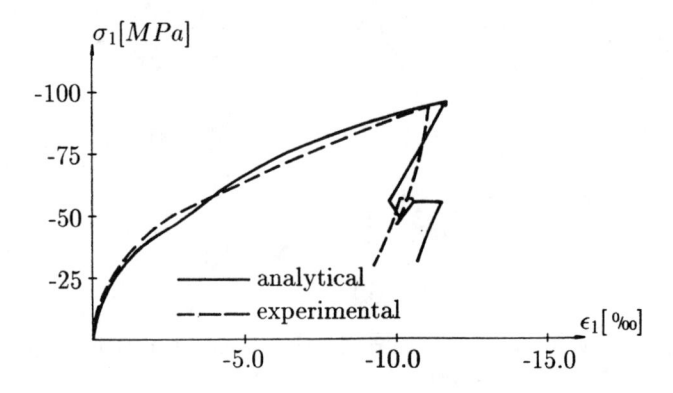

Figure 2.10: General three-dimensional nonproportional loading: σ_1-ϵ_1 diagram

Nevertheless, the assumption of proportional deviatoric stress and strain rates for general triaxial loading is a serious deficiency of the considered invariant-based hypoelastic constitutive model. This assumption was not corroborated by experimental results. Moreover, the combination of the criterion for loading and unloading in terms of the principal stresses with the constitutive equations in rate form, originally given in terms of octahedral stresses and strains, requires reformulation of the constitutive equations in terms of the principal stresses and strains. Because of the mentioned difficulties and inconsistencies, it was proposed in [Stankowski (1983)] to describe the deformational behavior of concrete for general triaxial loading within the framework of the theory of strain-hardening plasticity.

In [Shafer (1985)] the present material model was extended, within the framework of the theory of hypoelasticity, to general triaxial states of stress without reference to the hypothesis of proportional deviatoric stress and strain rates. This constitutive model also allows consideration of the post-peak behavior. However, although the mathematical formulation of the model is relatively complicated, deficiencies were found for the case of unloading [Eberhardsteiner (1991)]. An improvement of this material model concerning the criterion for loading and unloading would require considerable modifications. For a detailed description and evaluation reference to [Shafer (1985), Eberhardsteiner (1991)] is made.

Comments on Invariant-Based Hypoelastic Models

There are no fundamental differences between equivalent-uniaxial hypoelastic material models and invariant-based hypoelastic constitutive models. Both models yield the tangent material stiffness matrix or the tangent material compliance matrix by data fitting without the aid of complicated theoretical concepts. Hence, invariant-based hypoelastic constitutive models are characterized by the same advantages and disadvantages as equivalent-uniaxial material models. The main advantage is the simplicity of the formulation. The main disadvantage is the violation of fundamental physical principles such as the laws of thermodynamics [Chen (1982), Saleeb (1982)] and the form invariance condition [Bažant (1983)].

2.3.3 Models Based on the Theory of Plasticity

Summary of the Theoretical Foundations of the Theory of Plasticity

Fundamental assumptions In the following the basic relations of the theory of plasticity will be summarized. The brief survey follows [Simo (1988), Lubliner (1990)]. It starts with the basic equations for rate-dependent inelastic material behavior. These relations are then specialized for rate-independent material behavior. For a comprehensive introduction to the theory of plasticity reference to [Lubliner (1990)] is made.

In the theory of plasticity the dependence of the strain on the history of the material can be described by means of an array of internal variables $\boldsymbol{\xi}$, yielding

$$\varepsilon = \mathbf{f}(\boldsymbol{\sigma}, T, \boldsymbol{\xi}) \ . \tag{2.94}$$

(2.94) is more general than the inverse relation (2.6), because it is not always possible to express $\boldsymbol{\sigma}$ as a function of ε, T and $\boldsymbol{\xi}$. The internal variables describe irreversible material behavior. The evolution of the internal variables is expressed by means of rate equations which can formally be written as

$$\dot{\xi}_k = h_k^*(\boldsymbol{\sigma}, T, \boldsymbol{\xi}) \ . \tag{2.95}$$

In (2.95) h_k^* denotes an arbitrary function defining the evolution of the internal variable ξ_k. Similar to the strain in (2.94), the internal variables in (2.95) depend on the local state of the material, defined by $\boldsymbol{\sigma}$, T and $\boldsymbol{\xi}$. The internal variables serve as a means to describe hardening or softening of the material.

A fundamental assumption within the framework of the theory of small strains is the additive decomposition of the strain tensor into an elastic part ε^e and an inelastic part ε^a:

$$\varepsilon = \varepsilon^e(\boldsymbol{\sigma}, T) + \varepsilon^a(\boldsymbol{\xi}) \ . \tag{2.96}$$

The constitutive relations are given as

$$\boldsymbol{\sigma} = \mathbf{C} : \varepsilon^e = \mathbf{C} : (\varepsilon - \varepsilon^a) \ , \tag{2.97}$$

where $\mathbf{C}$ is the tensor of the elastic moduli.

Yield surface and flow rule If a material behaves elastically within a certain range of stress and temperature and inelastically beyond this range, then it is appropriate to define a function $f(\boldsymbol{\sigma}, T, \boldsymbol{\xi})$ such that $f(\boldsymbol{\sigma}, T, \boldsymbol{\xi}) < 0$ indicates elastic material behavior. Hence, in this case $\dot{\varepsilon}^a = \mathbf{0}$ holds.

$$f(\boldsymbol{\sigma}, T, \boldsymbol{\xi}) = 0 \tag{2.98}$$

defines a yield surface in the stress space. $f(\boldsymbol{\sigma}, T, \boldsymbol{\xi}) > 0$, together with $\dot{\varepsilon}^a \neq \mathbf{0}$, indicates inelastic material behavior. This type of formulation is denoted as stress space plasticity.

$\dot{\varepsilon}^a = \mathbf{0}$ does not necessarily imply that $\dot{\boldsymbol{\xi}} = \mathbf{0}$ in the elastic material region. E.g., the internal variables used for the description of strain-aging change with increasing age of the material even in the stress-free state. However, in general, the rates of the internal variables are assumed to vanish in the elastic region, i.e., $\dot{\boldsymbol{\xi}} = \mathbf{0}$ if $f(\boldsymbol{\sigma}, T, \boldsymbol{\xi}) \leq 0$. On the basis of this assumption (2.95) can be redefined as

$$\dot{\xi}_k = \phi(f) h_k(\boldsymbol{\sigma}, T, \boldsymbol{\xi}) \ , \tag{2.99}$$

where the scalar function $\phi(f)$ satisfies the conditions $\phi(f) = 0$ for $f \leq 0$ and $\phi(f) > 0$ for $f > 0$, and h_k is an appropriate function for the purpose of describing the evolution of the internal variable ξ_k.

2.3 Time-Independent Material Models for Concrete

Based on the assumption that $\varepsilon^a = \varepsilon^a(\boldsymbol{\xi})$, the rate of the inelastic strain is obtained with the help of (2.99) as

$$\dot{\varepsilon}^a_{ij} = \frac{\partial \varepsilon^a_{ij}}{\partial \xi_k} \dot{\xi}_k = \phi(f) g_{ij}(\boldsymbol{\sigma}, T, \boldsymbol{\xi}) , \qquad (2.100)$$

where

$$g_{ij} = \frac{\partial \varepsilon^a_{ij}}{\partial \xi_k} h_k . \qquad (2.101)$$

For simplicity, the components g_{ij} of the tensor function **g** are usually assumed to be derivable from a scalar function $g(\boldsymbol{\varepsilon}, T, \boldsymbol{\xi})$, denoted as flow potential, such that

$$g_{ij} = \frac{\partial g}{\partial \sigma_{ij}} . \qquad (2.102)$$

Hence,

$$\dot{\varepsilon}^a_{ij} = \phi(f) \frac{\partial g}{\partial \sigma_{ij}} . \qquad (2.103)$$

(2.103) is known as the flow rule.

Hardening and softening If the stress and the temperature at a point outside the yield suface ($\Rightarrow f > 0$) are held constant, the inelastic strain will increase. This increase will be bounded if hardening of the material takes place. It will be unbounded if softening occurs. Hardening is characterized by a the decrease of f from $f > 0$ to $f = 0$. Thus, for hardening, $\dot{f} < 0$. Softening, however, is characterized by an increase of f. Thus, for softening, $\dot{f} > 0$. Using (2.99), the rate of f is obtained as

$$\dot{f}(\boldsymbol{\sigma} = \text{const.}, T = \text{const.}, \boldsymbol{\xi}) = \frac{\partial f}{\partial \xi_k} \dot{\xi}_k = -\phi(f) H , \qquad (2.104)$$

where the plastic modulus H is defined as

$$H = -\frac{\partial f}{\partial \xi_k} h_k . \qquad (2.105)$$

$\phi(f)$ is non-negative. Consequently, hardening is characterized by $H > 0$ and softening by $H < 0$. If f only depends on $\boldsymbol{\sigma}$ and not on internal variables, then, following from (2.105), $H = 0$, representing the special case of a perfectly-plastic material. It is characterized by a yield surface with a fixed position in the stress space.

A material with rate-dependent behavior, for which a yield function can be formulated, is termed as viscoplastic. The inelastic strains are referred to as viscoplastic strains. They are denoted as $\boldsymbol{\varepsilon}^{vp}$.

Rate-independent plasticity If the loading causing inelastic deformations is sufficiently slow such that for the limiting case of a vanishing loading rate the stress point is prevented from leaving the yield surface in the direction of $f > 0$, then the rate dependency of the material will be negligible. Disregarding this dependency results in the special case of rate-independent plasticity. Consequently, the admissible values for the yield function are restricted to $f \leq 0$. Thus, inelastic loading is expressed by $f = 0$ and $\dot{f} = 0$. In this case the inelastic strains are termed as plastic strains. They are denoted as ε^p. The second condition for plastic loading, $\dot{f} = 0$, can be used for determination of $\phi(f)$. Ignoring the dependence of f on T and making use of (2.99), $\dot{f}(\sigma, \xi)$ is obtained from (2.98) as

$$\dot{f}(\sigma, \xi) = \frac{\partial f}{\partial \sigma_{ij}} \dot{\sigma}_{ij} + \phi(f) \frac{\partial f}{\partial \xi_k} h_k \,. \tag{2.106}$$

Substituting (2.105) into $\dot{f}(\sigma, \xi) = 0$, yields

$$\phi(f) = \frac{1}{H} \frac{\partial f}{\partial \sigma_{ij}} \dot{\sigma}_{ij} \,. \tag{2.107}$$

For the case of a perfectly-plastic material, (2.107) is an indeterminate expression, because both $(\partial f / \partial \sigma_{ij}) \dot{\sigma}_{ij} = \dot{f} = 0$ and $H = 0$. Inserting (2.107) into (2.99) yields

$$\dot{\xi}_k = \frac{1}{H} \frac{\partial f}{\partial \sigma_{ij}} \dot{\sigma}_{ij} h_k \,. \tag{2.108}$$

Substituting (2.107) into (2.103), where the superscript "a" has to be replaced by the superscript "p", gives

$$\dot{\varepsilon}^p_{ij} = \frac{1}{H} \frac{\partial f}{\partial \sigma_{kl}} \dot{\sigma}_{kl} \frac{\partial g}{\partial \sigma_{ij}} \,. \tag{2.109}$$

The left-hand side and the right-hand side of (2.108) and (2.109) contain one time derivative each. Hence, (2.108) and (2.109) are rate-independent relations. Time plays no role in rate-independent plasticity. Nevertheless, time is used as a means to order the sequence of loading.

(2.108) and (2.109) can be rewritten as

$$\dot{\xi}_k = \dot{\lambda} h_k \,, \tag{2.110}$$

$$\dot{\varepsilon}^p_{ij} = \dot{\lambda} \frac{\partial g}{\partial \sigma_{ij}} \,, \tag{2.111}$$

where $\dot{\lambda}$ is referred to as the consistency parameter. By analogy to $\phi(f)$ in (2.107), $\dot{\lambda}$ is defined as

$$\dot{\lambda} = \begin{cases} \dfrac{1}{H} \dfrac{\partial f}{\partial \sigma_{ij}} \dot{\sigma}_{ij} & \text{for } f = 0 \,, \\ 0 & \text{for } f < 0 \,. \end{cases} \tag{2.112}$$

2.3 Time-Independent Material Models for Concrete

The derivative $\partial g/\partial \sigma_{ij}$, occurring in (2.111), specifies the direction of the plastic flow. $\dot{\lambda}$ serves as the scaling parameter for the magnitude of $\dot{\varepsilon}^p_{ij}$. Hence, admissible values of $\dot{\lambda}$ are restricted to $\dot{\lambda} \geq 0$. Following from (2.112), the loading/unloading conditions are given as

$$f \leq 0, \qquad \dot{\lambda} \geq 0, \qquad \dot{\lambda} f = 0. \tag{2.113}$$

In optimization theory the relations (2.113) are denoted as the Kuhn-Tucker conditions [Luenberger (1984), Simo (1988)]. In addition, $\dot{\lambda}$ and $\dot{f}$ obey the consistency condition

$$\dot{\lambda} \dot{f} = 0. \tag{2.114}$$

(2.113) and (2.114) govern elastic loading/unloading and plastic loading. If $f < 0$, then, following from (2.113), $\dot{\lambda} = 0$ and, because of (2.111), $\dot{\varepsilon}^p = 0$. Hence, the material is behaving elastically. If $f = 0$ and $\dot{f} < 0$, then, because of (2.114) $\dot{\lambda} = 0$, indicating elastic unloading from a plastic state. If $f = 0$ and $\dot{f} = 0$, then following from (2.113) and (2.114), either $\dot{\lambda} > 0$ or $\dot{\lambda} = 0$. For $\dot{\lambda} > 0$, plastic loading occurs. $\dot{\lambda} = 0$ signals so-called neutral loading.

Hardening and softening for rate-independent plasticity Let it be assumed that at time t a stress point $\boldsymbol{\sigma}$ is lying on the yield surface, i.e., $f(\boldsymbol{\sigma}, \boldsymbol{\xi}) = 0$, and that the stress is kept constant. Then, at time $t + dt$ hardening, characterized by $\dot{f} < 0$, yields $f < 0$. Hence, at time $t + dt$ the stress point is in the elastic region. Because the stress was held constant, the yield surface expanded in the vicinity of $\boldsymbol{\sigma}$. Thus, hardening is associated with an expansion of the yield surface. Consequently, softening is associated with a contraction of the yield surface.

For a hardening material, characterized by $H > 0$, the loading/unloading conditions are given as

$$f < 0 \qquad\qquad\qquad \Rightarrow \quad \text{elastic loading/unloading,}$$

$$f = 0 \text{ and } \quad \frac{\partial f}{\partial \sigma_{ij}} \dot{\sigma}_{ij} < 0 \quad \Rightarrow \quad \text{elastic unloading from a plastic state,}$$

$$f = 0 \text{ and } \quad \frac{\partial f}{\partial \sigma_{ij}} \dot{\sigma}_{ij} = 0 \quad \Rightarrow \quad \text{neutral loading from a plastic state,}$$

$$f = 0 \text{ and } \quad \frac{\partial f}{\partial \sigma_{ij}} \dot{\sigma}_{ij} > 0 \quad \Rightarrow \quad \text{plastic loading.} \tag{2.115}$$

These conditions follow from (2.112). They are illustrated in Fig.2.11.

For a perfectly-plastic material the condition $\dot{f} = 0$ results in $(\partial f/\partial \sigma_{ij})\dot{\sigma}_{ij} = 0$. For such a material also $H = 0$. The right-hand side of the equation for $\dot{\lambda}$ (2.112) is an indeterminate expression.

For a softening material, characterized by $H < 0$, in case of $(\partial f/\partial \sigma_{ij})\dot{\sigma}_{ij} < 0$ it is impossible to decide whether elastic unloading from the yield surface or softening, associated with a contraction of the yield surface, occurs.

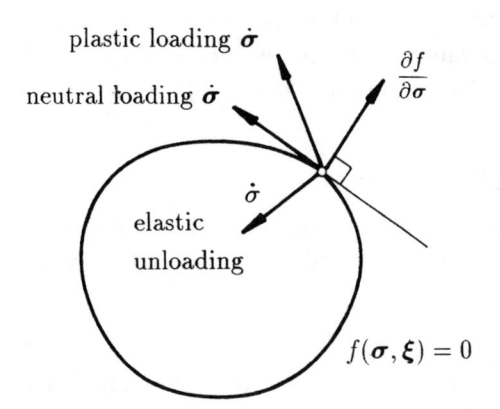

Figure 2.11: Loading criteria for a hardening material

The difficulties to distinguish between plastic loading and neutral loading for a perfectly-plastic material and between elastic unloading and softening for a softening material can be overcome by means of the loading/unloading conditions formulated in terms of the rate of trial stress [Simo (1988)]. To this end, λ is computed from the condition $\dot{f} = 0$, following from the consistency condition (2.114). Replacing in the expression for $\dot{f}$ in (2.106) $\phi(f)$ by $\dot{\lambda}$ and σ_{ij} by the expression on the right-hand side of (2.97), noting that $\dot{\varepsilon}^a = \dot{\varepsilon}^p$, and making use of the flow rule (2.111), yields

$$\dot{\lambda} = \frac{\dfrac{\partial f}{\partial \sigma} : \mathbf{C} : \dot{\varepsilon}}{\dfrac{\partial f}{\partial \sigma} : \mathbf{C} : \dfrac{\partial g}{\partial \sigma} + \dfrac{\partial f}{\partial \xi} : \mathbf{h}} . \qquad (2.116)$$

Introducing the rate of the trial stress, defined as

$$\dot{\sigma}^{Trial} = \mathbf{C} : \dot{\varepsilon} , \qquad (2.117)$$

and assuming the denominator of (2.116) to be greater than zero, the loading/unloading conditions can be defined alternatively as

$$f < 0 \qquad \Rightarrow \quad \text{elastic loading/unloading,}$$

$$f = 0 \quad \text{and} \quad \frac{\partial f}{\partial \sigma_{ij}} \dot{\sigma}^{Trial}_{ij} < 0 \quad \Rightarrow \quad \text{elastic unloading from a plastic state,}$$

$$f = 0 \quad \text{and} \quad \frac{\partial f}{\partial \sigma_{ij}} \dot{\sigma}^{Trial}_{ij} = 0 \quad \Rightarrow \quad \text{neutral loading from a plastic state,}$$

$$f = 0 \quad \text{and} \quad \frac{\partial f}{\partial \sigma_{ij}} \dot{\sigma}^{Trial}_{ij} > 0 \quad \Rightarrow \quad \text{plastic loading.} \qquad (2.118)$$

2.3 Time-Independent Material Models for Concrete

Fig. 2.12 contains the geometric interpretation of (2.116). The loading/unloading conditions (2.118) hold regardless of hardening, perfectly-plastic or softening material behavior.

However, the assumption of the denominator in (2.116) to be greater than zero restricts the admissible softening of a material. E.g., for the associated flow rule with $g \equiv f$, because of the positive definiteness of **C**, the first term in the denominator always has a positive value.

Assuming **h** to be of the form

$$\mathbf{h} = \mathbf{D} : \frac{\partial f}{\partial \boldsymbol{\xi}} , \quad (2.119)$$

where **D** denotes a matrix of suitably chosen hardening moduli, for the case of $g \equiv f$ and on the basis of the aforementioned assumption for the denominator of (2.116),

$$\frac{\partial f}{\partial \boldsymbol{\sigma}} : \mathbf{C} : \frac{\partial f}{\partial \boldsymbol{\sigma}} + \frac{\partial f}{\partial \boldsymbol{\xi}} : \mathbf{D} : \frac{\partial f}{\partial \boldsymbol{\xi}} > 0 . \quad (2.120)$$

Hence, (2.120) represents a restriction on the amount of allowable softening [Simo (1988)].

With respect to the expansion of the yield surface, basically there are two different mechanisms. The first one is a global expansion, characterized by the enlargement of the elastic domain, which is bounded by the yield surface. This mechanism is denoted as isotropic hardening. The second mechanism is a translation of the yield surface without change of the size of the enclosed elastic domain. This mechanism is known as kinematic hardening.

Isotropic hardening Mathematically, isotropic hardening is described by separating $f(\boldsymbol{\sigma}, \boldsymbol{\xi})$ into two parts. One of them depends on $\boldsymbol{\sigma}$ and the other one on $\boldsymbol{\xi}$, resulting in

$$f(\boldsymbol{\sigma}, \boldsymbol{\xi}) = F(\boldsymbol{\sigma}) - k(\boldsymbol{\xi}) , \quad (2.121)$$

where k denotes the yield stress depending on the internal variables $\boldsymbol{\xi}$. Usually, only one internal variable is taken into account. It is denoted as hardening parameter. This parameter is often defined as an equivalent plastic strain $\bar{\varepsilon}^p$. Its rate is given as

$$\dot{\kappa} \equiv \dot{\bar{\varepsilon}}^p = \sqrt{\dot{\boldsymbol{\varepsilon}}^p : \dot{\boldsymbol{\varepsilon}}^p} . \quad (2.122)$$

κ may also be defined as the plastic work per unit volume, W^p. The respective rate equation is

$$\dot{\kappa} \equiv \dot{W}^p = \boldsymbol{\sigma} : \dot{\boldsymbol{\varepsilon}}^p . \quad (2.123)$$

A law for isotropic hardening, in rate form, is given as

$$\dot{k}(\kappa) = H_i(\kappa) \dot{\kappa} , \quad (2.124)$$

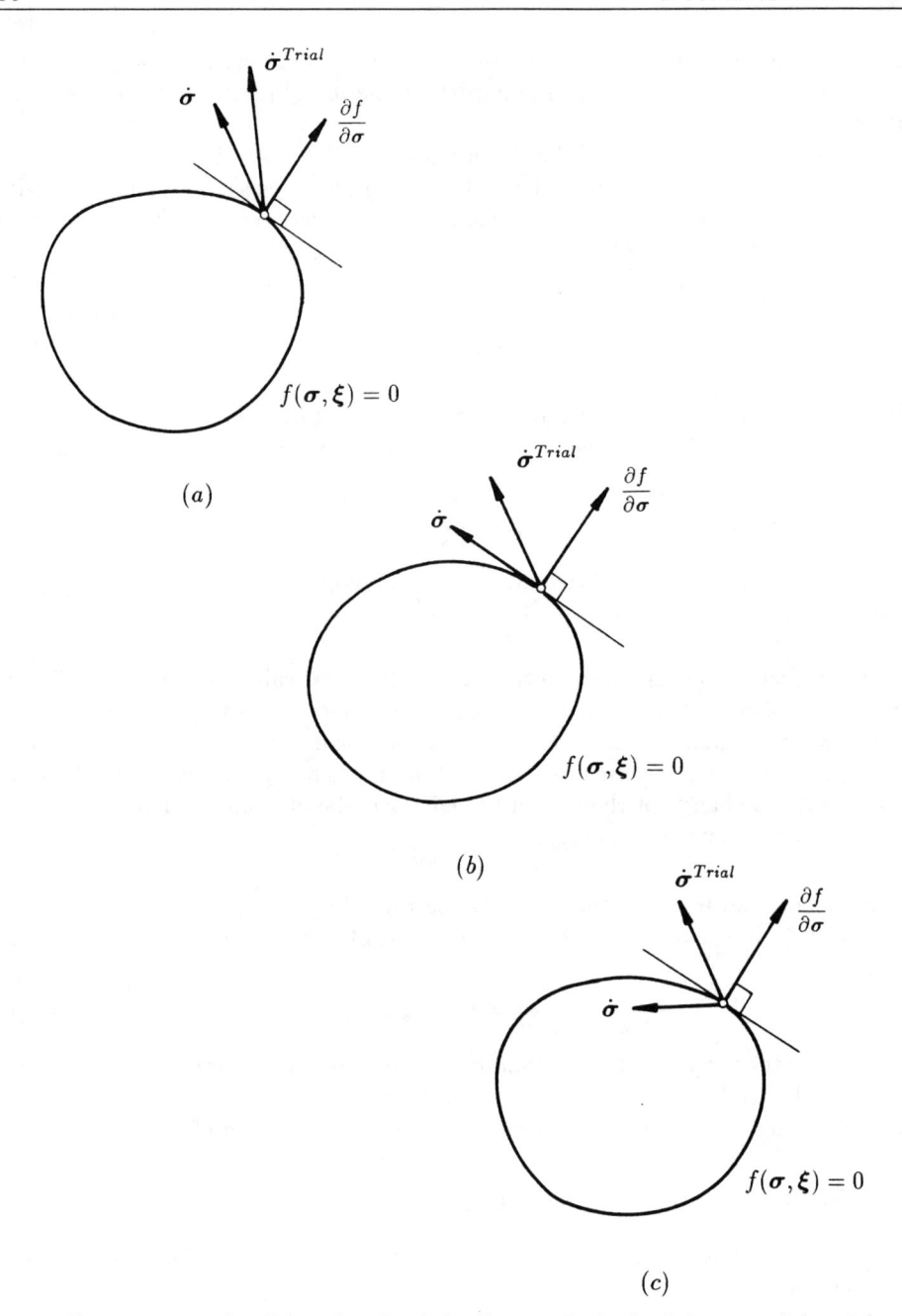

Figure 2.12: Conditions for plastic loading on the basis of the rate of the trial stress: (a) hardening material, (b) perfectly-plastic material, (c) softening material

2.3 Time-Independent Material Models for Concrete

where $H_i(\kappa)$ denotes the modulus of isotropic hardening.

Thus, isotropic hardening is accounted for by an increase of the value of the yield stress in consequence of an increase of the value of the hardening variable. The term isotropic hardening refers to the fact that, if $F(\boldsymbol{\sigma})$ is an isotropic function, then $f(\boldsymbol{\sigma}, \boldsymbol{\xi})$ will be an isotropic function even if plastic deformations have accumulated (Fig.2.13(a)). Therefore, the assumption of isotropic hardening does not allow consideration of hardening-induced anisotropic material behavior.

Kinematic hardening Kinematic hardening is based on the introduction of the so-called back stress $\boldsymbol{\rho}$ as an internal variable which can be interpreted as a reference point of the yield surface. Replacing the argument of F in (2.121) by $(\boldsymbol{\sigma} - \boldsymbol{\rho})$ and omitting the argument of k in this equation, yields

$$f(\boldsymbol{\sigma}, \boldsymbol{\xi}) = F(\boldsymbol{\sigma} - \boldsymbol{\rho}) - k . \tag{2.125}$$

Hence, the yield stress k is constant. A commonly used law for kinematic hardening in rate form is given as

$$\dot{\boldsymbol{\rho}} = H_k(\kappa) \dot{\boldsymbol{\varepsilon}}^p \tag{2.126}$$

where $H_k(\kappa)$ denotes the kinematic hardening modulus.

In case of plastic loading, the reference point $\boldsymbol{\rho}$ is translated from its original position at the origin of the principal stress space to a new position, causing the yield surface to perform a translation (Fig.2.13(b)). Hence, anisotropic behavior is induced in consequence of plastic deformation ($\rightarrow$ induced anisotropy).

Combined hardening Combination of (2.121) and (2.125) yields combined hardening (Fig.2.13(c)), characterized by

$$f(\boldsymbol{\sigma}, \boldsymbol{\rho}, \kappa) = F(\boldsymbol{\sigma} - \boldsymbol{\rho}) - k(\kappa) . \tag{2.127}$$

For combined hardening, the isotropic and the kinematic hardening modulus, $H_i(\kappa)$ and $H_k(\kappa)$, are sometimes derived from a common hardening modulus $H(\kappa)$ by introducing a parameter β, $0 \leq \beta \leq 1$, defining the percentage of isotropic hardening. Thus,

$$H_i(\kappa) = \beta H(\kappa) , \quad H_k(\kappa) = (1 - \beta) H(\kappa) . \tag{2.128}$$

In Fig.2.13 the different types of hardening are shown for the two-dimensional case. Moreover, specialization for one-dimensional loading and unloading is included.

Generalized hardening rules can be obtained, e.g., by introducing a second surface in the stress space, referred to as the bounding surface, such that the current yield surface, denoted as the loading surface, is constrained to move within the bounding surface. Hardening is assumed to depend on the distance between the current stress point on the loading surface and a stress point on the bounding surface, referred to as image stress point. This concept was reported originally in [Dafalias (1975)] to describe cyclic multiaxial loading in the plastic range.

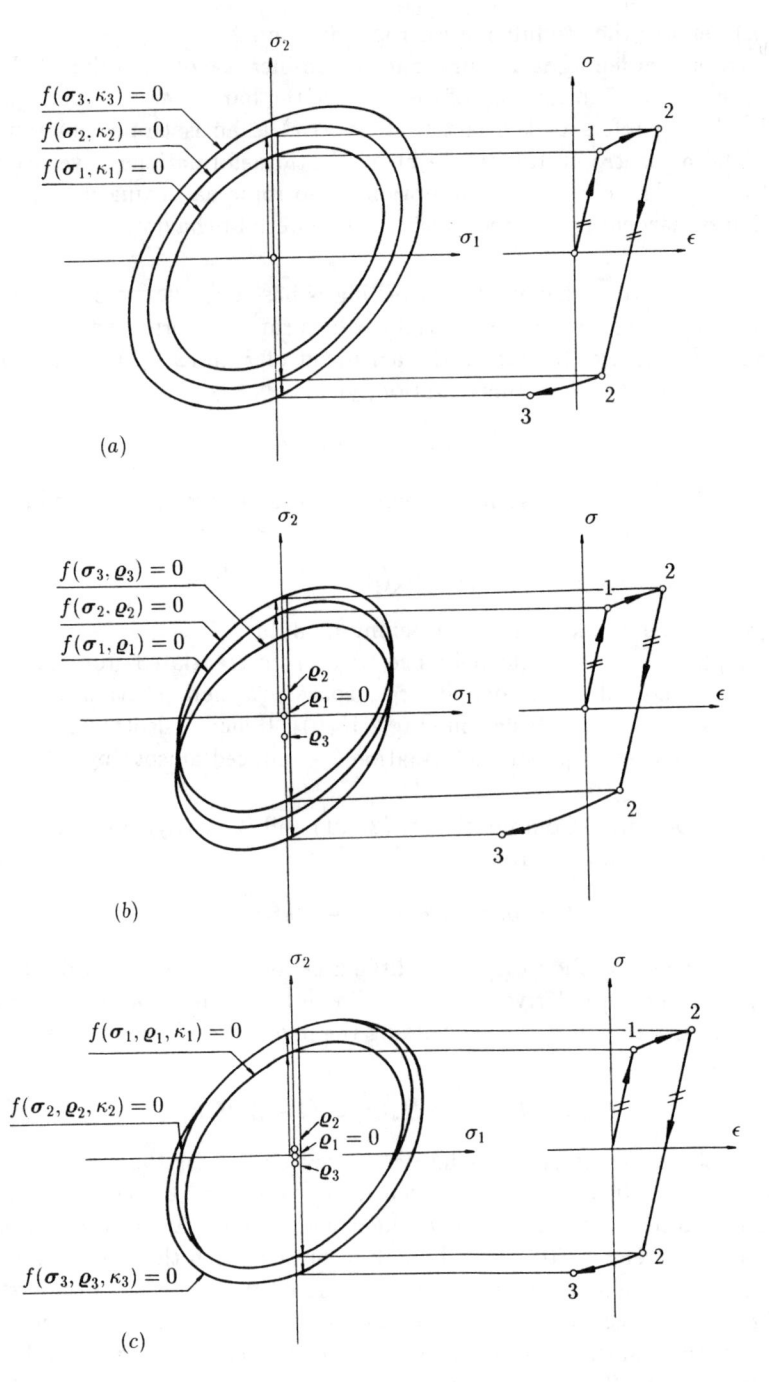

Figure 2.13: (a) Isotropic, (b) kinematic and (c) combined hardening

2.3 Time-Independent Material Models for Concrete

Drucker's stability postulate Hardening was defined by Drucker in terms of the work performed during a loading/unloading cycle [Drucker (1950)]. According to this definition, the term work-hardening has come to stay. Generalization of the inequality

$$d\sigma d\varepsilon^p \geq 0 , \qquad (2.129)$$

which is valid for a uniaxial state of stress, to a triaxial state of stress yields

$$d\sigma_{ij} d\varepsilon^p_{ij} \geq 0 , \qquad (2.130)$$

where the sign of equality holds for the special case of a perfectly-plastic material. (2.130) is known as Drucker's inequality. (2.129) is illustrated in Fig.2.14.

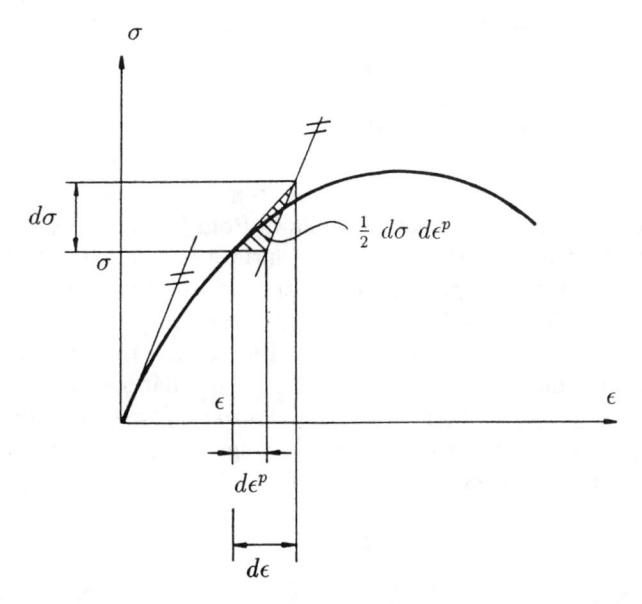

Figure 2.14: Graphic illustration of Drucker's inequality

The left-hand side of (2.129) can be interpreted as twice the work per unit volume, performed by the differential stress increment $d\sigma$ during a load cycle. Such a cycle consists of applying a differential increment of the external load (in addition to the current loads), producing a differential stress increment $d\sigma$, and of removing this differential increment of the external load. A material, characterized by a nonnegative work in a closed loading/unloading cycle according to (2.129) or (2.130), is stable under load (or stress) control. This means that it is possible to apply arbitrary differential stress increments. This is true, however, only for work-hardening materials. A softening material is characterized by $d\sigma < 0$, and $d\varepsilon > 0$. Hence, it is not possible to apply arbitrary differential stress increments. Consequently, such a material is unstable under load control. However, the material is stable under displacement (or strain) control.

The left-hand side of (2.130) can be rewritten as the scalar product

$$d\boldsymbol{\sigma} \cdot d\boldsymbol{\varepsilon}^p \geq 0 ,\qquad(2.131)$$

where the second-order tensors $d\boldsymbol{\sigma}$ and $d\boldsymbol{\varepsilon}^p$ are represented as vectors

$$\begin{aligned}d\boldsymbol{\sigma} &= \lfloor d\sigma_{11}\ d\sigma_{22}\ d\sigma_{33}\ d\tau_{12}\ d\tau_{23}\ d\tau_{31}\rfloor^T ,\\ d\boldsymbol{\varepsilon}^p &= \lfloor d\varepsilon^p_{11}\ d\varepsilon^p_{22}\ d\varepsilon^p_{33}\ d\gamma^p_{12}\ d\gamma^p_{23}\ d\gamma^p_{31}\rfloor^T .\end{aligned}\qquad(2.132)$$

Thus, the vectors $d\boldsymbol{\sigma}$ and $d\boldsymbol{\varepsilon}^p$ form an angle of less than or equal to $90°$, excluding plastic strain rates in the opposite direction of the stress rate.

A necessary but not sufficient condition for the validity of Drucker's stability postulate is obtained by replacing the stress differential $d\boldsymbol{\sigma}$ in (2.131) by a finite stress increment $(\boldsymbol{\sigma} - \boldsymbol{\sigma}^*)$, yielding

$$(\boldsymbol{\sigma} - \boldsymbol{\sigma}^*) \cdot d\boldsymbol{\varepsilon}^p \geq 0 . \qquad(2.133)$$

$\boldsymbol{\sigma}$ refers to a point on the yield surface; $\boldsymbol{\sigma}^*$ characterizes a point located either in the elastic region or on the yield surface such that loading from $\boldsymbol{\sigma}^*$ to $(\boldsymbol{\sigma} + d\boldsymbol{\sigma})$ is elastic for the portion from $\boldsymbol{\sigma}^*$ to $\boldsymbol{\sigma}$ and plastic from $\boldsymbol{\sigma}$ to $(\boldsymbol{\sigma} + d\boldsymbol{\sigma})$, producing a differential plastic strain increment $d\boldsymbol{\varepsilon}^p$. It follows from (2.133) that the rate of plastic strains must be perpendicular to the yield surface and that the latter must be convex. Fig.2.15 contains a graphic interpretation of the inequality (2.133) for the two-dimensional case. It is noted that both requirements are necessary but not sufficient for Drucker's stability postulate, because they do not say anything about the expansion or contraction of the yield surface, i.e., about hardening or softening.

A plastic strain rate, perpendicular to the yield surface $f = 0$, is proportional to the gradient of f in the stress space. Hence, f represents a plastic potential. Replacing g in (2.111) by f results in

$$\dot{\varepsilon}^p_{ij} = \dot{\lambda}\frac{\partial f}{\partial \sigma_{ij}} . \qquad(2.134)$$

(2.134) is known as the normality rule or the associated flow rule expressing the fact that for $g = f$ the flow rule is associated with the yield function. Consequently, a flow rule, characterized by $g \neq f$, is denoted as nonassociated flow rule.

Only for a smooth yield surface the direction of $\dot{\boldsymbol{\varepsilon}}^p$ is defined uniquely by (2.134). If the yield surface is not smooth, i.e., if it is composed of several surfaces bounded by edges, then at these edges the direction of $\dot{\boldsymbol{\varepsilon}}^p$ is not defined uniquely. Fig.2.16 contains an illustration of the analogous two-dimensional situation.

Koiter's generalized flow rule The generalized flow rule, proposed in [Koiter (1953)], serves as a remedy for the nonuniqueness of the direction of $\dot{\boldsymbol{\varepsilon}}^p$ at edges and corners. If, e.g., the yield surface consists of n smooth surfaces, given by $f_k(\boldsymbol{\sigma}, \boldsymbol{\xi}) = 0$, $k = 1, ..., n$, then the elastic region is defined as $f_k < 0$ for all $k = 1, ..., n$. The stress point $\boldsymbol{\sigma}$ is located on the yield surface if, at least for one k, $f_k = 0$. It is located on an edge, if $f_k = 0$ for at least two k. The generalized flow rule is given as

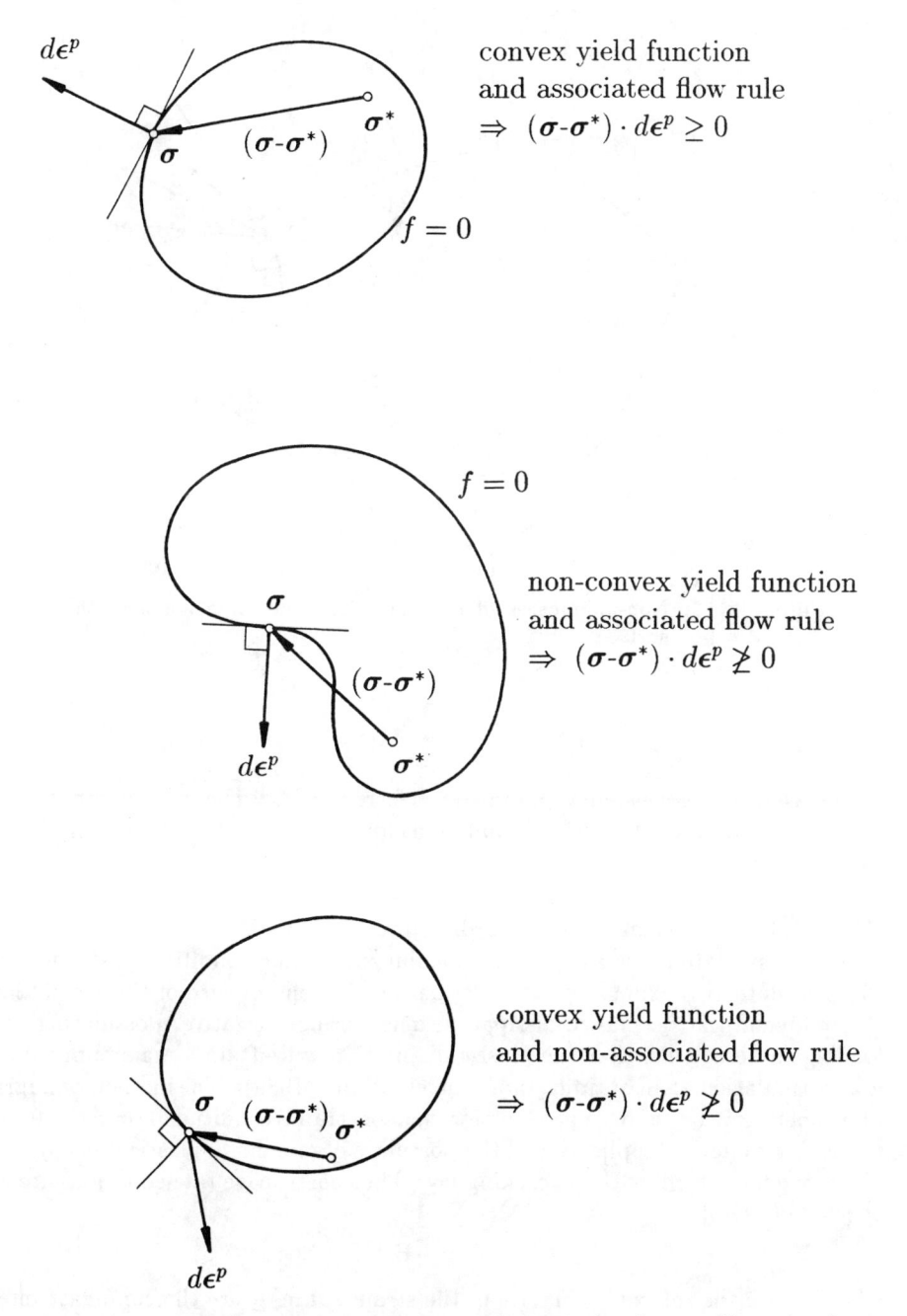

2.3 Time-Independent Material Models for Concrete

convex yield function
and associated flow rule
$\Rightarrow (\boldsymbol{\sigma}-\boldsymbol{\sigma}^*) \cdot d\boldsymbol{\epsilon}^p \geq 0$

non-convex yield function
and associated flow rule
$\Rightarrow (\boldsymbol{\sigma}-\boldsymbol{\sigma}^*) \cdot d\boldsymbol{\epsilon}^p \not\geq 0$

convex yield function
and non-associated flow rule
$\Rightarrow (\boldsymbol{\sigma}-\boldsymbol{\sigma}^*) \cdot d\boldsymbol{\epsilon}^p \not\geq 0$

Figure 2.15: Illustration of the associated flow rule for the two-dimensional case and of the convexity of the yield curve

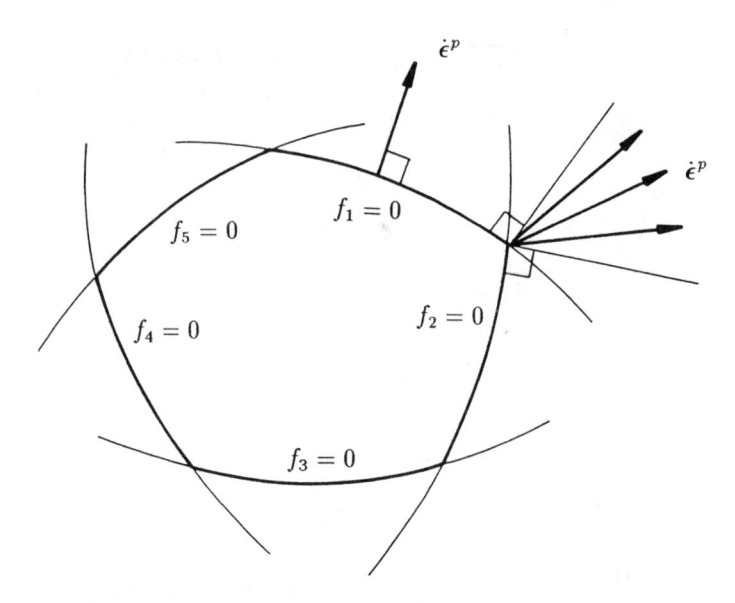

Figure 2.16: Non-uniqueness of the normality rule for a non-smooth yield surface

$$\dot{\varepsilon}^p_{ij} = \sum_k \dot{\lambda}_k \frac{\partial f_k}{\partial \sigma_{ij}} \, , \tag{2.135}$$

where $\dot{\lambda}_k$ is the consistency parameter referred to f_k. The summation over k in (2.135) is restricted to those yield functions for which $f_k = 0$, because only for them $\dot{\lambda}_k > 0$.

Postulate of maximum plastic dissipation The convexity of the yield function, the associated flow rule and the loading/unloading conditions also follow from the postulate of maximum plastic dissipation. In consequence of the second law of thermodynamics the plastic dissipation must be non-negative. Postulating plastic loading to occur such that the plastic dissipation will attain a maximum value, allows formulation of an optimization problem [Simo (1989)]. The method of Lagrange multipliers can be used to account for the constraint condition $f(\boldsymbol{\sigma}, \boldsymbol{\xi}) = 0$. For a hardening material, application of the postulate of maximum plastic dissipation also yields the basic form of the hardening law. The constitutive rate equations are given as [Simo (1989)]

$$\dot{\mathbf{q}} = -\mathbf{H}\dot{\boldsymbol{\alpha}} \, , \tag{2.136}$$

where $\boldsymbol{\alpha}$ are the internal variables in the strain space, $\mathbf{q}$ are the conjugate internal variables in the stress space and $\mathbf{H}$ denotes a matrix of the hardening tangent moduli. The postulate of maximum plastic dissipation implies a hardening law such that

2.3 Time-Independent Material Models for Concrete

$$\dot{\alpha}_k = \dot{\lambda}\frac{\partial f(\boldsymbol{\sigma},\mathbf{q})}{\partial q_k} . \tag{2.137}$$

It follows from this postulate that the hardening law is associated with the yield function. Hence, it is denoted as associated hardening law.

Constitutive rate equations The rate equations for the evolution of the plastic strains [(2.111) or (2.134)] and of hardening [(2.124) and (2.126)] and the consistency condition (2.114) can be combined with the constitutive rate equations, following from (2.97) with $\boldsymbol{\varepsilon}^a = \boldsymbol{\varepsilon}^p$ as

$$\dot{\boldsymbol{\sigma}} = \mathbf{C} : (\dot{\boldsymbol{\varepsilon}} - \dot{\boldsymbol{\varepsilon}}^p) , \tag{2.138}$$

to a relationship between the rate of the total strains and the stress rate. Formally, this relationship can be written as follows:

$$\dot{\boldsymbol{\sigma}} = \mathbf{C}^{ep} : \dot{\boldsymbol{\varepsilon}} . \tag{2.139}$$

In (2.139) $\mathbf{C}^{ep} \equiv \mathbf{C}_T$ denotes the tensor of the elasto-plastic material tangent moduli. $\mathbf{C}^{ep}$ is given as [Chen (1982)]

$$\mathbf{C}^{ep} = \mathbf{C} - \frac{1}{h}\mathbf{C} : \frac{\partial g}{\partial \boldsymbol{\sigma}} \otimes \frac{\partial f}{\partial \boldsymbol{\sigma}} : \mathbf{C} \tag{2.140}$$

with

$$h = \frac{\partial f}{\partial \boldsymbol{\sigma}} : \mathbf{C} : \frac{\partial g}{\partial \boldsymbol{\sigma}} + h_v . \tag{2.141}$$

h_v denotes the contribution of hardening to h. Hence, for ideal plasticity $h_v = 0$. In (2.140) $\otimes$ denotes the tensor product (e.g., $\mathbf{a} \otimes \mathbf{b} = \mathbf{C}$ is equivalent to $a_{ij}b_{kl} = C_{ijkl}$). For the special case of the associated flow rule (2.134), because of $g = f$, $\mathbf{C}^{ep}$ is symmetric.

Strain space plasticity The plasticity theory commonly employed for constitutive models of concrete belongs to the category of stress-space formulations based on Drucker's stability postulate. It is characterized by defining the yield surface and the loading conditions in the stress space. Drucker's postulate for stable material behavior under stress-controlled loading holds for a work-hardening material with a convex yield surface and an associated flow rule. Thus, problems will occur for strain- softening materials characterized by unstable material behavior under load control. A yield surface in the stress space is expanding as long as the material is hardening. It is contracting in the softening region of the material. For a yield surface in the strain space there is no such difference. Such a yield surface is expanding also in the softening region (Fig.2.17).

Substitution of (2.97) with $\boldsymbol{\varepsilon}^a = \boldsymbol{\varepsilon}^p$ into (2.98) and disregard of the temperature T results in the following expression for the yield surface in the strain space:

$$\hat{f}(\boldsymbol{\varepsilon},\boldsymbol{\varepsilon}^p,\boldsymbol{\xi}) = 0 . \tag{2.142}$$

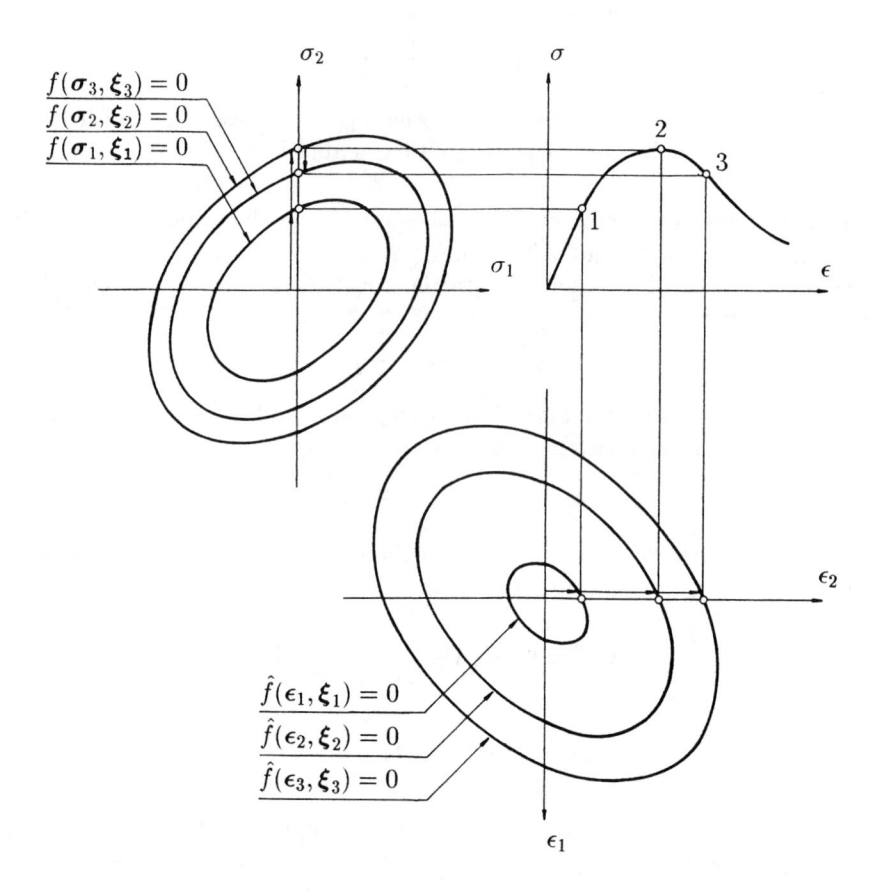

Figure 2.17: Yield surfaces in the stress space and corresponding yield surfaces in the strain space

The loading criteria in the strain space require the plastic strain rate $\dot{\epsilon}^p$ to be non-zero whenever the strain point on the yield surface in the strain space is moving "outwards", resulting in an expansion of the yield surface [Casey (1981)]. During loading, the yield surface in the strain space is always expanding (at least locally). The corresponding yield surface in the stress space, however, is either expanding or stationary or even contracting.

The flow rule in the strain space can be derived either from Il'yushin's postulate, requiring the work done in a closed cycle in the strain space to be non-negative or from the stability postulate in the strain space, given by Nguyen and Bui as [Lubliner (1990)]

$$\dot{\sigma}^p_{ij}\dot{\epsilon}_{ij} \geq 0 , \qquad (2.143)$$

where

$$\dot{\sigma}^p_{ij} = C_{ijkl}\dot{\epsilon}^p_{kl} \qquad (2.144)$$

2.3 Time-Independent Material Models for Concrete

is the rate of the so-called plastic stress σ_{ij}^p.

(2.143) is obtained by inserting the constitutive relations (2.97) with $\varepsilon^a = \varepsilon^p$ into

$$(\boldsymbol{\sigma} - \boldsymbol{\sigma}^*) : \dot{\boldsymbol{\varepsilon}}^p \geq 0 \ . \tag{2.145}$$

(2.145) follows directly from (2.133). (2.145) and (2.133) are based on such a stress history that the plastic strains associated with the stress states $\boldsymbol{\sigma}$ and $\boldsymbol{\sigma}^*$ are identical. The strains $\boldsymbol{\varepsilon}$ on the yield surface in the strain space are obtained as $\boldsymbol{\varepsilon} = \boldsymbol{\varepsilon}^* + \dot{\boldsymbol{\varepsilon}} dt$, where $\boldsymbol{\varepsilon}^*$ is either on or inside the yield surface in the strain space [Lubliner (1990)].

The inequality (2.143) holds for

$$\dot{\sigma}_{ij}^p = \dot{\lambda} \frac{\partial \hat{f}}{\partial \varepsilon_{ij}} \ . \tag{2.146}$$

The loading/unloading conditions in the strain space are summarized as [Casey (1981)]

$$\hat{f} < 0 \quad\Rightarrow\quad \text{elastic loading/unloading},$$

$$\hat{f} = 0 \text{ and } \frac{\partial \hat{f}}{\partial \varepsilon_{ij}} \dot{\varepsilon}_{ij} < 0 \quad\Rightarrow\quad \text{unloading from a plastic state},$$

$$\hat{f} = 0 \text{ and } \frac{\partial \hat{f}}{\partial \varepsilon_{ij}} \dot{\varepsilon}_{ij} = 0 \quad\Rightarrow\quad \text{neutral loading from a plastic state},$$

$$\hat{f} = 0 \text{ and } \frac{\partial \hat{f}}{\partial \varepsilon_{ij}} \dot{\varepsilon}_{ij} > 0 \quad\Rightarrow\quad \text{plastic loading}. \tag{2.147}$$

The loading conditions (2.147) hold regardless of hardening, perfectly-plastic or softening material behavior. With the help of

$$\frac{\partial \hat{f}}{\partial \boldsymbol{\varepsilon}} : \dot{\boldsymbol{\varepsilon}} = \frac{\partial \hat{f}}{\partial \boldsymbol{\sigma}} : \frac{\partial \boldsymbol{\sigma}}{\partial \boldsymbol{\varepsilon}} : \dot{\boldsymbol{\varepsilon}} = \frac{\partial \hat{f}}{\partial \boldsymbol{\sigma}} : \mathbf{C} : \dot{\boldsymbol{\varepsilon}} \ , \tag{2.148}$$

the equivalence of the loading conditions in the strain space (2.147) and of the loading conditions (2.118) derived on the basis of stress space plasticity can be shown. This was done in [Simo (1988)]. Previously, stress-space and strain-space based loading conditions were considered as non-equivalent [Casey (1983)]. Use of the loading conditions (2.147) together with (2.148) for a material model including compressive softening was reported in [Ohtani (1989)].

For constitutive modelling of concrete, strain space plasticity is not very popular. Only recently such material models have been proposed for concrete [Han (1987), Pekau (1992), Mizuno (1992)]. It is claimed that strain space plasticity is superior to stress space plasticity, provided the softening material region is taken into account. However, according to [Simo (1988)], the two formulations are equivalent.

Biaxial elastic-plastic constitutive model

A two-parameter biaxial elastic-plastic constitutive model was reported in [Epstein (1978)]. Because of the restriction to biaxial stress states, the yield surface degenerates to a yield curve given as

$$f(\boldsymbol{\sigma}, \boldsymbol{\xi}) = F(\boldsymbol{\sigma}) - \alpha k_c(\kappa_c) - (1-\alpha) k_t(\kappa_t) , \qquad (2.149)$$

where

$$\boldsymbol{\sigma} = \left\{ \begin{array}{c} \sigma_1 \\ \sigma_2 \end{array} \right\}, \qquad \boldsymbol{\xi} = \left\{ \begin{array}{c} \kappa_c \\ \kappa_t \end{array} \right\}. \qquad (2.150)$$

In (2.149) F denotes the part of f depending on $\boldsymbol{\sigma}$; $k_c(\kappa_c)$ is the yield stress for uniaxial compression depending on the hardening parameter κ_c; $k_t(\kappa_t)$ is the yield stress for uniaxial tension depending on the hardening parameter κ_t. α depends on the location of the stress point on the yield curve. In particular,

$$\begin{array}{llll} \alpha = 1 & \text{for} & \sigma_1 \leq 0 \wedge \sigma_2 \leq 0 & \text{(biaxial compression)}, \\ \alpha = 0 & \text{for} & \sigma_1 \geq 0 \wedge \sigma_2 \geq 0 & \text{(biaxial tension)}, \\ 0 < \alpha < 1 & \text{for} & \sigma_1 \sigma_2 < 0 & \text{(tension-compression)}. \end{array} \qquad (2.151)$$

In the tension-compression region α can be taken as the ratio of the compressive stress over the current yield stress for uniaxial compression $k_c(\kappa_c)$.

Initial and intermediate yield curve and ultimate load curve The form of the function f is chosen such that for the initial yield curve good correspondence with the curve defining the elastic limit for biaxial stress states (Fig.1.8) is achieved.

Hence, for the initial yield curve $k_c(\kappa_c = 0) = \sigma_{yc}^C$ and $k_t(\kappa_t = 0) = \sigma_{yt}^C$, where σ_{yc}^C and σ_{yt}^C are the elastic limits for uniaxial compression and tension, respectively. For loading beyond the elastic limit, the yield curve or parts of it are expanding until the ultimate load curve for biaxial stress states (Fig.1.8) is reached. In this case $k_c = \sigma_P^C$ and $k_t = \sigma_T^C$, where σ_P^C and σ_T^C denote the uniaxial compressive and tensile strength, respectively.

The initial yield curve, an intermediate yield curve (loading curve) and the ultimate load curve (bounding curve) are shown in Fig.2.18 together with the corresponding stress-strain curve and the stress - plastic strain curve for a uniaxial state of stress. (For the sake of brevity, these curves will be referred to as the uniaxial stress-strain curve and the uniaxial stress - plastic strain curve.)

A rough approximation of the experimentally obtained curves shown in Fig.1.8 is obtained by taking the von Mises yield function

$$f = \sqrt{J_2} - \frac{k_c(\kappa_c)}{\sqrt{3}} \qquad (2.152)$$

for the region of biaxial compression, straight lines for the tension-compression regions and a circle for the region of biaxial tension. In this case the functions F for the different regions of the biaxial constitutive model are given as

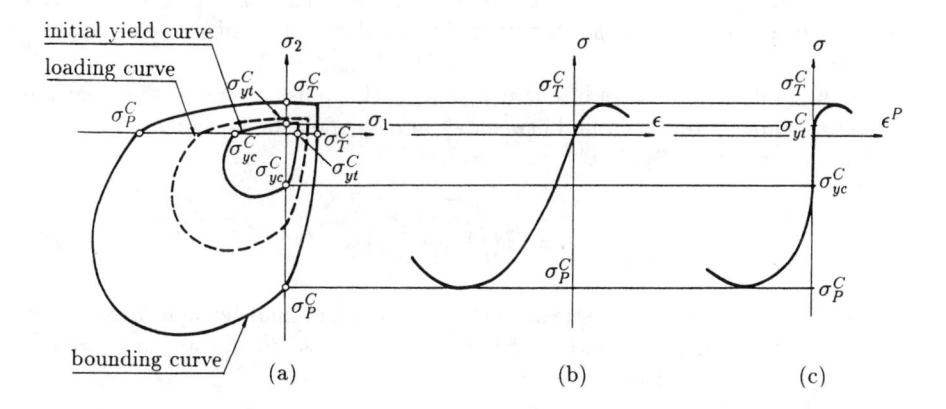

2.3 Time-Independent Material Models for Concrete

Figure 2.18: (a) Initial and intermediate yield curve and ultimate load curve, (b) corresponding uniaxial σ-ε diagram, (c) corresponding uniaxial σ-ε^p curve

$$F = \sqrt{\sigma_1^2 - \sigma_1\sigma_2 + \sigma_2^2} \quad \text{for} \quad \sigma_1 \leq 0 \quad \text{and} \quad \sigma_2 \leq 0 \,,$$
$$F = \sigma_1 + \sigma_2 \quad \text{for} \quad \sigma_1\sigma_2 < 0 \,,$$
$$F = \sqrt{\sigma_1^2 + \sigma_2^2} \quad \text{for} \quad \sigma_1 \geq 0 \quad \text{and} \quad \sigma_2 \geq 0 \,. \tag{2.153}$$

Flow rule The associated flow rule (2.134) is used for determination of the plastic strain rate.

Hardening law The rates of the hardening parameters κ_c and κ_t are defined as

$$\dot{\kappa}_c = \alpha\dot{\kappa} \,, \qquad \dot{\kappa}_t = (1-\alpha)\dot{\kappa} \,, \tag{2.154}$$

where $\dot{\kappa}$ is the rate of the equivalent plastic strain (2.122), and α is taken according to (2.151). Compressive plastic loading only affects κ_c whereas tensile plastic loading only affects κ_t. Therefore, hardening of the compressive region of the yield curve is separated from hardening of the tensile region of the yield curve. This is necessary for a realistic material model for concrete.

The hardening laws for the yield stresses κ_c and κ_t are determined from a uniaxial compression and a uniaxial tension test, respectively. From a uniaxial stress-strain diagram (Fig.2.18(b)) a stress - plastic strain curve can be extracted by subtracting the elastic strain from the total strain (Fig.2.18(c)). This allows representation of the stress beyond the elastic limit as a function of the plastic strain ε^p. In particular, for uniaxial compression and uniaxial tension one obtains

$$k_c(\varepsilon_c^p) = \sigma_{yc}^C + C_\varepsilon^p(\varepsilon_c^p) \,,$$
$$k_t(\varepsilon_t^p) = \sigma_{yt}^C + T_\varepsilon^p(\varepsilon_t^p) \,, \tag{2.155}$$

where ε_c^p and ε_t^p denote the plastic strain for uniaxial compression and tension, respectively, and C_ε^p and T_ε^p are functions for the description of the relation between ε^p and σ.

Since k_c and k_t are formulated in terms of the plastic strain, the relationship between ε_c^p and κ_c and the one between ε_t^p and κ_t must be specified.

It follows from (2.122) that

$$\dot{\kappa} = \dot{\varepsilon}_1^p \sqrt{1 + \left(\frac{\dot{\varepsilon}_2^p}{\dot{\varepsilon}_1^p}\right)^2} , \qquad (2.156)$$

where ε_1^p denotes the plastic strain, measured in a uniaxial tension or compression test, respectively, in the direction of the applied stress. $\dot{\varepsilon}_2^p/\dot{\varepsilon}_1^p$ can be expressed by means of the associated flow rule as

$$\frac{\dot{\varepsilon}_2^p}{\dot{\varepsilon}_1^p} = \frac{\frac{\partial f}{\partial \sigma_2}}{\frac{\partial f}{\partial \sigma_1}} . \qquad (2.157)$$

The desired relationship between the hardening parameter κ for the biaxial case and the plastic strain $\varepsilon^p \equiv \varepsilon_1^p$ based on experiments for the uniaxial case is then obtained by substituting (2.157) into (2.156). Hence, k_c can be reformulated in terms of κ_c instead of ε_c^p. Similarly, k_t can be reformulated in terms of κ_t instead of ε_t^p. (2.155) is then replaced by

$$\begin{aligned} k_c(\kappa_c) &= \sigma_{yc}^C + C_\kappa^p(\kappa_c) , \\ k_t(\kappa_t) &= \sigma_{yt}^C + T_\kappa^p(\kappa_t) , \end{aligned} \qquad (2.158)$$

where C_κ^p and T_κ^p are functions describing the σ-ε^p relation in uniaxial compression and tension in terms of κ_c and κ_t, respectively.

Failure criteria Failure and the respective mode of failure are determined with the help of the ultimate strength curve and the failure criterion described in subsection 2.2.2.

Constitutive rate equations Substitution of the associated flow rule (2.134) into the evolution equation for the hardening parameter (2.122) yields

$$\dot{\kappa} = \dot{\lambda} \sqrt{\frac{\partial f}{\partial \boldsymbol{\sigma}} : \frac{\partial f}{\partial \boldsymbol{\sigma}}} . \qquad (2.159)$$

Solving (2.159) for $\dot{\lambda}$ and inserting the result into (2.134) gives the flow rule in terms of $\dot{\kappa}$:

$$\dot{\varepsilon}^p = \frac{\dot{\kappa}}{\sqrt{\frac{\partial f}{\partial \boldsymbol{\sigma}} : \frac{\partial f}{\partial \boldsymbol{\sigma}}}} \frac{\partial f}{\partial \boldsymbol{\sigma}} . \qquad (2.160)$$

2.3 Time-Independent Material Models for Concrete

From the consistency condition (2.114) the relation

$$\dot{f} = \frac{\partial f}{\partial \boldsymbol{\sigma}} : \dot{\boldsymbol{\sigma}} + \frac{\partial f}{\partial k_c}\frac{\partial k_c}{\partial \kappa_c}\dot{\kappa}_c + \frac{\partial f}{\partial k_t}\frac{\partial k_t}{\partial \kappa_t}\dot{\kappa}_t = 0 \qquad (2.161)$$

is obtained for plastic loading. From (2.161) the rate of the hardening parameter $\dot{\kappa}$ can be derived in terms of the total strain rate $\dot{\boldsymbol{\varepsilon}}$ as

$$\dot{\kappa} = \frac{1}{h}\sqrt{\frac{\partial f}{\partial \boldsymbol{\sigma}} : \frac{\partial f}{\partial \boldsymbol{\sigma}}}\, \frac{\partial f}{\partial \boldsymbol{\sigma}} : \mathbf{C} : \dot{\boldsymbol{\varepsilon}}\,, \qquad (2.162)$$

where

$$h = \frac{\partial f}{\partial \boldsymbol{\sigma}} : \mathbf{C} : \frac{\partial f}{\partial \boldsymbol{\sigma}} + \sqrt{\frac{\partial f}{\partial \boldsymbol{\sigma}} : \frac{\partial f}{\partial \boldsymbol{\sigma}}}\left[\alpha^2 \frac{\partial C_k^p}{\partial \kappa_c} + (1-\alpha)^2 \frac{\partial T_k^p}{\partial \kappa_t}\right]\,. \qquad (2.163)$$

The derivation consists of the following steps:

(a) the constitutive rate equations (2.138) are inserted into (2.161),

(b) in the resulting relation $\dot{\boldsymbol{\varepsilon}}^p$ is expressed in terms of $\dot{\kappa}$, using (2.160),

(c) $\partial f/\partial k_c$ and $\partial f/\partial k_t$ are computed from (2.149),

(d) (2.154) is taken into account.

Inserting (2.162) into (2.160) and substituting the so-obtained relation into (2.138) results in the stress rate in terms of the total strain rate:

$$\dot{\boldsymbol{\sigma}} = \mathbf{C}^{ep} : \dot{\boldsymbol{\varepsilon}}\,, \qquad (2.164)$$

where the tensor of the elasto-plastic tangent moduli is given as

$$\mathbf{C}^{ep} = \mathbf{C} - \frac{1}{h}\mathbf{C} : \frac{\partial f}{\partial \boldsymbol{\sigma}} \otimes \frac{\partial f}{\partial \boldsymbol{\sigma}} : \mathbf{C}\,. \qquad (2.165)$$

Because of use of the associated flow rule, $\mathbf{C}^{ep}$ is symmetric.

An improved version of this constitutive model was presented in [Murray (1979)]. The improvements are twofold. Firstly, the original constitutive model is extended to a three-parameter model, taking into account that tensile failure in one direction hardly affects the tensile strength in the perpendicular direction. Thus, the tensile strength parameter $k_t(\kappa_t)$ is replaced by two independent tensile strength parameters $k_{t1}(\kappa_{t1})$ and $k_{t2}(\kappa_{t2})$. Consequently, the hardening parameter κ_t is decomposed into two hardening parameters, κ_{t1} and κ_{t2}. Secondly, improved functions for the different regions of the loading curve, characterized by a better agreement with experimentally obtained loading curves at limiting stages of biaxial states of stress (Fig.1.8) [Kupfer (1969), Kupfer (1973)], are developed. The improved functions are essential features of the model because, according to (2.157), these functions determine the ratio of the components of the plastic strain rates.

Triaxial elastic-plastic constitutive model

A relatively sophisticated triaxial plasticity-based constitutive model was proposed in [Han (1987)]. Earlier elastic-plastic models (e.g. [Chen (1975)]) are characterized by the affinity of the initial yield surface and of intermediate yield surfaces with the ultimate strength surface. Under tensile loading this assumption usually leads to an overestimation of the plastic strains whereas under confined compressive loading the plastic compressive strains are underestimated. For this reason in the model proposed in [Han (1987)] only the deviatoric sections of the loading surface are chosen affine to the corresponding deviatoric sections of the ultimate strength surface. The meridians of the loading surface are chosen such that in the zone of triaxial tension the initial yield surface coincides with the ultimate strength surface. It departs from the latter in the zone of mixed stresses resulting in an increase of the hardening region with increasing hydrostatic pressure (Fig.2.19). In contrast to the ultimate strength surface, the initial yield surface intersects the hydrostatic axis in the compressive region.

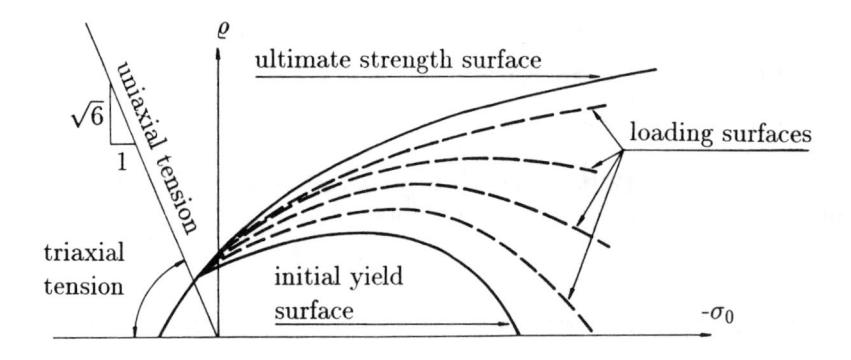

Figure 2.19: Meridional sections of the initial yield surface, of loading surfaces and of the ultimate strength surface

Ultimate strength surface The ultimate strength surface f_u is given as

$$f_u = \rho - \rho_u(\sigma_0, \theta) = 0 , \qquad (2.166)$$

where $\rho = ||\mathbf{s}||$ denotes the distance of the stress point $\boldsymbol{\sigma}$ from the hydrostatic axis and ρ_u is the distance of a corresponding stress point on the ultimate strength surface from this axis. The corresponding stress point has the same values of σ_0 and θ as $\boldsymbol{\sigma}$. A function for the ultimate strength surface such as the five-parameter ultimate strength surface, proposed in [Willam (1975)] and described previously (see (2.15),(2.16)), is recommended.

2.3 Time-Independent Material Models for Concrete

Loading surface Any loading surface can be expressed as

$$f = \rho - c(\sigma_0, \kappa)\rho_u(\sigma_0, \theta) = 0 , \qquad (2.167)$$

where c is a shape factor, depending on σ_0 and κ. The hardening parameter κ ($c_y \leq \kappa \leq 1$, where c_y denotes the initial value of c) is found to be dependent on the plastic work W^p. (In contrast to (2.123) κ is not equal to W^p.) $\kappa = 1$ indicates that the loading surface and, hence, the stress point has reached the ultimate strength surface.

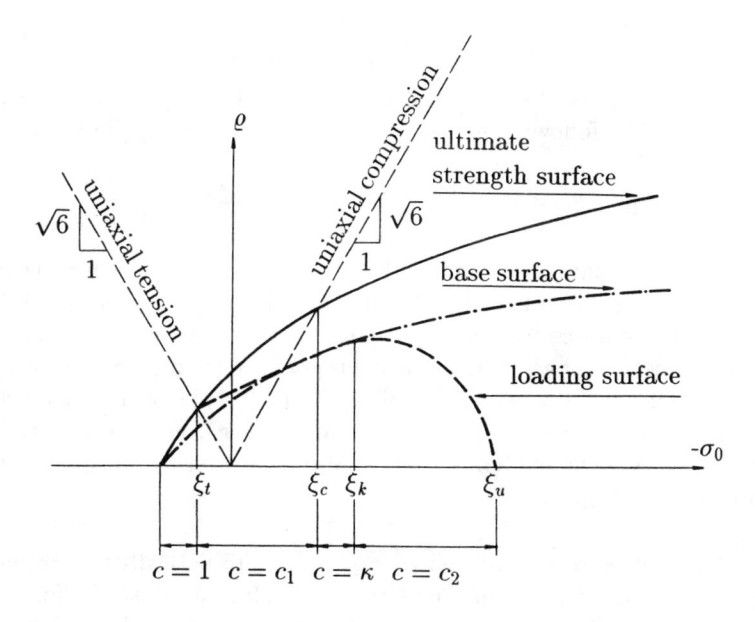

Figure 2.20: Definition of the yield surface

For the definition of $c(\sigma_0, \kappa)$ the hydrostatic axis is subdivided into four different regions (Fig.2.20):

$$\begin{array}{rl} \sigma_0 \geq \xi_t : & c = 1 , \\ \xi_t \geq \sigma_0 \geq \xi_c : & c = c_1(\sigma_0, \kappa) , \\ \xi_c \geq \sigma_0 \geq \xi_k : & c = \kappa , \\ \xi_k \geq \sigma_0 : & c = c_2(\sigma_0, \kappa) . \end{array} \qquad (2.168)$$

For $\sigma_0 \geq \xi_t$, i.e., for the tensile region, the initial yield surface coincides with the ultimate strength surface, reflecting the observed elastic brittle behavior under tensile loading. Hence, $c = 1$. For the part of the compressive region with a relatively low hydrostatic pressure, defined by $\xi_c \geq \sigma_0 \geq \xi_k$, c is equal to the hardening parameter κ. The intermediate region of mixed stresses, with $\xi_t \geq \sigma_0 \geq \xi_c$, is characterized by a decrease of c from $c = 1$ at $\sigma_0 = \xi_t$ to $c = \kappa$ at $\sigma_0 = \xi_c$. For this region, $c = c_1(\sigma_0, \kappa)$ is given as [Han (1985)]

$$c_1(\sigma_0, \kappa) = 1 + (1 - \kappa) \frac{\xi_t(2\xi_c - \xi_t) - 2\xi_c\sigma_0 + \sigma_0^2}{(\xi_c - \xi_t)^2} . \qquad (2.169)$$

Hence, for the two last material domains the hardening region increases with increasing hydrostatic pressure. For $\kappa = 1$ the loading surface coincides with the ultimate strength surface. In the part of the compressive region with a relatively large hydrostatic pressure, i.e., for $\sigma_0 \leq \xi_k$, $c_2 = \kappa$ at $\sigma_0 = \xi_k$, and $c_2 = 0$ at $\sigma_0 = \xi_u$, i.e., at the point of intersection of the loading surface with the hydrostatic axis, ξ_u is assumed as follows

$$\xi_u = \frac{A}{1 - \kappa} , \qquad (2.170)$$

where A is a material parameter. For $\kappa \to 1$ the mentioned point of intersection approaches infinity. The following expression was proposed for c_2 [Han (1985)]:

$$c_2(\sigma_0, \xi_u(\kappa)) = \kappa \frac{(\xi_u - \sigma_0)(\xi_u + \sigma_0 - 2\xi_k)}{(\xi_u - \xi_k)^2} . \qquad (2.171)$$

This material domain is characterized by a large hardening region, reflecting the experimentally observed ability of concrete to sustain relatively large plastic deformations under confined compressive loading. Again, for $\kappa = 1$ the loading surface coincides with the ultimate strength surface. It is proposed in [Han (1984), Han (1987)] to set $\xi_t = 0$ and $\xi_k = \xi_c = \sigma_P^G/3$. It follows from (2.167), which is illustrated in Fig.2.20, that multiplication of ρ_u by the shape factor c yields the initial yield surface and the subsequent loading surfaces. Multiplication of ρ_u by κ gives the so-called base surface (Fig.2.20).

Flow rule Application of the associated flow rule automatically determines the ratio between the octahedral normal and the octahedral shear strain for a given stress point on a particular loading surface. However, it is argued in [Han (1984), Han (1987)] that the use of the associated flow rule for concrete generally results in differences between the observed and the predicted material behavior. This is the reason why application of a nonassociated flow rule is proposed in [Han (1987)]. In this case the rate of the plastic strains does not depend on the loading surface. The nonassociated flow rule is deduced from a Drucker-Prager-type plastic potential, given as

$$g = \alpha I_1 + \sqrt{J_2} - k = 0 , \qquad (2.172)$$

where α and k are material parameters. The rate of the plastic strain is obtained from (2.111) as

$$\dot{\varepsilon}_{ij}^p = \dot{\lambda} \left(\alpha \delta_{ij} + \frac{1}{\sqrt{2}} \frac{s_{ij}}{\|\mathbf{s}\|} \right) . \qquad (2.173)$$

2.3 Time-Independent Material Models for Concrete

The first term on the right-hand side of (2.173) represents the volumetric part of $\dot{\varepsilon}^p_{ij}$. It only depends on the dilatancy factor α. In [Han (1987)] it is proposed to express α as a linear function of the hardening parameter κ, ranging from -0.6 - -0.7 for $\kappa = c_y$ to 0.1 - 0.28 for $\kappa = 1$, i.e., for the ultimate strength. In this way, the experimentally observed material behavior under compressive loading, characterized by plastic volume contraction followed by plastic volume expansion shortly before reaching the ultimate strength, is taken into account.

Hardening law The shape and the size of the loading surface are governed by the hardening parameter κ which depends on the plastic work. The rate of plastic work for a general three-dimensional stress state is given in (2.123). Specialization of (2.123) for a uniaxial compression test, characterized by $\boldsymbol{\sigma} = \lfloor -\bar{\sigma}\ 0\ 0 \rfloor^T$, where $\bar{\sigma}$ is the absolute value of the uniaxial compressive stress, yields

$$\dot{W}^p = \bar{\sigma}\dot{\bar{\varepsilon}}^p . \tag{2.174}$$

In (2.174), $\dot{\bar{\varepsilon}}^p$ denotes the absolute value of the rate of plastic strain in a uniaxial compression test. Assuming the rate of plastic work for a general three-dimensional stress state to be equal to the respective rate for a uniaxial compression test and making use of (2.111) yields

$$\dot{\bar{\varepsilon}}^p = \dot{\lambda}\frac{1}{\bar{\sigma}}\frac{\partial g}{\partial \boldsymbol{\sigma}} : \boldsymbol{\sigma} . \tag{2.175}$$

Inserting $\rho = \sqrt{2/3}\bar{\sigma}$, following from specialization of (2.14) for uniaxial compression ($\sigma_1 = -\bar{\sigma}, \sigma_2 = 0, \sigma_3 = 0$), into (2.167) and setting $c = \kappa$ (2.168$_3$) and $\rho_u = \rho_c$ in (2.167), results in

$$f = \sqrt{\frac{2}{3}}\bar{\sigma} - \kappa\rho_c(\sigma_0) . \tag{2.176}$$

Substitution of (2.176) into the condition $f = 0$ for plastic loading yields

$$\bar{\sigma} = \sqrt{\frac{3}{2}}\kappa\rho_c(\sigma_0) . \tag{2.177}$$

Integration of (2.175) gives $\bar{\varepsilon}^p$. Hence, from a stress - plastic strain curve (Fig.2.21) obtained from a uniaxial compression test, the stress $\bar{\sigma}$ and the corresponding plastic modulus $d\bar{\sigma}/d\bar{\varepsilon}^p$ can be obtained. $d\bar{\sigma}/d\bar{\varepsilon}^p$ is denoted as plastic base modulus H^p_b.

In the ρ-σ_0 diagram (Fig.2.20) $\bar{\sigma}$ is located on the loading path for uniaxial compression. This path is defined by $\rho = \sqrt{2/3}\bar{\sigma}$ and $\sigma_0 = -\bar{\sigma}/3$. The corresponding value of κ is obtained from the condition that the loading surface must pass through this point. The position of the loading surface and, hence, κ are determined by $\bar{\varepsilon}^p$. Thus, $\bar{\varepsilon}^p$ may be viewed as the primary hardening variable.

For the relation between $\bar{\varepsilon}^p$ and $\bar{\sigma}$, the function

$$\bar{\sigma}(\bar{\varepsilon}^p) = |\sigma^C_P|\left[c_y + (1-c_y)\sqrt{1-\left(\frac{\bar{\varepsilon}^p_u - \bar{\varepsilon}^p}{\bar{\varepsilon}^p_u}\right)^2}\right] \tag{2.178}$$

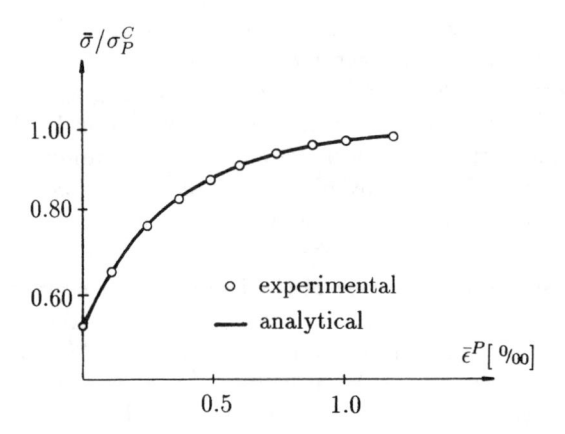

Figure 2.21: Stress - plastic strain curve for uniaxial compression

was proposed [Meschke (1991)], where $\bar{\varepsilon}_u^p$ is the value of $\bar{\varepsilon}^p$ at $\bar{\sigma} = |\sigma_P^C|$. (2.178) is obtained from fitting of data from a uniaxial compression test conducted by Kupfer et al. [Kupfer (1969)]. Fig.2.21 demonstrates the excellent agreement of the experimentally obtained $\bar{\sigma}/\sigma_P^C$-$\bar{\varepsilon}^p$ curve with the respective curve obtained from (2.178). From (2.178) the plastic base modulus is obtained as

$$H_b^p(\bar{\varepsilon}^p) = \frac{d\bar{\sigma}}{d\bar{\varepsilon}^p} = |\sigma_P^C|(1-c_y)\frac{\bar{\varepsilon}_u^p - \bar{\varepsilon}^p}{(\bar{\varepsilon}_u^p)^2}\frac{1}{\sqrt{1 - \left(\frac{\bar{\varepsilon}_u^p - \bar{\varepsilon}^p}{\bar{\varepsilon}_u^p}\right)^2}}. \tag{2.179}$$

Since $H_b^p(\bar{\varepsilon}^p)$ and the analogous quantity $H_b^p(\kappa)$ result from a uniaxial test, the respective expressions neither contain the influence of the hydrostatic stress nor the effect of the angle of similarity. The term plastic base modulus reflects the fact that H_b^p only serves as the basis for determination of a more realistic plastic modulus H^p for general three-dimensional stress states, taking the aforementioned effects into account. For this reason, a modification factor $M(\sigma_0, \theta)$ was introduced in [Han (1984)], yielding

$$H^p(\sigma_0, \theta, \kappa) = M(\sigma_0, \theta) H_b^p(\kappa) \tag{2.180}$$

with

$$M(\sigma_0, \theta) = \begin{cases} f_m & \text{if } 0 < f_m \leq 1, \\ 1 & \text{otherwise}, \end{cases} \tag{2.181}$$

where

$$f_m(\sigma_0, \theta) = \frac{-0.15}{(1.4 - \cos\theta)\left(\frac{1}{3} - \frac{\sigma_0}{\sigma_P^C}\right)\left(2.5 - \frac{\sigma_0}{\sigma_P^C}\right)}. \tag{2.182}$$

This modification factor is based on experimental results [Palaniswamy (1974)]. In Fig.2.22 M is plotted for a few selected values of θ as a function of the normalized hydrostatic stress.

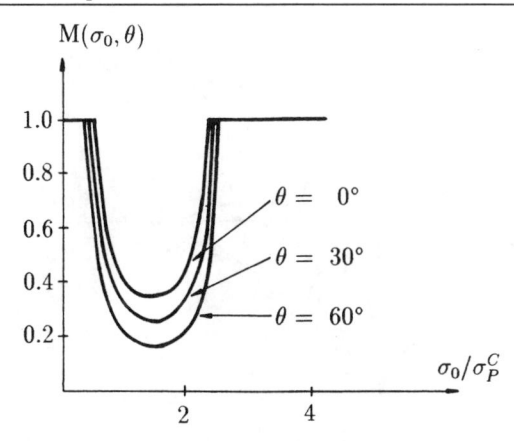

2.3 Time-Independent Material Models for Concrete

Figure 2.22: Modification factor $M(\sigma_0, \theta)$

Loading/unloading criterion Within the framework of the theory of plasticity (2.113) and (2.114) allow to distinguish between elastic loading/unloading and plastic loading. For the present constitutive model this criterion may lead to differences between the experimentally observed and the analytically predicted material behavior. This may be the case, e.g., for loading paths consisting of hydrostatic loading, followed first by partial hydrostatic unloading and deviatoric loading until the loading surface is hit and additional plastic deformations are predicted, and then by loading parallel to the hydrostatic axis (Fig.2.23) [Meschke (1991)]. In this case, hardening because of deviatoric loading yields an expansion of the loading surface, resulting in considerable differences between the experimentally obtained and the analytically predicted stress-strain response for the last part of the loading path, i.e., for loading parallel to the hydrostatic axis. The experimental results suggest a loading/unloading criterion in terms of the principal stresses [Stankowski (1985)], indicating additional plastic deformations if at least one of the principal stresses exceeds its previously attained maximum value (point 5 in Fig.2.23(a)). The loading/unloading criterion within the framework of the theory of plasticity, however, indicates plastic loading, if the current stress point attains the loading surface (point 6 in Fig.2.23(a)). Hence, in contrast to the loading/unloading criterion in terms of the principal stresses, the loading/unloading criterion within the framework of the theory of plasticity indicates elastic loading for the last part of the loading path (section 5-6 in Fig.2.23(a)). The mentioned difference will remain relatively small if, for the respective loading path, deviatoric loading is restricted to the elastic region, i.e., if no hardening under deviatoric loading takes place. In this case this difference is restricted to the section D-E of the loading path shown in Fig.2.23(b).

Constitutive rate equations For plastic loading the consistency condition (2.114) gives

$$\dot{f} = \frac{\partial f}{\partial \boldsymbol{\sigma}} : \dot{\boldsymbol{\sigma}} + \frac{\partial f}{\partial \bar{\varepsilon}^p} \dot{\bar{\varepsilon}}^p = 0 \ . \tag{2.183}$$

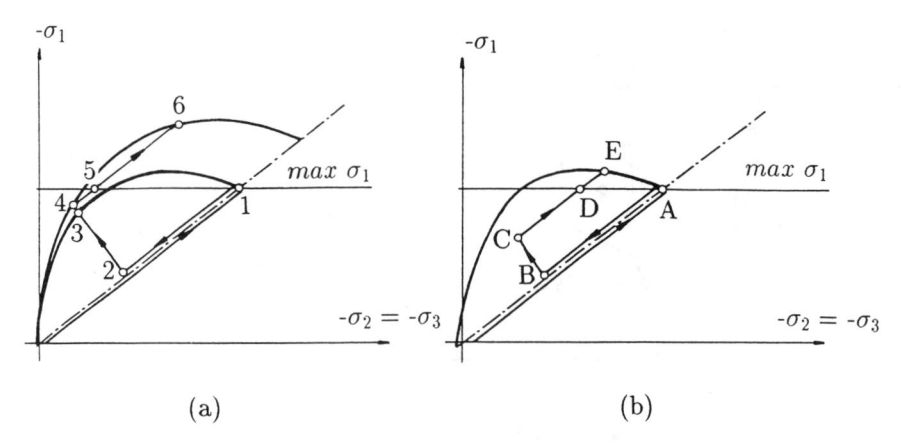

Figure 2.23: Prediction of plastic loading: (a) hardening under deviatoric loading, (b) no hardening under deviatoric loading

Substituting the constitutive rate equations (2.138) and the rate equation (2.175) for $\dot{\bar{\varepsilon}}^p$ into (2.183) and taking the nonassociated flow rule (2.111) into account, yields the consistency parameter λ in terms of the total strain rate as

$$\dot{\lambda} = \frac{1}{h} \frac{\partial f}{\partial \boldsymbol{\sigma}} : \mathbf{C} : \dot{\boldsymbol{\varepsilon}} \tag{2.184}$$

where

$$h = \frac{\partial f}{\partial \boldsymbol{\sigma}} : \mathbf{C} : \frac{\partial g}{\partial \boldsymbol{\sigma}} - \frac{\partial f}{\partial \bar{\sigma}} \frac{\partial \bar{\sigma}}{\partial \bar{\varepsilon}^p} \frac{1}{\bar{\sigma}} \frac{\partial g}{\partial \boldsymbol{\sigma}} : \boldsymbol{\sigma} \ . \tag{2.185}$$

The derivative of f with respect to $\bar{\sigma}$, occurring in (2.185), is obtained from (2.176) as

$$\frac{\partial f}{\partial \bar{\sigma}} = \sqrt{\frac{2}{3}} + \frac{1}{3} \frac{\partial \rho_c}{\partial \sigma_0} \kappa + \frac{1}{3} \rho_c \frac{\partial \kappa}{\partial \sigma_0} \ . \tag{2.186}$$

In (2.186) use of $\partial \sigma_0 / \partial \bar{\sigma} = -1/3$ has been made. According to (2.179), $\partial \bar{\sigma} / \partial \bar{\varepsilon}^p$, appearing in (2.185), is equal to the plastic base modulus H_b^p. Nevertheless, in [Han (1984)] $\partial \bar{\sigma} / \partial \bar{\varepsilon}^p$ was assumed to be equal to the plastic modulus H^p according to (2.180). In this way, irrespective of the inherent mathematical inconsistency, the dependence of the plastic modulus on the hydrostatic stress and the angle of similarity was taken into account.

Inserting (2.184) into the nonassociated flow rule (2.111) and substituting the result into the constitutive rate equations (2.138), the stress rate is obtained in terms of the rate of the total strain as

$$\dot{\boldsymbol{\sigma}} = \mathbf{C}^{ep} : \dot{\boldsymbol{\varepsilon}} \ , \tag{2.187}$$

where

2.3 Time-Independent Material Models for Concrete

$$\mathbf{C}^{ep} = \mathbf{C} - \frac{1}{h}\mathbf{C} : \frac{\partial g}{\partial \boldsymbol{\sigma}} \otimes \frac{\partial f}{\partial \boldsymbol{\sigma}} : \mathbf{C} \qquad (2.188)$$

denotes the tensor of the elastic-plastic material tangent moduli on the continuum level. Because of the use of a nonassociated flow rule this tensor is unsymmetric.

Failure criteria Failure and the respective mode of failure are determined on the basis of the ultimate strength surface and the failure criterion described in subsection 2.2.3.

Numerical evaluation An extensive numerical investigation of the described constitutive model is contained in [Meschke (1991)]. The evaluation of this material model is based on experiments conducted by Kupfer et al. [Kupfer (1969)], Schickert and Winkler [Schickert (1977)], Kotsovos and Newman [Kotsovos (1978)] and Scavuzzo et al. [Scavuzzo (1983)]. In the following, this evaluation is summarized on the basis of the results presented in [Meschke (1991)].

For one-dimensional and two-dimensional loading, in general, there is good agreement between theoretical and experimental results. For three-dimensional stress paths the situation is illustrated in Figs.2.24 - 2.26. In these figures the computed stress-strain curves are compared with corresponding experimentally obtained curves for the loading paths, shown in Fig.1.10. Values of σ_0 and τ_0 for selected stress points on these loadings paths are summarized in Table 2.1. These loading paths are the same as the ones considered for the three-dimensional equivalent-uniaxial model (Figs.2.4 - 2.6) and for the three-dimensional hypoelastic model (Figs.2.8 - 2.10). This allows comparing the predictive capabilities of the different material models.

For hydrostatic loading followed by loading in a deviatoric plane (Fig.1.10(a)), the computed strains agree reasonably well with the experimental ones as long as the absolute value of the hydrostatic pressure is less than $|\sigma_P^C|$ (Fig.2.24). It was shown in [Meschke (1991)] that the assumption of a linear dependence of the dilatancy factor α on κ is the reason for predicting a too soft material behavior for $\sigma_0 < \sigma_P^C$ (i.e., $|\sigma_0| > |\sigma_P^C|$), illustrated in the τ_0/σ_P-γ_0 diagram in Fig.2.24. For this reason, in [Meschke (1991)] a more sophisticated relation was chosen for the dilatancy factor α. Formally, this relation can be written as

$$\alpha = \alpha(c, \kappa, \sigma_0, \sigma_P^C, \theta) . \qquad (2.189)$$

It is based on experimentally obtained results from deviatoric loading paths with different values of the angle of similarity θ. These loading paths are starting from different hydrostatic stress levels. In addition, the influence of the uniaxial compressive strength of concrete on the plastic volumetric response is accounted for. This results in a better agreement of the analytically predicted stress-strain response with the one obtained from the experiment.

Fig.2.25 contains stress-strain diagrams resulting from the loading path shown in Fig.1.10(b). For the smaller value of the octahedral shear stress τ_0, the computed results agree well with the experimental results. However, for a similar loading path with a larger value of τ_0, significant differences between the computed and the experimentally obtained strains are found. The reason for this situation is the loading criterion based on the theory of plasticity, which indicates plastic loading too late. At the beginning of loading parallel to the hydrostatic axis, no coupling between σ_0 and γ_0 is predicted for both loading paths. However, as is evident from Fig.2.25(b), only for the experiment with the smaller value of τ_0 there is no such coupling.

For the star-type loading path of Fig.1.10(c), the analytically predicted σ_1-ε_1 diagram agrees well with the experimentally obtained σ_1-ε_1 diagram (Fig.2.26). This is not the case for the σ_2-ε_2 diagrams which are not shown in Fig.2.26.

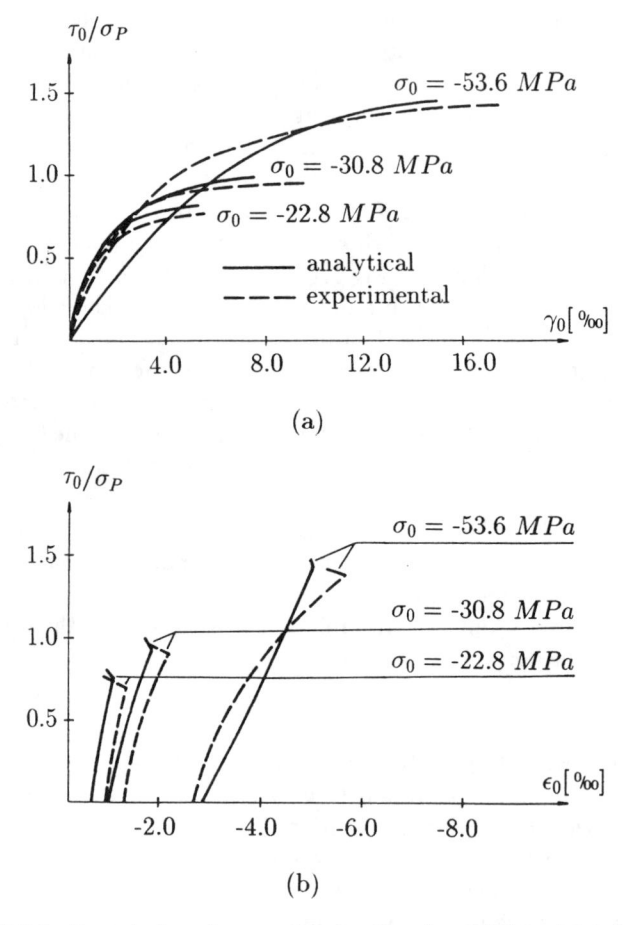

Figure 2.24: Deviatoric axisymmetric loading from different levels of hydrostatic loading: (a) τ_0/σ_P-γ_0 diagram, (b) τ_0/σ_P-ε_0 diagram

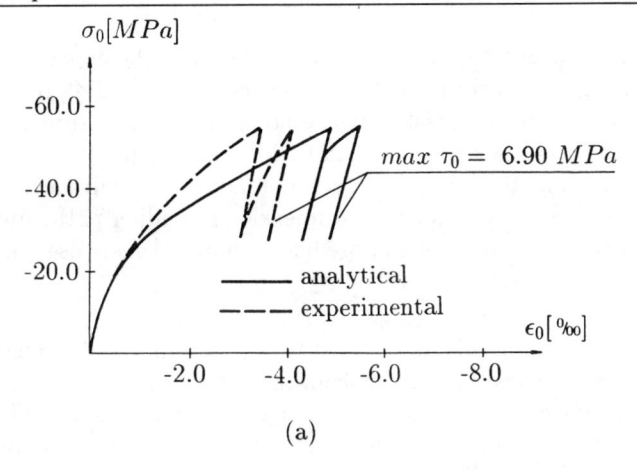

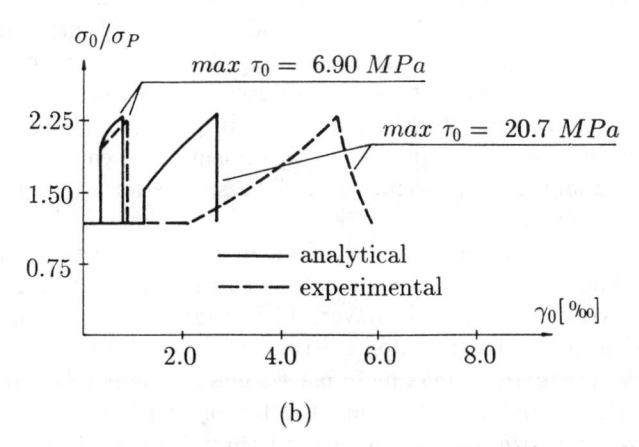

2.3 Time-Independent Material Models for Concrete

Figure 2.25: Nonproportional axisymmetric loading: (a) σ_0-ε_0 diagram, (b) σ_0/σ_P-γ_0 diagram

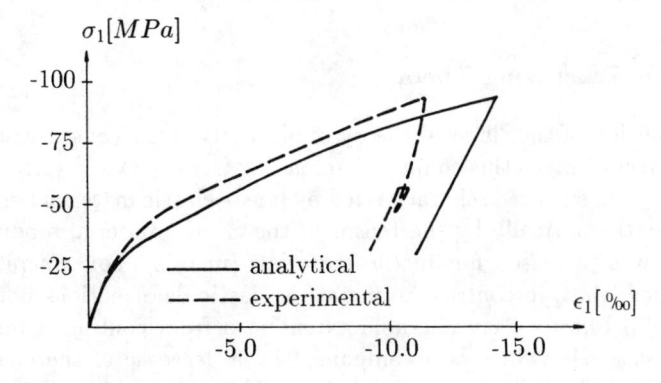

Figure 2.26: General three-dimensional nonproportional loading: σ_1-ε_1 diagram

The differences between the analytically predicted and the experimentally obtained σ_1-ε_1 diagram are caused by the loading criterion. According to this criterion, the material response will be elastic, if the stress point is within the previously attained maximum yield surface. Especially for loading paths with pronounced nonproportional loading this criterion does not agree well with experimental results. Nevertheless, for many other three-dimensional loading paths investigated in [Meschke (1991)] the agreement of the predicted material response with the corresponding experimental results is satisfactory.

Discussion An important advantage of the proposed material model for application to practical problems is that, apart from the definition of the ultimate strength surface, only material parameters are required which can be obtained from a conventional uniaxial test. In addition, the constitutive model can be adjusted to new experimental results by specifying improved functions for the dilatancy factor α and the modification factor M for the plastic modulus. Use of the relatively simple expressions for α and M, proposed in [Han (1984)], leads to differences between the analytically predicted and the experimentally observed material behavior for higher levels of hydrostatic pressure. However, as shown in [Meschke (1991)], an improvement of the function for α, yielding better agreement between the experimentally obtained and the analytically predicted material response for a wide range of stress states, is quite complicated.

Use of a nonassociated flow rule facilitates the modelling of the deformational material behavior. The dilatancy factor α is employed to adjust the volumetric plastic strain to the experimental results. However, the abandonment of the associated flow rule means that a necessary condition for the validity of both Drucker's stability postulate and the postulate of maximum plastic dissipation is violated. A practical consequence of this abandonment is that the tensor of the tangent material moduli becomes unsymmetric. It could be argued that there would be no need for a nonassociated flow rule if the loading surfaces were chosen properly. However, it is very difficult to construct loading surfaces together with a hardening law steering the expansion of the loading surface such that all theoretical requirements for the description of the material behavior of concrete are fulfilled.

2.3.4 Plastic Fracturing Theory

A basic assumption of the classical theory of plasticity is the constitutive rate equation (2.138). According to this equation, unloading from the yield surface and reloading up to the yield surface is characterized by linear-elastic material behavior which is described mathematically by the tensor of the elastic material moduli **C**. Originally, (2.138) was proposed for ductile materials (metals). For concrete, however, it can be argued, that, in contrast to metals, inelastic deformations not only result from plastic slip but to a considerable extent also from continued microcracking at higher stress levels, which is accompanied by a decrease of the elastic moduli. Especially in the softening region a significant stiffness degradation can be observed.

2.3 Time-Independent Material Models for Concrete

The plastic fracturing theory is an extension of the classical theory of plasticity. It accounts for the degradation of the elastic material moduli with increasing deformation. Hence, the rate constitutive equation is obtained from $\boldsymbol{\sigma} = \mathbf{C}\boldsymbol{\varepsilon}^e$ as

$$\dot{\boldsymbol{\sigma}} = \mathbf{C}\dot{\boldsymbol{\varepsilon}}^e + \dot{\mathbf{C}}\boldsymbol{\varepsilon}^e \;, \qquad (2.190)$$

where $\dot{\boldsymbol{\varepsilon}}^e = \dot{\boldsymbol{\varepsilon}} - \dot{\boldsymbol{\varepsilon}}^p$ and $\dot{\mathbf{C}}$ denotes the rate of degradation of the elastic material moduli. Introducing the fracturing stress rate

$$\dot{\boldsymbol{\sigma}}^{fr} = -\dot{\mathbf{C}}\boldsymbol{\varepsilon}^e \;, \qquad (2.191)$$

(2.190) can be rewritten as

$$\dot{\boldsymbol{\sigma}} = \mathbf{C}\dot{\boldsymbol{\varepsilon}}^e - \dot{\boldsymbol{\sigma}}^{fr} \;. \qquad (2.192)$$

As in the classical theory of plasticity, the rate of the plastic strain is deduced from a potential or loading function formulated in terms of the stress tensor and of hardening parameters. Similarly, the rate of the fracturing stress is derived from a potential function given in terms of the strain tensor and of hardening parameters.

In contrast to the classical theory of plasticity, which indicates elastic material behavior for neutral loading, the plastic fracturing theory predicts inelastic material behavior for this kind of loading. The latter prediction agrees better with the observed material response.

A detailed description of the plastic fracturing theory is contained in [Bažant (1979), Bažant (1980d)].

2.3.5 Endochronic Theory of Plasticity

The term "endochronic" expresses the description of the material behavior in terms of an intrinsic time. Its meaning is comparable to the equivalent plastic strain in the classical theory of plasticity. For concrete, two variables, serving as intrinsic times, govern the evolution of the plastic and the fracturing strains [Bažant (1978b)]. The theory can be regarded as a more complicated case of viscoplasticity, characterized by the dependence of the viscosity on the strain rate in addition to its dependence on the stress and the strain. The endochronic theory is especially suited for predicting the material response for nonproportional loading paths with pronounced rotations of the principal stress directions. However, hardly any experimental data exist for such complicated loading paths. In contrast to the previously discussed material models, the endochronic theory leads to nonlinear constitutive rate equations, i.e., the tangent moduli depend on the stress and the strain rate. A detailed description of the endochronic theory of plasticity is contained in [Bažant (1976), Bažant (1978b), Bažant (1978c), Bažant (1980c)].

2.4 Material Models for the Time-Dependent Behavior of Concrete

2.4.1 Introduction

A vast body of literature exists on time-dependent behavior of concrete. State-of-the-art reports with several hundreds of references can be found, e.g., in [ASCE (1982), Bažant (1982b), RILEM (1986)]. This section is restricted to an introduction to the mathematical modelling of time-dependent effects. It is aimed at consideration of such effects within the framework of finite element analyses. The introduction only covers the most popular methods. It is certainly not a comprehensive review on currently used models for creep, shrinkage and aging of concrete.

The time-dependent behavior of concrete structures in consequence of creep, shrinkage and aging of concrete may have a strong influence on the deformations and the stresses. Time-dependent strains may be twice or even three times as large as the instantaneous strains. For structures under sustained loads, for which loss of stability may occur, the load at failure may be reduced considerably with increasing time. Stresses resulting from nonuniform creep and shrinkage, caused, e.g., by a variable moisture content throughout the structure, may cause cracking. Occasionally, time-dependent effects have resulted in loss of serviceability or even failure, years after construction. However, frequently the stress redistributions caused by creep and shrinkage exert a favorable influence on the stress state in the sense that, e.g., restraints are reduced with increasing time.

Usually, the long-term behavior of concrete structures under service loads is of practical interest. In most cases the level of stress under service loads is sufficiently low such that linear material behavior may be assumed to hold. Hence, at any instant of time the relation between the strain and the stress may be considered as linear. Creep laws, defined in the codes [ACI-Committee 209, Chiorino (1984)], are based on the assumption of linear material behavior and, thus, on the applicability of the principle of superposition. Such creep laws are obviously unsuited to predict creep deformations occurring at high levels of sustained stress.

Creep laws may be established by means of integral-type and rate-type formulations. For the prediction of the time-dependent behavior of concrete structures, integral-type formulations are by far more popular. The creep laws, contained in the ACI and CEB/FIP codes [ACI-Committee 209, Chiorino (1984)], belong to this class of formulations. They provide a phenomenological description of creep. Such creep laws are obtained by curve-fitting of experimental results. The main advantage of these creep laws is their simplicity. Their disadvantage is an approximation of the real time-dependent behavior, which is often rather crude. An alternative to the ACI and CEB/FIP creep and shrinkage laws is the so-called BP-model [Bažant (1978), Bažant (1980b)]. It also belongs to the integral-type creep laws. However, its physical basis is more attractive. Moreover, there is a better agreement of analytically predicted and experimentally observed time-dependent effects.

The evolution of creep and shrinkage strains depends on intrinsic and extrinsic

2.4 Material Models for the Time-Dependent Behavior of Concrete

factors. Intrinsic factors are fixed when the concrete is cast. They, e.g., contain the composition parameters of the concrete mix. Extrinsic factors may change after the concrete has been cast. They consist of the parameters describing the environmental conditions such as temperature and humidity. Extrinsic factors can be subdivided into state variables and other variables. Temperature, age and pore humidity at a certain material particle are examples for state variables. They characterize the local state of a continuum. Hence, they can be used in constitutive equations. Environmental temperature and humidity as well as the size and the shape of a concrete member, although having a strong influence on creep and shrinkage, cannot be treated as state variables. These quantities rather represent boundary conditions. Therefore, the distribution of the temperature and the humidity in the concrete member should actually be determined before computation of creep and shrinkage strains. In order to avoid such a complicated mode of analysis, codes of practice consider the influence of the environmental temperature and humidity and of the shape and size of a concrete member on creep and shrinkage by means of empirical formulae, yielding relatively crude approximations of the actual situation.

Knowledge of both the time-dependent material properties and the history of the environmental conditions is a necessary requirement for more accurate predictions of the long-term behavior of concrete structures. Unfortunately, the required data are usually not available. Hence, frequently only standard values can be used. In general, this leads to relatively large differences between the experimentally observed and the analytically predicted time-dependent material behavior. However, these differences can be reduced considerably, if at least short-term shrinkage and creep data are available.

For the mathematical description of the time-dependent behavior of concrete it is convenient to decompose the total strain ε at time t as follows:

$$\begin{aligned} \varepsilon(t) &= \varepsilon^e(t) + \varepsilon^p(t) + \varepsilon^{cr}(t) + \varepsilon^{sh}(t) + \varepsilon^T(t) \\ &= \varepsilon^{in}(t) + \varepsilon^{cr}(t) + \varepsilon^{sh}(t) + \varepsilon^T(t) \\ &= \varepsilon^\sigma(t) + \varepsilon^0(t) \ . \end{aligned} \qquad (2.193)$$

In (2.193) $\varepsilon^{in}(t) = \varepsilon^e(t) + \varepsilon^p(t)$ denotes the instantaneous strain, with $\varepsilon^e(t)$ and $\varepsilon^p(t)$ standing for the elastic and the plastic part of $\varepsilon^{in}(t)$. ε^e and ε^p were introduced already when describing time-independent constitutive models. $\varepsilon^{cr}(t)$ is the creep strain and $\varepsilon^{sh}(t)$ is the shrinkage strain. Determination of $\varepsilon^{cr}(t)$ and $\varepsilon^{sh}(t)$ will be described in this section. $\varepsilon^T(t)$ denotes the temperature-induced strain, given as

$$\varepsilon^T = \int_{T_0}^{T} \alpha(T) dT \ , \qquad (2.194)$$

where T denotes the current temperature, α is the coefficient of thermal expansion and T_0 refers to an initial reference temperature. For $\alpha =$const., (2.194) yields

$$\varepsilon^T = \alpha \cdot (T - T_0) \ . \qquad (2.195)$$

$\varepsilon^{in}(t)$ and $\varepsilon^{cr}(t)$ can be combined to the stress-induced strain $\varepsilon^{\sigma}(t)$. The sum of $\varepsilon^{sh}(t)$ and $\varepsilon^{T}(t)$ gives the strain $\varepsilon^{0}(t)$ which does not depend on the stress. Decomposition of the total strain according to (2.193) is somewhat arbitrary. It leaves the impression that each part of the total strain can be related uniquely to a certain source. However, strictly speaking, this is not true. Problems, e.g., occur with stress-induced shrinkage and temperature-induced strains. In consequence of hydration there is an increase of temperature in a concrete specimen. Hence, it is not easy to determine $\varepsilon^{T}(t)$. Nevertheless, (2.193) usually serves as the basis to describe the time-dependent behavior of concrete at least in an approximate manner.

In the following, the shrinkage and creep laws of the ACI and CEB/FIP code will be described in detail because they are the most widely used shrinkage and creep laws. Moreover, as an alternative to the code laws, the BP-model will be presented. Its physical background is better than the one of the code laws. Also the degree of correspondence of the numerically predicted and the experimentally observed material behavior is better than for the code laws.

2.4.2 Material Models for Shrinkage of Concrete

It was pointed out in chapter 1 that the moisture distribution throughout the depth of a concrete specimen varies from the wetter inner region to the dryer outer region. Hence, drying shrinkage does not proceed uniformly throughout the depth of a concrete body. In addition, it was mentioned that the rate of shrinkage depends on the state of stress.

For these reasons, a realistic prediction of shrinkage strains would require determination of the moisture distribution throughout a concrete body to obtain the main parameters governing the process of shrinkage. However, mathematical models for shrinkage are usually characterized by assuming that shrinkage is uniform throughout the depth of a concrete body and that it is independent of the acting stresses. Hence, shrinkage is modelled as a volumetric process, i.e., the shrinkage strains are the same in any direction.

Shrinkage law according to the ACI-Committee 209 According to [ACI-Committee 209] the shrinkage law is given as

$$\varepsilon^{sh}(t) = K_H^{sh} \, K_{TH}^{sh} \, K_S^{sh} \, K_B^{sh} \, K_F^{sh} \, K_A^{sh} \, \frac{t - \tau_0}{f + t - \tau_0} \, \varepsilon_\infty^{sh} \, , \tag{2.196}$$

where t denotes the age of concrete and τ_0 stands for the age of concrete at the completion of curing. t and τ_0 are measured in days. ε_∞^{sh} is the ultimate shrinkage strain for $t \to \infty$, f is a constant and $K_H^{sh}, K_{TH}^{sh}, K_S^{sh}, K_B^{sh}, K_F^{sh}$ and K_A^{sh} are correction factors for non-standard conditions, describing the influence of

- the relative humidity of the surrounding environment (K_H^{sh}),
- the minimum thickness of the concrete member (K_{TH}^{sh}),

- the consistency of fresh concrete (K_S^{sh}),
- the cement content of concrete (K_B^{sh}),
- the content of fine aggregate particles (K_F^{sh}) and
- the air content of concrete (K_A^{sh})

on the shrinkage strains.

For standard conditions, characterized by an environmental humidity of 40%, a minimum thickness not exceeding 6 inches and a slump not exceeding 4 inches, each one of the correction factors is equal to one. For non-standard conditions they are given as

$$K_H^{sh} = \begin{cases} 1.40 - 0.010\, h_e, & 40\% \le h_e \le 80\%, \\ 3.00 - 0.030\, h_e, & 80\% \le h_e \le 100\%, \end{cases}$$

$$K_{TH}^{sh} = \begin{cases} 1.23 - 0.038\, d_{min}, & \text{for load durations} \le 1 \text{ year}, \\ 1.17 - 0.029\, d_{min}, & \text{for the ultimate value}, \end{cases}$$

$$K_S^{sh} = 0.89 + 0.041\, s,$$

$$K_B^{sh} = 0.75 + 0.034\, b,$$

$$K_F^{sh} = \begin{cases} 0.30 + 0.0140\, f, & \text{for } f \le 50\%, \\ 0.90 + 0.0020\, f, & \text{for } f \ge 50\%, \end{cases}$$

$$K_A^{sh} = 0.95 + 0.0080\, a. \tag{2.197}$$

In (2.197) h_e, f and a are the environmental relative humidity, the content of fine aggregates by weight, and the air content, all given in %; d_{min} and s denote the minimum thickness of the concrete member and the slump, both given in inches, and b stands for the number of 94-lb sacks of cement per cubic yard of concrete. For moist-cured concrete, $\varepsilon_\infty^{sh} = 0.0008$ and $f = 35$. For steam-cured concrete $\varepsilon_\infty^{sh} = 0.00073$ and $f = 55$.

Shrinkage law according to the CEB/FIP model code The shrinkage law of the CEB/FIP model code 1978 [CEB-FIP (1978)] is defined by graphs. Mathematical expressions serving as approximations of the graphical representations can be found in [Chiorino (1984)] where the shrinkage law is given as

$$\varepsilon^{sh}(t) = \varepsilon_1^{sh}\, \varepsilon_2^{sh}\, [\beta^{sh}(t) - \beta^{sh}(\tau_0)]. \tag{2.198}$$

In (2.198) the age t of concrete and its age τ_0 at the completion of curing are given in days; ε_1^{sh} and ε_2^{sh} are functions of the humidity h_e and the effective thickness d_{eff}, respectively. β^{sh} is a function depending on t and d_{eff}. The expressions for $\varepsilon_1^{sh}, \varepsilon_2^{h}$ and β^{sh} are given as follows:

$$\varepsilon_1^{sh} = \left(0.000775\, h_e^3 - 0.1565\, h_e^2 + 11.0325\, h_e - 303.25\right) \cdot 10^{-5},$$

$$\varepsilon_2^{sh} = \exp\left[0.00174\, d_{eff} - \frac{0.32}{d_{eff}} - \ln\left(\frac{d_{eff}^{0.251}}{1.9}\right)\right], \quad (2.199)$$

$$\beta^{sh}(t) = \left[\frac{t}{t + K_3(d_{eff})}\right]^{K_4(d_{eff})},$$

where

$$K_3(d_{eff}) = 11.8\, d_{eff} + 16,$$

$$K_4(d_{eff}) = \exp\left[-0.00257\, d_{eff} + \frac{0.32}{d_{eff}} + \ln\left(0.22\, d_{eff}^{0.4}\right)\right].$$

For a temperature, other than $20^\circ C$, t is replaced by

$$t = \frac{\alpha}{30} \sum_m \{[T(t_m) + 10]\, \Delta t_m\}, \quad (2.200)$$

where $T(t_m)$ is the average temperature in degrees centigrade within the time period Δt_m and α is a coefficient for the type of cement used. For the shrinkage law, α is set equal to 1. $\alpha \neq 1$ will only be employed for the creep law, if a slow hardening or a rapid hardening cement is used.

Shrinkage law according to the Bažant and Panula (BP) model The shrinkage law proposed by Bažant and Panula [Bažant (1978)] is based on fundamental results of diffusion theory, combined with empirical expressions. It is a semi-empirical law.

At the beginning of drying, the pore humidity is equal to 100%. Drying of concrete is caused by the movement of moisture through the concrete member to the surface such that the pore humidity finally attains the environmental humidity. The drying process which causes shrinkage can be modelled mathematically by nonlinear diffusion theory. The dependence of the diffusivity of concrete on the pore humidity is the source of the nonlinearity. Drying shrinkage is assumed to be proportional to the reduction of the pore humidity.

The basic results of diffusion theory are summarized as follows:

- The drying times of geometrically similar specimens of different sizes are proportional to the square of the thickness, resulting in a horizontal shift of the shrinkage curves in the logarithmic time scale for different member sizes (see Fig.1.17(d)). In contrast to the shrinkage law by Bažant and Panula, the ACI shrinkage law predicts a vertical shift of the shrinkage curves for different member sizes. The reason for such a shift is the correction factor K_{TH}^{sh} in (2.197).

- Drying shrinkage would tend to zero if, in the sense of a limiting case, the size of the cross-sectional area of a concrete member became infinitely large;

2.4 Material Models for the Time-Dependent Behavior of Concrete

- Initially, the shrinkage strain is proportional to the square root of the drying time. This theoretical result is confirmed by experimental data [Bažant (1987)].

On the basis of these results the semi-empirical shrinkage law is formulated as

$$\varepsilon^{sh}(t, \tau_0) = \varepsilon^{sh}_{\infty,0} k_h S(t - \tau_0) , \qquad (2.201)$$

where t is the age of concrete and τ_0 stands for its age at the start of drying. t and τ_0 are measured in days. $\varepsilon^{sh}_{\infty,0}$ denotes the ultimate shrinkage at 0% humidity and k_h is a function of humidity, given as

$$k_h = \begin{cases} 1 - \left(\dfrac{h_e}{100}\right)^3 , & \text{for } h_e \leq 98\%, \\ -0,20 , & \text{for } h_e > 98\%, \text{ to account for swelling} , \end{cases} \qquad (2.202)$$

and

$$S(t - \tau_0) = \left(1 + \dfrac{\tau^{sh}}{t - \tau_0}\right)^{-1/2} \qquad (2.203)$$

is an empirical time law, with

$$\tau^{sh} = 600 \left(\dfrac{k_s \, d_{eff}}{150}\right)^2 \dfrac{10}{C(\tau_0)} \qquad (2.204)$$

denoting the shrinkage half-time. In (2.204) C denotes the diffusivity at the start of drying in mm^2/day, k_s is a shape factor calculated from diffusion theory and d_{eff} is the effective thickness of the concrete member in mm, defined as the ratio of twice the volume to the surface exposed to drying. Numerical values of k_s for different geometric forms and empirical formulae for $\varepsilon^{sh}_{\infty,0}$ and C in terms of the strength and of composition parameters of the concrete mix are contained in [Bažant (1978)]. For short durations of drying, i.e., for $t - \tau_0 \ll \tau^{sh}$, the time law (2.203) is approximated as

$$S(t - \tau_0) \approx \left(\dfrac{t - \tau_0}{\tau^{sh}}\right)^{1/2} . \qquad (2.205)$$

2.4.3 Models for Creep of Concrete

Integral-Type Creep Laws

It follows from Fig.1.18(a) that up to about 40% of the uniaxial compressive strength σ_P^C the relationship between the stress and the creep strain is approximately linear. Within this range also the short-term behavior is approximately linear. Hence, the strain at time t can be represented as

$$\varepsilon(t) = J(t, t_0)\sigma + \varepsilon^0(t) , \qquad (2.206)$$

where $J(t, t_0)$ is the compliance function, t denotes the current age of concrete and t_0 stands for the age of concrete at loading. The stress σ is then kept constant from t_0 to t. Hence, $J(t, t_0)$ is the strain produced by the unit stress. Typical shapes of the compliance function for different ages of concrete at loading are shown in Fig.2.27. In Fig.2.27(a) J is plotted in terms of the age t of concrete and in Fig.2.27(b) in terms of the logarithm of the duration $(t - t_0)$ of the action of the load.

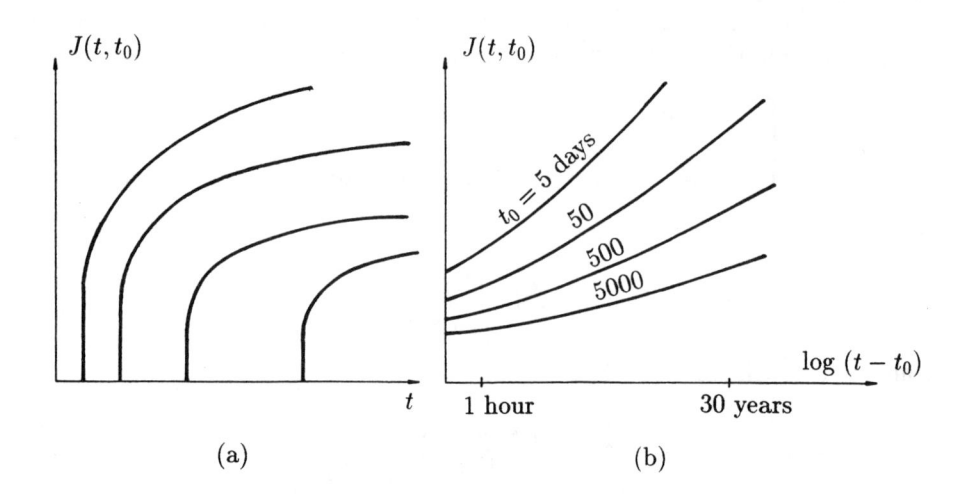

Figure 2.27: Compliance function $J(t, t_0)$ for different ages at loading: (a) J-t diagram, (b) J-$\log(t - t_0)$ diagram

(2.206) holds for a stress which is kept constant over the time period $(t - t_0)$. In concrete structures, however, because of creep, shrinkage and temperature changes, stresses may vary with time even if the load is held constant. Hence, (2.206) has to be extended to stress histories with varying stress. If a linear stress-strain relationship can be assumed to hold for both the instantaneous and the time-dependent strain, then the principle of superposition will be valid.

According to this principle the stress history is subdivided into differentials $d\sigma$ applied at different points of time (Fig.2.28). The contribution of a typical stress differential, applied at time t_0 (Fig.2.28), to the stress-dependent strain at time t is $J(t, t_0) d\sigma(t_0)$. Hence,

$$\varepsilon(t) = \int_0^t J(t, t_0) \, d\sigma(t_0) + \varepsilon^0(t) , \qquad (2.207)$$

where the integral represents the stress-dependent strain.

Experiments have shown that the principle of superposition can only be applied if [Bažant (1982b)]

- the stress does not exceed a value of about 40% of the uniaxial compressive strength of concrete,

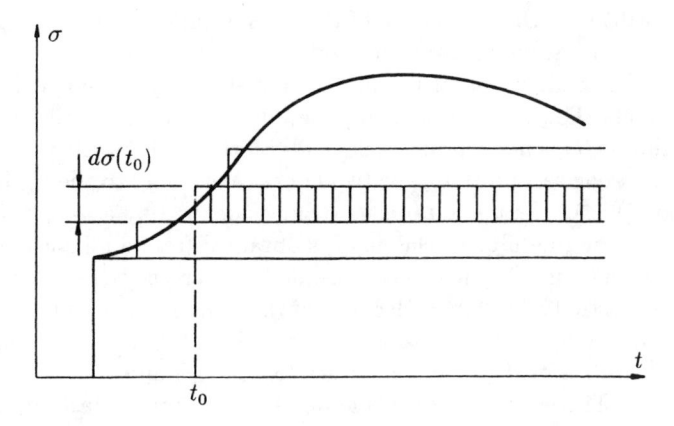

2.4 Material Models for the Time-Dependent Behavior of Concrete

Figure 2.28: Principle of superposition for variable load histories

- the strain does not decrease,
- the stress is not increased suddenly, long after initial loading,
- the distribution of the moisture content does not change significantly during creep.

(2.206) and (2.207) yield the total strain at time t. For time-dependent analysis of concrete structures it is convenient to separate the total strains into the instantaneous and the time-dependent strains. Hence, the compliance function is subdivided into the elastic compliance $1/E^C(t_0)$ and the creep compliance $C(t, t_0)$, respectively, yielding

$$J(t, t_0) = \frac{1}{E^C(t_0)} + C(t, t_0) \; . \tag{2.208}$$

The elastic compliance refers to the vertical parts of the J-t diagrams shown in Fig.2.27(a). It is noteworthy that because of aging of concrete the compliance function cannot be formulated in terms of $t - t_0$, i.e., of a single variable specifying the load duration. The creep compliance can be rewritten as

$$C(t, t_0) = \frac{\phi(t, t_0)}{E^C(t_0)} \; , \tag{2.209}$$

where $\phi(t, t_0) = J(t, t_0) E^C(t_0) - 1$ is the creep coefficient which defines the ratio of the creep strain $\varepsilon^{cr}(t)$ over the instantaneous elastic strain $\varepsilon^e(t)$.

(2.208) allows analysis of the short-term behavior on the basis of time-independent material models as described previously and of the creep strains by means of the creep compliance as

$$\varepsilon^{cr}(t) = \int_0^t C(t, t_0) \, d\sigma(t_0) \; . \tag{2.210}$$

It follows from (2.210) that the evaluation of the creep strain requires storage of the stress history. This is a disadvantage of integral-type creep laws.

Experimentally obtained values for the modulus of elasticity always include some time-dependent effects. This is the reason why the subdivision of the stress-dependent strain into $\varepsilon^e(t)$ and $\varepsilon^{cr}(t)$ is not unique. It depends on the load duration in the experiment, ranging from one second to one or two hours. According to the CEB-Design Manual [Chiorino (1984)], the instantaneous strain is defined by a load duration of 30 to 60 seconds. (If the modulus of elasticity is obtained from an experiment with a load duration of one hour, then it is recommended to increase the measured value of the modulus of elasticity by 25%). Because of different methods of subdividing the total strain into the instantaneous and the time-dependent part, it is important to use consistently determined numerical values for $E^C(t_0)$ and $C(t, t_0)$ in (2.208) or $\phi(t, t_0)$ in (2.209). The experimental value for the modulus of elasticity $E^C_{ex}(t_0)$, obtained on the basis of a load duration Δt, should be replaced by $E^C(t_0)$ following from (2.208) as

$$\frac{1}{E^C(t_0)} + C(t_0 + \Delta t, t_0) = \frac{1}{E^C_{ex}(t_0)} \ . \tag{2.211}$$

Alternative to (2.206) and (2.207), the stress at time t, caused by the strain $\varepsilon(t_0) - \varepsilon^0(t_0)$, which is applied at time t_0 and held constant up to the current time t, can be represented as

$$\sigma(t) = R(t, t_0)\left[\varepsilon(t_0) - \varepsilon^0(t_0)\right] \ , \tag{2.212}$$

where $R(t, t_0)$ denotes the relaxation function. $R(t, t_0)$ is defined as the stress at time t, caused by the unit of stress-producing strain applied at t_0. Since $\varepsilon^0(t_0)$ does not produce a stress, it must be subtracted in (2.212) from the total strain. If the stress varies with time, then, by analogy to (2.207), (2.212) must be replaced by

$$\sigma(t) = \int_0^t R(t, t_0)\left[d\varepsilon(t_0) - d\varepsilon^0(t_0)\right] \ . \tag{2.213}$$

On the basis of creep strains which are proportional to the stress, knowledge of either $J(t, t_0)$ or $R(t, t_0)$ suffices to define the creep properties. If $J(t, t_0)$ is known, the relaxation function can be obtained approximately as [Bažant (1979d)]

$$R(t, t_0) \approx \frac{1 - \Delta_0}{J(t, t_0)} - \frac{0.115}{J(t, t-1)}\left[\frac{J(t_0 + \Delta, t_0)}{J(t, t-\Delta)} - 1\right] \ . \tag{2.214}$$

In (2.214) $\Delta = (t - t_0)/2$, $0 < \Delta_0 < 0.02$; t and t_0 are given in days. Usually, Δ_0 is set equal to 0.008. Although (2.213) and (2.207) are equivalent, the time-dependent material behavior of concrete is almost exclusively described in terms of the compliance function. One of the reasons for this probably is that most often the physical conditions for concrete are closer to constant stress than to constant strain. Moreover, it is easier to perform creep tests for concrete than relaxation tests.

2.4 Material Models for the Time-Dependent Behavior of Concrete

In the following, various expressions for the compliance function or the creep coefficient will be summarized. They are either contained in the ACI code, or in the CEB-FIP model code or in the creep law proposed by Bažant and Panula. It should be kept in mind that the code laws were formulated mainly for the purpose of describing the time-dependent behavior of beam structures. Usually it is assumed that the numerical values of $C(t, t_0)$ are the same over the cross-section of a concrete member. This assumption only holds for sealed concrete members characterized by a uniform moisture content. Normally drying occurs. In this case a single compliance function for the description of the cross-sectional creep behavior only yields a mean value for the creep strains over the cross-section. Thus, such a compliance function obviously does not reflect a constitutive material property.

Creep law according to the ACI-Committee 209 In [ACI-Committee 209] the creep compliance function is formulated according to (2.209). The creep coefficient is given as

$$\phi(t, t_0) = \frac{(t - t_0)^{0.6}}{10 + (t - t_0)^{0.6}} \phi_\infty , \qquad (2.215)$$

where the points of time t and t_0 are measured in days and where ϕ_∞ denotes the ultimate creep coefficient at $t = \infty$, defined as

$$\phi_\infty = 2.35 \; K_{t_0}^{cr} \; K_H^{cr} \; K_{TH}^{cr} \; K_S^{cr} \; K_F^{cr} \; K_A^{cr} . \qquad (2.216)$$

The coefficients $K_{t_0}^{cr}$, K_H^{cr}, K_{TH}^{cr}, K_S^{cr}, K_F^{cr} and K_A^{cr} are correction factors accounting for the influence of

- the age of concrete at loading ($K_{t_0}^{cr}$),
- the relative humidity of the surrounding environment (K_H^{cr}),
- the minimum thickness of the concrete member (K_{TH}^{cr}),
- the consistency of fresh concrete (K_S^{cr}),
- the content of fine aggregate particles (K_F^{cr}) and
- the air content of concrete (K_A^{cr})

on the creep strains. For standard conditions, which are identical to the ones on which the ACI-Committee 209 shrinkage law is based, each one of the correction factors is equal to one. For non-standard conditions the proposed numerical expressions are given as

$$K_{t_0}^{cr} = \begin{cases} 1.25 \; t_0^{-0.118} , & \text{for moist-cured concrete } (t_0 \geq 7 \text{days}) , \\ 1.13 \; t_0^{-0.095} , & \text{for steam-cured concrete } (t_0 \geq 2 \text{days}) , \end{cases}$$

$$K_H^{cr} = 1.27 - 0.0067 \; h_e , \quad \text{for } h_e \geq 40\% ,$$

$$K_{TH}^{cr} = \begin{cases} 1.14 - 0.023 \; d_{min} , & \text{for } t - t_0 \leq 365 \text{ days} , \\ 1.10 - 0.017 \; d_{min} , & \text{for the ultimate value} , \end{cases}$$

$$K_S^{cr} = 0.82 + 0.067\,s,$$
$$K_F^{cr} = 0.88 + 0.0024\,f,$$
$$K_A^{cr} = \begin{cases} 1.0, & \text{for } a \le 6\% \\ 0.46 + 0.090\,a, & \text{for } a > 6\%. \end{cases} \qquad (2.217)$$

Creep law according to the CEB/FIP model code According to the CEB/FIP model code 1978 [CEB-FIP (1978)], the compliance function is assumed as

$$J(t, t_0) = \frac{1}{E^C(t_0)} + \frac{\phi(t, t_0)}{E_{28}^C}, \qquad (2.218)$$

where E_{28}^C refers to the modulus of elasticity of the concrete at the age of 28 days. The creep coefficient ϕ in (2.218) is a sum of three parts:

$$\phi(t, t_0) = \beta_a(t_0) + \phi_v\,\beta_v(t - t_0) + \phi_f\,[\beta_f(t) - \beta_f(t_0)], \qquad (2.219)$$

where $\beta_a(t_0)/E_{28}^C$ represents the initial creep strain which is present after the first 24 hours after application of the load, $\phi_v = 0.4$ denotes the coefficient for consideration of the reversible or delayed elastic creep strain, $\phi_f = \phi_{f1}\phi_{f2}$ is the coefficient for consideration of the irreversible creep strain, β_v is a function of time and β_f is a function of time and of the effective thickness d_{eff}. The age of concrete t and its age at loading t_0 are given in days. For temperatures different from 20°C, t is replaced by (2.200).

The coefficients are given in the form of graphs. Functions serving as approximations of the graphs are contained in Appendix D of [Chiorino (1984)]. They are given as

$$\beta_a(t_0) = 0.8\left[1 - \left(\frac{t_0}{t_0 + 47}\right)^{1/2.45}\right],$$

$$\beta_v(t - t_0) = \left[\frac{t - t_0}{t - t_0 + 328}\right]^{1/4.2},$$

$$\varphi_{f1} = 4.45 - 0.035\,h_e,$$

$$\varphi_{f2} = \exp\left[4.4 \cdot 10^{-5}\,d_{eff} - \frac{0.357}{d_{eff}} - \ln\left(\frac{d_{eff}^{0.1667}}{2.6}\right)\right],$$

$$\beta_f(t) = \left[\frac{t}{t + K_1(d_{eff})}\right]^{K_2(d_{eff})},$$

$$K_1(d_{eff}) = \exp\left[\frac{5.02}{d_{eff}} + \ln\left(6.95\,d_{eff}^{1.25}\right)\right],$$

$$K_2(d_{eff}) = \exp\left[0.00144\,d_{eff} - \frac{1.1}{d_{eff}} - \ln\left(1.005\,d_{eff}^{0.2954}\right)\right]. \qquad (2.220)$$

2.4 Material Models for the Time-Dependent Behavior of Concrete

Creep law according to the Bažant and Panula (BP) model In contrast to the two described creep laws, the BP-model distinguishes between basic creep and drying creep. The former is related to sealed specimens characterized by constant humidity. The latter refers to the additional creep in consequence of drying. It depends on the moisture reduction and, consequently, on the humidity as well as on the shape and the size of the concrete member.

In the simplified version of the BP-model [Bažant (1980b)] the compliance function is given as

$$J(t,t_0) = \frac{1}{E_a^C} + C_b(t,t_0) + C_d(t,t_0,\tau_0) . \qquad (2.221)$$

In (2.221) E_a^C is the asymptotic modulus of elasticity obtained from extrapolation of the compliance function J to an infinitely short load duration; C_b is the compliance following from basic creep; C_d denotes the mean additional compliance resulting from drying creep; τ_0 stands for the age of concrete at the start of drying. If E^C is known from an experiment with a load duration Δt (Δt is usually within the range from a few seconds up to one or two hours), then E_a^C can be obtained by substituting $1/E^C = J(t_0 + \Delta t, t_0)$ into (2.221) and solving for E_a^C. Because of $log(0) = -\infty$, $1/E_a^C$ defines the common horizontal asymptote of the compliance functions shown in Fig.2.27(b). Thus, it represents the true instantaneous strain produced by a unit stress.

The influence of the load duration on basic creep can be approximated to a relatively high degree of accuracy by the power law $(t-t_0)^n$. The influence of the age of concrete at loading on basic creep can be approximated by the power law $(1/t_0)^m$, resulting in the so-called double power law for C_b [Bažant (1976b), Bažant (1978)]:

$$C_b(t,t_0) = \frac{\phi_1}{E_a^C} \left(t_0^{-m} + \alpha\right) (t - t_0)^n , \qquad (2.222)$$

where E_a^C, ϕ_1, α, m and n are material parameters. For normal-weight and normal-strength concrete the material parameters for the simplified version of the BP-model are given as follows:

$$\frac{1}{E_a^C} = 10^{-6} \cdot \left(14.5 + \frac{3446.8}{(\sigma_{P,28}^C)^2}\right) ,$$

$$\phi_1 = 0.30 + \frac{152.2}{(\sigma_{P,28}^C)^{1.2}} ,$$

$$m = 0.28 + \frac{47.5}{\left(\sigma_{P,28}^C\right)^2} ,$$

$$n = 0.115 + 6.1 \cdot 10^{-7} \left(\sigma_{P,28}^C\right)^3 ,$$

$$\alpha = 0.05 , \qquad (2.223)$$

where the unit of the uniaxial compressive strength of concrete at the age of 28 days, $\sigma^C_{P,28}$, is MPa. Typical values of the material parameters are $\phi_1 \approx 2-4$, $m \approx 1/3$ and $n \approx 1/8$. E^C_a is about 50% larger than the conventional modulus of elasticity.

In the complete version of the BP-model the material parameters also depend on the composition of the concrete mix. It is noted that the values of the compliance function based on the double power law are very sensitive with respect to a change of the value of n. The factor $(t-t_0)^n$ in the expression for C_b is supported theoretically by the activation energy model for creep [Wittmann (1985)]. Apart from its relative simplicity, the advantages of the double power law for C_b are its wide range of applicability, extending from very short load durations to load durations of many years, and separation of the influence of concrete aging and creep. The double power law even yields a good approximation for the dynamic modulus of elasticity which is obtained by setting $t - t_0 = 10^{-7}$ days.

The term C_d in the expression for the compliance function is related to drying creep. It includes the effect of shrinkage on creep. This term is given as

$$C_d(t, t_0, \tau) = \frac{\phi'_d}{E^C_a} t_0^{-m/2} k_h \varepsilon^{sh}_{\infty,0} S_d(t, t_0) , \qquad (2.224)$$

where

$$k_h = 1 - \left(\frac{h_e}{100}\right)^{3/2},$$

$$\phi'_d = \phi_d \left(1 + \frac{t_0 - \tau_0}{10\tau^{sh}}\right)^{-1/2},$$

$$S_d(t, t_0) = \left(1 + \frac{3\tau^{sh}}{t - t_0}\right)^{-0.35}. \qquad (2.225)$$

h_e was defined during description of the shrinkage law according to the ACI-Committee 209. τ_0, τ^{sh} and $\varepsilon^{sh}_{\infty,0}$ were defined during presentation of the BP shrinkage law. (2.224) and (2.225) follow from basic results of diffusion theory, which have been discussed previously. ϕ_d can be computed from an empirical expression [Bažant (1978)], taking the composition of the concrete mix into account. If no pertinent information is available, it is recommended to choose $\phi_d = 0.0056$.

The complete BP-model is containing an additional term in the expression for the compliance function. This term is formally written as $-C_p(t, t_0, \tau_0)$. It represents the decrease of creep after the drying process has reached the final state. This term is only relevant for very thin cross-sections or at elevated temperatures.

In parts IV and V of [Bažant (1978)] the formulae for basic creep and drying creep were modified in order to allow consideration of temperature effects. These effects were taken into account in the material law for basic creep and drying creep. This was done by modifying the material parameters to account for the increase of the creep rate in consequence of elevated temperatures and for the decrease of creep because of the acceleration of hydration. In addition, the acceleration of drying and aging was taken into account.

2.4 Material Models for the Time-Dependent Behavior of Concrete

Further improvements of the double power law of the BP-model are reported in [Bažant (1985), Bažant (1985b)]. In [Bažant (1985)] a logarithmic double power law and in [Bažant (1985b)] a triple power law was proposed. In comparison to the double power law, these two power laws result in a better correspondence with experimental data for very long load durations. Extensions of the BP-model to high-strength concrete and to cyclic humidity are proposed in [Bažant (1984)] and [Bažant (1985c)], respectively.

Approximation of the Creep Compliance by a Dirichlet Series

An approximation of a given creep compliance function $C(t, t_0)$ can be obtained by its expansion into a series of exponential functions, i.e., into a Dirichlet series [Bažant (1973)]

$$C(t, t_0) \cong \sum_{r=1}^{s} a_r(t_0) \left[1 - e^{-(t-t_0)/\tau_r} \right], \qquad (2.226)$$

where $a_r(t_0)$ are coefficients depending on the age at loading and τ_r are constants denoted as relaxation times. The degree of accuracy of the so-obtained approximation of the creep compliance function depends on the number s of terms in the series expansion and on the proper choice of the relaxation times. In order to obtain a good approximation of the creep compliance function for a small value of s, the relaxation times should be chosen as [ASCE (1982)]

$$\frac{\tau_r}{\tau_{r-1}} = 10, \quad r = 2, \ldots, s, \qquad (2.227)$$

with

$$\tau_1 \leq 2t_1, \quad \tau_s \geq t_n/2. \qquad (2.228)$$

In (2.228) t_1 and t_n are the bounds of the time interval of interest for the time-dependent analysis. If aging is taken into account, τ_1 should be much smaller than the age of concrete t_0 at initial loading, e.g.,

$$\tau_1 \leq t_0/10. \qquad (2.229)$$

The conditions (2.228) and (2.229) result from the following properties of the exponential function:

$$1 - e^{-0.1} < 0.1, \quad 1 - e^{-10} \approx 1.$$

For a set of chosen relaxation times the coefficients $a_r(t_0)$ can be obtained by means of a least squares fit of the expression on the right-hand side of (2.226) [Kabir (1977)].

In [Kabir (1977)] (2.226) was extended to account for the influence of elevated constant temperatures T on creep by multiplying the exponent of (2.226) by the function $\phi(T)$, resulting in

$$C(t, t_0, T) \cong \sum_{r=1}^{s} a_r(t_0) \left[1 - e^{-\phi(T)(t-t_0)/\tau_r}\right] . \qquad (2.230)$$

This multiplication yields a horizontal shift of the creep curve with respect to the logarithmic time scale, representing the acceleration of creep at elevated temperatures.

It can be shown that the Dirichlet series expansion of the compliance function is equivalent to a rate-type constitutive relation corresponding to a Kelvin chain model [Bažant (1973)]. On the basis of the approximation (2.226) recursive algorithms can be developed [Kabir (1977), ASCE (1982)]. They are computationally advantageous because only information from the previous time step is required for determination of the incremental creep strains. Hence, in contrast to the integral-type creep law (2.207) or (2.210), there is no need to store the complete stress history. However, in view of the rapid progress in computer technology this advantage has become somewhat less important. It is no longer a real problem to store a complete stress history.

Rate-Type Creep Laws

Rate-type creep laws are given in the form of differential equations containing strain rates and stress rates. Such creep laws are an alternative to integral-type creep laws. Usually rate-type creep laws are based on rheological models. Simple rheological models are the Maxwell and the Kelvin unit (Fig.2.29(a) and (b)).

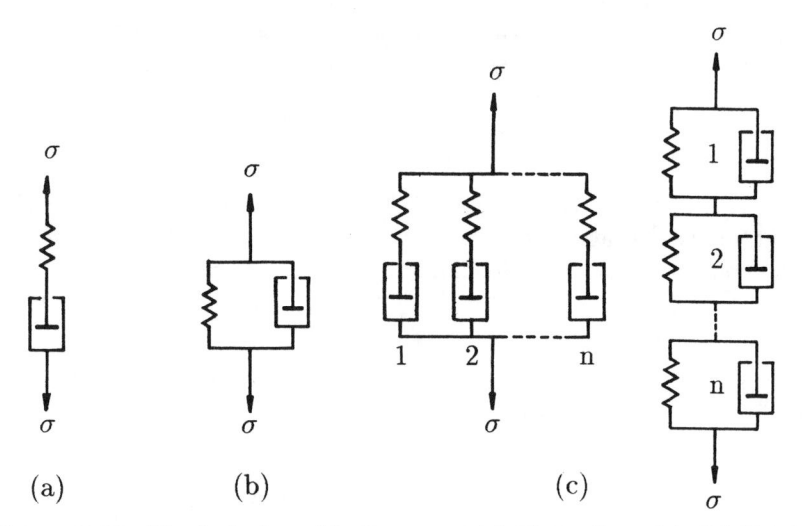

Figure 2.29: Rheological models for creep: (a) Maxwell model, (b) Kelvin model, (c) combined models

A Maxwell unit represents a series connexion of a spring with the modulus of elasticity E and a dashpot with the viscosity η. The respective constitutive relations

2.4 Material Models for the Time-Dependent Behavior of Concrete

are $\sigma_s = E\varepsilon_s$ and $\sigma_d = \eta\dot{\varepsilon}_d$. A Kelvin unit consists of a spring and a dashpot in parallel connexion. If aging of concrete is taken into account, the constitutive relations for the spring and the dashpot become

$$\dot{\sigma}_s = E(t)\dot{\varepsilon}_s \quad \text{and} \quad \sigma_d = \eta(t)\dot{\varepsilon}_d . \tag{2.231}$$

The differential equation for a Maxwell unit,

$$\dot{\varepsilon} = \frac{\dot{\sigma}}{E(t)} + \frac{\sigma}{\eta(t)} , \tag{2.232}$$

follows from $\dot{\varepsilon} = \dot{\varepsilon}_s + \dot{\varepsilon}_d$ and $\sigma = \sigma_s = \sigma_d$. The differential equation for the Kelvin unit is obtained from $\dot{\sigma} = \dot{\sigma}_s + \dot{\sigma}_d$ and $\varepsilon = \varepsilon_s = \varepsilon_d$ as

$$\ddot{\varepsilon} + \frac{E(t) + \dot{\eta}(t)}{\eta(t)}\dot{\varepsilon} = \frac{\dot{\sigma}}{\eta(t)} . \tag{2.233}$$

Maxwell and Kelvin units can be combined to more complicated rheological models, yielding rate-type creep laws in terms of the material parameters $E_r(t)$ and $\eta_r(t)$, $r = 1, \ldots, n$ (Fig.2.29(c)). The degree of accuracy of approximations of integral-type creep laws by such rheological models increases with increasing value for n.

A more recent proposal for a creep law is based on the solidification theory of creep [Bažant (1989)]. In this theory the effect of concrete aging on creep is related to the increase of the volume fraction of the load-bearing portion of the solidified matter. The constitutive model can be formulated as a rate-type creep law in the form of a Kelvin chain with age-independent elastic moduli and viscosities. It only requires four material parameters.

Extension of Creep and Shrinkage Laws to Multiaxial Stress States

The preceding equations for consideration of the time-dependent behavior of concrete only hold for uniaxial loading. The extension to multiaxial stress states is obtained by assuming isotropic material behavior for creep. The respective extension of (2.210) yields

$$\varepsilon^{cr}(t) = \int_0^t \mathbf{A} C(t,t_0) \, d\boldsymbol{\sigma}(t_0) . \tag{2.234}$$

For three-dimensional stress states,

$$\begin{aligned}
\boldsymbol{\varepsilon}^{cr}(t) &= \lfloor \varepsilon_{11}^{cr} \; \varepsilon_{22}^{cr} \; \varepsilon_{33}^{cr} \; 2\varepsilon_{12}^{cr} \; 2\varepsilon_{23}^{cr} \; 2\varepsilon_{31}^{cr} \rfloor^T , \\
d\boldsymbol{\sigma}(t) &= \lfloor d\sigma_{11} \; d\sigma_{22} \; d\sigma_{33} \; d\sigma_{12} \; d\sigma_{23} \; d\sigma_{31} \rfloor^T ,
\end{aligned} \tag{2.235}$$

$$\mathbf{A} = \begin{bmatrix} 1 & -\nu^{cr} & -\nu^{cr} & 0 & 0 & 0 \\ & 1 & -\nu^{cr} & 0 & 0 & 0 \\ & & 1 & 0 & 0 & 0 \\ & & & 2(1+\nu^{cr}) & 0 & 0 \\ & & & & 2(1+\nu^{cr}) & 0 \\ \text{sym.} & & & & & 2(1+\nu^{cr}) \end{bmatrix} . \tag{2.236}$$

For biaxial stress states, **A** is obtained by deleting the respective lines and columns in (2.236). Poisson's ratio ν^{cr} refers to creep of concrete. For sealed concrete specimens, ν^{cr} can be set equal to the value of Poisson's ratio for short-term loading. For unsealed concrete members, drying results in a variable moisture content over the depth of the specimen, accompanied by microcracking and visible cracking. For this case, the compliance function is usually taken as a mean value. The corresponding Poisson's ratio for creep then depends on the humidity distribution and on the size and the shape of the cross-section. It can drop almost to zero [Bažant (1982b)]. It is noteworthy that cracking caused by shrinkage induces some anisotropy. Thus, the assumption of isotropic material behavior in (2.236) is only a relatively crude approximation.

Shrinkage is modelled as a volumetric process. Hence, the extension to multiaxial conditions simply yields

$$\varepsilon^{sh}(t) = \lfloor \varepsilon^{sh} \quad \varepsilon^{sh} \quad \varepsilon^{sh} \quad 0 \quad 0 \quad 0 \rfloor^T, \tag{2.237}$$

where $\varepsilon^{sh} = \varepsilon^{sh}(t)$ is defined according to the employed shrinkage law.

2.4.4 Material Models for Concrete Aging

Because of progressive hydration, both the modulus of elasticity and the strength of concrete increase with increasing age. According to the ACI Committee 209 [ACI-Committee 209], aging is accounted for by the expressions

$$\sigma_P^C(t) = \sigma_{P,28}^C \frac{t}{a+bt}$$

$$E^C(t) = E_{28}^C \sqrt{\frac{t}{a+bt}}, \tag{2.238}$$

where the particular values for the constants a and b depend on the type of cement. For ordinary cement $a = 4$ and $b = 0.85$.

In the CEB/FIP model code [CEB-FIP (1978)] aging is modelled as

$$\sigma_P^C(t) = \sigma_P^C(t=\infty)\beta_c(t) = \frac{\sigma_{P,28}^C}{\beta_c(t=28)}\beta_c(t) = \frac{\sigma_{P,28}^C}{0.669}\beta_c(t), \tag{2.239}$$

where

$$\beta_c(t) = \left(\frac{t}{t+47}\right)^{1/2.45} \tag{2.240}$$

and

$$E^C(t) = \frac{E_{28}^C}{\beta_i(t)}, \tag{2.241}$$

with

2.4 Material Models for the Time-Dependent Behavior of Concrete

$$\beta_i(t) = \sqrt[3]{\frac{\beta_c(t=28)}{\beta_c(t)}} = 0.875 \left(\frac{t+47}{t}\right)^{1/7.35} . \tag{2.242}$$

In (2.238) - (2.242) the age of concrete, t, is given in days.

Consideration of aging in a time-dependent analysis requires application of a modified creep compliance [Van Zyl (1978), Walter (1988)]:

$$\bar{C}(t,t_0) = C(t,t_0) + \frac{1}{E^C(t_0)} - \frac{1}{E^C(t)} . \tag{2.243}$$

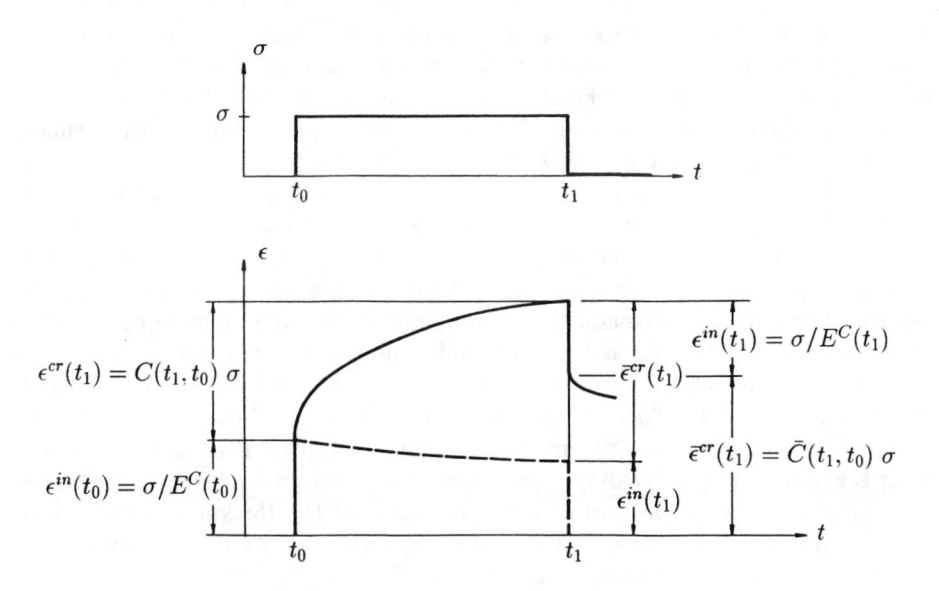

Figure 2.30: Separation of the strain into an instantaneous and a time-dependent part

On the basis of (2.243) the creep strain can be identified as follows: A load applied at time t_0 produces the stress σ which is kept constant until time t_1 (Fig.2.30). Then, the stress-induced strain at time t_1 is obtained as

$$\varepsilon(t_1) = \varepsilon^{in}(t_0) + \varepsilon^{cr}(t_1) = \left[\frac{1}{E^C(t_0)} + C(t_1,t_0)\right] \sigma . \tag{2.244}$$

Complete unloading at time t_1 results in an instantaneous decrease of the strain by $\sigma/E^C(t_1)$. Because of aging, $E^C(t_1) > E^C(t_0)$. Hence, the strain immediately after unloading is equal to the creep strain given as

$$\varepsilon(t_1) = \bar{\varepsilon}^{cr}(t_1) = \left[\frac{1}{E^C(t_0)} + C(t_1,t_0) - \frac{1}{E^C(t_1)}\right] \sigma = \bar{C}(t_1,t_0)\sigma . \tag{2.245}$$

In contrast to the ACI Committee 209 and the CEB/FIP model code, the influence of aging is accounted for in the expression for the creep compliance in the BP model. It is considered in the term $C_b(t, t_0)$.

Aging, with special emphasis on young concrete, was recently considered in [Meschke (1994)] within the framework of rate-type creep laws.

2.4.5 Discussion of Creep and Shrinkage Laws

The empirical shrinkage and creep laws in the ACI and CEB/FIP model code are simple tools for the modelling of the time-dependent behavior of concrete structures. However, the application of creep laws from codes will usually only yield crude approximations of the real time-dependent behavior. The respective compliance functions are mainly intended to obtain approximations of the time-dependent behavior of beam structures. These functions result in mean values of the structural response for the individual cross-sections of such concrete members. Hence, they do not allow prediction of the time-dependent stress and strain distribution over these cross-sections. Moreover, creep experiments, which are the basis of the code laws, are usually characterized by specimens with relatively small cross-sections and by relatively short load durations. Extrapolation to larger cross-sections and longer load durations contains additional uncertainties. From a physical point of view, environmental effects such as humidity and temperature are not treated correctly. Instead of introducing these effects as boundary conditions for determination of the pore humidity and the local temperature at each point of the concrete member, they are incorporated directly into the compliance function. In addition, the size effect is accounted for by scaling the ultimate creep and shrinkage coefficient instead of making a time shift proportional to the square of the thickness of the concrete member as would follow from diffusion theory. Hence, the ACI and CEB/FIP code laws underestimate the long-term shrinkage of thick concrete members. Yet, the use of these empirical laws greatly simplifies time-dependent analyses. The price for this simplification is the reduction of the accuracy of the results.

The code laws were designed primarily for beam structures under the usual conditions of environmental humidity. These laws are not well suited to approximate the time-dependent behavior of mass concrete or of sealed concrete members. The range of validity of the code laws is further restricted by the climatic conditions for which these laws have been developed, i.e., for moderate climates.

With respect to the CEB/FIP model code, the restriction of the influence of the humidity and of the size to the irreversible creep component is questionable. Apart from this, the introduction of a reversible creep component for an aging material has been questioned [Bažant (1979b)]. Moreover, the reversible creep component was obtained from creep recovery tests on the basis of the principle of superposition. Experiments have indicated, however, that the principle of superposition is not valid for decreasing strains [ASCE (1982)].

A comparison of creep and shrinkage curves according to the ACI Committee

209 and the CEB/FIP model code for standard conditions shows that the differences between corresponding curves are quite large. This can be explained, at least partially, by the different split of the time-dependent strains into creep strains and shrinkage strains. For this reason it is important to take the creep and the shrinkage law from the same code.

At least partially, the theoretical basis of the BP-model is better than the one of the ACI and the CEB/FIP model code. Moreover, the range of applicability of the BP-model is wider than the one of the two model codes. The BP-model distinguishes between basic creep and drying creep. It also considers the effect of temperature and properly accounts for the size effect in shrinkage and drying creep. The BP-model yields the correct limiting values of shrinkage and drying creep for mass concrete.

The price for these advantages of the BP-model is the relative great complexity of the resulting formulae. However, the gain in accuracy of the predicted time-dependent behavior usually justifies the additional effort. Based on 80 different data sets, the 95% confidence limits for the BP-model are ± 37%, whereas for the ACI and the CEB/FIP model code they are ± 77% and ± 92%, respectively [Bažant (1980b)].

Apart from different degrees of accuracy of different creep and shrinkage laws, it should be kept in mind that all of these laws are based on the assumption of isotropic time-dependent material behavior. This assumption may be an additional source of differences between the analytically predicted and the experimentally observed time-dependent structural behavior. Moreover, some time-dependent material properties show a relatively large statistical scatter. Also the history of the environmental conditions can only be estimated. Because of the mentioned deficiencies of the investigated material models for creep and shrinkage, stress analyses of concrete structures considering creep and shrinkage should include analyses based on the limiting values of the time-dependent material properties. This allows an assessment of the influence of time-dependent effects on the long-term serviceability of a concrete structure.

2.5 Material Models for Reinforcing and Prestressing Steel

2.5.1 Time-Independent Material Models

The state of stress in reinforcing and prestressing bars is one-dimensional. Hence, the material models are relatively simple. Early constitutive models were characterized by a bilinear-elastic or multilinear-elastic constitutive relationship. Such material models for reinforcing steel were applied together with nonlinear-elastic constitutive relations for concrete. This was only suitable for monotonic external loading.

For consideration of more general loading paths including unloading and reloading, elastic-plastic material models with isotropic and/or kinematic hardening have been developed. For a one-dimensional state of stress the yield condition is given as [Simo (1988)]

$$f(\sigma^S, \rho^S, \kappa^S) = |\sigma^S - \rho^S(\varepsilon^p)^S| - \sigma_Y^S(\kappa^S) = 0 , \qquad (2.246)$$

where σ^S denotes the uniaxial stress acting in the reinforcing (or prestressing) bar, ρ^S is the back stress of the reinforcing bar as a function of the plastic strain $(\varepsilon^p)^S$ and $\sigma_Y^S(\kappa^S)$ is the yield stress of the reinforcing bar, depending on the equivalent plastic strain $\kappa^S = \int |(\dot{\varepsilon}^p)^S| dt$.

Assuming constant moduli H_i^S and H_k^S for isotropic and kinematic hardening, respectively, the hardening laws for these two modes of hardening are obtained as

$$\sigma_Y^S(\kappa^S) = \sigma_{Y,0}^S + H_i^S \kappa^S ,$$
$$\dot{\rho}^S(\varepsilon^p)^S = H_k^S(\dot{\varepsilon}^p)^S , \qquad (2.247)$$

where $\sigma_{Y,0}^S$ is the initial yield stress. Defining one hardening modulus H^S for isotropic and kinematic hardening, the portion of isotropic and kinematic hardening can be recovered by introducing a material parameter β, $0 \leq \beta \leq 1$, such that

$$H_i^S = \beta H^S , \qquad H_k^S = (1-\beta) H^S . \qquad (2.248)$$

Fig.2.31 shows stress-strain diagrams for the reinforcing and the prestressing steel, based on linear-isotropic, linear-kinematic and linear-combined hardening rules. (Stress-strain diagrams for these three hardening rules were presented in subsection 2.3.3 (see Fig.2.13).) These material models could be refined by replacing the linear hardening laws by multilinear or nonlinear hardening laws. However, for reinforcing steel, in general, the only material parameters available are the initial tangent modulus, the yield strength and the tensile strength, allowing only specification of a linear hardening law.

For prestressing steel, stresses in the compressive range resulting from cyclic loading can be excluded. Hence, it is sufficient to describe the stress-strain relationship in the tensile region. For prestressing steel more information for the analytical specification of a stress-strain curve is available than for reinforcing steel. Usually, the initial tangent modulus E_0^Z, a defined elastic limit $\sigma_{0.01}^Z$ and a defined yield stress $\sigma_{0.1}^Z$ (or $\sigma_{0.2}^Z$), the tensile strength σ_T^Z and the corresponding strain at failure ε_U^Z are known material properties (Fig.1.23). With these material properties at hand, a trilinear stress-strain curve can be generated. Such stress-strain curves were reported in [Kang (1980), Van Greunen (1983)]. Alternatively, a smooth stress-strain curve can be constructed, yielding an improved approximation of the actual stress-strain diagram. In [Hofstetter (1987)] a linear stress-strain relation was assumed to hold up to 80 % of the elastic limit:

$$\sigma^Z = E_0^Z \varepsilon^Z , \qquad \sigma^Z \leq 0.80 \sigma_{0.01}^Z . \qquad (2.249)$$

For the nonlinear part of the stress-strain diagram the analytical expression

$$\sigma^Z(\varepsilon^Z) = \sum_{i=1}^{8} a_i \frac{1}{(\varepsilon^Z)^{(i-1)}} , \qquad \sigma^Z \geq 0.80 \sigma_{0.01}^Z , \qquad (2.250)$$

2.5 Material Models for Reinforcing and Prestressing Steel

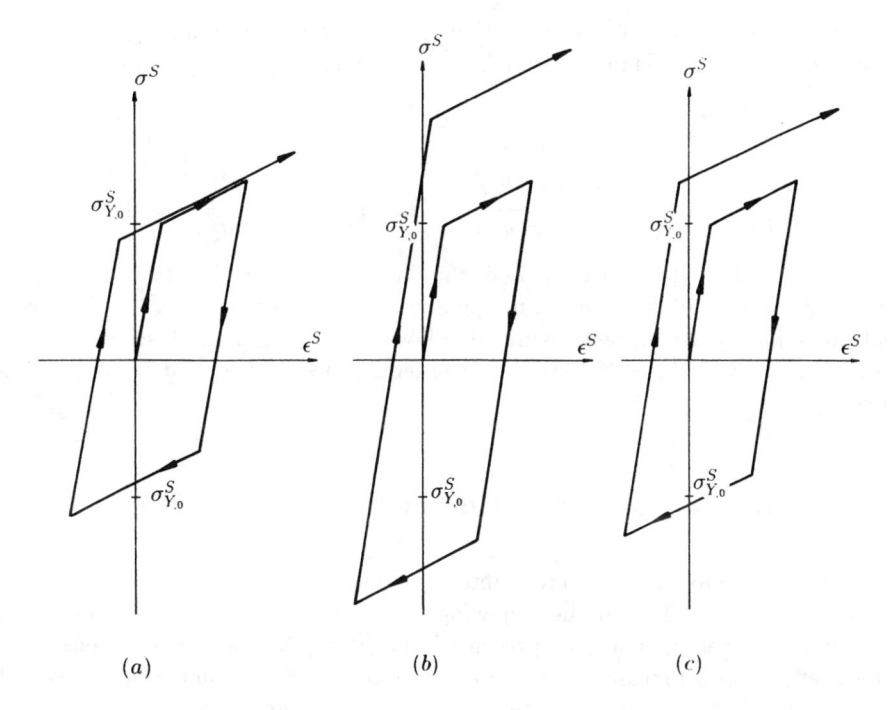

Figure 2.31: Stress-strain diagrams for reinforcing and prestressing steel with (a) linear-kinematic, (b) linear-isotropic and (c) linear-combined hardening

is employed. The parameters $a_i, i = 1, ..., 8$, are computed from the known values

$$\varepsilon^Z = \frac{\sigma^Z}{E_0^Z} \quad : \quad \sigma^Z = 0.80\sigma_{0.01}^Z, \quad \frac{d\sigma^Z}{d\varepsilon^Z} = E_0^Z, \quad \frac{d^2\sigma^Z}{(d\varepsilon^Z)^2} = 0,$$

$$\varepsilon^Z = \frac{\sigma_{0.01}^Z}{E_0^Z} + 0.0001 \quad : \quad \sigma^Z = \sigma_{0.01}^Z,$$

$$\varepsilon^Z = \frac{\sigma_{0.1}^Z}{E_0^Z} + 0.001 \quad : \quad \sigma^Z = \sigma_{0.1}^Z,$$

$$\varepsilon^Z = \varepsilon_u^Z \quad : \quad \sigma^Z = \sigma_T^Z, \quad \frac{d\sigma^Z}{d\varepsilon^Z} = 0, \quad \frac{d^2\sigma^Z}{(d\varepsilon^Z)^2} = 0. \quad (2.251)$$

Unloading and reloading are modelled as linear functions characterized by a slope equal to E_0^Z.

2.5.2 Time-Dependent Material Model for Prestressing Steel

As was pointed out in chapter 1, the time-dependent behavior of prestressing tendons rather corresponds to the time-dependent behavior observed in a relaxation test than in a creep test. In [Magura (1964)] it was proposed to compute the loss of stress resulting from relaxation from the expressions

$$\frac{\sigma^Z(t)}{\sigma^Z(t_0)} = 1, \quad \text{for} \quad \frac{\sigma^Z(t_0)}{\sigma^Z_{0.1}} < 0.55,$$

$$\frac{\sigma^Z(t)}{\sigma^Z(t_0)} = 1 - \frac{\log t}{10}\left(\frac{\sigma^Z(t_0)}{\sigma^Z_{0.1}} - 0.55\right), \quad \text{for} \quad \frac{\sigma^Z(t_0)}{\sigma^Z_{0.1}} \geq 0.55, \quad (2.252)$$

where t_0 denotes the time at prestressing and t is the elapsed time since prestressing. t_0 and t are measured in hours. Consequently, $\sigma^Z(t_0)$ is the stress in the prestressing steel immediately after prestressing. It follows from (2.252) that stress relaxation can be neglected, if the initial stress in the tendon does not exceed 55% of the yield stress $\sigma^Z_{0.1}$.

2.6 Models for Consideration of the Interface Behavior

The material behavior at concrete interfaces and at concrete-steel interfaces was described in section 1.4. In the following, mathematical models for consideration of the interface behavior will be presented. They may be used for the investigation of local effects in structures. For the analysis of concrete structures, however, the local material behavior at interfaces can only be considered in an average manner by appropriate modifications of the constitutive relations for the concrete and the reinforcing steel (see subsection 3.4.5).

2.6.1 Models for Consideration of Aggregate Interlock

Mathematical models for consideration of aggregate interlock of cracked concrete may be subdivided into two categories. The first category is characterized by considering the shear transfer capacity of cracked concrete by introducing a modified shear modulus. It is referred to as the smeared representation of aggregate interlock. The second category consists of models for a particular discrete crack. They contain relationships between the relative displacements, tangential and normal to the crack plane, and the forces transferred across the crack planes.

The smeared representation of aggregate interlock of cracked concrete is characterized by replacing the shear modulus G of the intact concrete by a modified shear modulus G_{AI} [Suidan (1973)]. The ratio

$$r = \frac{G_{AI}}{G} \quad (2.253)$$

is denoted as shear retention factor. This approach is especially attractive for the concept of smeared cracks. This concept will be described in detail in subsection 3.2.5. Especially in older constitutive models frequently a constant value was taken for r. Typically, r was set equal to 0.2.

2.6 Models for Consideration of the Interface Behavior

Experiments have indicated, however, that the ability of concrete to transfer shear forces across cracks decreases with increasing crack width (see subsection 1.4.2). In order to avoid the shortcomings of a constant shear retention factor, an expression for G_{AI} in terms of the strain ε_{11} perpendicular to the crack was proposed in [Cedolin (1977)]. It is given as

$$G_{AI} = \begin{cases} F\left(1 - \dfrac{\varepsilon_{11}}{\bar{\varepsilon}_{11}}\right), & \text{for } \varepsilon_{11} \leq \bar{\varepsilon}_{11}, \\ 0, & \text{for } \varepsilon_{11} > \bar{\varepsilon}_{11}. \end{cases} \quad (2.254)$$

In (2.254) F is a constant and $\bar{\varepsilon}_{11}$ is the limiting value of ε_{11}. This value is defined such that for strains ε_{11} exceeding $\bar{\varepsilon}_{11}$ aggregate interlock is negligible. Numerical tests [Cedolin (1977)] have shown that F can be chosen equal to one tenth of the modulus of elasticity of concrete whereas $\bar{\varepsilon}_{11}$ may be set equal to 0.004.

Models for consideration of aggregate interlock at a particular discrete crack are characterized by a constitutive relationship of the form

$$\mathbf{t} = \mathbf{f}(\mathbf{u}^{rel}) \quad (2.255)$$

or of the form

$$\dot{\mathbf{t}} = \mathbf{D}\dot{\mathbf{u}}^{rel}. \quad (2.256)$$

The traction vector $\mathbf{t}$ is given as a function of the vector of the relative displacements of the two adjacent crack faces, denoted as $\mathbf{u}^{rel}$. Both vectors are referred to a local coordinate system with axes parallel and normal to the crack planes. The matrix $\mathbf{D}$ in (2.256) is the so-called crack stiffness matrix. For a two-dimensional problem the components of $\mathbf{t}$ and $\mathbf{u}^{rel}$ are parallel to the local base vectors $\mathbf{e}_t$ and $\mathbf{e}_n$ (Fig.2.32), i.e.,

$$\mathbf{t} = \begin{Bmatrix} \sigma_{nn} \\ \tau_{nt} \end{Bmatrix}, \quad \mathbf{u}^{rel} = \begin{Bmatrix} u_n^{rel} \\ u_t^{rel} \end{Bmatrix}, \quad (2.257)$$

where σ_{nn} and τ_{nt} denote the normal stress and the shear stress transferred across the crack faces, and u_n^{rel} and u_t^{rel} are the relative displacements of the two adjacent crack faces, normal and parallel to the crack plane (Fig.2.32). This plane is defined by the normal vector $\mathbf{e}_n$. Hence, for a two-dimensional problem (2.255) can be written as

$$\sigma_{nn} = f_n(u_n^{rel}, u_t^{rel}), \quad \tau_{nt} = f_t(u_n^{rel}, u_t^{rel}). \quad (2.258)$$

From (2.258) the constitutive equations in rate form are obtained as

$$\begin{Bmatrix} \dot{\sigma}_{nn} \\ \dot{\tau}_{nt} \end{Bmatrix} = \begin{bmatrix} D_{nn} & D_{nt} \\ D_{tn} & D_{tt} \end{bmatrix} \begin{Bmatrix} \dot{u}_n^{rel} \\ \dot{u}_t^{rel} \end{Bmatrix}, \quad (2.259)$$

where

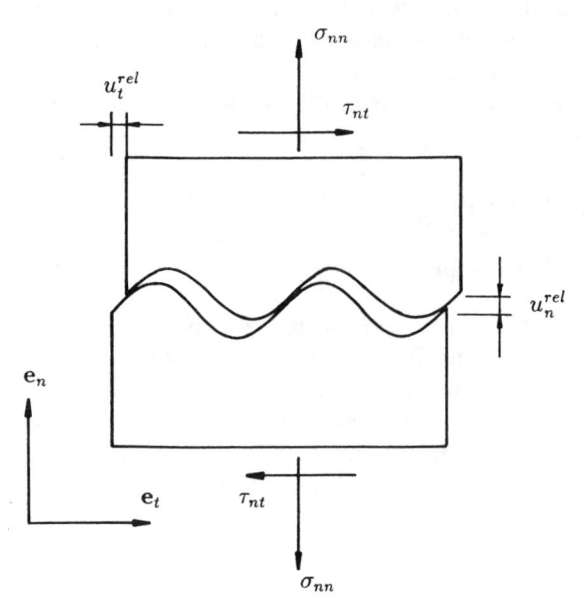

Figure 2.32: Model for aggregate interlock at a discrete crack

$$D_{nn} = \frac{\partial f_n}{\partial u_n^{rel}}, \quad D_{nt} = \frac{\partial f_n}{\partial u_t^{rel}}, \quad D_{tn} = \frac{\partial f_t}{\partial u_n^{rel}}, \quad D_{tt} = \frac{\partial f_t}{\partial u_t^{rel}}. \quad (2.260)$$

In general, $D_{nt} \neq D_{tn}$. Two basically different approaches for the formulation of the constitutive relations (2.258) are known. The first approach is based on empirical expressions, making use of experimental results (e.g. [Bažant (1980)]). The second approach relies on a theoretical model [Walraven (1981)]. In this model the aggregate particles are assumed to be rigid spheres of different size, protruding from a flat crack plane. The distribution of the aggregates is determined by means of statistical analysis. The hardened cement matrix between the aggregates is assumed to obey a rigid - perfectly plastic stress-strain law.

A summary of the constitutive relations for three empirical and two theoretically-based crack models can be found in [Feenstra (1991a)]. Further models for aggregate interlock are contained, e.g., in [Divakar (1987), Tassios (1987), Yoshikawa (1989)].

2.6.2 Models for Consideration of Bond

Bond slip is a consequence of the failure of the concrete in front of the lugs of the reinforcing bar. An empirical constitutive relationship between the local bond slip u_t^{rel} and the local bond stress τ^B was proposed in [Pochanart (1989)]. This relationship is shown in Fig.2.33.

It is characterized by an envelope for monotonic loading and a damage rule for cyclic loading. The envelope for monotonic loading can be subdivided into three

2.6 Models for Consideration of the Interface Behavior

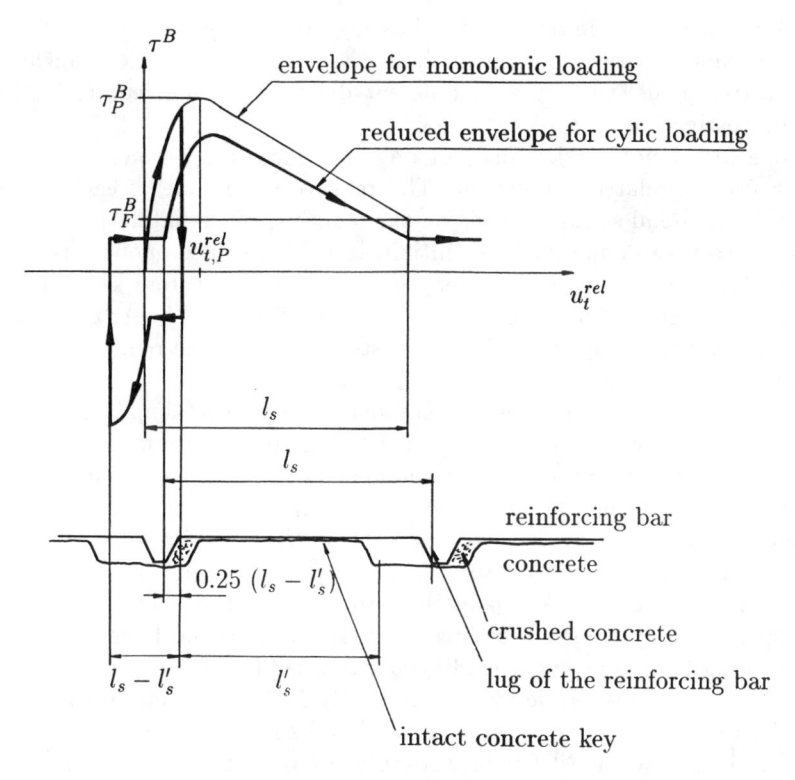

Figure 2.33: Model for a local bond stress - bond slip relationship

parts. The first part stretches from the origin to the bond strength τ_P^B. The formula for the bond stress τ^B as a function of u_t^{rel} is given as

$$\tau^B = \tau_P^B \left[1 - \left(\frac{u_{t,P}^{rel} - u_t^{rel}}{u_{t,P}^{rel}} \right)^3 \right]. \tag{2.261}$$

$u_{t,P}^{rel}$ denotes the bond slip corresponding to τ_P^B. The second part of the curve is characterized by a linear decrease of the bond stress with increasing bond slip until τ^B is equal to the frictional bond stress τ_F^B between the reinforcing steel and the concrete. At $\tau^B = \tau_F^B$, the local bond slip is assumed to be equal to the clear spacing l_s between the lugs of the deformed bar. Hence, when τ_B has been reduced to τ_F^B, the concrete keys between the lugs of a reinforcing bar have been destroyed completely. The third part of the curve is parallel to the abscissa. It is solely determined by the frictional bond stress τ_F^B.

In [Pochanart (1989)] τ_P^B and τ_F^B are expressed in terms of the ratio of the spacing of the lugs l_s over the height of the lugs l_h of a deformed bar with a diameter of one inch. With increasing ratio l_s/l_h, τ_P^B and τ_F^B are decreasing. The local bond slip corresponding to τ_P^B is given in terms of the bearing pressure against the steel lugs. The respective value increases linearly with the bearing pressure. Generalization of this constitutive model to bars with different diameters would require further experimental investigations.

The envelope for monotonic loading can be subdivided into a bond strength - related and a friction-related component. The relative magnitude of each one of the two parts of the bond stress depends on the length l'_s of the intact part of the concrete key between two adjacent lugs. Initially, the frictional component is equal to zero. With increasing bond slip, the length of the intact concrete key between two adjacent lugs is reduced until the local bond slip is equal to l_s. At this point, the bond resistance is reduced to the frictional stress between the reinforcing steel and the concrete.

The reduction of the bond strength in consequence of cyclic loading is accounted for by a reduced envelope. The latter is defined by an appropriate modification of the parameters of the envelope for monotonic loading. Unloading is characterized by an almost constant value of bond slip. Hence, in Fig.2.33 the unloading path is approximated by a vertical line. If there is no slip in the opposite direction, then the reloading path will coincide with the unloading path until the envelope of the previous load cycle is reached. A slip in the opposite direction will occur, if the bond stress, generated by reversed loading, exceeds the frictional bond stress τ_F^B. The reduction of τ_F^B caused by cyclic loading is accounted for by reducing the value for τ_F^B from the previous load cycle by approximately 18%. For continued reversed loading, the frictional part of the bond stress - bond slip curve, which is parallel to the abscissa, is followed until the lugs reach the adjacent concrete keys. From this point on the constitutive relation will follow a new reduced envelope. In order to account for the crushed concrete between the remaining concrete keys and the lugs of the reinforcing steel, the mentioned point is assumed to be located at a distance from the face of the concrete key of one quarter of the damaged length of the concrete key $l_s - l'_s$. Further details can be found in [Pochanart (1989)].

A review of several constitutive relations between the local bond stress and the local bond slip is contained in [CEB (1991)].

3 The Finite Element Method for RC and PC Structures

3.1 Review of the Finite Element Method

3.1.1 Introduction

In general, the mechanical behavior of concrete structures is nonlinear. There are two kinds of nonlinearity: material nonlinearity and geometric nonlinearity. Material nonlinearity is almost always present. It is characterized by nonlinear stress-strain relations for the intact concrete and by local material failure. Geometric nonlinearity must be considered, if the magnitudes of the displacements of structural members are such that the nonlinear terms in the strain-displacement equations are not negligibly small. This may be the case for thin concrete slabs and shells. If geometric nonlinearity must be taken into account, for RC and PC slabs and shells it is usually admissible to use a theory which is restricted to displacements of the same order of magnitude as the thickness of the slab or the shell. Such a theory is denoted as one of small displacements and moderately large rotations.

3.1.2 Principle of Virtual Displacements

The principle of virtual displacements is a fundamental variational principle of continuum mechanics. It may serve as the starting point for the derivation of a finite element formulation which is suited for linear as well as nonlinear structural analyses. In the following, the term "virtual displacement" will be defined.

Let $\mathbf{u}(\mathbf{x})$ be the displacement field of a body in terms of the position vector $\mathbf{x}$ of an arbitrary particle. Let

$$\mathbf{u}^\alpha(\mathbf{x}, \alpha) = \mathbf{u}(\mathbf{x}) + \alpha \boldsymbol{\eta}(\mathbf{x}) \tag{3.1}$$

be a field of fictitious displacements. The function $\mathbf{u}^\alpha(\mathbf{x}, \alpha)$ must satisfy the same differentiability condition and the same boundary conditions as the function $\mathbf{u}(\mathbf{x})$. The function $\boldsymbol{\eta}(\mathbf{x})$ must be kinematically admissible, i.e., it must be chosen such that these conditions can be satisfied. α is a small, real but otherwise arbitrary number. $\alpha \boldsymbol{\eta}(\mathbf{x})$ is called the variation of $\mathbf{u}(\mathbf{x})$ or the virtual displacement. Frequently, it is denoted as $\delta \mathbf{u}(\mathbf{x})$, i.e.,

$$\delta \mathbf{u}(\mathbf{x}, \alpha) = \alpha \boldsymbol{\eta}(\mathbf{x}) \ . \tag{3.2}$$

Differentiating (3.1) with respect to α and taking into account that (3.1) also holds for the limiting case of $\alpha \to 0$, yields

$$\left.\frac{\partial \mathbf{u}^\alpha}{\partial \alpha}\right|_{\alpha=0} = \eta(\mathbf{x}) \ . \tag{3.3}$$

Multiplying (3.3) by α results in an alternative definition of the variation of $\mathbf{u}(\mathbf{x})$:

$$\delta \mathbf{u}(\mathbf{x}, \alpha) = \alpha \left[\frac{\partial \mathbf{u}^\alpha}{\partial \alpha}\right]_{\alpha=0} . \tag{3.4}$$

In a similar manner the variation of the function $f(\mathbf{u})$ is obtained as

$$\begin{aligned}
\delta f(\mathbf{u}) &= \alpha \left[\frac{\partial f(\mathbf{u}^\alpha)}{\partial \alpha}\right]_{\alpha=0} = \alpha \left[\frac{\partial f(\mathbf{u}+\alpha\eta)}{\partial \alpha}\right]_{\alpha=0} \\
&= \alpha \left[\frac{\partial f(\mathbf{u}^\alpha)}{\partial \mathbf{u}^\alpha} \cdot \frac{\partial \mathbf{u}^\alpha}{\partial \alpha}\right]_{\alpha=0} = \alpha \frac{\partial f(\mathbf{u})}{\partial \mathbf{u}} \cdot \eta = \frac{\partial f(\mathbf{u})}{\partial \mathbf{u}} \cdot \delta \mathbf{u} \ .
\end{aligned} \tag{3.5}$$

It follows from (3.5) that δf can be computed by multiplying the directional derivative of $f(\mathbf{u})$ in the direction of $\alpha\eta$, defined as [Marsden (1988)]

$$D_\eta f(\mathbf{u}) = \left[\frac{\partial f(\mathbf{u}+\alpha\eta)}{\partial \alpha}\right]_{\alpha=0} = \frac{\partial f(\mathbf{u})}{\partial \mathbf{u}} \cdot \eta \ , \tag{3.6}$$

by α. With the help of (3.6) the linear approximation $f_\ell(\mathbf{u})$ of the function $f(\mathbf{u})$ at a given point $\mathbf{u}_0$ in the direction of the vector $\eta = \mathbf{u} - \mathbf{u}_0$ is obtained as

$$f_\ell(\mathbf{u}) = f(\mathbf{u}_0) + D_\eta f(\mathbf{u}_0) \ . \tag{3.7}$$

The principle of virtual displacements represents a weak form of the equilibrium conditions and the static boundary conditions. It can be formulated in terms of variables which are referred to the initial (undeformed) configuration. This formulation is denoted as the Lagrangian formulation. Alternatively, the principle of virtual displacements can be formulated in terms of variables which are referred to the current (deformed) configuration. This formulation is known as the Eulerian formulation.

The Eulerian formulation of the principle of virtual displacements is given by the equation (see, e.g., [Malvern (1969)])

$$g(\mathbf{u}, \delta\mathbf{u}) = -\int_V \boldsymbol{\sigma} : \delta\boldsymbol{\varepsilon} \ dV + \int_V \mathbf{b} : \delta\mathbf{u} \ dV + \int_{S^\sigma} \mathbf{t} : \delta\mathbf{u} \ dS^\sigma = 0 \ . \tag{3.8}$$

$\boldsymbol{\sigma}$ denotes the Cauchy stress tensor; $\boldsymbol{\varepsilon}$ is the linearized (or infinitesimal) strain tensor, also referred to as Cauchy strain tensor; $\mathbf{b}$ denotes the volume forces and $\mathbf{t}$ the surface tractions; $\mathbf{u}$ is the displacement vector; V is the volume and S is the surface of the body in the current configuration; S^σ is the part of S, on which surface tractions are prescribed; $\delta\mathbf{u}$ is the vector of the virtual displacements; $\delta\boldsymbol{\varepsilon}$ is the tensor of the virtual Cauchy strains. In a Cartesian coordinate system the components of the Cauchy strain tensor are given as

3.1 Review of the Finite Element Method

$$\varepsilon_{ij} = \frac{1}{2}\left(\frac{\partial u_i}{\partial x_j} + \frac{\partial u_j}{\partial x_i}\right), \qquad (3.9)$$

where x_i, $i = 1, 2, 3$, denotes the components of the position vector $\mathbf{x}$ of the particle under consideration in the current configuration. By means of (3.5) and (3.6) the components of the virtual Cauchy strain tensor are computed as

$$\begin{aligned}\delta\varepsilon_{ij} &= \alpha\left\{\frac{\partial}{\partial\alpha}\left[\frac{1}{2}\left(\frac{\partial u_i^\alpha}{\partial x_j} + \frac{\partial u_j^\alpha}{\partial x_i}\right)\right]\right\}_{\alpha=0} \\ &= \alpha\frac{1}{2}\left(\frac{\partial \eta_i}{\partial x_j} + \frac{\partial \eta_j}{\partial x_i}\right) = \frac{1}{2}\left(\frac{\partial(\delta u_i)}{\partial x_j} + \frac{\partial(\delta u_j)}{\partial x_i}\right). \qquad (3.10)\end{aligned}$$

For the analysis of deformable solids, however, the Lagrangian formulation is usually preferred. This formulation is given by the equation (see, e.g., [Malvern (1969)])

$$G(\mathbf{u},\delta\mathbf{u}) = -\int_{V_0}\mathbf{S}:\delta\mathbf{E}\,dV_0 + \int_{V_0}\bar{\mathbf{b}}:\delta\mathbf{u}\,dV_0 + \int_{S_0^\sigma}\bar{\mathbf{t}}:\delta\mathbf{u}\,dS_0^\sigma = 0. \qquad (3.11)$$

$\mathbf{S}$ denotes the 2nd Piola-Kirchhoff stress tensor; $\mathbf{E}$ is the Langrangian strain tensor, also known as the Green strain tensor; $\bar{\mathbf{b}}$ denotes the volume forces and $\bar{\mathbf{t}}$ the surface tractions; V_0 is the volume and S_0 is the surface of the body in the initial configuration; S_0^σ is the part of S_0, on which surface tractions are prescribed. In a Cartesian coordinate system the components of the Lagrangian strain tensor are given as

$$E_{IJ} = \frac{1}{2}\left(\frac{\partial u_I}{\partial X_J} + \frac{\partial u_J}{\partial X_I} + \frac{\partial u_L}{\partial X_I}\frac{\partial u_L}{\partial X_J}\right), \qquad (3.12)$$

where X_I, $I = 1, 2, 3$, denotes the components of the position vector $\mathbf{X}$ of the considered particle, referred to the initial configuration. Hence,

$$\mathbf{x} = \mathbf{X} + \mathbf{u}. \qquad (3.13)$$

By means of (3.5) and (3.6) the components of the virtual Lagrangian strain tensor are computed as

$$\delta E_{IJ} = \frac{1}{2}\left[\frac{\partial(\delta u_I)}{\partial X_J} + \frac{\partial(\delta u_J)}{\partial X_I} + \frac{\partial(\delta u_L)}{\partial X_I}\frac{\partial u_L}{\partial X_J} + \frac{\partial u_L}{\partial X_I}\frac{\partial(\delta u_L)}{\partial X_J}\right]. \qquad (3.14)$$

The relationship between the two stress tensors $\boldsymbol{\sigma}$ and $\mathbf{S}$ is given as [Malvern (1969)]

$$\boldsymbol{\sigma} = \frac{1}{J}\mathbf{F}\,\mathbf{S}\,\mathbf{F}^T, \qquad (3.15)$$

where

$$\mathbf{F} = \frac{\partial\mathbf{x}}{\partial\mathbf{X}} \qquad (3.16)$$

denotes the deformation gradient and

$$J = Det\,\mathbf{F} \,. \tag{3.17}$$

The matrix of the coefficients of $\mathbf{F}$ is the Jacobian matrix. Its determinant J is the Jacobian determinant. In the initial, undeformed configuration $\mathbf{x} = \mathbf{X}$ and, consequently, $J = 1$. In case of a unique relationship between the coordinates $\mathbf{X}$ and $\mathbf{x}$ of a material point, $\mathbf{F}$ is invertible, i.e., $J \neq 0$. Since the deformed configuration is evolving continuously from the initial configuration, it follows that $J > 0$ always holds.

The relations between the volume force vectors $\mathbf{b}$ and $\bar{\mathbf{b}}$ and the surface traction vectors $\mathbf{t}$ and $\bar{\mathbf{t}}$ are given as (see, e.g., [Malvern (1969)])

$$\mathbf{b} = \frac{1}{J}\bar{\mathbf{b}} \tag{3.18}$$

and

$$\mathbf{t}dS = \bar{\mathbf{t}}dS_0 \,, \tag{3.19}$$

respectively. The relationship betwwen dS and dS_0 is given as

$$\mathbf{n}dS = J\mathbf{F}^{-T} : \mathbf{n}_0 dS_0 \,, \tag{3.20}$$

where $\mathbf{n}$ and $\mathbf{n}_0$ are the unit vectors normal to dS and dS_0, respectively.

Substitution of (3.13) into (3.16) yields $\mathbf{F}$ in terms of the displacements as

$$\mathbf{F} = 1 + \frac{\partial \mathbf{u}}{\partial \mathbf{X}} \,. \tag{3.21}$$

If the partial derivatives of $\mathbf{u}$ with respect to $\mathbf{X}$ are small in comparison to 1, then the differences between $\mathbf{S}$ and $\boldsymbol{\sigma}$, $\bar{\mathbf{b}}$ and $\mathbf{b}$, $\bar{\mathbf{t}}$ and $\mathbf{t}$, V_0 and V and S_0 and S become negligible. Moreover, the quadratic terms in the expression for E_{IJ} (3.12) and the third and the fourth term on the right-hand side of (3.14) can be neglected. If the displacements and their partial derivatives are small, then, because of $\mathbf{x} \approx \mathbf{X}$, the difference between (3.10) and the reduced form of (3.14) disappears. Consequently, the differences between (3.11) and (3.8) disappear.

3.1.3 Linearization of the Mathematical Formulation of the Principle of Virtual Displacements

An essential feature of the strategy for the solution of nonlinear problems in the field of structural mechanics is the stepwise (or incremental) application of the external load. Another important characteristic of this solution strategy is the iterative determination of the nonlinear structural response for each load increment. Hence, this solution strategy is referred to as an incremental-iterative procedure.

In general, Newton's method is used for the mentioned iteration. A typical step of this iteration consists of

3.1 Review of the Finite Element Method

(a) the replacement of the nonlinear equation (3.11) by its linear approximation and

(b) the solution of the linearized equation to obtain an improved approximation of the solution of the nonlinear equation.

The iteration procedure is continued until a specified convergence criterion is met.

Let the structural response be known up to load increment n, i.e., let the displacements $\mathbf{u}_n$ resulting from given forces $\bar{\mathbf{b}}_n$ and $\bar{\mathbf{t}}_n$ be known. Then, the equation

$$G(\mathbf{u}_n, \delta\mathbf{u}) = -\int_{V_0} \mathbf{S}_n : \delta\mathbf{E}_n \, dV_0 + \int_{V_0} \bar{\mathbf{b}}_n : \delta\mathbf{u} \, dV_0 + \int_{S_0^\sigma} \bar{\mathbf{t}}_n : \delta\mathbf{u} \, dS_0^\sigma = 0 \qquad (3.22)$$

is fulfilled. This relation represents the mathematical formulation of the principle of virtual displacements referred to the initial configuration. (In contrast to $\delta\mathbf{u}$, $\delta\mathbf{E}_n$ contains the subscript n because of the dependence of $\delta\mathbf{E}_n$ on $\mathbf{u}_n$, following from (3.14).)

In the next step the load increment $n+1$ is applied, yielding the relation

$$G(\mathbf{u}_{n+1}, \delta\mathbf{u}) = -\int_{V_0} \mathbf{S}_{n+1} : \delta\mathbf{E}_{n+1} \, dV_0 + \int_{V_0} \bar{\mathbf{b}}_{n+1} : \delta\mathbf{u} \, dV_0 + \int_{S_0^\sigma} \bar{\mathbf{t}}_{n+1} : \delta\mathbf{u} \, dS_0^\sigma = 0 \, . \qquad (3.23)$$

In (3.23) the displacements

$$\mathbf{u}_{n+1} = \mathbf{u}_n + \Delta\mathbf{u}_{n+1} \qquad (3.24)$$

are not known, because the incremental displacements $\Delta\mathbf{u}_{n+1}$ are not known. (3.23) will be solved by means of a Newton iteration. To this end, $G(\mathbf{u}_{n+1}, \delta\mathbf{u})$ is replaced by its linear approximation, yielding, by analogy to (3.7),

$$G_\ell(\mathbf{u}_{n+1}, \delta\mathbf{u}) = G(\mathbf{u}_n, \delta\mathbf{u}) + D_{\Delta\mathbf{u}} G(\mathbf{u}_n, \delta\mathbf{u}) \, . \qquad (3.25)$$

If, for the sake of simplicity of the presentation, it is assumed that the loads do not depend on the displacements, then only the first term on the right-hand side of (3.23) will have to be replaced by its linear approximation. In this case, the linear approximation of (3.23) gives

$$\begin{aligned} G_\ell(\mathbf{u}_{n+1}, \delta\mathbf{u}) &= -\int_{V_0} \mathbf{S}_n : \delta\mathbf{E}_n \, dV_0 \\ &\quad - \int_{V_0} \left[\left(\frac{\partial \mathbf{S}_n}{\partial \mathbf{E}_n} : D_{\Delta\mathbf{u}} \mathbf{E}_n \right) : \delta\mathbf{E}_n + \mathbf{S}_n : D_{\Delta\mathbf{u}}(\delta\mathbf{E}_n) \right] dV_0 \\ &\quad + \int_{V_0} \bar{\mathbf{b}}_{n+1} : \delta\mathbf{u} \, dV_0 + \int_{S_0^\sigma} \bar{\mathbf{t}}_{n+1} : \delta\mathbf{u} \, dS_0^\sigma \, . \end{aligned} \qquad (3.26)$$

In the following, the directional derivatives $D_{\Delta u}\mathbf{E}_n$ and $D_{\Delta u}(\delta\mathbf{E}_n)$ will be written as $\Delta\mathbf{E}_{n+1}$ and $\Delta\delta\mathbf{E}_{n+1}$, respectively. $\Delta\mathbf{E}_{n+1}$ is obtained by replacing $\delta\mathbf{u}$ in (3.14) by $\Delta\mathbf{u}_{n+1}$ (suppressing the subscript $n+1$), which yields

$$\Delta E_{IJ} = \frac{1}{2}\left[\frac{\partial(\Delta u_I)}{\partial X_J} + \frac{\partial(\Delta u_J)}{\partial X_I} + \frac{\partial(\Delta u_L)}{\partial X_I}\frac{\partial u_L}{\partial X_J} + \frac{\partial u_L}{\partial X_I}\frac{\partial(\Delta u_L)}{\partial X_J}\right]. \quad (3.27)$$

This result follows from the equivalence (up to the scalar factor α) of the variation and the linearization of a function (see equations (3.5) and (3.6)). $\Delta\delta\mathbf{E}_n$ is obtained as the directional derivative of $\delta\mathbf{E}_n$, which follows from (3.14) as

$$\Delta\delta E_{IJ} = \frac{1}{2}\left[\frac{\partial\delta u_L}{\partial X_I}\frac{\partial\Delta u_L}{\partial X_J} + \frac{\partial\Delta u_L}{\partial X_I}\frac{\partial\delta u_L}{\partial X_J}\right]. \quad (3.28)$$

In (3.27) and (3.28) the total displacements $\mathbf{u}$ are referred to the load increment n whereas the incremental displacements $\Delta\mathbf{u}$ are referred to the load increment $n+1$. In order to simplify the notation, the respective subscripts were omitted in (3.27) and (3.28).

In (3.26)

$$\frac{\partial\mathbf{S}_n}{\partial\mathbf{E}_n} : D_{\Delta u}\mathbf{E}_n = \mathbf{C}_n : \Delta\mathbf{E}_{n+1} = \Delta\mathbf{S}_{n+1} \quad (3.29)$$

represents the linear approximation of the incremental stresses; $\mathbf{C}_n$ is the tensor of the material tangent moduli for the known equilibrium state n, referred to the initial configuration.

Within the framework of the Newton iteration, the linearized equation (3.26) is solved for the incremental displacements by setting $G_\ell(\mathbf{u}_{n+1},\delta\mathbf{u}) = 0$, yielding

$$\int_{V_0}[(\mathbf{C}_n:\Delta\mathbf{E}_{n+1}):\delta\mathbf{E}_n + \mathbf{S}_n:\Delta\delta\mathbf{E}_{n+1}]dV_0 =$$

$$\int_{V_0}\bar{\mathbf{b}}_{n+1}:\delta\mathbf{u}\,dV_0 + \int_{S_0^\sigma}\bar{\mathbf{t}}_{n+1}:\delta\mathbf{u}\,dS_0^\sigma - \int_{V_0}\mathbf{S}_n:\delta\mathbf{E}_n\,dV_0, \quad (3.30)$$

where use of (3.29) has been made. The left-hand side of (3.30) contains the terms which depend on the unknown incremental displacements. The right-hand side of (3.30) represents the difference of the virtual work of the external and the internal forces. At the beginning of load step $n+1$, i.e., after application of the load increment $n+1$, this difference is characterized by the fact that the virtual work of the external forces is referred to the total load after application of load step $n+1$, whereas the virtual work of the internal forces is based on the known equilibrium configuration before application of this load step.

The iteration procedure is continued until (3.23) is satisfied up to a prescribed tolerance.

3.1 Review of the Finite Element Method

For the vast majority of concrete structures the displacements are so small that it is not necessary to distinguish between material moduli referred to the initial and to the current configuration. For geometric linearity, because of $\mathbf{E} \equiv \boldsymbol{\varepsilon}$ and $\mathbf{S} \equiv \boldsymbol{\sigma}$, (3.29) is replaced by

$$\frac{\partial \boldsymbol{\sigma}_n}{\partial \boldsymbol{\varepsilon}_n} : \mathbf{D}_{\Delta u} \boldsymbol{\varepsilon}_n = \mathbf{C}_n : \Delta \boldsymbol{\varepsilon}_{n+1} = \Delta \boldsymbol{\sigma}_{n+1} . \tag{3.31}$$

The expression for the incremental strains,

$$\Delta \varepsilon_{ij} = \frac{1}{2}\left(\frac{\partial \Delta u_i}{\partial x_j} + \frac{\partial \Delta u_j}{\partial x_i}\right) , \tag{3.32}$$

follows from (3.9). Because of $\delta \mathbf{E} \equiv \delta \boldsymbol{\varepsilon}$, $\Delta \delta \mathbf{E} = \mathbf{0}$. Hence, (3.30) reduces to

$$\int_{V_0} (\mathbf{C}_n : \Delta \boldsymbol{\varepsilon}_{n+1}) : \delta \boldsymbol{\varepsilon} \, dV_0 =$$

$$\int_{V_0} \bar{\mathbf{b}}_{n+1} : \delta \mathbf{u} \, dV_0 + \int_{S_0^\sigma} \bar{\mathbf{t}}_{n+1} : \delta \mathbf{u} \, dS^\sigma - \int_{V_0} \boldsymbol{\sigma}_n : \delta \boldsymbol{\varepsilon} \, dV_0 . \tag{3.33}$$

The effectiveness of an algorithm for the iterative solution of a nonlinear problem depends on the rate of convergence of the iteration. A measure for this rate is the order of convergence p, which is to be determined from [Luenberger (1984)]

$$\lim_{k \to \infty} \frac{|r_{k+1} - r^*|}{|r_k - r^*|^p} = \beta . \tag{3.34}$$

In (3.34), r_k, $k = 0, 1, 2, \ldots, \infty$, is a sequence of real numbers, converging to the limit r^*, and β is a constant. According to (3.34), p only depends on the properties of the sequence for $k \to \infty$. The larger the value of p, the greater the rate of convergence. According to (3.34), e.g., the sequences

$$r_k = a^k \quad \text{and} \quad r_k = a^{2^k} , \quad 0 < a < 1 , \tag{3.35}$$

converge to zero with an order of convergence of one and two, respectively.

The Newton iteration is distinguished by an order of convergence of at least two [Luenberger (1984)] (also denoted as quadratic rate of convergence to the solution), provided its starting values are sufficiently close to the solution and the nonlinear equation is linearized exactly. However, the term "sufficiently close" cannot be specified *a priori*. If divergence occurs, then the size of the respective load increment must be reduced.

The presented mode of the Lagrangian formulation of the principle of virtual displacements is also referred to as the total Lagrangian formulation of this principle because the variables are total quantities in the sense of being referred to the initial, undeformed configuration of the structure. An alternative to this mode of the Lagrangian formulation of the principle of virtual displacements is the updated Lagrangian formulation of this principle, for which the variables are referred to the known equilibrium configuration of the previous load step [Bathe (1975), Bathe (1982)].

3.1.4 Finite Element Discretization

In matrix notation, (3.22) can be rewritten as

$$-\int_{V_0} (\delta \mathbf{E}_n)^T \mathbf{S}_n \, dV_0 + \int_{V_0} (\delta \mathbf{u})^T \bar{\mathbf{b}}_n \, dV_0 + \int_{S_0^\sigma} (\delta \mathbf{u})^T \bar{\mathbf{t}}_n \, dS_0^\sigma = 0 , \qquad (3.36)$$

where

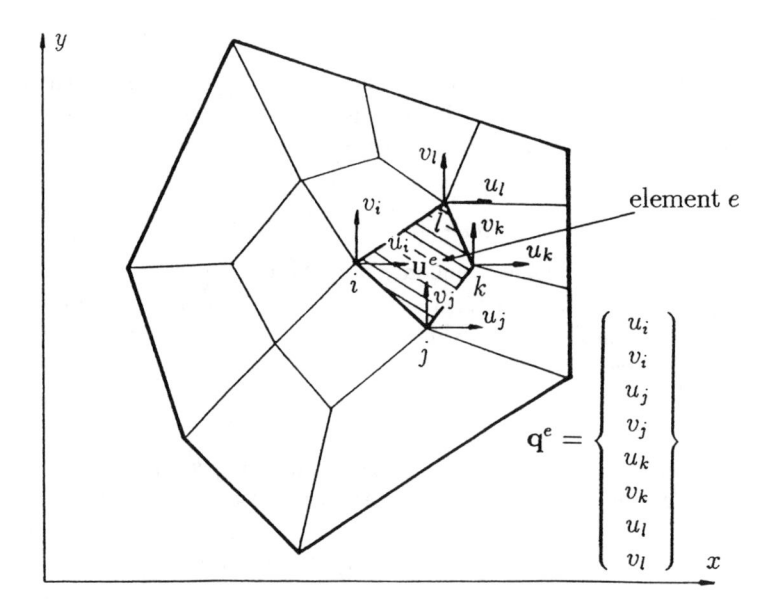

$$\begin{aligned}
\mathbf{E}^T &= \lfloor E_{11} \quad E_{22} \quad E_{33} \quad 2E_{12} \quad 2E_{23} \quad 2E_{31} \rfloor , \\
\mathbf{S}^T &= \lfloor S_{11} \quad S_{22} \quad S_{33} \quad S_{12} \quad S_{23} \quad S_{31} \rfloor , \\
\mathbf{u}^T &= \lfloor u_1 \quad u_2 \quad u_3 \rfloor , \\
\bar{\mathbf{b}}^T &= \lfloor \bar{b}_1 \quad \bar{b}_2 \quad \bar{b}_3 \rfloor , \\
\bar{\mathbf{t}}^T &= \lfloor \bar{t}_1 \quad \bar{t}_2 \quad \bar{t}_3 \rfloor .
\end{aligned} \qquad (3.37)$$

Figure 3.1: Finite element discretization

A characteristic feature of the finite element method is the discretization procedure. It consists of subdividing the considered domain into a number of finite elements. Fig.3.1 illustrates the discretization of a panel by means of quadrilateral finite elements. At an arbitrary point of the finite element e the displacement vector $\mathbf{u}^e$ is approximated as

$$\mathbf{u}^e = \mathbf{N}^e \mathbf{q}^e , \qquad (3.38)$$

3.1 Review of the Finite Element Method

where $\mathbf{q}^e$ is the vector of the displacements at the node points of element e and $\mathbf{N}^e$ is a matrix of suitably chosen shape functions.

Substitution of (3.38) into (3.14) yields the virtual Lagrangian strains for element e as

$$\delta \mathbf{E}^e = \mathbf{B}^e \, \delta \mathbf{q}^e \,, \qquad (3.39)$$

where

$$\mathbf{B}^e = \mathbf{B}_0^e + \mathbf{B}_L^e \,. \qquad (3.40)$$

$\mathbf{B}_0^e$ represents the part of $\mathbf{B}^e$ which is independent of $\mathbf{u}$, resulting from the first and the second term on the right-hand side of (3.14). $\mathbf{B}_L^e$ denotes the displacement-dependent part of $\mathbf{B}^e$, resulting from the third and the fourth term on the right-hand side of (3.14). Thus, for the special case of geometric linearity, $\mathbf{B}^e$ reduces to $\mathbf{B}^e = \mathbf{B}_0^e$.

If the structure to be analyzed was discretized by only one finite element, substitution of (3.38) and (3.39) into (3.36) would yield

$$(\delta \mathbf{q}^e)^T \left[-\int_{V_0^e} \mathbf{B}_n^{e^T} \mathbf{S}_n^e \, dV_0^e + \int_{V_0^e} \mathbf{N}^{e^T} \bar{\mathbf{b}}_n^e \, dV_0^e + \int_{S_0^\sigma} \mathbf{N}^{e^T} \bar{\mathbf{t}}_n^e \, dS_0^\sigma \right] = 0 \,, \qquad (3.41)$$

where $\delta \mathbf{q}^e$ is the vector of the virtual displacements of the node points of element e.

If, as is usually the case, the structure is subdivided into more than one finite element, the element vectors

$$\mathbf{f}_{in,n}^e = \int_{V_0^e} \mathbf{B}_n^{e^T} \mathbf{S}_n^e \, dV_0^e \,,$$

$$\mathbf{f}_{ex,n}^e = \int_{V_0^e} \mathbf{N}^{e^T} \bar{\mathbf{b}}_n^e \, dV_0^e + \int_{S_0^\sigma} \mathbf{N}^{e^T} \bar{\mathbf{t}}_n^e \, dS_0^\sigma \,, \qquad (3.42)$$

representing the internal and external generalized nodal forces of element e, must be assembled to global vectors $\mathbf{f}_{in}$ and $\mathbf{f}_{ex}$. These vectors represent the internal and external generalized nodal forces of the entire structure [Zienkiewicz (1989,1991)]. By analogy to (3.41),

$$\delta \mathbf{q}^T \left[-\mathbf{f}_{in,n} + \mathbf{f}_{ex,n} \right] = 0 \,, \qquad (3.43)$$

where $\delta \mathbf{q}$ is the global vector of virtual nodal displacements. (3.43) must hold for arbitrary virtual displacements $\delta \mathbf{q}$. Consequently,

$$-\mathbf{f}_{in,n} + \mathbf{f}_{ex,n} = \mathbf{0} \,. \qquad (3.44)$$

(3.44) represents the equilibrium conditions for the discretized structure.

For the special case of geometric linearity, $\mathbf{B}_n^e$ and $\mathbf{S}_n^e$ in (3.41) and (3.42) degenerate to $\mathbf{B}_0^e$ and to $\boldsymbol{\sigma}_n^e$, respectively. Hence, the expression for $\mathbf{f}_{in,n}^e$ in (3.42) has to be replaced by

$$\mathbf{f}_{in,n}^e = \int_{V_0^e} \mathbf{B}_0^{e^T} \boldsymbol{\sigma}_n^e \, dV_0^e \,. \qquad (3.45)$$

3.1.5 Newton Iteration

The solution for load step $n+1$ by means of a Newton iteration is based on linearization of the mathematical formulation of the principle of virtual displacements, i.e., on (3.30). In matrix notation, (3.30) can be rewritten as

$$\int_{V_0} \left[(\delta \mathbf{E}_n)^T \mathbf{C}_n \Delta \mathbf{E}_{n+1} + (\Delta \delta \mathbf{E}_{n+1})^T \mathbf{S}_n \right] dV_0 =$$

$$\int_{V_0} (\delta \mathbf{u})^T \bar{\mathbf{b}}_{n+1} \, dV_0 + \int_{S_0^\sigma} (\delta \mathbf{u})^T \bar{\mathbf{t}}_{n+1} \, dS_0^\sigma - \int_{V_0} (\delta \mathbf{E}_n)^T \mathbf{S}_n \, dV_0 \, . \quad (3.46)$$

The incremental vectors in (3.46) are defined analogous to (3.37).

If the structure was discretized by only one finite element, then, by analogy to (3.39), $\delta \mathbf{E}_n$ and $\Delta \mathbf{E}_{n+1}$ could be written as

$$\begin{aligned} \delta \mathbf{E}_n^e &= \mathbf{B}_n^e \, \delta \mathbf{q}^e \, , \\ \Delta \mathbf{E}_{n+1}^e &= \mathbf{B}_n^e \, \Delta \mathbf{q}_{n+1}^e \, . \end{aligned} \quad (3.47)$$

Linearization of $\mathbf{E}$, yielding $\Delta \mathbf{E}$ (3.27), is performed at the known solution of load step n. Hence, the incremental displacements in (3.27) refer to load step $n+1$, whereas the total displacements in this equation refer to load step n. Consequently, in (3.47) $\mathbf{B}^e$ is related to load step n. Substitution of (3.40) into (3.47) and of the resulting equations into the first term on the left-hand side of (3.46) yields

$$(\delta \mathbf{q}^e)^T \left[\mathbf{K}_0^e + \mathbf{K}_{L,n}^e \right] \Delta \mathbf{q}_{n+1}^e \, , \quad (3.48)$$

where

$$\mathbf{K}_0^e = \int_{V_0} (\mathbf{B}_0^e)^T \, \mathbf{C}_n \, \mathbf{B}_0^e \, dV_0 \quad (3.49)$$

denotes the element stiffness matrix for the special case of geometric linearity and

$$\mathbf{K}_{L,n}^e = \int_{V_0} \left[(\mathbf{B}_0^e)^T \, \mathbf{C}_n \, \mathbf{B}_{L,n}^e + (\mathbf{B}_{L,n}^e)^T \, \mathbf{C}_n \, \mathbf{B}_0^e + (\mathbf{B}_{L,n}^e)^T \, \mathbf{C}_n \, \mathbf{B}_{L,n}^e \right] dV_0 \quad (3.50)$$

represents the influence of geometric nonlinearity. $\mathbf{K}_{L,n}^e$ is denoted as initial displacement matrix or large displacement matrix. According to (3.28), the expression for $\Delta \delta \mathbf{E}_{n+1}$ contains both the virtual displacements and the incremental displacements. For a single finite element, the second term on the left-hand side of (3.46) results in

$$(\delta \mathbf{q}^e)^T \, \mathbf{K}_{\sigma,n}^e \, \Delta \mathbf{q}_{n+1}^e \, , \quad (3.51)$$

where $\mathbf{K}_{\sigma,n}^e$ is denoted as initial stress matrix.

Making use of (3.47) to (3.51) and of (3.42), discretization of (3.46) for a structure consisting of only one finite element yields

$$(\delta \mathbf{q}^e)^T \left[\mathbf{K}_0^e + \mathbf{K}_{L,n}^e + \mathbf{K}_{\sigma,n}^e \right] \Delta \mathbf{q}_{n+1}^e = (\delta \mathbf{q}^e)^T \left[\mathbf{f}_{ex,n+1}^e - \mathbf{f}_{in,n}^e \right] \, , \quad (3.52)$$

3.1 Review of the Finite Element Method

where

$$\mathbf{K}_0^e + \mathbf{K}_{L,n}^e + \mathbf{K}_{\sigma,n}^e = \mathbf{K}_n^e \qquad (3.53)$$

is the element tangent stiffness matrix.

If, as is usually the case, the structure is subdivided into more than one element, the element tangent stiffness matrices and the element load vectors must be assembled to the global tangent stiffness matrix $\mathbf{K}_n$ and the global load vectors $\mathbf{f}_{in,n}$ and $\mathbf{f}_{ex,n+1}$. By analogy to (3.52),

$$\delta\mathbf{q}^T \mathbf{K}_n \Delta\mathbf{q}_{n+1} = \delta\mathbf{q}^T [\mathbf{f}_{ex,n+1} - \mathbf{f}_{in,n}] \qquad (3.54)$$

must hold for arbitrary virtual nodal displacements $\delta\mathbf{q}$. Consequently,

$$\mathbf{K}_n \Delta\mathbf{q}_{n+1} = \mathbf{f}_{ex,n+1} - \mathbf{f}_{in,n} \;. \qquad (3.55)$$

For the special case of geometric linearity, $\delta\mathbf{E}_n$ and $\Delta\mathbf{E}_{n+1}$ in (3.46) are to be replaced by $\delta\varepsilon$ and $\Delta\varepsilon_{n+1}$, respectively. Moreover, because of $\Delta\delta\mathbf{E}_{n+1} = \mathbf{0}$, the second term on the left-hand side of (3.46) must vanish. Hence, $\mathbf{K}_n^e$ degenerates to $\mathbf{K}_0^e$. Consequently, $\mathbf{K}_n$ degenerates to $\mathbf{K}_0$.

Because of linearization of (3.11) the resulting incremental displacements are only approximations of the actual incremental displacements. For this reason, $\Delta\mathbf{q}_{n+1}$ is rewritten as $\Delta\mathbf{q}_{1,n+1}$. This notation shall indicate that $\Delta\mathbf{q}_{1,n+1}$ results from the first iteration step of the Newton iteration. For this iteration step the total incremental nodal displacements $\Delta\mathbf{q}_{n+1}^{(1)}$ are equal to the incremental nodal displacements as obtained from (3.55). Hence,

$$\Delta\mathbf{q}_{n+1}^{(1)} = \Delta\mathbf{q}_{1,n+1} \;. \qquad (3.56)$$

The total nodal displacements follow as

$$\mathbf{q}_{n+1}^{(1)} = \mathbf{q}_n + \Delta\mathbf{q}_{n+1}^{(1)} \;. \qquad (3.57)$$

Depending on the employed algorithm, either the total strains $\varepsilon_{n+1}^{(1)}$ and the total stresses $\sigma_{n+1}^{(1)}$, corresponding to the first iteration step of increment $n+1$, are computed from (3.57), or the incremental strains $\Delta\varepsilon_{n+1}^{(1)}$ and the incremental stresses $\Delta\sigma_{n+1}^{(1)}$ are determined from (3.56) and added to the final strains ε_n and the final stresses σ_n from the previous load increment n.

Use of the updated **B**-matrices and stresses for determination of the vector of the internal nodal forces $\mathbf{f}_{in,n+1}^{(1)}$ (analogous to determination of $\mathbf{f}_{in,n}^e$ according to (3.42)) results in

$$\mathbf{r}_{n+1}^{(1)} = -\mathbf{f}_{in,n+1}^{(1)} + \mathbf{f}_{ex,n+1} \;, \qquad (3.58)$$

where $r_{n+1}^{(1)}$ is the so-called residual vector. (3.58) replaces (3.44). (In contrast to $f_{in,n+1}^{(1)}$, no superscript has been added to $f_{ex,n+1}$, because $f_{ex,n+1}$ is a vector of prescribed nodal forces, which remains unchanged during the Newton iteration. However, this is not true for displacement-dependent loads.) The Euclidian norm of r, i.e. $||r|| = \sqrt{r_i r_i}$, can be used to formulate a criterion for termination of the Newton iteration. If $||r_{n+1}^{(1)}|| > c$, where c denotes a specified tolerance, the Newton iteration is continued by linearizing the mathematical formulation of the principle of virtual displacements at the previously obtained state, characterized by the nodal displacements $q_{n+1}^{(1)}$. This leads to

$$K_{n+1}^{(1)} \Delta q_{2,n+1} = r_{n+1}^{(1)}. \qquad (3.59)$$

Solution of (3.59) for $\Delta q_{2,n+1}$ allows computation of improved total incremental nodal displacements

$$\Delta q_{n+1}^{(2)} = \Delta q_{1,n+1} + \Delta q_{2,n+1} \qquad (3.60)$$

and of improved total nodal displacements

$$q_{n+1}^{(2)} = q_n + \Delta q_{n+1}^{(2)}. \qquad (3.61)$$

Subsequently, either the total strains and stresses, $\varepsilon_{n+1}^{(2)}$ and $\sigma_{n+1}^{(2)}$, are computed from the nodal displacements (3.61) or the incremental strains and stresses, $\Delta\varepsilon_{n+1}^{(2)}$ and $\Delta\sigma_{n+1}^{(2)}$, are determined from the total incremental nodal displacements (3.60) and added to the final (converged) strains and stresses from the previous load increment, ε_n and σ_n, to obtain $\varepsilon_{n+1}^{(2)}$ and $\sigma_{n+1}^{(2)}$. For an arbitrary iteration step i, (3.59) is to be replaced by

$$K_{n+1}^{(i-1)} \Delta q_{i,n+1} = r_{n+1}^{(i-1)}. \qquad (3.62)$$

The total incremental nodal displacements are given as

$$\Delta q_{n+1}^{(i)} = \sum_{j=1}^{i} \Delta q_{j,n+1}. \qquad (3.63)$$

The total nodal displacements are obtained as

$$q_{n+1}^{(i)} = q_n + \Delta q_{n+1}^{(i)}. \qquad (3.64)$$

The residual vector reads as follows:

$$r_{n+1}^{(i)} = -f_{in,n+1}^{(i)} + f_{ex,n+1}. \qquad (3.65)$$

3.1 Review of the Finite Element Method

Use of the total incremental nodal displacements $\Delta \mathbf{q}_{n+1}^{(i)}$ (instead of the incremental nodal displacements $\Delta \mathbf{q}_{i,n+1}$ from the iteration step i) for the computation of incremental strains and stresses and addition of these total incremental quantities to the final (converged) total quantities from the previous load step (instead of addition of the incremental quantities from iteration step i to the non-converged quantities from the previous iteration step) is an essential feature of path-dependent problems. It is noteworthy that computation of incremental stresses from incremental strains corresponding to individual iteration steps would lead to questionable results, if for a particular load increment and for a certain material point, e.g., elastic-plastic loading was indicated for iteration step i, whereas elastic unloading was signalled for iteration step $i+1$.

The iteration is continued until the norm of the residual vector is less than the prescribed tolerance c. For the special case of a system with only one degree of freedom, the Newton iteration is illustrated in Fig.3.2.

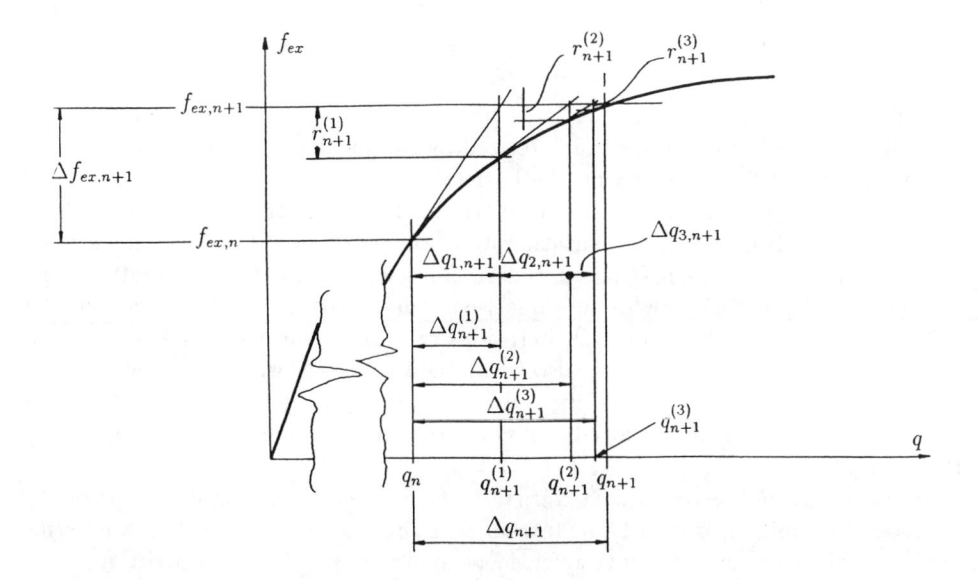

Figure 3.2: Newton iteration

The Newton iteration for nonlinear finite element analyses can be summarized as follows:

1. Determination of incremental nodal displacements by means of the incremental form of the equations of equilibrium for the discretized continuum ((3.55) for the first iteration step, (3.62) for the other iteration steps).

2. Update of the strains, the plastic strains, the internal variables and the stresses according to the employed constitutive model.

3. Use of these updates to compute the residual vector (3.65). If the Euclidean norm of this vector (or any other suitable norm, such as, e.g., the energy norm [Zienkiewicz (1989,1991)]) is smaller than a specified tolerance, the next load increment will be applied. Otherwise the iteration is continued. In any case, the continuation of the analysis is symbolized by the return to item 1.

An advantage of the Newton iteration is the quadratic rate of convergence to the solution, provided the correct tangent stiffness matrix is employed. A drawback of the Newton iteration is the possibility of divergence of the iteration, if the load increment is not chosen sufficiently small.

3.1.6 Alternatives to the Newton Iteration

Several modifications of the Newton iteration have been proposed. They can be classified as modified Newton iterations and quasi-Newton iterations. Both iterations are characterized by replacing the tangent stiffness matrix $\mathbf{K}_{n+1}^{(i)}$ by another matrix. In modified Newton iterations $\mathbf{K}_{n+1}^{(i)}$ is updated only at selected iteration steps. Usually, such updatings are restricted to the first iteration step of the individual load steps, resulting in $\mathbf{K}_{n+1}^{(i)} = \mathbf{K}_{n+1}^{(1)}$, $i \geq 2$.

Quasi-Newton iterations are characterized by computing an approximation of the actual tangent stiffness matrix on the basis of information from previous iteration steps. An example for such an iteration is the BFGS algorithm [Hinton (1992)]. Although the advantage of a quadratic rate of convergence of the Newton iteration is lost if the exact tangent stiffness matrix is not employed at each iteration step, application of modified Newton and quasi-Newton iterations can be attractive.

The rate of convergence of such iterations can be improved by means of a line search procedure. The latter can also be applied to the Newton iteration to avoid divergence of the iteration. Information on modified and quasi-Newton iterations as well as on line search procedures can be found in [Zienkiewicz (1989,1991), Hinton (1992)].

For ultimate load analyses the external loads are usually increased proportionally to a specified reference load $\mathbf{f}_{ex,0}$. Introducing the load parameter λ, the external loads and their increments, both referred to load step $n+1$, can be written as

$$\mathbf{f}_{ex,n+1} = \lambda_{n+1}\mathbf{f}_{ex,0} \quad \text{and} \quad \Delta\mathbf{f}_{ex,n+1} = \Delta\lambda_{n+1}\mathbf{f}_{ex,0} . \tag{3.66}$$

Substitution of (3.66_1) into (3.55) yields

$$\mathbf{K}_n \Delta \mathbf{q}_{1,n+1} = \lambda_{1,n+1}\mathbf{f}_{ex,0} - \mathbf{f}_{in,n} , \tag{3.67}$$

where the first subscript of $\Delta \mathbf{q}_{1,n+1}$ and $\lambda_{1,n+1}$ indicates the first iteration step.

3.1 Review of the Finite Element Method

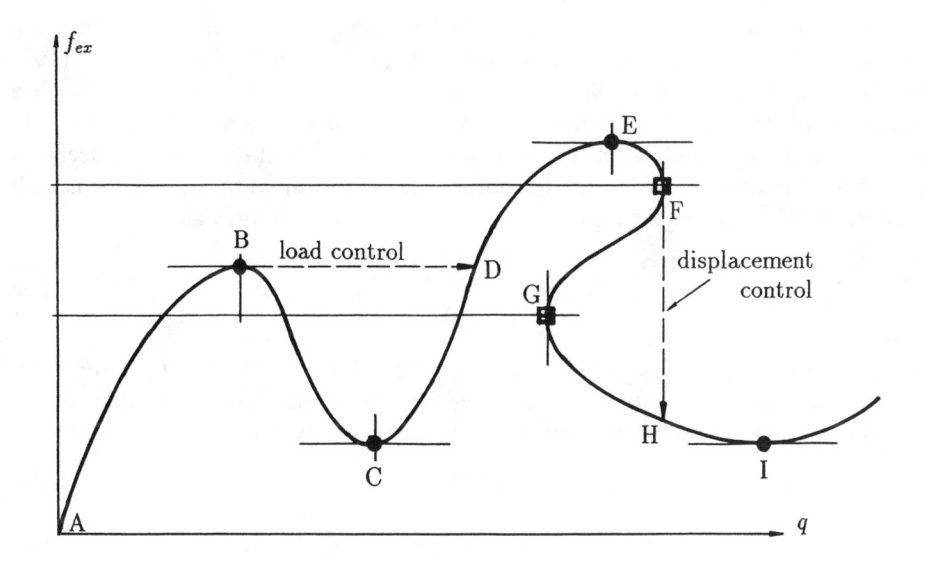

Figure 3.3: General load-displacement path including limit points

● limit point under load control
□ limit point under displacement control

The Newton iteration illustrated in Fig.3.2 is based on keeping the external loads constant during the iteration for one load increment. Thus, since λ_{n+1} is prescribed, the number of unknowns is equal to the number of nodal displacements m. Such a solution procedure, characterized by computing the displacements for a prescribed value of the load parameter, is denoted as a load-controlled solution scheme. When the first extremum on the load-displacement path is reached (point B in Fig.3.3), the tangent stiffness matrix $\mathbf{K}_n$ becomes singular. In the vicinity of such a point (a so called limit point), the tangent stiffness matrix is ill-conditioned. For the load-displacement path shown in Fig.3.3, a solution based on load control would jump from point B to point D instead of descending to point C.

This deficiency can be overcome by a displacement-controlled solution procedure characterized by computing the load parameter for a single prescribed displacement component. Thus, again, the number of unknowns is equal to m. In this way the load-displacement path shown in Fig. 3.3 could be traced until point F. When this point is reached, the displacement-controlled solution would jump to point H instead of proceeding to point G.

A load-controlled problem can be recast as a displacement-controlled problem only for relatively simple cases such as, e.g., for a single point load. For this case the corresponding displacement component is prescribed whereas the magnitude of the point load must be determined.

In order to overcome the deficiencies of load-controlled and displacement-controlled methods, so-called arc-length methods [Ramm (1981), Crisfield (1981), Fafard (1993)] have been developed. They allow tracing general load-displacement paths such as the one shown in Fig.3.3. Such methods can be viewed as generalized displacement-controlled iterative solution procedures. Both the incremental nodal displacements and the load parameter are treated as unknowns, resulting in $m+1$ unknowns. The unknown total incremental values can be written in vector form as

$$\Delta \mathbf{a}_{n+1}^{(i)} = \left\{ \begin{array}{c} \Delta \mathbf{q}_{n+1}^{(i)} \\ \Delta \lambda_{n+1}^{(i)} \end{array} \right\} . \tag{3.68}$$

The system of algebraic equations (3.55) or (3.62) for computation of the incremental nodal displacements consists of only m equations. Therefore, it has to be supplemented by an additional relation. The latter is given as

$$(\Delta \mathbf{a}_{n+1}^{(i)})^T \Delta \mathbf{a}_{n+1}^{(i)} = (\Delta S)^2 . \tag{3.69}$$

(3.69) constrains the "arc-length" of the tangent to the load-displacement path in the space of m nodal displacement components and one load parameter to a prescribed value ΔS. Geometrically, (3.69) can be interpreted as a hypersphere in the $m+1$ dimensional space.

Several versions of arc-length methods have been developed [Crisfield (1991), Hinton (1992)]. One of them is characterized by omitting the load parameter in (3.69), which forces the norm of $\Delta \mathbf{q}_{n+1}^{(i)}$ to be equal to ΔS. In another version, denoted as the normal plane method, (3.69) is used as a constraint only in the first iteration step. In the subsequent iteration steps the displacement increments are constrained to a hyperplane, perpendicular to the tangent of the first iteration step, requiring

$$(\Delta \mathbf{a}_{1,n+1})^T \Delta \mathbf{a}_{i,n+1} = 0 , \quad \text{for } i \geq 2 , \tag{3.70}$$

where

$$\Delta \mathbf{a}_{i,n+1} = \left\{ \begin{array}{c} \Delta \mathbf{q}_{i,n+1} \\ \Delta \lambda_{i,n+1} \end{array} \right\} \tag{3.71}$$

denotes the incremental values associated with the iteration step i. Because the external loads are not held constant during the iteration, the right-hand side of (3.62) must be supplemented by the incremental change of the external loads, yielding

$$\mathbf{K}_{n+1}^{(i-1)} \Delta \mathbf{q}_{i,n+1} = \mathbf{r}_{n+1}^{(i-1)} + \Delta \lambda_{i,n+1} \mathbf{f}_{ex,0} . \tag{3.72}$$

Direct addition of the constraint equation (3.69) or (3.70) to (3.72) would destroy both the symmetry and the banded form of the coefficient matrix. In order to preserve these desireable properties of the coefficient matrix, (3.72) is split into the following two parts:

$$\mathbf{K}_{n+1}^{(i-1)} \Delta \mathbf{q}_{i,n+1}^{I} = \mathbf{r}_{n+1}^{(i-1)} \quad \text{and} \quad \mathbf{K}_{n+1}^{(i-1)} \Delta \mathbf{q}_{i,n+1}^{II} = \mathbf{f}_{ex,0} . \tag{3.73}$$

3.1 Review of the Finite Element Method

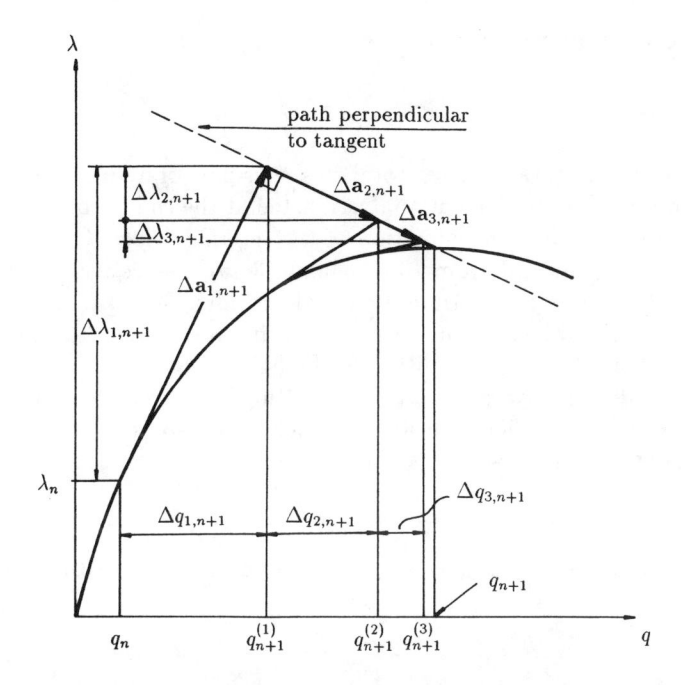

Figure 3.4: Illustration of an arc-length method

$\Delta \mathbf{q}_{i,n+1}$ is then obtained as

$$\Delta \mathbf{q}_{i,n+1} = \Delta \mathbf{q}^I_{i,n+1} + \Delta \lambda_{i,n+1} \Delta \mathbf{q}^{II}_{i,n+1} , \qquad (3.74)$$

where $\Delta \lambda_{i,n+1}$ is yet to be determined. Substitution of (3.74) into the constraint condition (3.70) finally yields

$$\Delta \lambda_{i,n+1} = \frac{-\Delta \mathbf{q}^T_{1,n+1} \Delta \mathbf{q}^I_{i,n+1}}{\Delta \mathbf{q}^T_{1,n+1} \Delta \mathbf{q}^{II}_{i,n+1} + \Delta \lambda_{1,n+1}} . \qquad (3.75)$$

The solution consists of

$$\begin{aligned} \mathbf{q}^{(i)}_{n+1} &= \mathbf{q}^{(i-1)}_{n+1} + \Delta \mathbf{q}_{i,n+1} , \\ \lambda^{(i)}_{n+1} &= \lambda^{(i-1)}_{n+1} + \Delta \lambda_{i,n+1} . \end{aligned} \qquad (3.76)$$

It can be interpreted geometrically as the point of intersection of the tangent at the known solution from iteration step $i - 1$ with the hyperplane perpendicular to the tangent of the first iteration step. For the special case of only one nodal displacement the solution procedure is illustrated in Fig.3.4.

The use of arc-length methods in finite element analyses of concrete structures is reported, e.g., in [Crisfield (1983), Lam (1992), Foster (1992)].

3.2 Update of the Stresses

3.2.1 Introduction

The update of the stress state for given total strains is part of item 2 of the Newton iteration for a nonlinear finite element analysis, listed at the end of subsection 3.1.5. The stresses are required for determination of the internal forces. On the element level these forces are obtained from (3.42_1) and (3.45), respectively. The global internal forces are required for determination of the residual vector (3.65). Since the procedure of the stress update is the same for all iteration steps, the superscript denoting the iteration step will be omitted in the following.

For time-independent analyses the strains resulting from creep and shrinkage of concrete are disregarded. Hence, the additive decomposition of the total strains according to (2.193) degenerates to

$$\varepsilon = \varepsilon^e + \varepsilon^p + \varepsilon^T . \tag{3.77}$$

For the special case of linear elasticity, because of $\varepsilon^p = 0$, the constitutive relations are given as

$$\sigma = \mathbf{C} : \varepsilon^e = \mathbf{C} : (\varepsilon - \varepsilon^T) . \tag{3.78}$$

For nonlinear constitutive models a general form of rate constitutive equations is

$$\dot{\sigma} = \mathbf{C}_T : \dot{\varepsilon}^e = \mathbf{C}_T : (\dot{\varepsilon} - \dot{\varepsilon}^p - \dot{\varepsilon}^T) , \tag{3.79}$$

where $\mathbf{C}_T$ denotes the tensor of the elastic tangent material moduli. For elasticity-based material models $\varepsilon^p = 0$. For elastic-plastic models $\mathbf{C}_T = \mathbf{C}$.

In order to keep the notation as simple as possible, temperature-induced strains will not be taken into account in the following. Hence, for constitutive models based on the theory of elasticity the total strains are equal to the elastic strains. For material models within the framework of the theory of elasto-plasticity the total strains are decomposed additively into elastic and plastic strains, i.e.,

$$\varepsilon = \varepsilon^e + \varepsilon^p . \tag{3.80}$$

Since for elastic-plastic constitutive models $\mathbf{C}_T = \mathbf{C} = \text{const.}$, the relations between the total stress and the total elastic strain follow from (3.79) as

$$\sigma = \mathbf{C} : \varepsilon^e = \mathbf{C} : (\varepsilon - \varepsilon^p) . \tag{3.81}$$

Let the vector of the total incremental nodal displacements $\Delta \mathbf{q}_{n+1}$ and the vector of the total nodal displacements $\mathbf{q}_{n+1}$ be known for a particular iteration step of load increment $n+1$. Then, in order to obtain the contribution of element e to the vector of internal forces according to (3.42_1) or (3.45), the stresses corresponding to the total strains must be computed. The total strains can be determined from the total nodal displacements.

3.2 Update of the Stresses

Within the framework of the finite element method the integration over the domain associated with an individual element is performed numerically. The numerical integration is based on the evaluation of the integrand at a number of points (sampling points or integration points) located in the element, followed by multiplication of the function values by weight coefficients and summation of the so-obtained numerical values over all integration points of a particular element [Zienkiewicz (1989,1991), Bathe (1982)].

Thus, for the evaluation of the integrals in (3.42_1) and in (3.45), respectively, the stresses must be updated at each integration point of a particular finite element e.

At first, for a particular integration point the total strains ε_{n+1} or $\mathbf{E}_{n+1}$ are computed from the total displacements by replacing $\mathbf{q}^e$ in (3.38) by $\mathbf{q}^e_{n+1}$ and substituting the so-obtained relation for the displacements $\mathbf{u}^e_{n+1}$ either into (3.9) (if geometric nonlinearity is disregarded) or into (3.12) (if geometric nonlinearity is considered). For small strains there is no conceptual difference between geometric linearity and geometric nonlinearity as far as the update of the stresses is concerned. In the following, the notation for geometric linearity will be used.

The total incremental strains are the difference between (non-converged) total strains of the actual load step $n+1$ and the total strains corresponding to the final (converged) displacements of the previous load step n. The procedure for the update of the stresses either in terms of the total incremental strains or the total strains depends on the type of the employed constitutive model.

Once the stresses at a particular integration point are known, the states of stress and strain have to be checked for material failure. Detection of material failure and modelling of the post-failure material behavior can be viewed as part of the stress update. Hence, it will also be described in this subsection.

3.2.2 Nonlinear Elastic Constitutive Models

For constitutive models with a nonlinear total (or secant) stress-strain relation (see subsection 2.3.2), there is a unique relationship between the total strains and the total stresses (see, e.g., (2.35)):

$$\sigma_{n+1} = \sigma_{n+1}(\varepsilon_{n+1}) . \tag{3.82}$$

Hence, e.g., for the equivalent-uniaxial constitutive model for biaxial stress states, formulated for principal directions, the total strains in these directions are needed for computation of the principal stresses by means of (2.35). This requires transformation of the total strains from the employed coordinate directions to principal directions. Finally, the principal stresses are transformed to the employed coordinate directions.

3.2.3 Hypoelastic Constitutive Models

In general, hypoelastic constitutive models are characterized by a path-dependent relationship between the strain rate and the stress rate. The actual stresses are obtained as

$$\sigma_{n+1} = \sigma_n + \Delta\sigma_{n+1} , \qquad (3.83)$$

where σ_n denotes the converged values of the stresses of the previous load step and $\Delta\sigma_{n+1}$ are the incremental stresses resulting from the incremental strains $\Delta\varepsilon_{n+1}$ which have accumulated during the iteration for the current load increment $n+1$. Since the tensor of the elastic tangent material moduli, $\mathbf{C}_T$, depends on the current state of stress or strain (see equations (2.5)), the incremental stress is obtained by integration of the rate constitutive equations (3.79) (with $\dot\varepsilon^p = 0$ and $\dot\varepsilon^T = 0$) as

$$\Delta\sigma_{n+1} = \int_{t_n}^{t_{n+1}} \mathbf{C}_T : \dot\varepsilon \, dt . \qquad (3.84)$$

The integration in (3.84) is performed numerically, employing, e.g., the generalized trapezoidal rule or the generalized midpoint rule [Bathe (1982), Ortiz (1985)]. The former is given as

$$x_{n+1} = x_n + [(1-\alpha)\dot x_n + \alpha \dot x_{n+1}]\Delta t_{n+1} \qquad (3.85)$$

and the latter as

$$x_{n+1} = x_n + \dot x_{n+\alpha}\Delta t_{n+1} . \qquad (3.86)$$

In (3.85) and (3.86), $\Delta t_{n+1} = t_{n+1} - t_n$ and $0 \le \alpha \le 1$. In (3.86), $\dot x_{n+\alpha} = \dot x(t_n + \alpha \Delta t_{n+1})$.

For $\alpha = 0$ and $\alpha = 1$ both integration methods yield identical results. Setting $\alpha = 0$ results in the explicit or forward Euler integration method. Setting $\alpha = 1$ yields the implicit or backward Euler integration scheme. For $\alpha = 1/2$, (3.85) yields the (implicit) trapezoidal rule and (3.86) the (implicit) midpoint rule. In contrast to forward and backward integration, which are of first-order accuracy with respect to Δt_{n+1}, for $\alpha = 1/2$ both integration schemes are characterized by second-order accuracy. First-order accuracy requires the numerically integrated values at t_{n+1} to agree with the exact values to within second-order terms in Δt_{n+1}. Alternatively, it requires that [Ortiz (1985)]

$$\left.\frac{dx_{n+1}}{d(\Delta t_{n+1})}\right|_{\Delta t_{n+1}\to 0} = \dot x_n . \qquad (3.87)$$

Second-order accuracy requires the numerically integrated values at t_{n+1} to agree with the exact values to within third-order terms in Δt_{n+1}. Alternatively, it requires that [Ortiz (1985)]

$$\left.\frac{d^2 x_{n+1}}{d(\Delta t_{n+1})^2}\right|_{\Delta t_{n+1}\to 0} = \ddot x_n . \qquad (3.88)$$

3.2 Update of the Stresses

For rate-independent material models (3.84) is rewritten as

$$\Delta\boldsymbol{\sigma}_{n+1} = \int_{\boldsymbol{\varepsilon}_n}^{\boldsymbol{\varepsilon}_{n+1}} \mathbf{C}_T : d\boldsymbol{\varepsilon} \ . \tag{3.89}$$

For such constitutive models the load steps take the role of the time steps in rate-dependent material models.

Explicit numerical integration of (3.89) yields

$$\Delta\boldsymbol{\sigma}_{n+1} \approx \mathbf{C}_{T,n} : \Delta\boldsymbol{\varepsilon}_{n+1} \ . \tag{3.90}$$

In order to reduce the error resulting from this numerical integration scheme, the strain increment $\Delta\boldsymbol{\varepsilon}_{n+1}$ may be subdivided into a number of subincrements. In this case the integral in (3.84) is approximated as

$$\Delta\boldsymbol{\sigma}_{n+1} \approx \sum_{k=0}^{\ell-1} \mathbf{C}_{T,n+k/\ell} : \frac{\Delta\boldsymbol{\varepsilon}_{n+1}}{\ell} \ , \tag{3.91}$$

where ℓ denotes the number of subincrements. (3.91) is known as the subincrementation method [Zienkiewicz (1989,1991), Bathe (1982)]. The errors resulting from application of (3.90) and (3.91) for the numerical integration of the constitutive rate equations are illustrated in Fig.3.5.

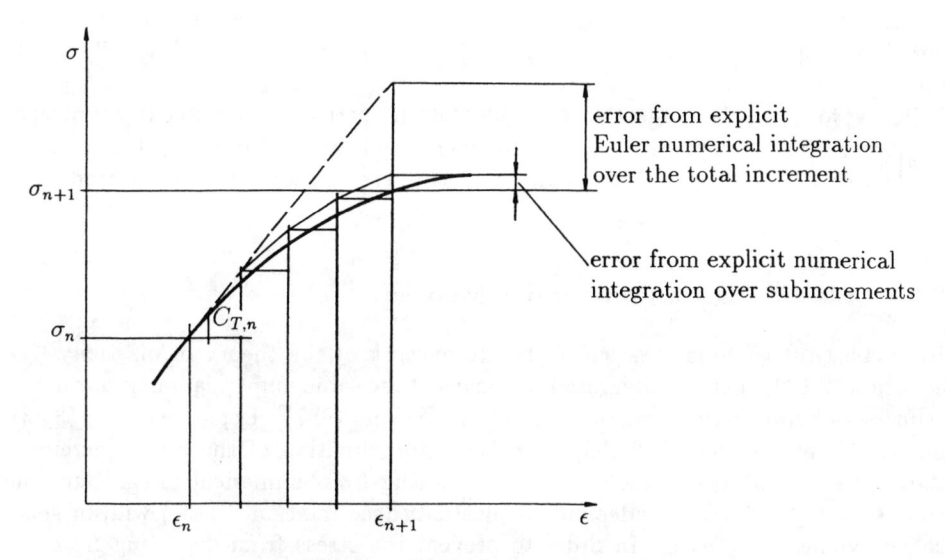

Figure 3.5: Errors resulting from explicit integration

An alternative scheme for explicit numerical integration is the Runge-Kutta method [Zienkiewicz (1989,1991)]. It is characterized by computing the stress $\boldsymbol{\sigma}_{n+1/2}$, corresponding to the strain $\boldsymbol{\varepsilon}_n + \Delta\boldsymbol{\varepsilon}_{n+1}/2$, as

$$\boldsymbol{\sigma}_{n+1/2} = \boldsymbol{\sigma}_n + \Delta\boldsymbol{\sigma}_{n+1/2} \,, \quad \text{with} \quad \Delta\boldsymbol{\sigma}_{n+1/2} = \mathbf{C}_{T,n} : \frac{\Delta\boldsymbol{\varepsilon}_{n+1}}{2} \,, \tag{3.92}$$

followed by computation of $\mathbf{C}_{T,n+1/2} = \mathbf{C}(\boldsymbol{\sigma}_{n+1/2})$ and determination of the stress increment

$$\Delta\boldsymbol{\sigma}_{n+1} = \mathbf{C}_{T,n+1/2} : \Delta\boldsymbol{\varepsilon}_{n+1} \,. \tag{3.93}$$

This method allows to estimate the error from the numerical integration by computing

$$\mathbf{r} = \Delta\boldsymbol{\sigma}_{n+1} - 2\Delta\boldsymbol{\sigma}_{n+1/2} \,. \tag{3.94}$$

If $||\mathbf{r}||$ exceeds a prescribed tolerance, then the size of the increments should be reduced.

Implicit integration methods, characterized by $\alpha > 0$ in (3.85) and (3.86), represent an alternative to explicit integration schemes. For both the generalized trapezoidal rule and the generalized midpoint rule the choice of $\alpha = 1/2$ is of interest because of the aforementioned higher-order accuracy.

In addition to the accuracy of the integration, the numerical stability of an integration scheme is important. An integration scheme is unconditionally stable, if perturbations applied to the solution at t_n are attenuated by the algorithm for computation of the solution at t_{n+1} for any size of the time step Δt_{n+1}. For conditionally stable algorithms such perturbations are attenuated only for time steps which are smaller than a critical value. It was shown in [Ortiz (1985)] that, with regards to numerical stability, the generalized midpoint rule is superior to the generalized trapezoidal rule. The generalized midpoint rule is unconditionally stable for $\alpha \geq 1/2$.

Hence, for implicit integration methods the time steps (or load steps) can be chosen larger than for explicit methods. However, application of an implicit integration method requires an iterative procedure for the stress update at the integration point level.

3.2.4 Elastic-Plastic Constitutive Models

Rate constitutive equations within the framework of the theory of plasticity (see equation (2.139)) can be integrated by means of the same numerical integration procedures as hypoelastic constitutive relations [Sloan (1987)]. Departing from (3.83), numerical integration of (2.139) yields an approximation of the stress increment $\Delta\boldsymbol{\sigma}_{n+1}$. However, because of the error resulting from numerical integration, the condition $f \leq 0$ for rate-independent plasticity (see relation 2.113$_1$) will, in general, be violated for $\boldsymbol{\sigma}_{n+1}$. In order to prevent the stress from departing from the yield surface in the direction of $f > 0$, algorithms were developed which allow exact enforcement of the condition $f = 0$ on the discretized level, i.e., within the framework of finite element analysis. They are known as return mapping algorithms or as closest point projection methods. The first designation expresses the fact that a stress state violating the yield condition is returned to the yield surface. The second

3.2 Update of the Stresses

designation refers to associated plasticity. The returned stress point is interpreted as the closest point projection of the stress, violating the condition $f \leq 0$, onto the yield surface with respect to the energy norm.

The return mapping algorithm [Simo (1985), Simo (1988b), Mitchell (1988)] basically consists of two major steps. In the first step the elastic trial stress σ^{Trial}, also referred to as the elastic predictor, is computed. Provided the condition $f \leq 0$ is violated by the trial stress, in the second step, also referred to as the plastic corrector, σ^{Trial} is returned to the yield surface.

In the context of the return mapping algorithm the constitutive equations and the discrete form of the loading/unloading conditions are both formulated at the "end" of load step $n+1$, or, equivalently, at the discrete time t_{n+1}. Thus, the constitutive equations follow from (3.81) as

$$\sigma_{n+1} = \mathbf{C} : (\varepsilon_{n+1} - \varepsilon^p_{n+1}) , \tag{3.95}$$

whereas the loading/unloading conditions are obtained from (2.113) as

$$f_{n+1}(\sigma_{n+1}, \mathbf{q}_{n+1}) \leq 0 , \quad \Delta\lambda_{n+1} \geq 0 , \quad \Delta\lambda_{n+1} f_{n+1}(\sigma_{n+1}, \mathbf{q}_{n+1}) = 0 , \tag{3.96}$$

where

$$\Delta\lambda_{n+1} = \dot{\lambda}\Delta t_{n+1} . \tag{3.97}$$

(A confusion of the vector of internal variables $\mathbf{q}_{n+1}$ with the vector of total nodal displacements seems to be impossible, because this subsection deals with the stress update at a particular integration point. Thus, the total nodal displacements do not appear in this subsection.)

The nature of the response is determined with the help of the elastic trial stress σ^{Trial}_{n+1}, assuming that the load step $n+1$ yields an elastic response. Hence, the trial state is obtained by "freezing" the plastic strains and the internal variables at the final (converged) values of load step n, yielding

$$\sigma^{Trial}_{n+1} = \mathbf{C} : (\varepsilon_{n+1} - \varepsilon^p_n) ,$$
$$\varepsilon^{p,Trial}_{n+1} = \varepsilon^p_n ,$$
$$\mathbf{q}^{Trial}_{n+1} = \mathbf{q}_n . \tag{3.98}$$

Substitution of (3.98_1) and (3.98_3) into the yield function (3.96_1) gives the trial value of the yield function

$$f^{Trial}_{n+1} = f^{Trial}_{n+1}(\sigma^{Trial}_{n+1}, \mathbf{q}_n) . \tag{3.99}$$

It is shown in [Simo (1988)] that convexity of the yield surface implies

$$f_{n+1} \leq f^{Trial}_{n+1} . \tag{3.100}$$

Thus, if

$$f^{Trial}_{n+1} < 0 , \quad \text{then} \quad f_{n+1} < 0 \Rightarrow \Delta\lambda_{n+1} = 0 . \tag{3.101}$$

Consequently, in this case the current load step results in elastic response as was assumed for the formulation of the trial state. Hence, the trial state is equal to the final state.

If, on the other hand, $f_{n+1}^{Trial} > 0$, then plastic loading will occur. In this case the plastic strains and the internal variables must be updated. Application of an implicit integration scheme according to (3.86) with $\alpha = 1$ to the flow rule (2.134) yields

$$\varepsilon_{n+1}^p = \varepsilon_n^p + \Delta\varepsilon_{n+1}^p \quad \text{with} \quad \Delta\varepsilon_{n+1}^p = \Delta\lambda_{n+1}\frac{\partial f_{n+1}}{\partial \boldsymbol{\sigma}_{n+1}} . \quad (3.102)$$

For constitutive models accounting for hardening, the implicit Euler integration scheme is also used for integration of the hardening law (see, e.g., equations (2.124), (2.126) or (2.137)). For the associated hardening law (2.137),

$$\boldsymbol{\alpha}_{n+1} = \boldsymbol{\alpha}_n + \Delta\boldsymbol{\alpha}_{n+1} \quad \text{with} \quad \Delta\boldsymbol{\alpha}_{n+1} = \Delta\lambda_{n+1}\frac{\partial f_{n+1}}{\partial \mathbf{q}_{n+1}} \quad (3.103)$$

is obtained.

Once the plastic strains ε_{n+1}^p are known, the stresses can be computed directly from (3.95). Alternatively, they can be computed from

$$\boldsymbol{\sigma}_{n+1} = \boldsymbol{\sigma}_{n+1}^{Trial} - \mathbf{C} : \Delta\varepsilon_{n+1}^p , \quad (3.104)$$

which follows from (3.95), using (3.98_1) and (3.102_1). Substitution of (3.102_2) into (3.104) yields

$$\boldsymbol{\sigma}_{n+1} = \boldsymbol{\sigma}_{n+1}^{Trial} - \Delta\lambda_{n+1}\mathbf{C} : \frac{\partial f_{n+1}}{\partial \boldsymbol{\sigma}_{n+1}} . \quad (3.105)$$

The return mapping of $\boldsymbol{\sigma}_{n+1}^{Trial}$ to the yield surface according to (3.105), resulting in $\boldsymbol{\sigma}_{n+1}$, is illustrated in Fig.3.6. This figure shows that the implicit Euler integration scheme is unconditionally stable because the trial stress is projected onto the yield surface in the direction of the gradient $\partial f_{n+1}/\partial \boldsymbol{\sigma}_{n+1}$ instead in the direction of $\partial f_n/\partial \boldsymbol{\sigma}_n$ as would be the case for the explicit Euler integration scheme. Therefore, from the point of numerical stability, there are no restrictions of the size of the load steps. However, the implicit Euler integration scheme is only of first-order accuracy. For this reason, the size of the load steps should not be chosen too large.

In the following, the update of the plastic strains and the internal variables for load step $n + 1$ will be described for the case of plastic loading. Because of the application of a backward Euler integration scheme, $\Delta\varepsilon_{n+1}^p$ and $\Delta\boldsymbol{\alpha}_{n+1}$ must be determined iteratively. For this purpose Newton iterations will be performed.

For plastic loading, characterized by $f_{n+1}^{Trial} > 0 \Rightarrow \Delta\lambda_{n+1} > 0$, ($3.96_3$) yields

$$f_{n+1}(\boldsymbol{\sigma}_{n+1}, \mathbf{q}_{n+1}) = 0 . \quad (3.106)$$

The expressions for the plastic strains (3.102) and the internal variables (3.103) can be rewritten as

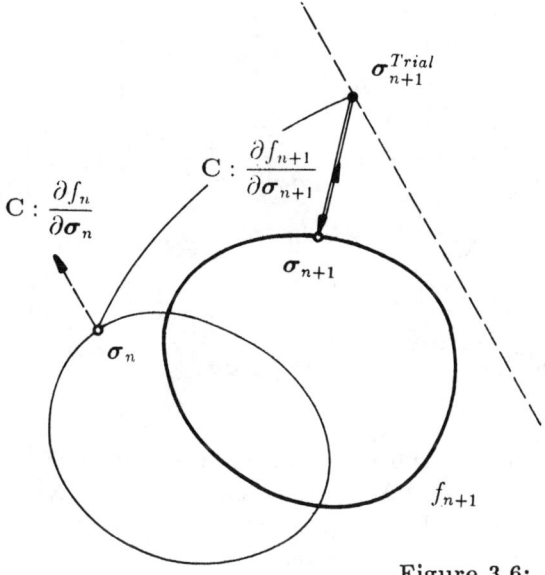

3.2 Update of the Stresses

Figure 3.6: Return mapping algorithm

$$-\varepsilon^p_{n+1} + \varepsilon^p_n + \Delta\lambda_{n+1}\frac{\partial f_{n+1}}{\partial \boldsymbol{\sigma}_{n+1}} = 0 ,$$

$$-\boldsymbol{\alpha}_{n+1} + \boldsymbol{\alpha}_n + \Delta\lambda_{n+1}\frac{\partial f_{n+1}}{\partial \mathbf{q}_{n+1}} = 0 . \tag{3.107}$$

The trial state provides the starting values for the Newton iterations. For plastic loading, (3.106) is violated, yielding the residual $r_{f,n+1} = f_{n+1}(\boldsymbol{\sigma}^{Trial}_{n+1}, \mathbf{q}_n) > 0$. For step k of the Newton iteration, the respective residual is given as

$$r^{(k)}_{f,n+1} = f^{(k)}_{n+1}(\boldsymbol{\sigma}^{(k)}_{n+1}, \mathbf{q}^{(k)}_{n+1}) . \tag{3.108}$$

Similarly, for iteration step k, the residuals of (3.107) are obtained as

$$\mathbf{r}^{(k)}_{\varepsilon,n+1} = -(\varepsilon^p_{n+1})^{(k)} + \varepsilon^p_n + \Delta\lambda^{(k)}_{n+1}\frac{\partial f^{(k)}_{n+1}}{\partial \boldsymbol{\sigma}^{(k)}_{n+1}} ,$$

$$\mathbf{r}^{(k)}_{\alpha,n+1} = -\boldsymbol{\alpha}^{(k)}_{n+1} + \boldsymbol{\alpha}_n + \Delta\lambda^{(k)}_{n+1}\frac{\partial f^{(k)}_{n+1}}{\partial \mathbf{q}^{(k)}_{n+1}} . \tag{3.109}$$

Within the framework of the Newton iteration, (3.106) and (3.107) are replaced by their respective linear approximations at the known values $\Delta\lambda^{(k)}_{n+1}$, $\boldsymbol{\sigma}^{(k)}_{n+1}$ and $\mathbf{q}^{(k)}_{n+1}$ of iteration step k:

$$r^{(k)}_{f,n+1} + \frac{\partial f^{(k)}_{n+1}}{\partial \boldsymbol{\sigma}^{(k)}_{n+1}}\Delta\boldsymbol{\sigma}_{k+1,n+1} + \frac{\partial f^{(k)}_{n+1}}{\partial \mathbf{q}^{(k)}_{n+1}}\Delta\mathbf{q}_{k+1,n+1} = 0 ,$$

$$r_{\varepsilon,n+1}^{(k)} - \Delta\varepsilon_{k+1,n+1}^p + \Delta\lambda_{k+1,n+1} \frac{\partial f_{n+1}^{(k)}}{\partial \sigma_{n+1}^{(k)}}$$

$$+ \Delta\lambda_{n+1}^{(k)} \left[\frac{\partial^2 f_{n+1}^{(k)}}{\partial (\sigma_{n+1}^{(k)})^2} \Delta\sigma_{k+1,n+1} + \frac{\partial^2 f_{n+1}^{(k)}}{\partial \sigma_{n+1}^{(k)} \partial q_{n+1}^{(k)}} \Delta q_{k+1,n+1} \right] = 0 ,$$

$$r_{\alpha,n+1}^{(k)} - \Delta\alpha_{k+1,n+1} + \Delta\lambda_{k+1,n+1} \frac{\partial f_{n+1}^{(k)}}{\partial q_{n+1}^{(k)}}$$

$$+ \Delta\lambda_{n+1}^{(k)} \left[\frac{\partial^2 f_{n+1}^{(k)}}{\partial (q_{n+1}^{(k)})^2} \Delta q_{k+1,n+1} + \frac{\partial^2 f_{n+1}^{(k)}}{\partial q_{n+1}^{(k)} \partial \sigma_{n+1}^{(k)}} \Delta\sigma_{k+1,n+1} \right] = 0 .$$

(3.110)

By analogy to the notation used in subsection 3.1.5 in the context of the Newton iteration for the nodal displacements, $\Delta\sigma_{k+1,n+1}$, $\Delta q_{k+1,n+1}$ and $\Delta\lambda_{k+1,n+1}$ are incremental quantities associated with the iteration step $k+1$. These quantities refer to a particular integration point. The corresponding total incremental quantities $\Delta\sigma_{n+1}^{(k+1)}$, $\Delta q_{n+1}^{(k+1)}$ and $\Delta\lambda_{n+1}^{(k+1)}$ are obtained by summation over the number of iteration steps, i.e.,

$$\Delta\sigma_{n+1}^{(k+1)} = \sum_{\ell=1}^{k+1} \Delta\sigma_{\ell,n+1} , \quad \Delta q_{n+1}^{(k+1)} = \sum_{\ell=1}^{k+1} \Delta q_{\ell,n+1} , \quad \Delta\lambda_{n+1}^{(k+1)} = \sum_{\ell=1}^{k+1} \Delta\lambda_{\ell,n+1} .$$

(3.111)

In the following, the equations (3.110) will be reduced such that they can be solved for $\Delta\lambda_{k+1,n+1}$. Noting that the updates on the integration point level are carried out for known total strains ε_{n+1}, the latter are fixed quantities. Hence, linearization of (3.95) yields

$$\Delta\varepsilon_{k+1,n+1}^p = -\mathbf{C}^{-1} : \Delta\sigma_{k+1,n+1} .$$

(3.112)

Substitution of (3.112) and of

$$\Delta\alpha_{k+1,n+1} = -\mathbf{H}^{-1} : \Delta q_{k+1,n+1} ,$$

(3.113)

following from (2.136), into (3.110) gives

$$r_{f,n+1}^{(k)} + \lfloor \frac{\partial f_{n+1}^{(k)}}{\partial \sigma_{n+1}^{(k)}} \quad \frac{\partial f_{n+1}^{(k)}}{\partial q_{n+1}^{(k)}} \rfloor \left\{ \begin{array}{c} \Delta\sigma_{k+1,n+1} \\ \Delta q_{k+1,n+1} \end{array} \right\} = 0 ,$$

$$\left\{ \begin{array}{c} r_{\varepsilon,n+1}^{(k)} \\ r_{\alpha,n+1}^{(k)} \end{array} \right\} + \mathbf{A}_{n+1}^{(k)} \left\{ \begin{array}{c} \Delta\sigma_{k+1,n+1} \\ \Delta q_{k+1,n+1} \end{array} \right\} + \Delta\lambda_{k+1,n+1} \left\{ \begin{array}{c} \frac{\partial f_{n+1}^{(k)}}{\partial \sigma_{n+1}^{(k)}} \\ \frac{\partial f_{n+1}^{(k)}}{\partial q_{n+1}^{(k)}} \end{array} \right\} = 0 ,$$

(3.114)

3.2 Update of the Stresses

where

$$\mathbf{A}_{n+1}^{(k)} = \begin{bmatrix} \mathbf{C}^{-1} + \Delta\lambda_{n+1}^{(k)} \dfrac{\partial^2 f_{n+1}^{(k)}}{\partial(\boldsymbol{\sigma}_{n+1}^{(k)})^2} & \Delta\lambda_{n+1}^{(k)} \dfrac{\partial^2 f_{n+1}^{(k)}}{\partial\boldsymbol{\sigma}_{n+1}^{(k)}\partial\mathbf{q}_{n+1}^{(k)}} \\ \Delta\lambda_{n+1}^{(k)} \dfrac{\partial^2 f_{n+1}^{(k)}}{\partial\mathbf{q}_{n+1}^{(k)}\partial\boldsymbol{\sigma}_{n+1}^{(k)}} & \mathbf{H}^{-1} + \Delta\lambda_{n+1}^{(k)} \dfrac{\partial^2 f_{n+1}^{(k)}}{\partial(\mathbf{q}_{n+1}^{(k)})^2} \end{bmatrix} . \quad (3.115)$$

Substitution of (3.114_2) into (3.114_1) yields

$$\Delta\lambda_{k+1,n+1} = \dfrac{r_{f,n+1}^{(k)} - \lfloor \dfrac{\partial f_{n+1}^{(k)}}{\partial\boldsymbol{\sigma}_{n+1}^{(k)}} \quad \dfrac{\partial f_{n+1}^{(k)}}{\partial\mathbf{q}_{n+1}^{(k)}} \rfloor (\mathbf{A}_{n+1}^{(k)})^{-1} \left\{ \begin{array}{c} r_{\varepsilon,n+1}^{(k)} \\ r_{\alpha,n+1}^{(k)} \end{array} \right\}}{\lfloor \dfrac{\partial f_{n+1}^{(k)}}{\partial\boldsymbol{\sigma}_{n+1}^{(k)}} \quad \dfrac{\partial f_{n+1}^{(k)}}{\partial\mathbf{q}_{n+1}^{(k)}} \rfloor (\mathbf{A}_{n+1}^{(k)})^{-1} \left\{ \begin{array}{c} \dfrac{\partial f_{n+1}^{(k)}}{\partial\boldsymbol{\sigma}_{n+1}^{(k)}} \\ \dfrac{\partial f_{n+1}^{(k)}}{\partial\mathbf{q}_{n+1}^{(k)}} \end{array} \right\}} . \quad (3.116)$$

Substitution of $\Delta\lambda_{k+1,n+1}$ into (3.114_2) yields $\Delta\boldsymbol{\sigma}_{k+1,n+1}$ and $\Delta\mathbf{q}_{k+1,n+1}$. Inserting $\Delta\boldsymbol{\sigma}_{k+1,n+1}$ into (3.112) and $\Delta\mathbf{q}_{k+1,n+1}$ into (3.113) gives $\Delta\boldsymbol{\varepsilon}_{k+1,n+1}^p$ and $\Delta\boldsymbol{\alpha}_{k+1,n+1}$. The respective total incremental quantities are obtained by summation over the number of iteration steps according to (3.111).

The iteration is continued until the residuals of (3.108) and (3.109) are smaller than prescribed tolerances. The result of the iteration at the integration point level consists of the total incremental consistency parameter $\Delta\lambda_{n+1}$, the plastic strains $\boldsymbol{\varepsilon}_{n+1}^p$, the internal variables $\boldsymbol{\alpha}_{n+1}$ and the stresses $\boldsymbol{\sigma}_{n+1}$, corresponding to the displacements from iteration step i of load step $n+1$ (see subsection 3.1.5).

The relations (3.106) to (3.116) describe a general iterative solution procedure for the stress update. This procedure must be employed for each integration point. In order to improve the efficiency of the iterative determination of the stress update, it should be reduced, if possible, to the iterative solution of a scalar nonlinear equation for $\Delta\lambda_{n+1}$ [Simo (1985), Hofstetter (1993)].

Once the stresses corresponding to the actual nodal displacements are known, the tangent material stiffness matrix $\mathbf{C}_{T,n+1}^{(i)}$ can be computed. This matrix is needed for the next iteration step to determine the incremental displacements. It is emphasized that $\mathbf{C}_{T,n+1}^{(i)}$ is not equal to the tangent material stiffness matrix $\mathbf{C}^{ep}$ (2.140) derived from the constitutive rate equations. If $\mathbf{C}^{ep}$ was used in a finite element analysis, the quadratic rate of asymptotic convergence, which is a characteristic feature of the Newton iteration, would be lost. For relatively large load steps the rate of convergence would be considerably smaller than the one obtained with $\mathbf{C}_{T,n+1}^{(i)}$. Hence, for each load step considerably more iteration steps would be necessary to satisfy the global equilibrium conditions. The reason for the deterioration of the rate of convergence was reported in [Simo (1985)]. In contrast to the continuum-based plasticity theory, computational plasticity is characterized by finite load steps.

Hence, the elastic-plastic tangent material moduli must not be derived from the continuum-based constitutive rate equations. They rather must be deduced through linearization of the equations used for determination of the stress update. This mode of derivation of the elastic-plastic tangent material moduli is consistent with the employed algorithm for computation of the stress update. Consequently, the resulting elastic-plastic tangent material moduli are referred to as consistent elastic-plastic tangent material moduli.

Derivation of the consistent elastic-plastic tangent material moduli for the general return mapping algorithm [Simo (1988), Simo (1988b)] requires determination of the total differentials $d\boldsymbol{\sigma}_{n+1}, d\boldsymbol{\varepsilon}^p_{n+1}, d\boldsymbol{\alpha}_{n+1}$ and df_{n+1}. They are obtained from the constitutive equations (3.95), the flow rule (3.102$_2$), the hardening law (3.103$_2$) and the yield condition (3.106) as follows:

$$d\boldsymbol{\sigma}_{n+1} = \mathbf{C} : \left(d\boldsymbol{\varepsilon}_{n+1} - d\boldsymbol{\varepsilon}^p_{n+1} \right) ,$$

$$d\boldsymbol{\varepsilon}^p_{n+1} = d\lambda_{n+1} \frac{\partial f_{n+1}}{\partial \boldsymbol{\sigma}_{n+1}} + \Delta\lambda_{n+1} \left(\frac{\partial^2 f_{n+1}}{\partial \boldsymbol{\sigma}^2_{n+1}} : d\boldsymbol{\sigma}_{n+1} + \frac{\partial^2 f_{n+1}}{\partial \boldsymbol{\sigma}_{n+1} \partial \mathbf{q}_{n+1}} : d\mathbf{q}_{n+1} \right) ,$$

$$d\boldsymbol{\alpha}_{n+1} = d\lambda_{n+1} \frac{\partial f_{n+1}}{\partial \mathbf{q}_{n+1}} + \Delta\lambda_{n+1} \left(\frac{\partial^2 f_{n+1}}{\partial \mathbf{q}^2_{n+1}} : d\mathbf{q}_{n+1} + \frac{\partial^2 f_{n+1}}{\partial \mathbf{q}_{n+1} \partial \boldsymbol{\sigma}_{n+1}} : d\boldsymbol{\sigma}_{n+1} \right) ,$$

$$df_{n+1} = \frac{\partial f_{n+1}}{\partial \boldsymbol{\sigma}_{n+1}} : d\boldsymbol{\sigma}_{n+1} + \frac{\partial f_{n+1}}{\partial \mathbf{q}_{n+1}} : d\mathbf{q}_{n+1} = 0 . \qquad (3.117)$$

In (3.117) all quantities represent final (converged) values of the Newton iteration at the integration point level, referring to iteration step i of load step $n+1$ (see subsection 3.1.5). (In order to simplify the notation, the superscript (i) has been omitted in this subsection.) Substitution of (3.117$_2$) into (3.117$_1$) yields

$$d\boldsymbol{\sigma}_{n+1} = \boldsymbol{\Xi}_{n+1} : \left[d\boldsymbol{\varepsilon}_{n+1} - d\lambda_{n+1} \frac{\partial f_{n+1}}{\partial \boldsymbol{\sigma}_{n+1}} - \Delta\lambda_{n+1} \frac{\partial^2 f_{n+1}}{\partial \boldsymbol{\sigma}_{n+1} \partial \mathbf{q}_{n+1}} : d\mathbf{q}_{n+1} \right] , \qquad (3.118)$$

where

$$\boldsymbol{\Xi}_{n+1} = \left[\mathbf{C}^{-1} + \Delta\lambda_{n+1} \frac{\partial^2 f_{n+1}}{\partial \boldsymbol{\sigma}^2_{n+1}} \right]^{-1} . \qquad (3.119)$$

Inserting (3.118) into (3.117$_4$) gives

$$\left[\Delta\lambda_{n+1} \frac{\partial f_{n+1}}{\partial \boldsymbol{\sigma}_{n+1}} : \boldsymbol{\Xi}_{n+1} : \frac{\partial^2 f_{n+1}}{\partial \boldsymbol{\sigma}_{n+1} \partial \mathbf{q}_{n+1}} - \frac{\partial f_{n+1}}{\partial \mathbf{q}_{n+1}} \right] : d\mathbf{q}_{n+1}$$

$$+ d\lambda_{n+1} \frac{\partial f_{n+1}}{\partial \boldsymbol{\sigma}_{n+1}} : \boldsymbol{\Xi}_{n+1} : \frac{\partial f_{n+1}}{\partial \boldsymbol{\sigma}_{n+1}}$$

$$= \frac{\partial f_{n+1}}{\partial \boldsymbol{\sigma}_{n+1}} : \boldsymbol{\Xi}_{n+1} : d\boldsymbol{\varepsilon}_{n+1} . \qquad (3.120)$$

Inserting (3.118) into (3.117$_3$) and making use of (3.113) yields

3.2 Update of the Stresses

$$\left[-\mathbf{H}^{-1} - \Delta\lambda_{n+1}\frac{\partial^2 f_{n+1}}{\partial \mathbf{q}_{n+1}^2} + \Delta\lambda_{n+1}^2 \frac{\partial^2 f_{n+1}}{\partial \mathbf{q}_{n+1}\partial \boldsymbol{\sigma}_{n+1}} : \boldsymbol{\Xi}_{n+1} : \frac{\partial^2 f_{n+1}}{\partial \boldsymbol{\sigma}_{n+1}\partial \mathbf{q}_{n+1}}\right] : d\mathbf{q}_{n+1}$$

$$+ \left[-\frac{\partial f_{n+1}}{\partial \mathbf{q}_{n+1}} + \Delta\lambda_{n+1}\frac{\partial^2 f_{n+1}}{\partial \mathbf{q}_{n+1}\partial \boldsymbol{\sigma}_{n+1}} : \boldsymbol{\Xi}_{n+1} : \frac{\partial f_{n+1}}{\partial \boldsymbol{\sigma}_{n+1}}\right] d\lambda_{n+1}$$

$$= \Delta\lambda_{n+1}\frac{\partial^2 f_{n+1}}{\partial \mathbf{q}_{n+1}\partial \boldsymbol{\sigma}_{n+1}} : \boldsymbol{\Xi}_{n+1} : d\boldsymbol{\varepsilon}_{n+1} \ . \tag{3.121}$$

Solution of (3.120) and (3.121) for $d\lambda_{n+1}$ and $\Delta\lambda_{n+1}d\mathbf{q}_{n+1}$ gives

$$\left\{\begin{array}{c} d\lambda_{n+1} \\ \Delta\lambda_{n+1}d\mathbf{q}_{n+1} \end{array}\right\} = \mathbf{Z}_{n+1}^{-1}\left\{\begin{array}{c} \frac{\partial f_{n+1}}{\partial \boldsymbol{\sigma}_{n+1}} : \boldsymbol{\Xi}_{n+1} : d\boldsymbol{\varepsilon}_{n+1} \\ \frac{\partial^2 f_{n+1}}{\partial \mathbf{q}_{n+1}\partial \boldsymbol{\sigma}_{n+1}} : \boldsymbol{\Xi}_{n+1} : d\boldsymbol{\varepsilon}_{n+1} \end{array}\right\}, \tag{3.122}$$

where $\mathbf{Z}_{n+1}$ is a $[2 \times 2]$ matrix. Its coefficients are obtained as

$$z_{11} = \frac{\partial f_{n+1}}{\partial \boldsymbol{\sigma}_{n+1}} : \boldsymbol{\Xi}_{n+1} : \frac{\partial f_{n+1}}{\partial \boldsymbol{\sigma}_{n+1}} \ ,$$

$$z_{12} = \frac{\partial f_{n+1}}{\partial \boldsymbol{\sigma}_{n+1}} : \boldsymbol{\Xi}_{n+1} : \frac{\partial^2 f_{n+1}}{\partial \boldsymbol{\sigma}_{n+1}\partial \mathbf{q}_{n+1}} - \frac{1}{\Delta\lambda_{n+1}}\frac{\partial f_{n+1}}{\partial \mathbf{q}_{n+1}} \ ,$$

$$z_{21} = z_{12} \ ,$$

$$z_{22} = -\frac{1}{\Delta\lambda_{n+1}^2}\mathbf{H}^{-1} - \frac{1}{\Delta\lambda_{n+1}}\frac{\partial^2 f_{n+1}}{\partial \mathbf{q}_{n+1}^2} + \frac{\partial^2 f_{n+1}}{\partial \mathbf{q}_{n+1}\partial \boldsymbol{\sigma}_{n+1}} : \boldsymbol{\Xi}_{n+1} : \frac{\partial^2 f_{n+1}}{\partial \boldsymbol{\sigma}_{n+1}\partial \mathbf{q}_{n+1}} \ . \tag{3.123}$$

Finally, substitution of (3.122) into (3.118) yields

$$\mathbf{C}_{T,n+1} = \frac{d\boldsymbol{\sigma}_{n+1}}{d\boldsymbol{\varepsilon}_{n+1}} = \boldsymbol{\Xi}_{n+1} - \sum_{i=1}^{2}\sum_{j=1}^{2} z_{ij,n+1}^{-1}\mathbf{N}_{i,n+1} \otimes \mathbf{N}_{j,n+1} \ , \tag{3.124}$$

where $z_{ij,n+1}^{-1}$ denotes the coefficients of the matrix $\mathbf{Z}_{n+1}^{-1}$ and

$$\mathbf{N}_{1,n+1} = \boldsymbol{\Xi}_{n+1} : \frac{\partial f_{n+1}}{\partial \boldsymbol{\sigma}_{n+1}} \qquad \mathbf{N}_{2,n+1} = \boldsymbol{\Xi}_{n+1} : \frac{\partial^2 f_{n+1}}{\partial \mathbf{q}_{n+1}\partial \boldsymbol{\sigma}_{n+1}} \ . \tag{3.125}$$

Obviously, the consistent elastic-plastic tangent material moduli (3.124) differ from the continuum-based elastic-plastic tangent material moduli according to (2.140) where, for the special case of an associated flow rule, g is to be replaced by f. However, if the time step Δt_{n+1} (or the load step) was infinitesimally small, then there would be no difference between the continuum-based tangent material moduli and the consistent tangent material moduli. In this case $\Delta\lambda_{n+1} \to 0$. Consequently, in (3.119) $\boldsymbol{\Xi} \to \mathbf{C}$. Setting $\Delta\lambda_{n+1} = 0$ and $\boldsymbol{\Xi}_{n+1} = \mathbf{C}$ in (3.120) and (3.121), $d\lambda_{n+1}$ can be computed. Substituting the obtained expression for $d\lambda_{n+1}$ into (3.118) and setting $\Delta\lambda_{n+1} = 0$ and $\boldsymbol{\Xi}_{n+1} = \mathbf{C}$, yields the continuum-based elastic-plastic tangent material moduli.

For the derivation of the consistent elastic-plastic tangent material moduli, the flow rule and the hardening law were assumed to be associated with the yield function. Hence, the tangent material stiffness matrix (3.124) is symmetric. Since hardening laws are usually not deduced on the basis of the postulate of maximum plastic dissipation (see subsection 2.3.3), most of the employed plasticity models have non-associated hardening laws (see, e.g., [Hofstetter (1993), Simo (1993)]). Contrary to the requirements to obtain a symmetric consistent elastic-plastic tangent material stiffness matrix, only an associated flow rule is needed in order to obtain a symmetric continuum-based elastic-plastic tangent material stiffness matrix.

Use of consistent elastic-plastic tangent material moduli is an essential feature for efficient finite element formulations based on the theory of plasticity. For a typical load step, use of consistent elastic-plastic tangent material moduli can save up to three quarters of the number of iteration steps required otherwise to satisfy global equilibrium within a prescribed tolerance [Simo (1985)]. Especially for large-scale three-dimensional problems savings of computer time are important.

Drawbacks of return mapping algorithms are the necessity to compute second derivatives in the expression for the matrix $\mathbf{A}_{n+1}^{(k)}$ (see equation (3.115)) and the costly Newton iteration at the integration point level, including the inversion of $\mathbf{A}_{n+1}^{(k)}$ and the multiplication of matrices with vectors. However, for many constitutive models the second drawback may be reduced considerably. In particular, very efficient algorithms may be developed for constitutive models with yield surfaces which do not depend on the third invariant of the stress tensor. For such models the curves of intersection of the yield surface with deviatoric planes are circles. Making use of the split of the material response into a volumetric and a deviatoric part, the previously described Newton iteration at the integration point level can often be reduced to the solution of a single nonlinear scalar equation in terms of the consistency parameter (see, e.g., for the von Mises yield criterion [Simo (1985)] and for a cap model [Hofstetter (1993)]). For the von Mises yield surface the respective algorithm is also known as the radial return mapping algorithm, because, as a consequence of the cylindrical yield surface, the trial stress is returned to the yield surface in the radial direction.

The return mapping algorithm for three-dimensional stress states can easily be modified for axisymmetric and plane strain problems. For the former, $\varepsilon_{13} = \varepsilon_{23} = 0$, for the latter, in addition, $\varepsilon_{33} = 0$. For plane stress problems, however, the required modifications are not so straightforward. Return mapping of a trial stress satisfying the conditions for plane stress, $\sigma_{13} = \sigma_{23} = \sigma_{33} = 0$, would yield a stress point which violates these conditions. For this reason, return mapping algorithms for plane stress conditions have been proposed in [Simo (1986)]. These algorithms are based on formulations of the constitutive equations in the subspace resulting from the plane stress conditions.

The return mapping algorithm can be extended to yield surfaces consisting of several smooth surfaces intersecting in a non-smooth fashion. In this case (3.102_2) is replaced by

3.2 Update of the Stresses

$$\Delta \varepsilon_{n+1}^p = \sum_j \Delta \lambda_{j,n+1} \frac{\partial f_{j,n+1}}{\partial \sigma_{n+1}} . \qquad (3.126)$$

(3.126) follows from Koiter's generalized flow rule, given in (2.135). In (3.126) the subscript j denotes the yield function. Consequently, the summation extends over the smooth yield surfaces of which the non-smooth yield surface is composed.

In contrast to plasticity models with a single, smooth yield surface, for multi-surface non-smooth plasticity models a positive trial value for the yield function j, i.e., $f_{j,n+1}^{Trial} > 0$, does not require the corresponding incremental consistency parameter $\Delta \lambda_{j,n+1}$ to be greater than zero [Simo (1988b)]. The condition for the yield function j to be active is $\Delta \lambda_{j,n+1} > 0$. In addition, for non-smooth strain-hardening or strain-softening plasticity the location of the intersection of two adjacent yield functions is not known *a priori*. Return mapping algorithms for non-smooth multi-surface plasticity were proposed in [Simo (1988b)].

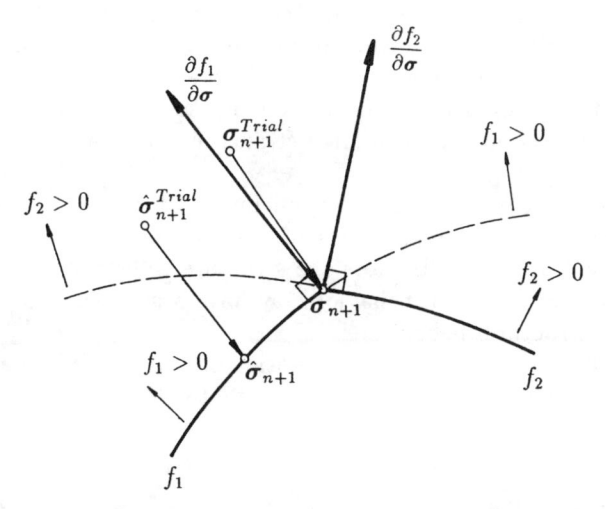

Figure 3.7: Return mapping for a non-smooth yield surface

Fig.3.7 shows two different elastic trial stress states, σ_{n+1}^{Trial} and $\hat{\sigma}_{n+1}^{Trial}$, which are returned to a non-smooth yield surface. For both trial stresses, $f_{1,n+1}^{Trial} > 0$ and $f_{2,n+1}^{Trial} > 0$. If the projection of σ_{n+1}^{Trial} to the corner formed by f_1 and f_2 yields $\Delta \lambda_{1,n+1} > 0$ and $\Delta \lambda_{2,n+1} > 0$, then σ_{n+1}^{Trial} is indeed returned to the corner. If the projection of $\hat{\sigma}_{n+1}^{Trial}$ to the corner, however, yields $\Delta \lambda_{1,n+1} > 0$ and $\Delta \lambda_{2,n+1} < 0$, then $\hat{\sigma}_{n+1}^{Trial}$ is not returned to the corner. Rather, it is returned to f_1.

3.2.5 Detection of Material Failure and Modelling of Post-Failure Material Behavior

Introduction

Computation of the state of strain and the state of stress at a particular integration point must be followed by a check against material failure. If failure is indicated, the type of failure must be identified. Moreover, a constitutive model for the description of the post-failure behavior has to be established. In chapter 1 the different modes of concrete failure were described. In chapter 2 mathematical models for the ultimate strength curve for biaxial stress states (subsection 2.2.2) and for the ultimate strength surface for triaxial stress states (subsection 2.2.3) were presented. Identification of the type of failure was discussed. In the post-peak region of stress-strain diagrams, i.e., after the ultimate strength envelope has been reached, the material behavior of concrete predominantly subjected to compression differs strongly from the behavior of concrete predominantly subjected to tension. In the first case, at least for relatively low levels of confinement, concrete is gradually softening until the ultimate strain is reached. Then, crushing occurs. In the second case, localization of deformations results in the formation of discrete cracks. The term "localization of deformations" means concentration of large strains in small subdomains of the considered domain.

Consideration of the post-peak behavior of concrete is of interest not only from an academic point of view. In certain parts of a structure the concrete strains may enter the post-peak region well before the structure collapses. Consequently, material modelling of the post-peak behavior may have a significant influence on the predictions of the structural behavior.

In what follows, different methods for consideration of concrete failure and of the post-failure behavior of concrete in the context of finite element analysis will be reviewed.

Modelling of Concrete Crushing

Failure of concrete predominantly subjected to compression is frequently determined by means of an ultimate strength envelope as described in subsections 2.2.2 and 2.2.3. When the stress point reaches this envelope, crushing of concrete is assumed. It is characterized by the complete loss of the load-carrying capacity of the material. Thus, both the stresses and the tangent material stiffness matrix are set equal to zero, i.e.,

$$\sigma_{n+1} = \mathbf{0} \, , \quad \mathbf{C}_{T,n+1} = \mathbf{0} \, . \tag{3.127}$$

However, if the crushed concrete in a structure is confined by intact material, then a residual strength and a residual stiffness of the crushed material will be preserved. Hence, the assumption of a complete loss of the load-carrying capacity of the material is conservative. Moreover, modelling of failure of the concrete, predominantly subjected to compression, on the basis of the ultimate strength envelope neglects the softening branch of the constitutive relations, i.e., the region between the strain at peak stress and the ultimate strain at crushing. On the basis of the ultimate strength envelope, crushing is signalled at smaller strains (in magnitude) than indicated in experiments. Hence, determination of crushing of concrete by means of an ultimate strength envelope is on the safe side.

A complete description of the behavior of concrete in the post-peak region requires not only knowledge of the ultimate (peak) strength but also of the ultimate strain and corresponding stress at crushing. For three-dimensional states of stress, however, the envelope of the ultimate strains is not available. For biaxial stress states such an envelope could be constructed by interpolation between known values of ultimate strains, which have been determined, e.g., by Kupfer [Kupfer (1969), Kupfer (1973)] for various ratios of principal stresses. However, in general, test results in the softening region are not reliable because the observed dependence of the results on the testing conditions is much more pronounced than in the pre-peak region [Yamaguchi (1991)].

If it is necessary to consider the post-peak softening behavior under predominantly compressive stresses, the problem of unreliable experimental data can be reduced by assuming the ultimate strain envelope as an isotropic expansion of the strain envelope at ultimate strength [Pekau (1992)]. A simple elastic-plastic constitutive model for compressive softening of concrete, based on the Drucker-Prager yield surface, was proposed in [Ohtani (1989)]. The contraction of the yield surface is governed by a damage parameter, expressed in terms of the hydrostatic pressure, which reflects the decrease of softening with increasing confining pressure. However, the available experimental data in the softening range are neither adequate nor sufficiently accurate to allow a critical evaluation of the proposed model.

Because of lack of experimental data the softening response in compression is frequently treated as perfectly-plastic material behavior until crushing occurs [Walter (1988)].

Modelling of Concrete Cracking

Mathematical models for concrete cracking include the prediction for crack initiation, the prediction of the evolution of the crack and a method of crack representation. Initiation of cracking is usually determined by a strength criterion. The evolution of a crack is described by a criterion based on fracture mechanics.

Fracture mechanics distinguishes between three different types of cracks in a plate (Fig.3.8). Mode I cracks are caused by in-plane tensile stresses acting perpendicular to the crack plane. Mode II cracks result from in-plane shear stresses, whereas mode III cracks are caused by out-of-plane shear stresses. For general solids the distinction between mode II and mode III is abandoned.

There are two different models for representation of cracks. They are denoted as the discrete crack model and the smeared crack model. The former is characterized by accounting for the displacement discontinuity across the crack. In the latter

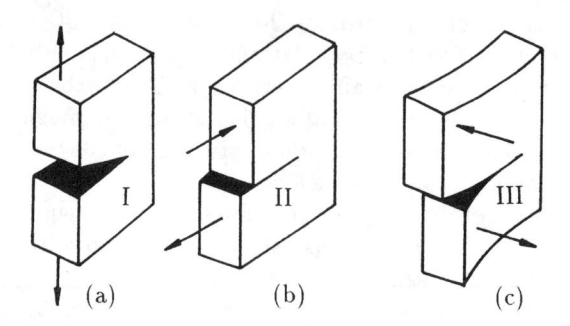

Figure 3.8: Different types of cracks: (a) mode I, (b) mode II, (c) mode III

the cracked material is treated as a continuum, assigning characteristic material properties to the cracked parts of the structure. Typically, the smallest cracked subregion is equal to the region associated with a sampling point for numerical integration over the volume of a finite element. The frequently used term "cracked integration point" refers to such a region.

Depending on the given problem, either the discrete crack model or the smeared crack model bears more resemblance with physical reality. If the structural behavior is governed by a few dominant cracks with large crack widths, the discrete crack model will be the better constitutive model. The smeared crack model, however, often allows a better approximation of the mechanical behavior of properly designed structures or structural elements made of reinforced concrete. This behavior is frequently characterized by many cracks with small crack widths in consequence of a suitable arrangement of the reinforcing steel.

Discrete crack models The first attempts to model cracking within the framework of the finite element method consisted of introducing discrete cracks with a predetermined location and orientation [Ngo (1967)]. This model was only applicable to problems with a few dominant cracks of *a priori* known location and orientation. Later on the cracks were allowed to extend along the common boundaries of adjacent finite elements. If the average value of the largest principal stress along the common boundary of two elements exceeds the tensile strength, a crack is modelled by separating the common nodes of the two elements as shown in Fig.3.9(a). The stresses perpendicular to the crack are set equal to zero. (This procedure will obviously be problematic, if the directions of the largest principal stress varies strongly along the common boundary of two elements.) However, the computed stresses at the crack tip are mesh-dependent. The smaller the finite elements, the larger the computed tensile stress at the crack tip. Hence, the finer the finite element mesh, the greater the likelihood of the prediction of crack propagation. This lack of objectivity of the numerical results with respect to mesh refinement was pointed out in [Bažant (1976c)]. Methods to obtain objective results were proposed in [Bažant (1979c)]. These methods are based on fracture mechanics. The most popu-

3.2 Update of the Stresses

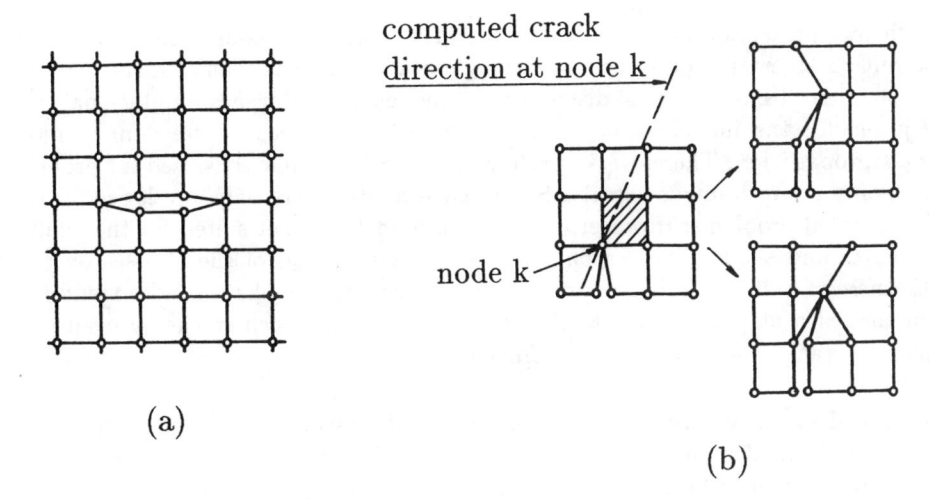

Figure 3.9: Discrete crack models: (a) cracks are following the common boundaries of adjacent finite elements, (b) common boundaries of finite elements are adjusted to follow the computed crack direction

lar method is to formulate the evolution of a crack in terms of the specific fracture energy G_f which is considered as a material parameter. G_f is the energy dissipated during the complete opening of a crack of unit area. When the crack is completely open, no tensile stresses are transferred across the crack faces.

Separation of nodes of finite elements is accomplished either by introducing new nodes when cracking of concrete is signalled or by providing double nodes from the outset. Double nodes are connected by stiff linkage elements. In case of cracking, the double nodes are separated by reducing the stiffness of the linkage elements. The residual stiffness of the linkage elements, which decreases with increasing crack width, is a consequence of the ability of cracked concrete to transfer forces across cracks by means of aggregate interlock and dowel action. Thus, the material behavior in the vicinity of cracks can be described with a relatively high degree of accuracy. However, because of the restrictions imposed on the location and the direction of cracks, this approach often yields too stiff numerical predictions of the structural response.

In order to eliminate this deficiency, the nodes of finite elements adjacent to a crack are relocated such that the boundaries of the elements coincide with the computed crack direction (Fig.3.9(b)) [Miguel (1990)]. An alternative method was proposed in [Grootenboer (1981)]. A feature of this method is the use of triangular hybrid elements. These elements are characterized by replacing the continuous interpolations of the displacements and the stresses within the elements by discontinuous interpolations, if cracking is signalled by the analysis. In this way the introduction of new nodes and, thus, the redefinition of the mesh is avoided. However, discontinuous interpolations require the introduction of additional degrees of freedom. One of them represents a slip parallel to the crack. Two additional degrees of freedom permit consideration of a linearly varying crack width within an element.

The main disadvantage of discrete crack models is the large computational effort required for the introduction of new nodes or new degrees of freedom to model crack propagation. Therefore, discrete crack models are mainly suited for problems with only a few dominant cracks. So far, such models were only developed for two-dimensional problems. However, these crack models are not suited for the analysis of plates and shells. The reason for this is the variation of the stresses over the thickness of such surface structures. Hence, in general, cracks will not extend from one face of a plate or a shell to the other. Modelling of such cracks by means of a discrete crack concept seems to be impossible.

Smeared crack models The cracks are usually assumed to be smeared over the region associated with a sampling point for numerical integration over the volume of a finite element. An advantage of this approach is the possibility to consider arbitrary directions of cracks without the need of redefining the mesh or introducing additional degrees of freedom. Moreover, smeared crack models for two-dimensional problems can be extended to three-dimensional problems without major difficulties. Because of the convenient implementation of smeared crack models into finite element programs such crack models have become by far more popular than discrete crack models.

For smeared crack models, displacement continuity is also preserved for the cracked material. This assumption is in conflict with the discontinuity of the displacements at cracks. It may lead to stress-locking, i.e., to overestimating the stiffness and the ultimate load. This has not been observed for discrete crack models [Rots (1989)]. Maintaining displacement continuity for the cracked concrete results in the transfer of relatively large strains to adjacent elements. This is incompatible with the observed strain localization. Because of the distribution of the cracks over finite regions, smeared crack models do not permit a reliable description of the material behavior in the vicinity of cracks. However, in general, they are well suited for the prediction of the global structural behavior.

Several smeared crack models have been reported in the literature. Most of these models are based on the theory of elasticity. They can be subdivided into fixed orthogonal crack models, fixed non-orthogonal crack models and rotating crack models. Recently, smeared crack models were proposed, which are based on the theory of plasticity. They are characterized by a Rankine yield criterion with an appropriate softening law.

In the crack models based on the theory of elasticity concrete is treated as an orthotropic material with reduced (or zero) strength perpendicular to the direction of the crack. The direction of the crack is assumed to be normal to the direction of the maximum principal tensile stress at a load level at which the corresponding stress point attains the ultimate strength envelope. The term "fixed crack" reflects the assumption that the direction of the crack does not change in the further course of the analysis. Fixed orthogonal crack models only allow the formation of additional cracks at an integration point in directions perpendicular to already existing cracks. Thus, for two-dimensional problems, at a single integration point at most two crack directions are possible. For three-dimensional problems, the maximum value of possible crack directions at an integration point is three. However, the assumption of orthogonal cracks is inconsistent with the consideration of shear stresses, from aggregate interlock, in the crack planes. These shear stresses cause a rotation of the principal stress axes from the position of the principal material axes (axes of orthotropy) defined at the initiation of cracking. The formation of non-orthogonal cracks was clearly shown in experiments conducted at the University of Toronto [Vecchio (1982), Collins (1985), Vecchio (1986), Bhide (1987), Stevens (1987)] (see Figs.1.36 and 1.37).

The mechanical inconsistency of fixed orthogonal crack models and the experimental results which do not support such constitutive models have led to the development of fixed non-orthogonal crack models. They allow consideration of a larger number of cracks at a single integration point in directions which need not be orthogonal to already existing cracks. However, fixed non-orthogonal crack models are more complicated than fixed orthogonal crack models, because information on all cracks at an integration point must be stored. Moreover, the opening of new cracks at an integration point is frequently accompanied by the closure of already existing cracks, which must be accounted for properly. In order to reduce the computational complexity, the number of cracks at an integration point is usually limited by allowing the formation of a new crack only if the deviation of its direction from the directions of already existing cracks is greater than a chosen threshold angle.

Rotating crack models are motivated by the experimental evidence that an existing crack will tend to close, if a new crack is formed in a rotated direction. In such models only the latest crack is taken into account. Its direction is perpendicular to the instant direction of the principal tensile stress. Hence, the crack direction rotates with the rotation of this principal stress. Consequently, there are no shear stresses in the crack planes. Implementation of a rotating crack model into a finite element program is much simpler than implementation of a fixed crack model.

Crack models based on the theory of plasticity yield plastic strains as the primary results. These strains allow identification of cracks and of their directions. Such constitutive models permit a unified treatment of the compressive and tensile behavior of concrete within the theory of plasticity. This offers computational advantages. Plasticity-based crack models are closely related to the rotating crack concept. Nevertheless, there is a fundamental difference between these two kinds of crack models. Plasticity-based crack models are usually developed from rate constitutive equations. Rotating crack models, representing elasticity-based crack models, however, are generally derived from total stress-strain relations.

Comparative studies of different types of smeared crack models can be found in [Crisfield (1989), Rots (1989), Feenstra (1993a), Welscher (1993)]. A large number of different versions of the four described kinds of smeared crack models was proposed. It is impossible to comment on all existing crack models. In the following, only a few representative crack models will be described. Reference to other crack models will be made.

Fixed orthogonal crack models. In early models within the smeared crack concept [Kabir (1977), Mang (1983), Owen (1983)] at most two cracks for two-dimensional stress states and three cracks for three-dimensional stress states were taken into account at a particular integration point. By analogy to discrete crack models, the initiation of cracking was assumed to occur in mode I. The direction of the first crack was assumed to be normal to the direction of the maximum principal tensile stress at a load level at which the corresponding stress point attains the ultimate strength envelope. This principal tensile stress was set equal to zero. The smeared crack concept is illustrated in Fig.3.10 for a quadrilateral isoparametric element with 8 node points and 9 integration points. The subregion associated with one of these integration points is assumed to be cracked. The direction of the crack is normal to the principal direction 1. Within the framework of elasticity-based crack models, at cracked integration points concrete is modelled as an orthotropic material (see equation (2.40)). The direction normal to the crack coincides with one of the principal axes of orthotropy.

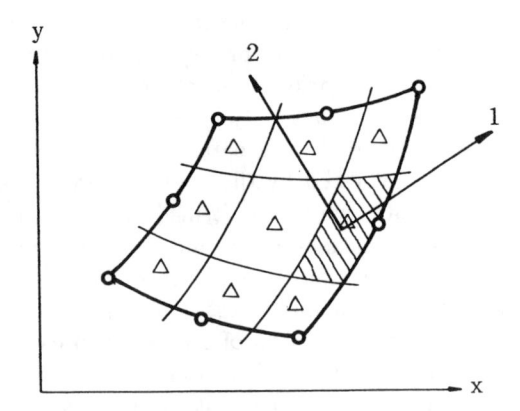

Figure 3.10: Crack at an integration point smeared over the subregion associated with this point

Frequently, a total stress-strain relation is used for the tensile region. Thus, if the crack model is combined with a nonlinear elastic material model for the compressive region, the elastic moduli E_1 and E_2 in (2.38) become secant moduli, i.e., $E_1 \equiv E_{S1}$ and $E_2 \equiv E_{S2}$. If a crack is detected, then the secant modulus of concrete in the direction normal to the crack will be set equal to zero. Hence, the coefficients in the respective row and column of the secant material stiffness matrix are replaced by zeros. Accordingly, the tangent modulus of concrete, referred to this direction, is set equal to zero. The coefficients in the respective row and column of the tangent material stiffness matrix are replaced by zeros. E.g., for a crack perpendicular to the local direction 1 (which is equal to one of the principal directions at the initiation of

3.2 Update of the Stresses

cracking), the constitutive equations for plane stress conditions follow from (2.40) and (2.39) as

$$\boldsymbol{\sigma} = \left\{ \begin{array}{c} \sigma_{11} \\ \sigma_{22} \\ \tau_{12} \end{array} \right\} = \left[\begin{array}{ccc} 0 & 0 & 0 \\ 0 & E_{S2} & 0 \\ 0 & 0 & G_{AI} \end{array} \right] \left\{ \begin{array}{c} \varepsilon_{11} \\ \varepsilon_{22} \\ \gamma_{12} \end{array} \right\}. \qquad (3.128)$$

In (3.128) E_{S2} denotes the secant modulus of concrete in the direction of the crack (which is equal to the other principal direction at crack initiation). Hence, at crack initiation $\tau_{12} = 0$. However, if loading continues after cracking, then, because of aggregate interlock (and for reinforced concrete also because of dowel action) shear forces will be transmitted across the crack, yielding a mixed mode (I and II) crack. In the smeared crack concept the shear transfer capacity of cracked concrete is accounted for by a modified shear modulus G_{AI} (see section 2.6).

If G_{AI} is assumed to be constant, then the corresponding rate constitutive equations, yielding the tangent material stiffness matrix required for the Newton iteration, will be analogous to (2.40):

$$\left\{ \begin{array}{c} \dot{\sigma}_{11} \\ \dot{\sigma}_{22} \\ \dot{\tau}_{12} \end{array} \right\} = \left[\begin{array}{ccc} 0 & 0 & 0 \\ 0 & E_{T2} & 0 \\ 0 & 0 & G_{AI} \end{array} \right] \left\{ \begin{array}{c} \dot{\varepsilon}_{11} \\ \dot{\varepsilon}_{22} \\ \dot{\gamma}_{12} \end{array} \right\}. \qquad (3.129)$$

In (3.129) E_{T2} denotes the tangent modulus of concrete in the direction parallel to the crack.

However, the assumption of a constant value for G_{AI} neglects the decreasing capacity of concrete to carry shear forces across cracks with increasing crack width (see subsection 1.4.2). In chapter 4 it will be demonstrated numerically that the assumption of a constant modified shear modulus leads to an overestimation of the load-carrying capacity.

In order to avoid the shortcomings of a constant modified shear modulus, a variable modified shear modulus, depending on the strain ε_{11} perpendicular to the crack (see (2.254)), was proposed in [Cedolin (1977)]. For a variable modified shear modulus, however, the relation between $\dot{\tau}_{12}$ and $\dot{\gamma}_{12}$ according to (3.129) does not hold. Rather, also $\dot{\varepsilon}_{11}$ is contained in the equation for $\dot{\tau}_{12}$, yielding a non-symmetric tangent material stiffness matrix. Nevertheless, in order to preserve the symmetry of the tangent material stiffness matrix, the term which violates this symmetry is frequently ignored. Thus, even for a variable modified shear modulus the tangent material stiffness matrix contained in (3.129) is employed. Because the total stresses are computed from the total strains according to (3.128), this inconsistency only affects the rate of convergence of the Newton iteration. If, however, the stresses were updated according to (3.89) the use of inconsistent tangent moduli would yield wrong stresses.

The rate constitutive equations for three-dimensional stress states contain three tangent shear moduli, G_{T12}, G_{T13} and G_{T23}. E.g., for a crack plane normal to the local direction 1 (Fig.3.11(a)), the tangent shear moduli G_{T12} and G_{T13} referring to the planes formed by the axes 1, 2 and 1, 3, respectively, will be replaced by the modified shear modulus G_{AI}, i.e.,

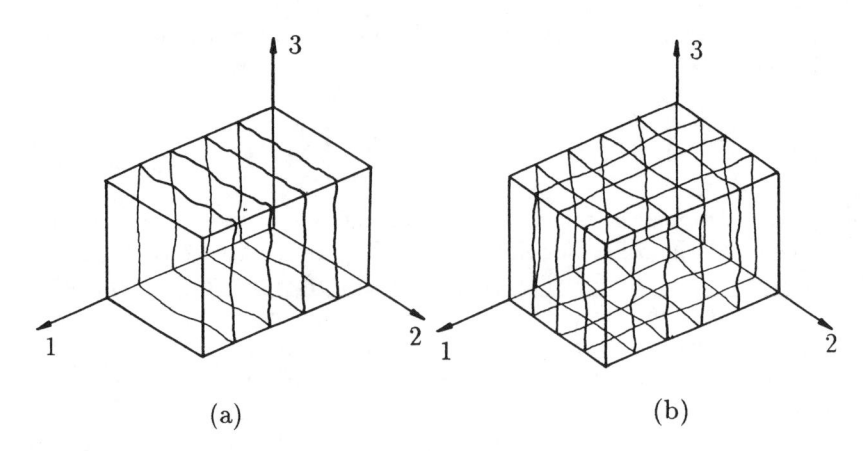

Figure 3.11: Examples for cracking in case of three-dimensional stress states: (a) one crack plane, (b) two orthogonal crack planes

$$G_{T12} = G_{AI}, \quad G_{T13} = G_{AI}. \tag{3.130}$$

If a second crack plane is formed, which, e.g., is normal to the locacl direction 2 (Fig.3.11(b)), then, according to [Owen (1983)], the tangent shear moduli are set equal to

$$G_{T13} = G_{AI}, \quad G_{T23} = G_{AI}, \quad G_{T12} = \begin{cases} 0.5\ G_{T13}, \\ 0.5\ G_{T23}, \end{cases} \quad \text{if} \quad G_{T23} < G_{T13}. \tag{3.131}$$

According to (3.131), the tangent shear modulus G_{T12} is further reduced to account for the second crack band in this plane. A third crack plane will be treated analogous to the second crack plane.

Stresses and strains in (3.128) are referred to a local coordinate system. It is defined by the directions normal and parallel to the crack (directions 1 and 2 in Fig.3.10). Stress rates and strain rates in (3.129) are referred to the same coordinate system. Stresses and strains and their rates are transformed to the directions of the global coordinate system by means of the transformation law for second-order tensors (2.47). Accordingly, the secant stiffness matrix in (3.128) and the tangent stiffness matrix in (3.129) are transformed to the global coordinate system by means of the transformation law for fourth-order tensors (2.44).

The second crack at an integration point is restricted to be orthogonal to the first crack. In the case of a three-dimensional stress state the third crack at an integration point is restricted to be orthogonal to the first and the second crack. If, for plane stress conditions, a second crack is formed at an integration point, or if, for a three-dimensional stress state, a third crack is formed at such a point, then aggregate interlock becomes the only means for a stress transfer at cracks. Ignoring aggregate interlock causes numerical problems in consequence of a singular material stiffness matrix. The latter will occur, if all finite elements containing a certain node

3.2 Update of the Stresses

point are cracked in the same direction and if no residual load-carrying capacity is left for the concrete. For this reason, the use of the modified shear modulus can also be viewed as a convenient way to avoid numerical problems.

The material model also accounts for the possibility of crack closure. A crack is assumed to be closed, if the decreasing tensile strain normal to the crack becomes equal to zero. Crack closure may be followed by compressive loading normal to the crack plane. However, the shear stiffness will not regain its original value. This is taken into account by using the modified shear modulus even for closed cracks.

The assumption of a complete release of the tensile stress normal to the crack immediately after the initiation of cracking is not a close approximation of the real behavior. Displacement-controlled experiments on specimens subjected to uniaxial tension rather indicate a strain-softening behavior in the post-peak region (see subsection 1.2.2), characterized by strain localization. In this region of material behavior the strains are increasing rapidly within a microcrack band. Its width is gradually decreasing. In the remaining part of the specimen the strains are decreasing, indicating unloading.

Load-displacement diagrams from ultimate load analyses based on the assumption of a sudden stress release at the onset of cracking usually contain a crack plateau. It is characterized by an increase of the displacements at practically constant loads. Experiments on concrete structures have shown that the crack plateaus are less pronounced than have been predicted numerically. Apart from this, the assumption of a sudden drop of stress is also disadvantageous from a numerical point of view. The stress, released instantly, must be redistributed to uncracked regions of concrete or to the reinforcement. Last, but not least, criticism concerning the mesh dependence of the results has been raised [Bažant (1976c)]. As mentioned previously, the smaller the finite elements, the larger the stresses at the integration point next to the crack tip. Thus, using a strength-based fracture criterion with a sudden stress release at the onset of cracking, the likelihood of predicting crack propagation for a fine mesh is greater than for a coarse mesh.

The deficiency of mesh-dependent results can be overcome by extending the strength-based fracture criterion by a model for the evolution of the crack, which is based on fracture mechanics [Bažant (1979c)]. (An extensive review of various smeared crack models based on fracture mechanics can be found in [RILEM (1989)].) The most popular method is characterized by introducing the specific fracture energy G_f as an additional material parameter.

Unobjective fracture criteria represent a fundamental deficiency of pertinent material models. Nevertheless, numerical results with distinct mesh dependence following from such a criterion were only observed for plain or lightly reinforced concrete structures. For moderately and heavily reinforced concrete structures this kind of mesh dependency is usually small.

A simple uniaxial material model for concrete in tension, which includes the softening material region, is obtained by approximating the stress-strain diagram of Fig.1.4(c) by a bilinear curve as shown in Fig.3.12. For comparison, the respective stress-strain relation without consideration of softening is included in this figure.

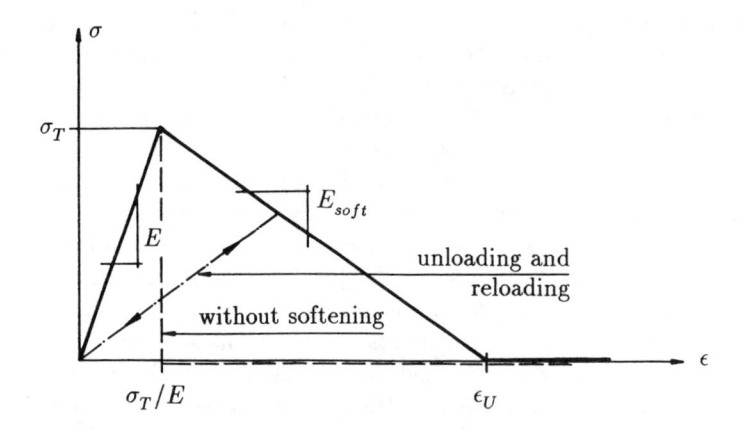

Figure 3.12: Bilinear stress-strain relationship for concrete in tension

The softening modulus E_{soft} has a negative value. The strain ε_U at which the tensile stress is released completely, is associated with the complete opening of the crack. Unloading and reloading is modelled by a secant. It is defined by the point on the softening branch of the stress-strain diagram at which unloading begins and by the origin. This representation of unloading and reloading is a crude approximation of the observed material behavior reflected by the stress-elongation diagram in Fig.1.5(b).

If cracking perpendicular to the direction 1 is detected, then, instead of setting $\sigma_{11} = 0$ according to (3.128), the value of σ_{11} is computed from the stress-strain equations for the softening region, given as

$$\left\{\begin{array}{c}\sigma_{11}\\ \sigma_{22}\\ \tau_{12}\end{array}\right\} = \left[\begin{array}{ccc}\mu E & 0 & 0\\ 0 & E_{S2} & 0\\ 0 & 0 & G_{AI}\end{array}\right]\left\{\begin{array}{c}\varepsilon_{11}\\ \varepsilon_{22}\\ \gamma_{12}\end{array}\right\}. \quad (3.132)$$

In (3.132) $\mu = \mu(\varepsilon_{11})$ is a reduction factor for the secant material stiffness normal to the crack, following from the employed stress-strain relation for uniaxial tension (Fig.3.12). The respective term on the main diagonal of the tangent material stiffness matrix is set equal to the softening modulus E_{soft}. Alternatively, this term is set equal to a very small positive value in order to preserve the positive definiteness of the tangent material stiffness matrix. If the stresses in the softening region are computed from total stress-strain relations, this inconsistency of the tangent material stiffness matrix will only affect the rate of convergence of the Newton iteration. Since the tangent material stiffness matrix is not correct, (3.89) must not be used for the stress update. Within the framework of material models for which the update of the stresses requires integration of the rate constitutive equations according to (3.89), such as, e.g., hypoelastic models, rate-type crack models must be used. Obviously, in such cases it is indispensable to employ the correct tangent material stiffness matrix.

3.2 Update of the Stresses

The secant material stiffness perpendicular to the crack, appearing in (3.132), is reduced gradually with increasing strain in this direction. In some crack models the Poisson effect is considered by means of non-zero off-diagonal terms in the row and the column referring to the direction perpendicular to the crack. These terms are reduced appropriately [Bažant (1983b)].

The question arises, which value should be taken for the softening modulus, e.g., in a linear softening material model. Fig.1.4(c) shows that the stress-strain relationship in the softening region depends on the length of the specimen perpendicular to the crack. Stress-elongation diagrams, however, are independent of the specimen length (see Fig.1.4(c)). This follows from concentration of the strains in a narrow band and from simultaneous unloading in the remaining part of the specimen, i.e., from strain localization. Thus unlike the modulus of elasticity, the softening modulus is not a material property. Rather, this modulus is obtained from an experiment as the average value of two parts of a specimen. One of them is the part of strain localization. The other one is the part where unloading occurs. For increasing length of the specimen, the ratio of the width of the crack band over the length of the specimen is decreasing. Consequently, the softening modulus becomes larger in magnitude. Hence, application of the same value of the softening modulus for finite element meshes with elements of different size yields mesh-dependent results.

The dependence of the computed structural response on the finite element mesh can be eliminated (apart from the conventional discretization error) by formulating suitable average material properties for the regions associated with the integration points. In this way, the inhomogeneous cracked material, attributed to an integration point and composed of regions of strain localization as well as unloading, is homogenized. This concept uses the specific fracture energy G_f as a material parameter [Hillerborg (1976), Bažant (1983b)]. For concrete, typical values of G_f are ranging from 50 Nm/m^2 to 200 Nm/m^2. After the initiation of cracking, which is signalled by the strength criterion, the release of the tensile stress is controlled by the specific fracture energy such that the energy released per unit area of the completely open crack is independent of the employed mesh.

A fixed orthogonal crack model yielding objective results was proposed in [Dahlblom (1990)]. This constitutive model represents a reformulation and extension of the fictitious crack model, originally proposed for the modelling of discrete cracks [Hillerborg (1976)]. The term "fictitious crack" indicates that prior to the formation of a single macrocrack the strains within a band of microcracks are lumped to a single fictitious macrocrack of crack width w_n.

A specimen of length L, subjected to uniaxial tension, is considered. It is assumed that cracking is initiated when the tensile stress reaches the tensile strength. The latter represents a material parameter of the fictitious crack model. The subregion of unloading of the softening specimen is assumed to extend over the whole length of the specimen. Consequently, the subregion of strain localization is assumed to be infinitely thin. This is in contrast to the crack band model [Bažant (1983b)] for which the width of the microcrack band associated with strain localization is regarded as a material property. The subregion of unloading of the specimen is described by linear elastic material behavior (Fig.3.13(a)) characterized by the modulus of elasticity E. The subregion of strain localization is characterized by the additional elongation w_n, denoted as crack width (Fig.3.13(b)). These assumptions are motivated by the experimental observation that softening is localized to a narrow band (Fig.1.4(b)). In the remaining part of the specimen the strains are even decreasing.

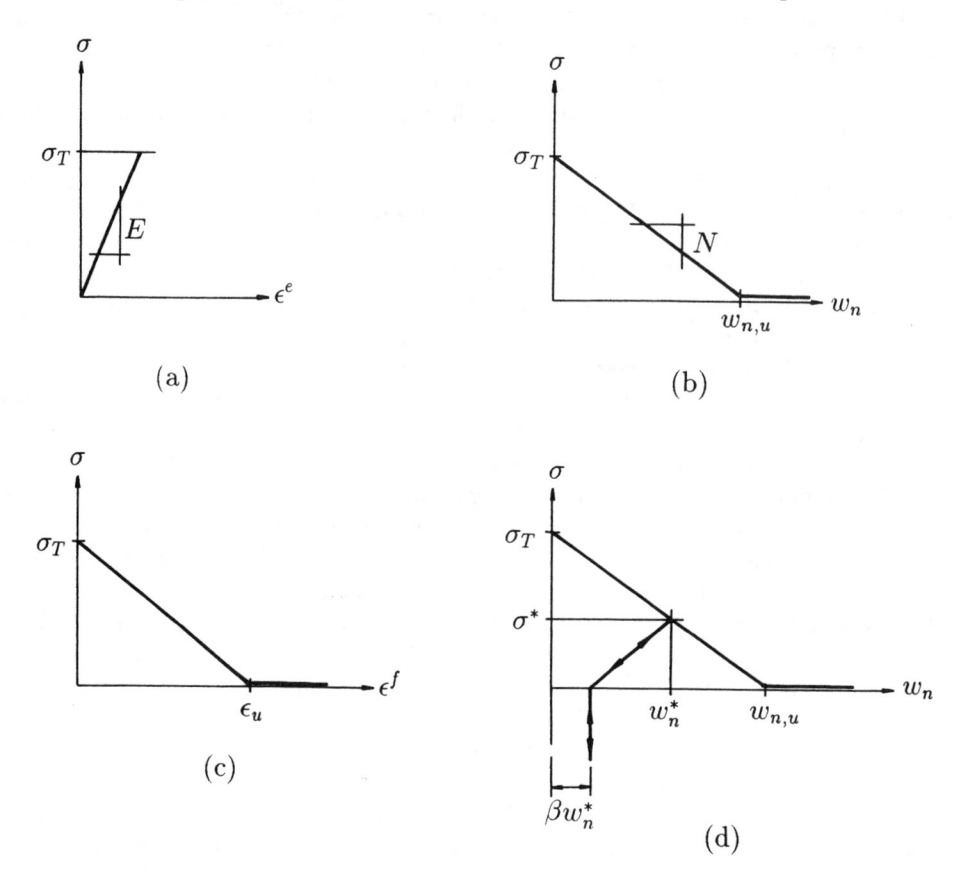

Figure 3.13: Fictitious crack model: (a) stress-strain relation for the pre-peak material domain and the subregion of unloading of the specimen in the softening material domain, (b) stress-elongation relation for the softening material domain, (c) stress-fracture strain diagram, (d) unloading and reloading

In contrast to the previously described crack models, where the rate of the total strain is also employed for the mechanical description of the cracked material, it has become popular to decompose the total strains ε in the softening domain into the strains of the material between the cracks, ε^{co}, and the crack (or fracture) strains, ε^f, [Bažant (1980), de Borst (1985)]:

$$\varepsilon = \varepsilon^{co} + \varepsilon^f . \tag{3.133}$$

3.2 Update of the Stresses

The advantage of this concept is that the behavior of the intact concrete between the cracks and the behavior of the cracks can be modelled separately, allowing combination of smeared crack models with nonlinear-elastic or elastic-plastic material models as well as models for the time-dependent behavior of concrete [de Borst (1987)]. For a specimen subjected to uniaxial tension, ε^{co} is computed from the stress-strain relation for linear elasticity yielding $\varepsilon^{co} = \varepsilon^e$, whereas ε^f is obtained by distributing the fictitious crack width w_n uniformly over the whole length of the specimen. Approximation of the relationship between the tensile stress perpendicular to the crack, σ, and the crack width w_n by a straight line of slope $N < 0$ yields (Fig.3.13(b))

$$w_n = \frac{\sigma - \sigma_T}{N} . \tag{3.134}$$

Alternatives to the linear σ-w_n relationship are bilinear or exponential σ-w_n relations, allowing a closer fit of experimental data (see Fig.1.4(a)). For the sake of simplicity of the presentation, a linear σ-w_n diagram will be used in the following. In this case the total elongation u of the specimen is obtained as

$$u = \varepsilon^e L + w_n , \quad \text{where} \quad \varepsilon^e = \frac{\sigma}{E} . \tag{3.135}$$

The energy D, dissipated when the tensile stress is released completely, i.e., for $\sigma = 0$ at $w_n = w_{n,u}$ (Fig. 3.13(b)), is obtained as

$$D = G_f A , \tag{3.136}$$

where A denotes the cross-sectional area of the specimen. The specific fracture energy is obtained as

$$G_f = \frac{1}{2} \sigma_T w_{n,u} . \tag{3.137}$$

The specific fracture energy G_f is the second material parameter of the fictitious crack model. It guarantees objective results with respect to mesh refinement. For known values of σ_T and G_f, $w_{n,u}$ and N follow from (3.134) and (3.137) (with $\sigma = 0$ at $w = w_{n,u}$) as

$$w_{n,u} = \frac{2G_f}{\sigma_T} , \quad N = -\frac{\sigma_T^2}{2G_f} . \tag{3.138}$$

Substituting (3.134) into (3.135) and dividing the so-obtained relation by the length of the specimen, L, gives the strain

$$\varepsilon = \frac{\sigma}{E} + \frac{\sigma - \sigma_T}{NL} \tag{3.139}$$

which is assumed to be uniformly distributed over the length of the specimen. Comparison of (3.139) with (3.133) shows that

$$\varepsilon^f = \frac{\sigma - \sigma_T}{NL} . \tag{3.140}$$

Substitution of (3.138$_2$) into (3.139) yields

$$\varepsilon = \sigma \left(\frac{1}{E} - \frac{2G_f}{\sigma_T^2 L} \right) + \frac{2G_f}{\sigma_T L} . \qquad (3.141)$$

Setting $\sigma = 0$ in (3.141) gives the strain ε_u at complete release of the tensile stress:

$$\varepsilon_u = \frac{w_{n,u}}{L} = \frac{2G_f}{\sigma_T L} . \qquad (3.142)$$

The definition of ε_u as the ratio of $w_{n,u}/L$ reflects the character of this strain as an average value. At $w_n = w_{n,u}$, because of $\sigma = 0$, the elastic strain is zero. Hence, ε_u represents the fracture strain (Fig.3.13(c)). From (3.141) the constitutive rate equation is obtained as

$$\dot{\sigma} = E_T \dot{\varepsilon} , \qquad (3.143)$$

where the tangent modulus E_T is given as

$$E_T = \frac{E}{1 - \dfrac{\lambda}{L}} \quad \text{with} \quad \lambda = \frac{2G_f E}{\sigma_T^2} = -\frac{E}{N} . \qquad (3.144)$$

The parameter λ is computed from material parameters of the specimen. Hence, it is also a material parameter. Because of its physical dimension as a length, λ is often referred to as the characteristic length of a material. Typical values of λ for standard concrete are ranging from 0.4 m to 0.8 m [Ottosen (1986)].

For material softening, $E_T < 0$. Hence, following from (3.144),

$$\lambda > L . \qquad (3.145)$$

The stress-strain relations in the softening region depend on the length L of the specimen in the direction normal to the crack (see (3.141) to (3.144)). This size-dependence of the σ-ε relationship can be explained by the fact that the fracture energy required to create a completely open crack depends on the area separated in consequence of cracking. In a direct tension test this is the cross-sectional area. Clearly, the fracture energy does not depend on the length of the specimen. However, in a smeared crack model the dissipated strain energy is described by means of the theory of a cracked continuum. Consequently, it is considered as a quantity per unit of volume. If the softening modulus was considered as a material parameter with a constant value, then, the shorter the specimen, the smaller the dissipated strain energy. This would make no physical sense. It follows that the length of the specimen must enter the σ-ε relationship in order to obtain an objective result for the dissipated strain energy. In other words, the dissipated strain energy must be the same for specimens of different lengths.

3.2 Update of the Stresses

Within the framework of the finite element method, L is related to the size of the finite elements in the direction normal to the crack plane. Hence, (3.145) restricts the element size in this direction. This restriction guarantees stable material behavior under displacement control. It follows from (3.144) that an increase of the value of L results in an increase of the absolute value of E_T. The limiting case, $L = \lambda$, corresponds to a sudden release of the tensile stress at crack initiation.

Shear loading in the crack plane is accounted for by introducing the tangential crack displacement w_t in the direction of the shear stress τ acting in the crack plane. A very simple relationship between w_n and w_t, containing the essential characteristics of shear loading in crack planes, is given as [Dahlblom (1990)]

$$w_t = \frac{\tau}{G_s} w_n , \tag{3.146}$$

where the so-called shear slip modulus G_s represents a material property. By setting $G_s = 3.8$ MPa, a set of experimentally obtained data was fitted reasonably well [Dahlblom (1990)].

Crack opening, which is equivalent to loading in the post-peak region, is characterized by $\dot{w}_n > 0$ and $w_n = w_n^*$, where w_n^* is the previously attained maximum value of the crack width. For the case of crack opening, the constitutive rate equations are obtained from (3.134) and (3.146) as

$$\left\{ \begin{array}{c} \dot{w}_n \\ \dot{w}_t \end{array} \right\} = \left[\begin{array}{cc} \frac{1}{N} & 0 \\ \frac{\tau}{G_s N} & \frac{w_n}{G_s} \end{array} \right] \left\{ \begin{array}{c} \dot{\sigma} \\ \dot{\tau} \end{array} \right\}. \tag{3.147}$$

For a constant shear stress in the crack plane, resulting in $\dot{\tau} = 0$, the off-diagonal term $\tau/(G_s N)$ in the coefficient matrix in (3.147) yields an increase of the tangential crack displacement w_t, which is proportional to the increase of the crack width w_n (see subsection 1.4.2). Coupling between $\dot{w}_n$ and $\dot{\tau}$, however, is not included in this constitutive model. The relations (3.147) are relatively simple constitutive rate equations for a crack. They may be replaced by more refined constitutive relations.

The equations (3.134) - (3.147) were derived for crack widths w_n satisfying the condition $w_n \leq w_{n,u}$, which holds for opening cracks. For $w_n > w_{n,u}$, the crack is completely open. In this case the ability to transfer tensile stresses across crack planes is lost completely. Thus, $\sigma = 0$ and $\dot{\sigma} = 0$. Following from $\dot{\sigma} = N\dot{w}_n$ in (3.147), $N = 0$ in this case.

Crack closing, which is equivalent to unloading in the post-peak region, is characterized by $\dot{w}_n < 0$. In this case a linear relationship between the stress and the crack width is assumed between the point (σ^*, w_n^*) at which unloading begins and the point defined by zero normal stress and the corresponding (residual) crack width (Fig.3.13(d)). The residual crack width is given as $w_n = \beta w_n^*$ with $0 \leq \beta \leq 1$. The limiting values $\beta = 0$ and $\beta = 1$ correspond to the limiting cases of a fully recoverable and a totally irrecoverable crack width. For the case of $\beta = 1$, the fracture strain remains constant during unloading and reloading. Consequently, the crack is inactive. In the present model this situation is taken into account by linear elasticity. In

[Dahlblom (1990)] it is reported that $\beta = 0.2$ has resulted in good correspondence between the computed material behavior and the experimental results contained in [Reinhardt (1984)].

For unloading with a tensile stress still acting across the crack planes, the stress - crack width relation follows from Fig.3.13(d) as

$$w_n = \left[\beta + (1-\beta)\frac{\sigma}{\sigma^*}\right] w_n^* . \tag{3.148}$$

Differentation of (3.146) and (3.148) with respect to time yields the constitutive rate equations for unloading with a tensile stress still acting across the crack planes as

$$\left\{\begin{array}{c} \dot{w}_n \\ \dot{w}_t \end{array}\right\} = \left[\begin{array}{cc} (1-\beta)\frac{w_n^*}{\sigma^*} & 0 \\ (1-\beta)\frac{w_n^*}{\sigma^*}\frac{\tau}{G_s} & \frac{w_n}{G_s} \end{array}\right] \left\{\begin{array}{c} \dot{\sigma} \\ \dot{\tau} \end{array}\right\} . \tag{3.149}$$

A comparison of the rate equations for crack opening (equation (3.147)) with the rate equations for crack closing with $\sigma > 0$ (equation (3.149)) shows that by setting

$$\frac{1}{N} = (1-\beta)\frac{w_n^*}{\sigma^*} \tag{3.150}$$

in (3.147) the relations (3.149) are obtained.

If, during unloading, σ changes from a tensile stress to a compressive stress, i.e., if the crack is closed during unloading, the residual crack width will remain constant, i.e., $w_n = \beta w_n^*$ (Fig.3.13(d)). Thus, $\dot{w}_n = 0$, yielding $\dot{w}_t = (w_n/G_s)\dot{\tau}$. Hence, (3.147) will also be valid for a closed crack, if N is set equal to ∞.

In order to be able to apply the described constitutive model to structural analysis by the finite element method, its formulation must be extended to multiaxial states of stress. In the following, a three-dimensional state of stress will be considered. A local coordinate system is defined in the same way as for the previously described fixed orthogonal crack model. The coordinate axes are denoted by the subscripts 1, 2 and 3. Each normal stress is related to the respective crack width ($\sigma_{11} \rightarrow w_{11}$, $\sigma_{22} \rightarrow w_{22}$, $\sigma_{33} \rightarrow w_{33}$). Each shear stress corresponds to the respective tangential crack displacement ($\sigma_{12} \rightarrow w_{12}$, $\sigma_{21} \rightarrow w_{21}$, ..., $\sigma_{31} \rightarrow w_{31}$). Thus, a crack displacement vector $\mathbf{w}^f$ and a corresponding traction vector $\mathbf{t}^f$ can be assigned to each crack plane. $\mathbf{w}^f$ contains the crack width and the two tangential crack displacements. $\mathbf{t}^f$ consists of the axial stress normal to the crack plane and the two shear stresses acting in this plane. E.g., for a crack plane perpendicular to the local direction 1, $\mathbf{w}^f$ and $\mathbf{t}^f$ are given as

$$\mathbf{w}^f \equiv \mathbf{w}_1^f = \left\{\begin{array}{c} w_{11} \\ w_{12} \\ w_{13} \end{array}\right\} , \quad \mathbf{t}^f \equiv \mathbf{t}_1^f = \left\{\begin{array}{c} \sigma_{11} \\ \sigma_{12} \\ \sigma_{13} \end{array}\right\} . \tag{3.151}$$

From the equations (3.134) and (3.146) the stress-displacement relations are obtained as

3.2 Update of the Stresses

$$w_{ii} = \frac{\sigma_{ii} - \sigma_T}{N}, \qquad w_{ij} = \frac{w_{ii}}{G_s}\sigma_{ij}, \qquad \text{no summation over } i, \; i \neq j. \qquad (3.152)$$

The crack strains are given as

$$\varepsilon^f_{ii} = \frac{w_{ii}}{L_i}, \qquad \varepsilon^f_{ij} = \frac{1}{2}\left(\frac{w_{ij}}{L_i} + \frac{w_{ji}}{L_j}\right), \qquad \text{no summation over } i, \; i \neq j, \qquad (3.153)$$

where L_i, $i = 1,2,3$, are equivalent lengths. An equivalent length is defined as the length over which the crack width and the tangential crack displacements are smeared uniformly. L_i is measured parallel to the local coordinate axis i.

In the finite element method the equivalent length L_i represents the maximum length of a subregion of a finite element in a certain direction. The subregion is associated with a "cracked integration point". L_i is measured in the local direction i, normal to the considered crack plane. Fig.3.14 refers to the definition of equivalent lengths for two-dimensional triangular elements with one integration point per element and two-dimensional quadrilateral elements with four integration points per element.

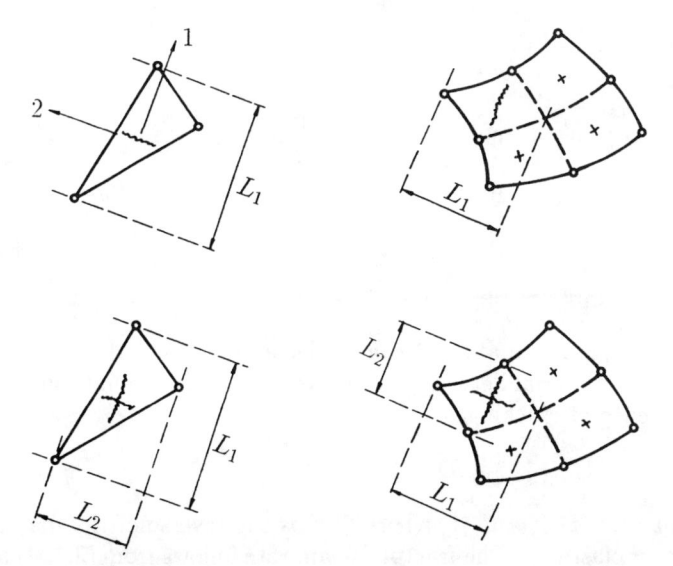

Figure 3.14: Definition of equivalent lengths L_i for two-dimensional finite elements

Substituting (3.152) into (3.153) yields

$$\varepsilon^f_{ii} = J_i\,(\sigma_{ii} - \sigma_T), \qquad 2\varepsilon^f_{ij} = K_{ij}\sigma_{ij}, \qquad \text{no summation over } i, \; i \neq j, \qquad (3.154)$$

where

$$J_i = \frac{1}{N L_i}, \qquad K_{ij} = \frac{1}{G_s}\left(\frac{w_{ii}}{L_i} + \frac{w_{jj}}{L_j}\right). \qquad (3.155)$$

$((3.154_1)$) is found to be similar to (3.140).) The total strains are computed analogous to (3.133). For the sake of simplicity, the strains between the cracks are assumed to be elastic, i.e., $\boldsymbol{\varepsilon}^{co} = \boldsymbol{\varepsilon}^{e}$. Hence, the total strains referred to the local coordinate system are obtained as

$$\boldsymbol{\varepsilon} = \mathbf{D}\boldsymbol{\sigma} - \boldsymbol{\sigma}_T \mathbf{J} , \qquad (3.156)$$

where

$$\boldsymbol{\varepsilon}^T = \lfloor \varepsilon_{11}\ \varepsilon_{22}\ \varepsilon_{33}\ 2\varepsilon_{12}\ 2\varepsilon_{23}\ 2\varepsilon_{31} \rfloor ,$$
$$\boldsymbol{\sigma}^T = \lfloor \sigma_{11}\ \sigma_{22}\ \sigma_{33}\ \sigma_{12}\ \sigma_{23}\ \sigma_{31} \rfloor ,$$
$$\mathbf{J}^T = \lfloor J_1\ J_2\ J_3\ 0\ 0\ 0 \rfloor$$

and

$$\mathbf{D} = \begin{bmatrix} \frac{1}{E} + J_1 & -\frac{\nu}{E} & -\frac{\nu}{E} & 0 & 0 & 0 \\ -\frac{\nu}{E} & \frac{1}{E} + J_2 & -\frac{\nu}{E} & 0 & 0 & 0 \\ -\frac{\nu}{E} & -\frac{\nu}{E} & \frac{1}{E} + J_3 & 0 & 0 & 0 \\ 0 & 0 & 0 & \frac{1}{G} + K_{12} & 0 & 0 \\ 0 & 0 & 0 & 0 & \frac{1}{G} + K_{23} & 0 \\ 0 & 0 & 0 & 0 & 0 & \frac{1}{G} + K_{31} \end{bmatrix} . $$
$$(3.157)$$

The material compliance matrix $\mathbf{D}$ can be decomposed into the compliance matrix $\mathbf{D}^{co}$ for the intact material between the cracks, and an additional compliance matrix $\mathbf{D}^{f}$ in consequence of cracking, i.e.,

$$\mathbf{D} = \mathbf{D}^{co} + \mathbf{D}^{f} . \qquad (3.158)$$

In the present case, $\mathbf{D}^{co} = \mathbf{C}^{-1}$, where $\mathbf{C}^{-1}$ is the inverse of the material stiffness matrix for linear elasticity. The fracture strain rate follows from (3.154) and (3.155_2) as

$$\dot{\varepsilon}^f_{ii} = J_i \dot{\sigma}_{ii} , \quad 2\dot{\varepsilon}^f_{ij} = \frac{1}{G_s}\sigma_{ij}\left(J_i\dot{\sigma}_{ii} + J_j\dot{\sigma}_{jj}\right) + K_{ij}\dot{\sigma}_{ij} ,$$
$$\text{no summation over } i,\ i \neq j . \qquad (3.159)$$

Thus, the constitutive rate equations, referred to the local coordinate system, are given as

$$\dot{\boldsymbol{\varepsilon}} = \mathbf{D}_T \dot{\boldsymbol{\sigma}} , \qquad (3.160)$$

where

3.2 Update of the Stresses

$$\mathbf{D}_T = \begin{bmatrix} \dfrac{1}{E}+J_1 & -\dfrac{\nu}{E} & -\dfrac{\nu}{E} & 0 & 0 & 0 \\ -\dfrac{\nu}{E} & \dfrac{1}{E}+J_2 & -\dfrac{\nu}{E} & 0 & 0 & 0 \\ -\dfrac{\nu}{E} & -\dfrac{\nu}{E} & \dfrac{1}{E}+J_3 & 0 & 0 & 0 \\ J_1\dfrac{\sigma_{12}}{G_s} & J_2\dfrac{\sigma_{12}}{G_s} & 0 & \dfrac{1}{G}+K_{12} & 0 & 0 \\ 0 & J_2\dfrac{\sigma_{23}}{G_s} & J_3\dfrac{\sigma_{23}}{G_s} & 0 & \dfrac{1}{G}+K_{23} & 0 \\ J_1\dfrac{\sigma_{31}}{G_s} & 0 & J_3\dfrac{\sigma_{31}}{G_s} & 0 & 0 & \dfrac{1}{G}+K_{31} \end{bmatrix}$$
(3.161)

Analogous to the aforementioned decomposition of the compliance material matrix $\mathbf{D}$, the tangent material compliance matrix $\mathbf{D}_T$ can be decomposed into the tangent compliance matrix $\mathbf{D}_T^{co}$ for the intact material between the cracks and an additional tangent material compliance matrix $\mathbf{D}_T^f$ in consequence of cracking, i.e.,

$$\mathbf{D}_T = \mathbf{D}_T^{co} + \mathbf{D}_T^f . \qquad (3.162)$$

In the present case $\mathbf{D}_T^{co} = \mathbf{C}^{-1}$. Combination of an elasticity-based crack model with a plasticity-based material model can be found in [de Borst (1985), de Borst (1987)]. Because of the nonlinear interaction between cracking and elastic-plastic material behavior between the cracks, an iterative algorithm is employed. It handles the crack model and the material model consecutively in an iterative manner. Unfortunately, this algorithm is not always robust [Rots (1989)].

Formulations for plane strain, plane stress and axisymmetry can be obtained by means of specialization of (3.156) and (3.160). If there is only one crack at an integration point and if this crack is, e.g., normal to the direction 1, then $L_2 = L_3 \to \infty$, yielding $J_2 = J_3 = K_{23} = 0$ (see (3.155)).

Inversion of $\mathbf{D}_T$ yields the tangent material stiffness matrix. The unsymmetry of this matrix follows from the unsymmetry of the tangent material compliance matrix, which is a consequence of the coupling term $\tau/(G_s N)$ in the coefficient matrix in (3.147). If this term is ignored, i.e., if the rate relation for $\dot{w}_t$ in (3.149) is replaced by $\dot{w}_t = (w_n/G_s)\dot{\tau}$, then $\mathbf{D}_T$ and, thus, $\mathbf{D}_T^{-1}$ will be symmetric.

All of the hitherto presented constitutive equations are referred to principal axes of orthotropy. Thus, within the framework of a finite element formulation the stress tensor, the strain tensor, the material stiffness tensor and the material compliance tensor must be transformed to the global coordinate system. This is accomplished with the help of the transformation laws for second-order and fourth-order tensors, given in (2.47) and (2.44), respectively.

No assumption is necessary for the modified shear modulus G_{AI} appearing, e.g., in (3.128). Rather, G_{AI} follows directly from the constitutive relations (3.156). These equations yield

$$2\varepsilon_{12} = \left(\frac{1}{G} + K_{12}\right)\sigma_{12} . \tag{3.163}$$

(It is unessential that (3.128) refers to a plane state of stress whereas (3.156) is related to a three-dimensional state of stress.) Analogous to the third equation in (3.128), (3.163) can formally be written as

$$\sigma_{12} = G_{AI}\gamma_{12} . \tag{3.164}$$

Making use of (3.163) and of (3.155$_2$), G_{AI} is obtained as

$$G_{AI} = \frac{G}{1 + \dfrac{G}{G_s}\left(\dfrac{w_{11}}{L_1} + \dfrac{w_{22}}{L_2}\right)} . \tag{3.165}$$

If there is only one crack at an integration point and if this crack is, e.g., normal to the direction 1, then in (3.165) $L_2 \to \infty$ and $w_{22} = 0$. At the initiation of cracking, $w_{11} = 0$. Hence, following from (3.165), $G_{AI} = G$. However, with increasing crack width w_{11} the value of G_{AI} is decreasing. For $w_{11} \to \infty$, $G_{AI} \to 0$.

For unloading of a crack perpendicular to the direction i, with a tensile stress still acting across the crack plane, i.e., with $\sigma_{ii} > 0$, the term $1/N$ in (3.155$_1$) is expressed by means of (3.150) (replacing w_n^* by w_{ii}^* and σ^* by σ_{ii}^*). This gives

$$J_i = (1 - \beta)\frac{w_{ii}^*}{\sigma_{ii}^* L_i} , \quad \text{no summation over } i . \tag{3.166}$$

The coefficients K_{ij} are still computed from (3.155$_2$).

For a closed crack with $\sigma_{ii} < 0$, $J_i = 0$. This follows from setting $N = \infty$ in (3.155$_1$). The coefficients K_{ij} are still computed from (3.155$_2$), setting $w_{ii} = \beta w_{ii}^*$. Hence, for closed cracks the original tangent material stiffnesses are only recovered for the relations between normal stresses and normal strains. The shear stiffnesses remain at reduced levels. Only for totally recoverable cracks ($\to \beta = 0$) the original shear stiffnesses are regained when these cracks are closed. Finally, for a completely open crack perpendicular to the direction i, $J_i = \infty$, following from (3.155$_1$) with $N = 0$. State changes within a given strain increment $\Delta\varepsilon$, such as changes from the pre-peak tensile region to the tensile softening region, or from the softening region to a completely open crack, or from reloading to softening, can be treated by subdividing $\Delta\varepsilon$ into two parts. One of them refers to the region before the change of state. The other one is related to the region after this state change [Rots (1989)].

The restriction of fixed orthogonal crack models to orthogonal crack directions at an integration point is a deficiency of these constitutive models. In fact, this restriction is inconsistent with the treatment of aggregate interlock. The shear stresses in the crack planes, generated by continued loading after cracking, and the normal stresses in the directions of the principal axes of orthotropy, defined as the crack directions, will cause a deviation of the principal stress axes from the principal axes of orthotropy. The principal stresses in these directions may exceed the tensile

3.2 Update of the Stresses

strength. This is possible because only the stresses acting in directions which are perpendicular to already existing cracks are checked for cracking. Tensile stresses in other directions, which are larger than the tensile strength, indicate crack initiation in these directions. However, such crack initiations are ignored. For this reason, fixed orthogonal crack models may not only result in the prediction of a too stiff structural response and in the overestimation of the ultimate load-carrying capacity, but also in computed crack distributions which differ considerably from the observed crack pattern.

Overestimation of both the stiffness and the ultimate load may be the consequence of an improper choice of the modified shear modulus [Crisfield (1989)]. Especially for relatively large, constant values of the shear retention factor, relatively great differences between the observed and the computed structural behavior may appear. As a remedy the modified shear modulus can be reduced to a very small number. (However, in order to avoid numerical problems it must not be reduced to zero.) In this case a lower bound of the actual load-carrying capacity will be obtained. Still, the crack patterns remain unrealistic. E.g., experimentally investigated panels [Vecchio (1982), Bhide (1987)], shown in Figs.1.36 and 1.37, have experienced secondary cracks, forming acute angles with the primary cracks. Such secondary cracks could obviously not be predicted by means of fixed orthogonal crack models.

Nevertheless, in spite of the conceptional shortcomings of fixed orthogonal crack models, they have been used sucessfully for ultimate load analyses of reinforced and prestressed concrete structures. The reason for the apparent success of these material models is that their mechanical deficiencies frequently do not have a significant influence on the structural behavior. If, e.g., cracks are formed primarily in only one direction or if secondary (and tertiary) cracks are nearly perpendicular to already existing cracks, the assumptions on which fixed orthogonal crack models are based, will be adequate. This is reflected by the good quality of the computed results.

Fixed non-orthogonal crack models. Motivated by the mechanical inconsistencies of fixed orthogonal crack models and by experimental results [Vecchio (1982), Bhide (1987)], the restriction of secondary and tertiary crack planes at a particular integration point to be perpendicular to already existing crack planes has been abandoned [de Borst (1985), Cervenka (1985), Rots (1989), Barzegar (1989)]. This has resulted in fixed non-orthogonal crack models.

Non-orthogonal crack models are considerably more complex than orthogonal crack models. Firstly, for each "cracked integration point" information about all previously formed cracks has to be stored. Secondly, at an integration point the crack planes are not defined just by one local coordinate system. Rather, a local coordinate system is assigned to each crack plane. The local fracture strains and the local tractions are referred to this coordinate system. Then, analogous to orthogonal crack models, constitutive equations for the intact material between the cracks and constitutive relations between the local fracture strains and the local stresses are formulated. Transformation of the local fracture strains for a single crack to the global coordinate system follows from (2.47). The total fracture strains ε^f at a particular integration point, referred to the global coordinate system, are computed by summation over the contributions from all cracks at the respective integration point.

On principle, an arbitrary number of cracks at a particular integration point can be taken into account. Hence, the formation of a new crack at a point could be assumed whenever a principal stress exceeds the tensile strength, regardless of the direction of this principal stress. However, consideration of new cracks even after small rotations of the principal stress axes would be computationally inefficient, because the information on all existing cracks has to be stored. Moreover, as new cracks are formed at a material point, existing cracks often tend to close. Hence, for a relatively large number of cracks at a single integration point it would be difficult to treat the state changes properly. They involve unloading and reloading and closure of existing cracks. Consequently, the number of cracks at an integration point is often limited by introducing a threshold angle. A new crack at a material point will only be considered if the angles between the normal vector of the new crack plane and the normal vectors of already existing crack planes are larger than the chosen threshold angle [de Borst (1985)]. However, for relatively large values of the threshold angle the same objection holds as for orthogonal crack models. If the angle formed by the direction of the largest principal stress and the normal to an existing crack plane is smaller than the chosen threshold angle, then the respective principal tensile stress may be greater than the tensile strength without causing crack initiation. Small threshold angles or, equivalently, a relatively large number of cracks at a single integration point can be taken into account more easily if elastic unloading of cracks ($\to \beta = 1$ in (3.148)) is assumed. In this case a crack becomes inactive immediately after reversal of the rate of the normal strain perpendicular to the crack plane. An inactive crack has no influence on the constitutive behavior. Information on inactive cracks is only stored to check for re-opening of such cracks in subsequent load increments.

Contrary to the structural response computed by means of fixed orthogonal crack models, the structural response resulting from non-orthogonal crack models was found to be insensitive with respect to the value chosen for the modified shear modulus [Cervenka (1985), Barzegar (1989), Wang (1990)]. The reason for this insensitivity is the limitation of the increase of the shear stresses in the crack planes by the formation of new cracks.

The main problem with non-orthogonal crack models is that their implementation into a finite element program is difficult. Different assumptions for the formation of new cracks, such as different choices of the value of the threshold angle, and for the modelling of unloading and reloading of opening cracks (elastic unloading or secant unloading) as well as of the behavior of closed cracks (recovery of the original shear stiffness or retention of a reduced shear stiffness) may influence the results considerably. Unless crack closure is modelled as elastic unloading, it will be difficult to consider state changes of existing cracks at a single integration point properly. Further research is required for the development of fixed non-orthogonal crack models applicable to structural analyses by means of the finite element method.

3.2 Update of the Stresses

Rotating crack models. In order to avoid the mechanical shortcomings of fixed orthogonal crack models, resulting from the rotation of the principal stress axes, the concept of rotating cracks was proposed [Cope (1980), Gupta (1984), Balakrishnan (1988)]. Originally, it was based on a relationship between total strains and stresses [Cope (1980)] similar to (3.128). However, the local coordinate axes 1, 2, 3, defining the principal axes of crack-induced orthotropy, were not assumed to be fixed after crack initiation. Rather, they were allowed to follow the rotation of the principal stress axes. Hence, the local axes 1, 2, 3, aligned with the directions perpendicular to the crack planes, also coincide with the principal stress axes after crack initiation. Consequently, multiple cracks at a material point are restricted to orthogonal planes. The shear stresses and the shear strains in the local coordinate system are always equal to zero. E.g., for the special case of a biaxial stress state with a crack plane perpendicular to the local direction 1, (3.128) is reduced to

$$\left\{ \begin{array}{c} \sigma_1 \\ \sigma_2 \end{array} \right\} = \left[\begin{array}{cc} E_{S1} & 0 \\ 0 & E_{S2} \end{array} \right] \left\{ \begin{array}{c} \varepsilon_1 \\ \varepsilon_2 \end{array} \right\}, \qquad (3.167)$$

where, contrary to (3.128), the secant modulus E_{S1} in the direction normal to the crack is not equal to zero, permitting consideration of tension softening. It follows from (3.167) that also for the cracked material the principal stress axes coincide with the principal strain axes. This property, known as the coaxiality of the stress tensor and strain tensor, represents a necessary condition for the form invariance of crack-induced orthotropic material behavior [Bažant (1983)] (see the comments on equivalent-uniaxial models in subsection 2.3.2). The principal stresses can then be computed from the principal strains, e.g., on the basis of the equivalent-uniaxial models described in subsection 2.3.2, combined with the failure criteria presented in section 2.2.

From the condition of coaxiality of the principal stresses and the principal strains, the tangent shear stiffness of the cracked material, G_T, can easily be determined. For convenience, this will be shown only for the special case of a biaxial stress state. Let it be assumed that the principal stresses σ_1, σ_2 and the principal strains $\varepsilon_1, \varepsilon_2$ are known for load step n. The incremental shear stress $\Delta\tau_{12}$ and the incremental engineering shear strain $\Delta\gamma_{12}$, both referred to the principal axes for load step n, yield a rotation of the principal stress axes and of the principal strain axes, denoted as $\Delta\theta_\sigma$ and $\Delta\theta_\varepsilon$, respectively. These rotations are obtained from the relations

$$\tan 2\Delta\theta_\sigma = \frac{\Delta\tau_{12}}{\sigma_{11} - \sigma_{22}}, \quad \tan 2\Delta\theta_\varepsilon = \frac{\Delta\gamma_{12}}{2(\varepsilon_{11} - \varepsilon_{22})}, \qquad (3.168)$$

where $\sigma_{11} = \sigma_1 + \Delta\sigma_{11}$ and $\sigma_{22} = \sigma_2 + \Delta\sigma_{22}$. Coaxiality at the end of load step $n+1$ requires $\Delta\theta_\sigma$ to coincide with $\Delta\theta_\varepsilon$. Substitution of $\Delta\tau_{12} = G_T \Delta\gamma_{12}$ into the condition $\tan 2\Delta\theta_\sigma = \tan 2\Delta\theta_\varepsilon$ yields

$$G_T = \frac{\sigma_{11} - \sigma_{22}}{2(\varepsilon_{11} - \varepsilon_{22})}. \qquad (3.169)$$

Hence, there is no need for the introduction of an empirical expression for the shear modulus of cracked concrete. Unlike as for non-orthogonal crack models, only the most recent cracks need to be taken into account. They are parallel to the actual principal axes. Information on previously formed cracks does not have to be stored. The advantage of this approach is its computational simplicity. However, it does not reflect the physical nature of cracking because cracks obviously do not rotate at a point [Bažant (1983)]. Nevertheless, disregard of previously formed cracks can be justified by the experimental evidence that upon formation of a new crack at a material point the previously formed cracks at this point tend to close. Thus, rotating crack models can be viewed as a special case of fixed non-orthogonal crack models. It is obtained by setting the threshold angle equal to zero and by deactivating the previously formed cracks. Application of the concept of strain decomposition allows to combine rotating crack models with elastic-plastic and time-dependent material models.

Crack models based on the theory of plasticity. Combination of smeared crack models based on the theory of elasticity (fixed orthogonal crack models, fixed non-orthogonal crack models, rotating crack models) with constitutive models for concrete in compression, based on the theory of plasticity, may lead to numerical problems for multiaxial stress states. With regards to a plane state of stress, this may be the case for tension in one principal direction and compression in the other. For such mixed stress states both cracking and crushing may occur at a particular material point. However, the local iteration on different constitutive models for cracking and crushing may result in an oscillating numerical process [Feenstra (1993a)]. This is the reason why a unified treatment of the material behavior of concrete at a particular material point is preferable to the use of two different theories. An additional motivation for using an elastic-plastic material model for both the compressive and the tensile region is the availability of efficient and robust return mapping algorithms.

Within the framework of the theory of plasticity the maximum tensile stress criterion (Rankine criterion),

$$f_i = \sigma_i - \sigma_T = 0 , \qquad (3.170)$$

serves as the yield surface. In (3.170) σ_i denotes a principal stress; σ_T is the tensile strength of concrete. For three-dimensional stress states, $i = 1, 2, 3$. For plane stress states, $i = 1, 2$. Fig. 3.15 shows the yield curve for such a state of stress. For simplicity, the tensile strength of concrete can be neglected in (3.170) [Crisfield (1989)]. If this is not done, an appropriate softening law must be specified [Feenstra (1993a), Welscher (1993)]. If σ_T is defined as a function of a single internal variable κ_T, i.e.,

$$\sigma_T = \sigma_T(\kappa_T) , \qquad (3.171)$$

then the softening model will be referred to as isotropic. Since κ_T is a measure for the accumulated damage of the material, it is also denoted as the internal damage parameter.

3.2 Update of the Stresses

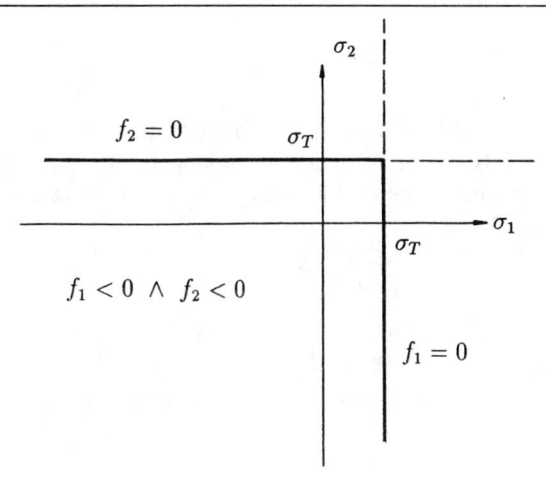

Figure 3.15: Maximum tensile stress criterion (Rankine criterion) for a plane state of stress

In [Feenstra (1993a)] $\sigma_T(\kappa_T)$ is given either by the linear softening law

$$\sigma_T(\kappa_T) = \bar{\sigma}_T \left(1 - \frac{\kappa_T}{\kappa_T^u}\right) \tag{3.172}$$

or by the exponential softening law

$$\sigma_T(\kappa_T) = \bar{\sigma}_T \exp\left(-\frac{\kappa_T}{\kappa_T^u}\right), \tag{3.173}$$

where $\bar{\sigma}_T = \sigma_T(\kappa_T = 0)$ denotes the tensile strength and κ_T^u is the value of κ_T, for which $\sigma_T(\kappa_T) = 0$, indicating that the capacity of concrete to carry tensile stresses has been reduced to zero. $\dot{\kappa}_T$ is defined in terms of the rate of the specific plastic work for a general state of stress and strain,

$$\dot{W}^p = \boldsymbol{\sigma} : \dot{\boldsymbol{\varepsilon}}^p, \tag{3.174}$$

which is assumed to be equal to the rate of the specific plastic work

$$\dot{\bar{W}}^p = \sigma_T(\kappa_T)\dot{\kappa}_T \tag{3.175}$$

done by the unaxial tensile stress σ_T on the rate of the internal variable κ_T. The specific plastic work W^p, corresponding to (3.174) and (3.175), is obtained as

$$W^p = \int_{t=0}^{t=\infty} \dot{W}^p dt, \quad \bar{W}^p = \int_{t=0}^{t=\infty} \dot{\bar{W}}^p dt = \int_{\kappa=0}^{\kappa=\infty} \sigma_T(\kappa_T) d\kappa_T. \tag{3.176}$$

The plastic work per unit volume, W^p (or $\bar{W}^p$), is related to the specific fracture energy G_f by

$$W^p = \frac{G_f}{L}, \qquad (3.177)$$

where L denotes an equivalent length. (3.177) is based on the assumption of a uniform distribution of W^p over the equivalent length. By analogy to (3.144) and (3.145), for a linear softening model the equivalent length must obey the inequality

$$\frac{2G_f E}{\bar{\sigma}_T^2} > L. \qquad (3.178)$$

In [Welscher (1993)] the internal variable κ_T has the physical meaning of the crack strain ε^f. According to (3.140), ε^f depends on L.

Substitution of (3.171) into (3.170) yields the maximum tensile stress criterion with isotropic softening:

$$f_i = \sigma_i - \sigma_T(\kappa_T) = 0. \qquad (3.179)$$

The yield surfaces of (3.179) are defined in the principal stress space. Consequently, in [Welscher (1993)] the return mapping algorithm is also formulated in the principal stress space. The principal trial stresses $\sigma_{i,n+1}^{Trial}$ and the corresponding principal directions $\mathbf{n}_{i,n+1}^{Trial}$ are obtained by solving the following eigenproblem:

$$\boldsymbol{\sigma}_{n+1}^{Trial} \mathbf{n}_{i,n+1}^{Trial} = \sigma_{i,n+1}^{Trial} \mathbf{n}_{i,n+1}^{Trial}, \qquad (3.180)$$

where $\boldsymbol{\sigma}_{n+1}^{trial}$ are the trial stresses defined in (3.98$_1$). For isotropic yield functions the principal directions of the trial stresses are equal to the ones of the stresses returned to the yield surface. Therefore, the return mapping of the trial stresses and the computation of the consistent tangent moduli can be performed in the principal stress space with fixed principal axes. Subsequently, the principal stresses are transformed back to the global coordinate system [Welscher (1993), Meschke (1994)]. Following the respective algorithms proposed in [Simo (1988b)], an iterative procedure is used to determine which yield functions of the non-smooth yield surface are active.

Instead of referring the yield functions f_i to principal stress directions, in [Feenstra (1993a), Feenstra (1993b)] the maximum tensile stress criterion (3.170) is formulated in terms of a global coordinate system x, y, z. For the special case of plane stress

$$f = \frac{1}{2}\left(\mathbf{a}^T\boldsymbol{\sigma} + \sqrt{\boldsymbol{\sigma}^T\mathbf{P}\boldsymbol{\sigma}}\right) - \sigma_T(\kappa_T) = 0, \qquad (3.181)$$

where

$$\mathbf{a}^T = \lfloor 1 \ 1 \ 0 \rfloor,$$
$$\boldsymbol{\sigma}^T = \lfloor \sigma_x \ \sigma_y \ \tau_{xy} \rfloor,$$
$$\mathbf{P} = \begin{bmatrix} 1 & -1 & 0 \\ -1 & 1 & 0 \\ 0 & 0 & 4 \end{bmatrix}. \qquad (3.182)$$

3.2 Update of the Stresses

Models with isotropic softening are useful if cracking occurs in one direction only, as is the case for mixed biaxial states of stress. However, for biaxial stress states with tension in both principal directions, the computed material behavior may deviate from the experimentally observed material behavior. For a biaxial stress state, for which the tensile stress σ_1 is already in the softening region whereas the tensile stress σ_2 is still in the pre-peak region, the reduction of the tensile stress $\sigma_1(\kappa_T)$ in consequence of the accumulated internal variable $\kappa_T > 0$ also yields a reduction of the tensile strength in the perpendicular direction. However, the reduction of the tensile strength in the direction parallel to existing cracks is not supported by experimental evidence. In this case either a kinematic softening rule should be chosen [Feenstra (1993b)] or internal variables depending on the direction of the accumulated damage should be introduced [Welscher (1993)].

3.2.6 Implementation of Material Models

In general, commercial nonlinear finite element programs offer the possibility to add user-defined subroutines. Within the framework of the finite element method the constitutive model is implemented at the integration point level. Thus, implementation of a nonlinear constitutive law into a computer program for nonlinear finite element analysis is restricted to step 2 of the global solution procedure summarized in subsection 3.1.5. Therefore, it is easy to design the skeleton of a subroutine for a user-defined material law. Fig.3.16 shows the skeleton of the subroutine HYPELA of the finite element program MARC [MARC (1992)]. HYPELA was primarily designed for adding a user-defined hypoelastic material model to MARC. However, since internal variables may be considered, it is also possible to define elastic-plastic constitutive models.

The arguments of HYPELA are subdivided into input and output variables. Input variables are known when HYPELA is called. Output variables must be updated in HYPELA. They are returned to the calling subroutine.

Input variables are

- the number of the current finite element, assigned by the user, and the internal number of the respective finite element, N(1) and N(2),

- the identification number for the current material, MATS,

- the number of the integration point in the current element, NN,

- for elements, which are subdivided into several layers over the thickness, the number of the current layer, KC,

- the number of rows or columns of the tangent material matrix, NGENS,

- the number of normal stress components, NDI,

- the number of shear stress components, NSHEAR,

```
      SUBROUTINE HYPELA(CT,G,E,DE,S,TEMP,DTEMP,NGENS,N,NN,KC,MATS,
     *                  NDI,NSHEAR)
C
C* * * * * *
C
C     USER SUBROUTINE TO DEFINE YOUNG'S MODULUS AND POISSON'S RATIO
C     AS FUNCTION OF STRESS IN NON-LINEAR ELASTIC SMALL STRAIN
C     MATERIAL.
C
C     CT        STRESS STRAIN LAW TO BE FORMED BY USER
C     G         CHANGE IN STRESS DUE TO TEMPERATURE EFFECTS
C     E         TOTAL STRAIN
C     DE        INCREMENT OF STRAIN
C     S         STRESS - SHOULD BE UPDATED BY USER
C     TEMP      STATE VARIABLES
C     DTEMP     INCREMENT OF STATE VARIABLES
C     NGENS     SIZE OF STRESS - STRAIN LAW
C     N         ELEMENT NUMBER
C     NN        INTEGRATION POINT NUMBER
C     KC        LAYER NUMBER
C     MATS      MATERIAL I.D.
C     NDI       NUMBER OF DIRECT COMPONENTS
C     NSHEAR    NUMBER OF SHEAR COMPONENTS
C
C* * * * * *
C
      IMPLICIT REAL*8 (A-H,O-Z)
C
      DIMENSION CT(NGENS,NGENS),G(NGENS),E(NGENS),DE(NGENS),S(NGENS)
      DIMENSION TEMP(1),DTEMP(1),N(2)
C
C USER CODING
C
      RETURN
      END
```

Figure 3.16: Skeleton of a user-defined subroutine for the definition of a constitutive law

- the final (converged) values of the total strains referring to the previous increment n, denoted as E,
- the final (converged) values of the total stress referring to the previous increment n, denoted as S,
- the final (converged) values of the total internal variables referring to the previous increment n, denoted as TEMP, and
- the total incremental strains for the current increment $n+1$, denoted as DE.

Output variables are

- the total stresses referring to increment $n+1$, denoted as S,
- the total incremental internal variables for increment $n+1$, denoted as DTEMP,
- the tangent material stiffness matrix, denoted as CT, and
- if present, initial stresses, denoted as G.

3.3 Formulations for Time-Dependent Problems

3.3.1 Introduction

Consideration of time-dependent effects requires discretization with respect to time in addition to spatial discretization. This requirement is met by subdividing the total time interval under consideration into a sufficiently large number of time steps

$$\Delta t_i = t_i - t_{i-1} , \qquad (3.183)$$

where t_i, $i = 1, \ldots, m$, denotes discrete points of time. Because of the time dependence of the material behavior of concrete, the state of stress in a structure is time-dependent even if the external loads and the temperature are kept constant.

Fig.3.17 shows the stress history corresponding to a given load history. The figure illustrates continuous changes of stress at constant loads within the time intervals as well as discontinuous stress changes resulting from load increments at discrete points of time.

Usually, algorithms for consideration of the time-dependent behavior of concrete structures are based on the additive decomposition of the total strains according to (2.193). Hence, the total strain at $t = t_i$ is given as

$$\varepsilon(t_i) = \varepsilon^e(t_i) + \varepsilon^p(t_i) + \varepsilon^{cr}(t_i) + \varepsilon^{sh}(t_i) + \varepsilon^T(t_i) . \qquad (3.184)$$

The difference between the additive decomposition of the strains for consideration of time-independent material behavior (see equation (3.77)) and of time-dependent constitutive behavior is reflected by the creep strains and shrinkage strains in (3.184). Hence, in the constitutive relations (3.78) and (3.81) the elastic strains must be computed from (3.184) if creep and shrinkage are taken into account.

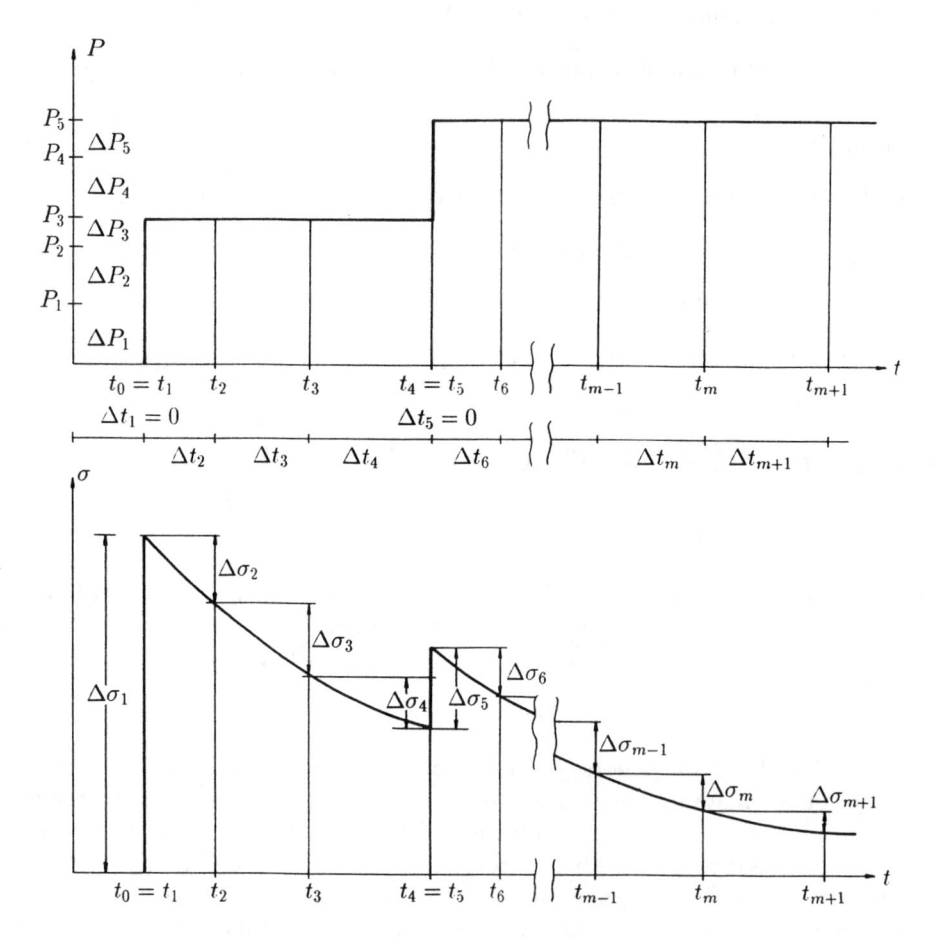

Figure 3.17: Load history and corresponding stress history for a given time discretization

For the formulation of the global solution algorithm within the framework of the finite element method it makes no fundamental difference whether changes of the state of stress and the state of strain are caused by the application of static external loads or by the time-dependence of the material. The latter requires consideration of creep strains and shrinkage strains. Similar to temperature-induced strains, creep strains and shrinkage strains are treated as initial strains. These strains are

3.3 Formulations for Time-Dependent Problems

computed on the integration point level. Thus, the modifications of an existing computer program for nonlinear time-independent finite element analysis to allow consideration of time-dependent material behavior are restricted to step 2 of the global solution procedure summarized in subsection 3.1.5. Once the time-dependent strains are known, the elastic strains and, subsequently, the stresses are updated analogous to the respective updates for time-independent analyses. Thereafter, a check for material failure is made.

The solution procedure for rate-type creep laws is similar to the one for rate-independent plasticity. The rate law for the creep strains is integrated by means of a suitable numerical integration scheme. However, for concrete the creep strains are usually not computed from a rate-type creep law. It is by far more popular to compute creep strains from one of the creep laws described in section 2.4. For this reason, only algorithms based on a creep compliance function will be considered in the following. Moreover, only creep compliances which are independent of the acting stresses will be taken into account. Creep strains determined from such creep compliances are proportional to the acting stresses. Several algorithms are well suited for the computation of creep strains. In contrast to computation of such stress-dependent strains, computation of stress-independent strains such as shrinkage strains and temperature-induced strains is straightforward.

Algorithms for determination of creep strains are based on the approximation of a gradually changing stress history as shown in Fig.3.17 by a piecewise constant stress history with stress changes restricted to discrete points of time. Application of external loads is restricted to the time points t_i, $i = 1, \ldots, m$. In the analysis such load increments are treated as time steps of zero length. E.g., in Fig.3.17 external loads are applied at the time points t_1 and t_5. Consequently, $\Delta t_1 = 0$ and $\Delta t_5 = 0$. Algorithms for computation of creep strains differ by

- the way of consideration of the stress history (algorithms with or without storage of the complete stress history) and by

- their numerical accuracy (first-order and second-order algorithms).

For sufficiently low load levels the rate of the creep strains decreases with increasing time after the application of the load. Therefore, the length of the time steps may increase with increasing time after loading. In [Bažant (1982b)] time steps are suggested which are constant in the logarithmic time scale. If an algorithm of second-order accuracy is employed, then four time steps per decade, with an initial time step of about 0.1 day, will yield sufficiently accurate results.

In contrast to algorithms for consideration of time-dependent material behavior in nonlinear finite element programs, effective modulus methods are particularly simple tools for obtaining crude approximations of time-dependent changes of stresses and strains in concrete structures. The term effective-modulus indicates that the modulus of elasticity in the constitutive relations for linear elasticity is replaced by an effective modulus to compute the changes of stresses and strains with the help of a finite element program based on linear elasticity. Obviously, cracking of concrete induced by creep and shrinkage is not included in such approximate analyses.

3.3.2 Algorithms for the Computation of Creep Strains

Algorithms with storage of the complete stress history

An algorithm of first-order accuracy with respect to the chosen time steps is obtained by approximating the actual stress history by a piecewise constant stress history, admitting stress changes $\Delta\boldsymbol{\sigma}_i \equiv \Delta\boldsymbol{\sigma}(t_i)$ only at discrete time points t_i, $i = 1, \ldots, m$. The stresses are assumed to be equal to their respective values at the beginning of the time step. Consequently, the history integral in (2.234) is approximated by the sum

$$\boldsymbol{\varepsilon}^{cr}(t_m) = \mathbf{A} \sum_{i=1}^{m-1} C(t_m, t_i)\Delta\boldsymbol{\sigma}(t_i) , \quad m \geq 2 . \tag{3.185}$$

Fig.3.17 illustrates a situation which is characterized by $t_0 = t_1$ and $\boldsymbol{\sigma}(t_0) \equiv \boldsymbol{\sigma}_0 = \mathbf{0}$. For such a situation $\Delta\boldsymbol{\sigma}_1 = \boldsymbol{\sigma}_1 - \boldsymbol{\sigma}_0$ represents the stress resulting from initial loading.

According to (3.185), determination of the creep strains at time t_m gives

$$\boldsymbol{\varepsilon}^{cr}(t_m) = \mathbf{A}\left[C(t_m, t_1)\Delta\boldsymbol{\sigma}(t_1) + C(t_m, t_2)\Delta\boldsymbol{\sigma}(t_2) + \ldots \right.$$
$$\left. \ldots + C(t_m, t_{m-1})\Delta\boldsymbol{\sigma}(t_{m-1})\right]. \tag{3.186}$$

By analogy, computation of the creep strains at time t_{m+1} yields

$$\boldsymbol{\varepsilon}^{cr}(t_{m+1}) = \mathbf{A}\left[C(t_{m+1}, t_1)\Delta\boldsymbol{\sigma}(t_1) + C(t_{m+1}, t_2)\Delta\boldsymbol{\sigma}(t_2) + \ldots \right.$$
$$\left. \ldots + C(t_{m+1}, t_{m-1})\Delta\boldsymbol{\sigma}(t_{m-1}) + C(t_{m+1}, t_m)\Delta\boldsymbol{\sigma}(t_m)\right]. \tag{3.187}$$

The evaluation of the creep strains at t_{m+1} is based only on known values of the stresses. Therefore, the respective algorithm represents an explicit solution procedure. Hence, computation of the incremental creep strains

$$\Delta\boldsymbol{\varepsilon}^{cr}_{m+1} = \boldsymbol{\varepsilon}^{cr}(t_{m+1}) - \boldsymbol{\varepsilon}^{cr}(t_m) , \tag{3.188}$$

accumulating in the time interval $\Delta t_{m+1} = t_{m+1} - t_m$, requires knowledge of the stress history from t_1 up to t_m. Thus, in a finite element analysis all previous stress changes $\Delta\boldsymbol{\sigma}(t_1), \Delta\boldsymbol{\sigma}(t_2), \ldots, \Delta\boldsymbol{\sigma}(t_m)$ at each integration point of each finite element must be stored.

The accuracy of this algorithm can be improved by replacing the explicit solution procedure for the approximation of the history integral in (2.234) by an implicit rule of second-order accuracy with respect to the chosen time steps. On the basis of the trapezoidal rule (3.185) is replaced by

$$\boldsymbol{\varepsilon}^{cr}(t_m) = \mathbf{A} \sum_{i=1}^{m} \frac{1}{2}[C(t_m, t_{i-1}) + C(t_m, t_i)]\Delta\boldsymbol{\sigma}(t_i) , \quad m \geq 2 . \tag{3.189}$$

3.3 Formulations for Time-Dependent Problems

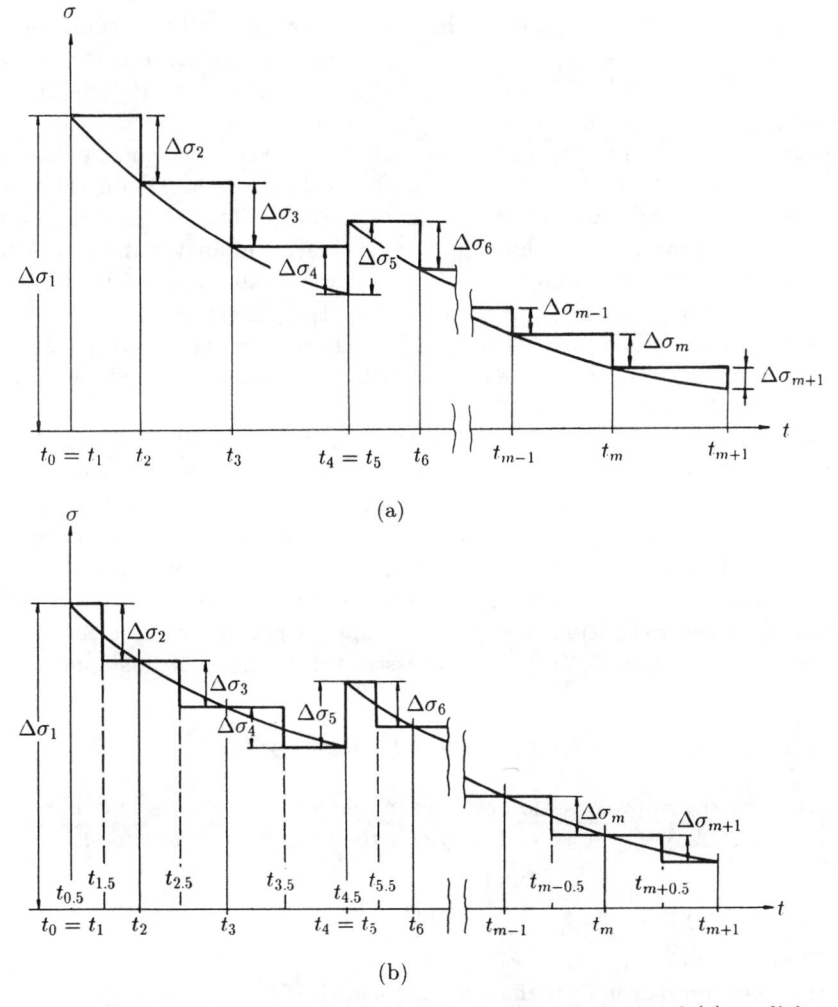

Figure 3.18: Approximation of a stress history on the basis of (a) explicit integration and (b) implicit integration

On the basis of the midpoint rule (3.185) is replaced by

$$\varepsilon^{cr}(t_m) = \mathbf{A} \sum_{i=1}^{m} C(t_m, t_{i-1/2}) \Delta \boldsymbol{\sigma}(t_i) , \quad m \geq 2 . \tag{3.190}$$

Fig.3.18 illustrates a situation which is characterized by $t_0 = t_1$ and $\sigma_0 = 0$. For such a situation $\Delta \boldsymbol{\sigma}_1$ represents the stress resulting from initial loading. Fig.3.18(a) shows the approximation of a stress history on the basis of the explicit integration scheme according to (3.185). Fig.3.18(b) illustrates the approximation of the same stress history on the basis of the implicit midpoint rule according to (3.190). Combined with an expansion of the creep compliance function into a Dirichlet series, the explicit numerical integration scheme was applied in [Kabir (1977), Van Zyl (1978), Kang (1980), Van Greunen (1983)]. The implicit midpoint rule was used in [Walter (1988)].

$\Delta\boldsymbol{\sigma}_{m+1}$ is not involved in the approximation of the creep strains at time t_{m+1} by means of (3.187). Hence, (3.187) represents an explicit numerical integration scheme. However, $\Delta\boldsymbol{\sigma}_{m+1}$ appears in the approximations of $\varepsilon^{cr}(t_{m+1})$ according to (3.189) and (3.190). Consequently, these equations represent implicit numerical integration schemes. Hence, for the case of implicit numerical integration schemes the creep compliance enters into the tangent stiffness matrix. Therefore, the approximations based on the implicit numerical integration schemes are more expensive. However, they are more accurate than the approximation by means of the explicit numerical integration scheme.

Algorithms requiring storage of the complete stress history have two drawbacks. Firstly, depending on the size of the problem, such storage may be computationally expensive. Secondly, computation of the creep strains requires lengthy summations. However, the advent of powerful workstations has reduced the significance of these drawbacks. An advantage of such algorithms is that any creep compliance function can be employed directly. Hence, there is no need to compute the coefficients of a Dirichlet series expansion, serving as an approximation of the creep compliance function (see equation (2.226)), and to assess the accuracy of this approximation.

Algorithms without storage of the complete stress history

Substituting the approximation of a given creep function by a Dirichlet series according to (2.226), e.g., into (3.185) yields the creep strains at time t_m:

$$\varepsilon^{cr}(t_m) \;=\; \mathbf{A}\sum_{i=1}^{m-1}\sum_{r=1}^{s} a_r(t_i)\left[1 - e^{-(t_m - t_i)/\tau_r}\right]\Delta\boldsymbol{\sigma}(t_i) \quad m \geq 2 \,. \quad (3.191)$$

Writing the summation over the time steps in (3.191) out in full, yields

$$\varepsilon^{cr}(t_m) \;=\; \mathbf{A}\sum_{r=1}^{s}\Delta\boldsymbol{\sigma}(t_1)a_r(t_1)\left[1 - e^{-(t_m - t_1)/\tau_r}\right]$$

$$+\; \mathbf{A}\sum_{r=1}^{s}\Delta\boldsymbol{\sigma}(t_2)a_r(t_2)\left[1 - e^{-(t_m - t_2)/\tau_r}\right] + \ldots$$

$$\ldots \;+\; \mathbf{A}\sum_{r=1}^{s}\Delta\boldsymbol{\sigma}(t_{m-1})a_r(t_{m-1})\left[1 - e^{-(t_m - t_{m-1})/\tau_r}\right]. \quad (3.192)$$

By analogy to (3.192), the creep strains at time t_{m+1} are given as

$$\varepsilon^{cr}(t_{m+1}) \;=\; \mathbf{A}\sum_{r=1}^{s}\Delta\boldsymbol{\sigma}(t_1)a_r(t_1)\left[1 - e^{-(t_{m+1} - t_1)/\tau_r}\right]$$

3.3 Formulations for Time-Dependent Problems

$$+ \; \mathbf{A} \sum_{r=1}^{s} \Delta\boldsymbol{\sigma}(t_2) a_r(t_2) \left[1 - e^{-(t_{m+1}-t_2)/\tau_r}\right] + \ldots$$

$$\ldots \; + \; \mathbf{A} \sum_{r=1}^{s} \Delta\boldsymbol{\sigma}(t_{m-1}) a_r(t_{m-1}) \left[1 - e^{-(t_{m+1}-t_{m-1})/\tau_r}\right]$$

$$+ \; \mathbf{A} \sum_{r=1}^{s} \Delta\boldsymbol{\sigma}(t_m) a_r(t_m) \left[1 - e^{-(t_{m+1}-t_m)/\tau_r}\right]. \tag{3.193}$$

Subtraction of (3.192) from (3.193) results in the incremental creep strains $\Delta\varepsilon^{cr}_{m+1}$:

$$\Delta\varepsilon^{cr}_{m+1} \;=\; \mathbf{A} \left[\sum_{r=1}^{s} a_r(t_1) \Delta\boldsymbol{\sigma}(t_1) e^{-(t_m-t_1)/\tau_r}\right] \left[1 - e^{-\Delta t_{m+1}/\tau_r}\right]$$

$$+ \; \mathbf{A} \left[\sum_{r=1}^{s} a_r(t_2) \Delta\boldsymbol{\sigma}(t_2) e^{-(t_m-t_2)/\tau_r}\right] \left[1 - e^{-\Delta t_{m+1}/\tau_r}\right] + \ldots$$

$$\ldots \; + \; \mathbf{A} \left[\sum_{r=1}^{s} a_r(t_{m-1}) \Delta\boldsymbol{\sigma}(t_{m-1}) e^{-(t_m-t_{m-1})/\tau_r}\right] \left[1 - e^{-\Delta t_{m+1}/\tau_r}\right]$$

$$+ \; \mathbf{A} \left[\sum_{r=1}^{s} a_r(t_m) \Delta\boldsymbol{\sigma}(t_m)\right] \left[1 - e^{-\Delta t_{m+1}/\tau_r}\right]. \tag{3.194}$$

With the help of the abbreviations

$$\begin{aligned} a_r^*(t_1) &= \Delta\boldsymbol{\sigma}(t_1) a_r(t_1) \\ a_r^*(t_m) &= a_r^*(t_{m-1}) e^{-\Delta t_m/\tau_r} + \Delta\boldsymbol{\sigma}(t_m) a_r(t_m) \end{aligned} \tag{3.195}$$

(3.194) can be reformulated as

$$\Delta\varepsilon^{cr}_{m+1} = \mathbf{A} \sum_{r=1}^{s} a_r^*(t_m) \left[1 - e^{-\Delta t_{m+1}/\tau_r}\right]. \tag{3.196}$$

Use of (3.196) avoids storage of the complete stress history. The latter is contained in the coefficients $a_r^*(t_m)$ which are computed by means of the recursive formula (3.195). This algorithm was applied in [Kabir (1977), Van Zyl (1978), Kang (1980), Van Greunen (1983)]. Because of the recursive formulation, the algorithm is economic as far as the required storage capacity is concerned. However, it requires determination of the coefficients $a_r(t_i)$ for all discrete points of time t_i, $i = 1, \ldots, m$. The numerical accuracy could be improved by substituting (2.226) into (3.189) or (3.190) instead of into (3.185), i.e., by a reformulation based on the trapezoidal or the midpoint rule. The disadvantage of such a formulation is that the current stress change, which is not known, appears in the approximation of the incremental creep strains.

Effective modulus methods

Effective modulus methods are characterized by considering the time interval between loading at t_0 and the current point of time t as a single time step $\Delta t = t - t_0$. The external loads are assumed to be constant in the entire time interval. The strains at time t follow from (2.206) and (2.207) as

$$\varepsilon(t) = J(t, t_0)\sigma(t_0) + \int_{t_0^+}^{t} J(t, \tau) d\sigma(\tau) + \varepsilon^0(t) \ . \tag{3.197}$$

The first term on the right-hand side of (3.197) represents the strain produced by the stress which has been applied at t_0. This stress is considered to be constant in the time interval Δt. The superscript + in t_0^+ indicates that the time interval (t_0^+, t), defined by the integration limits of the integral in (3.197), does not contain the discrete point of time t_0 at which the stress $\sigma(t_0)$ has been applied. Consequently, the second term on the right-hand side of (3.197) refers to the strain resulting from the change of stress after application of $\sigma(t_0)$. Substitution of (2.208) into (3.197), use of (2.209) and disregard of aging of concrete, i.e., setting $E^C(t) = E^C(t_0) = E =$ const., yields [Trost (1967)]

$$\varepsilon(t) = \frac{1 + \phi(t, t_0)}{E}\sigma(t_0) + \frac{\sigma(t) - \sigma(t_0)}{E}[1 + \phi(t, t_0)\rho(t, t_0)] + \varepsilon^0(t) \ , \tag{3.198}$$

where $\rho(t, t_0)$, termed as relaxation coefficient, is given as

$$\rho(t, t_0) = \frac{\int_{t_0^+}^{t} \phi(t, \tau) d\sigma(\tau)}{\phi(t, t_0)[\sigma(t) - \sigma(t_0)]} \ . \tag{3.199}$$

It is noted that $\rho \leq 1$. ρ accounts for the decrease of the creep capacity of concrete for stress changes occurring at $t > t_0$. Thus, ρ is actually a correction factor for consideration of aging. Therefore, it is more appropriate to refer to ρ as the aging coefficient. Rewriting (3.198) as

$$\Delta\sigma(t) = E_{eff}\left[\Delta\varepsilon(t) - \Delta\bar{\varepsilon}(t) - \varepsilon^0(t)\right] \ , \tag{3.200}$$

where $\Delta\sigma(t) = \sigma(t) - \sigma(t_0)$, $\Delta\varepsilon(t) = \varepsilon(t) - \varepsilon(t_0)$ and

$$E_{eff} = \frac{E}{1 + \rho(t, t_0)\phi(t, t_0)} \quad \text{and} \quad \Delta\bar{\varepsilon}(t) = \frac{\sigma(t_0)}{E}\phi(t, t_0) \ , \tag{3.201}$$

allows computation of the change of stress $\Delta\sigma(t)$ in the time interval Δt on the basis of linear-elastic finite element analysis by replacing the modulus of elasticity by E_{eff} and treating the strains $\Delta\bar{\varepsilon}(t) + \varepsilon^0(t)$ as initial strains. Different ages of concrete in a structure can be considered by adjusting E_{eff} according to the age of the respective concrete at loading.

Bažant [Bažant (1972)] has extended this method by accounting for the variability of the elastic modulus in consequence of aging. $\sigma(t)$ is assumed to vary linearly with the relaxation function $R(t, t_0)$. Thus,

3.3 Formulations for Time-Dependent Problems

$$\sigma(t) = aE(t_0) + bR(t, t_0) , \tag{3.202}$$

where a and b are arbitrary constants. Because of $R(t_0, t_0) = E(t_0)$,

$$\varepsilon(t_0) = a + b . \tag{3.203}$$

Inserting

$$d\sigma(\tau) = b\frac{\partial R(\tau, t_0)}{\partial \tau} d\tau , \tag{3.204}$$

obtained from (3.202), into (3.197) yields

$$\varepsilon(t) = J(t, t_0)\sigma(t_0) + b \int_{t_0^+}^{t} J(t, \tau) \frac{\partial R(t, t_0)}{\partial \tau} d\tau + \varepsilon^0(t) . \tag{3.205}$$

The integral in (3.205) is computed by setting $\varepsilon(t) = 1$ and $\varepsilon^0(t) = 0$ in (3.197). In this case, $\sigma(t_0) = E(t_0)$ and, following from (2.212), $d\sigma(\tau) = (\partial R(\tau, t_0))/(\partial \tau) \, d\tau$. Consequently, (3.197) becomes

$$1 = J(t, t_0)E(t_0) + \int_{t_0^+}^{t} J(t, \tau) \frac{\partial R(t, t_0)}{\partial \tau} d\tau . \tag{3.206}$$

Making use of (3.206), (2.208) and (2.209), (3.205) can be written as

$$\varepsilon(t) = \varepsilon(t_0) + a\phi(t, t_0) + \varepsilon^0(t) . \tag{3.207}$$

The starting point for the derivation of (3.207) was (3.202). It follows that a stress history which depends linearly on the relaxation function yields a strain history which depends linearly on the creep coefficient. Frequently, the assumption of a strain history according to (3.207) is a good approximation of the actual strain history. Computation of the constants a and b from (3.202) and (3.203) allows formulation of a relationship similar to (3.200), where E_{eff} is given as

$$E_{eff} = \frac{E(t_0) - R(t, t_0)}{\phi(t, t_0)} . \tag{3.208}$$

Replacing E in (3.201) by $E(t_0)$ allows writing the expression for the effective modulus in a form similar to (3.201). The aging coefficient is then obtained as

$$\rho(t, t_0) = \frac{E(t_0)}{E(t_0) - R(t, t_0)} - \frac{1}{\phi(t, t_0)} . \tag{3.209}$$

For a given creep compliance function $J(t, t_0)$ an approximation of the relaxation function can be determined by means of (2.214).

It is repeated that effective modulus methods result in crude approximations of time-dependent changes of stresses and strains. The strains are obtained from a linear-elastic stress-strain relationship. Therefore, nonlinear effects resulting from cracking of concrete in consequence of creep and shrinkage cannot be considered by such methods.

3.3.3 Algorithm for Consideration of Relaxation of the Prestressing Steel

An algorithm accounting for the time-dependent loss of the tendon stress because of relaxation of the prestressing steel was reported in [Kang (1980)]. Fig.3.19 refers to this algorithm which was employed in [Van Zyl (1978), Van Greunen (1983), Walter (1988)]. The algorithm is based on the empirical formula (2.252). The evaluation of the loss of the tendon stress because of relaxation of the prestressing steel is based on the assumption of a constant tendon strain within a single time step.

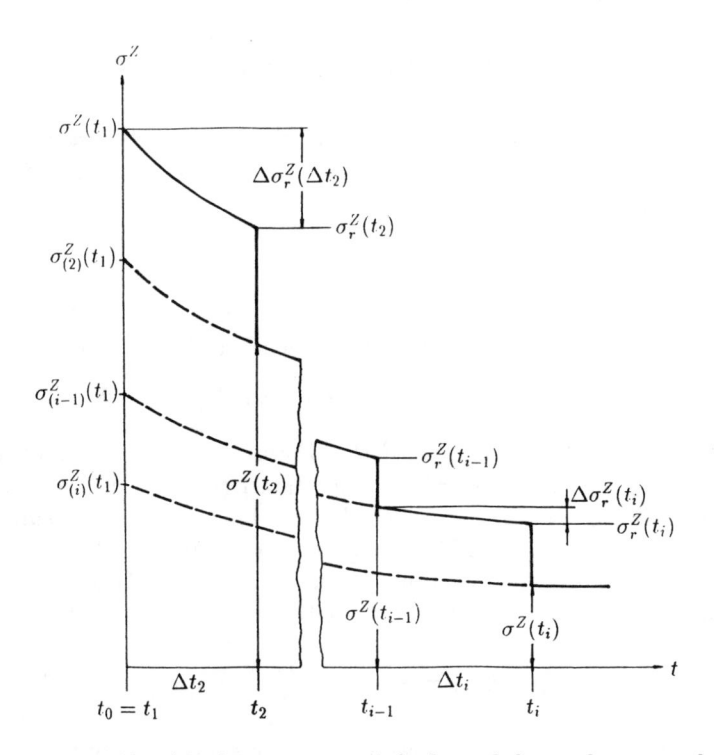

Figure 3.19: Algorithmic treatment of the loss of the tendon stress because of relaxation of the prestressing steel

It is assumed that the tendon stress σ^Z at the point of time t_1, at the beginning of the first nonzero time step, $\Delta t_2 = t_2 - t_1$, is known. It consists of the stress resulting from prestressing and of stresses generated by changes of the tendon strain after prestressing, caused by the application of external loads. These stresses are part of the result from nonlinear finite element analysis for time t_1. Assuming the tendon strain to be constant within the time step Δt_2, the tendon stress σ_r^Z at time t_2 at the end of the first nonzero time step is computed by means of (2.252). The subscript r in σ_r^Z indicates that for computation of the tendon stress at t_2 so far only the loss of the stress from relaxation of the prestressing steel has been taken into account. Subsequently, the changes of the tendon stress in consequence of creep and shrinkage of concrete are computed, yielding the final tendon stress $\sigma^Z(t_2)$.

(2.252) relates the initial tendon stress at prestressing to the stress at the current point of time. However, because of other changes of the tendon stress, considered at discrete points of time, (2.252) cannot be applied directly for computation of the loss of the tendon stress because of relaxation of the prestressing steel. In Fig.3.19 these points are the end points of the time intervals. For the purpose of applying (2.252) for the time interval Δt_2, a fictitious initial tendon stress $\sigma^Z_{(2)}(t_1)$ is computed by inserting the value of $\sigma^Z(t_2)$ for $\sigma^Z(t)$ in (2.252) and solving for $\sigma^Z(t_0) \equiv \sigma^Z_{(2)}(t_1)$. The so-obtained value for the fictitious initial tendon stress is then inserted in (2.252) for $\sigma^Z(t_0)$. Then, (2.252) is solved for $\sigma^Z(t) \equiv \sigma^Z_r(t_3)$. The total loss of the tendon stress because of relaxation of the prestressing steel, measured from the point of time of application of the prestressing force until the current point of time, i.e., from t_1 to t_i, is obtained as

$$\sum_{j=2}^{i} \Delta \sigma^Z_r(\Delta t_j) = \sum_{j=2}^{i} \left[\sigma^Z(t_{j-1}) - \sigma^Z_r(t_j) \right] . \tag{3.210}$$

3.4 Finite Element Modelling of Reinforced and Prestressed Concrete

3.4.1 Introduction

Within the framework of the finite element method reinforced concrete can be represented either

(a) by superposition of the material models for the constituent parts of this composite material, i.e., for concrete and for the reinforcing steel, or

(b) by a constitutive model for the composite material, considering reinforced concrete as a continuum on the macrolevel.

Because of their wider range of applicability, models of type (a) are more popular. The finite element method is well suited for the superposition of the material models for the constituent parts of a composite material. Material models of this type can be employed for virtually all kinds of reinforced concrete structures. Depending on the type of the problem to be solved, concrete is represented by three-dimensional elements, shell or plate elements, plane stress or plane strain elements, axisymmetric elements or beam elements. The reinforcement is modelled either

(a1) by separate truss or beam elements ($\rightarrow$ discrete representation of the reinforcement) or

(a2) by separate elements of identical type as the concrete elements, which are superimposed on the latter ($\rightarrow$ embedded representation) or (a3) by distribution of the reinforcement to thin layers of equivalent thickness ($\rightarrow$ distributed representation).

Superposition of concrete and reinforcing steel to represent reinforced concrete requires constitutive models to account for bond and dowel action on the concrete-steel interface. However, often perfect bond between concrete and steel is assumed to be preserved up to the ultimate load. Moreover, in most finite element analyses dowel action is disregarded.

Perfect bond up to structural failure is the most popular assumption for all types of representation of the reinforcement. For perfect bond the displacements of the reinforcing bars coincide with the displacements of the adjacent concrete. Bond slip between concrete and steel remains small as long as the concrete sections between the lugs of deformed bars do not fail (see subsection 1.4.1). Therefore, at least for static loading, frequently the influence of bond slip on the overall structural behavior is relatively small for a relatively large range of the load intensity. A satisfactory degree of correspondence between computed and measured displacements of reinforced concrete structures often can be obtained with material models neglecting bond slip [Stevens (1991)]. This may not hold for dynamic loading, for which usually progressive deterioration of bond is observed.

Especially for analytical investigations of structural details, such as joints of members, an adequate representation of bond behavior and dowel action may be necessary. Discrete representation of the reinforcement allows modelling of bond and dowel action by means of special bond elements connecting adjacent nodes of concrete and steel elements. The distributed representation and the embedded representation of the reinforcement, however, do not permit the use of bond elements, because the displacements of concrete and steel at the interface are presupposed to be the same. Consequently, the effect of bond slip can only be accounted for implicitly by modifying the constitutive relations for concrete or steel.

In constitutive models of type (b) the material behavior of reinforced concrete on the macrolevel is described such as if this composite material was a single material. Constitutive models of this type are essentially based on the results of experiments on reinforced concrete panels. Vecchio and Collins, e.g., have developed the so-called "modified compression field theory" [Vecchio (1986), Vecchio (1988)] which is based on their experiments [Vecchio (1982)]. Since reinforced concrete is treated as a single material, neither the reinforcement nor the steel-concrete interaction needs to be modelled separately. However, currently such models are only available for biaxial stress states [Cervenka (1985), Vecchio (1986), Stevens (1987), Vecchio (1988)]. Moreover, the reinforcement must be distributed uniformly.

These restrictions are the reason for concentration on material models of type (a). Constitutive models for concrete, reinforcing steel and prestressing steel were presented in chapter 2. The implementation of nonlinear material models into a finite element program was described in previous sections of this chapter. In this section, finite elements for concrete and for the reinforcement will be presented. In addition, constitutive models to account for the behavior of the concrete-steel interface will be addressed briefly.

3.4.2 Finite Elements for Concrete

Three-dimensional Elements

Popular three-dimensional finite elements for discretization of concrete structures are isoparametric linear 8-node and quadratic 20-node brick elements. The parametric concept is characterized by mapping so-called "parent elements" of simple geometric form onto elements of the desired geometric shape. The parent elements are defined in a system of normalized coordinates ξ_1, ξ_2, ξ_3. The actual finite elements are embedded in a global Cartesian system of reference. In this way, finite elements of a more general geometric shape are obtained.

Isoparametric elements are a special case of parametric elements. They are characterized by identical interpolation functions for the parametric mapping of parent elements and for the description of the element displacements. Hence, for isoparametric elements the coordinates $\mathbf{x}^e$ of an arbitrary point of element e are obtained by replacing the vector of the nodal displacements $\mathbf{q}^e$ in (3.38) by the vector of the coordinates of the node points of this element. The isoparametric mapping is illustrated in Fig.3.20. The requirement of uniqueness of the mapping function $\mathbf{f}$ places some restrictions on $\mathbf{x}^e = \mathbf{f}(\boldsymbol{\xi}^e)$ [Bathe (1982), Zienkiewicz (1989,1991)].

The displacement vector of an arbitrary point of a three-dimensional brick element e is given as

$$\mathbf{u}^e = \lfloor u_1 \; u_2 \; u_3 \rfloor^T . \tag{3.211}$$

The vector of the nodal displacements reads as

$$\mathbf{q}^e = \lfloor u_1^{(1)} \; u_2^{(1)} \; u_3^{(1)} \; u_1^{(2)} \; \ldots \; u_3^{(n)} \rfloor^T . \tag{3.212}$$

The matrix of the shape functions $\mathbf{N}^e$, employed for both the mapping of the parent elements and the description of the element displacements, is given as

$$\mathbf{N}^e = \begin{bmatrix} \mathbf{I} N^{(1)} & \mathbf{I} N^{(2)} & \ldots & \mathbf{I} N^{(n)} \end{bmatrix} , \tag{3.213}$$

where $\mathbf{I}$ represents a 3×3 unit matrix and n denotes the number of nodes of element e. The subscripts in (3.212) label the coordinate directions. The superscripts on the right-hand side of (3.212) and (3.213) denote the node points of the element. The shape functions $N^{(i)}$ are formulated in terms of the local coordinates. For the linear isoparametric 8-node serendipity element they are given as

$$N^{(i)} = \frac{1}{8}(1 + \xi_1^{(i)}\xi_1)(1 + \xi_2^{(i)}\xi_2)(1 + \xi_3^{(i)}\xi_3) , \qquad i = 1, 2, \ldots, 8 . \tag{3.214}$$

For the quadratic isoparametric 20-node serendipity element the shape functions for the corner nodes are given as

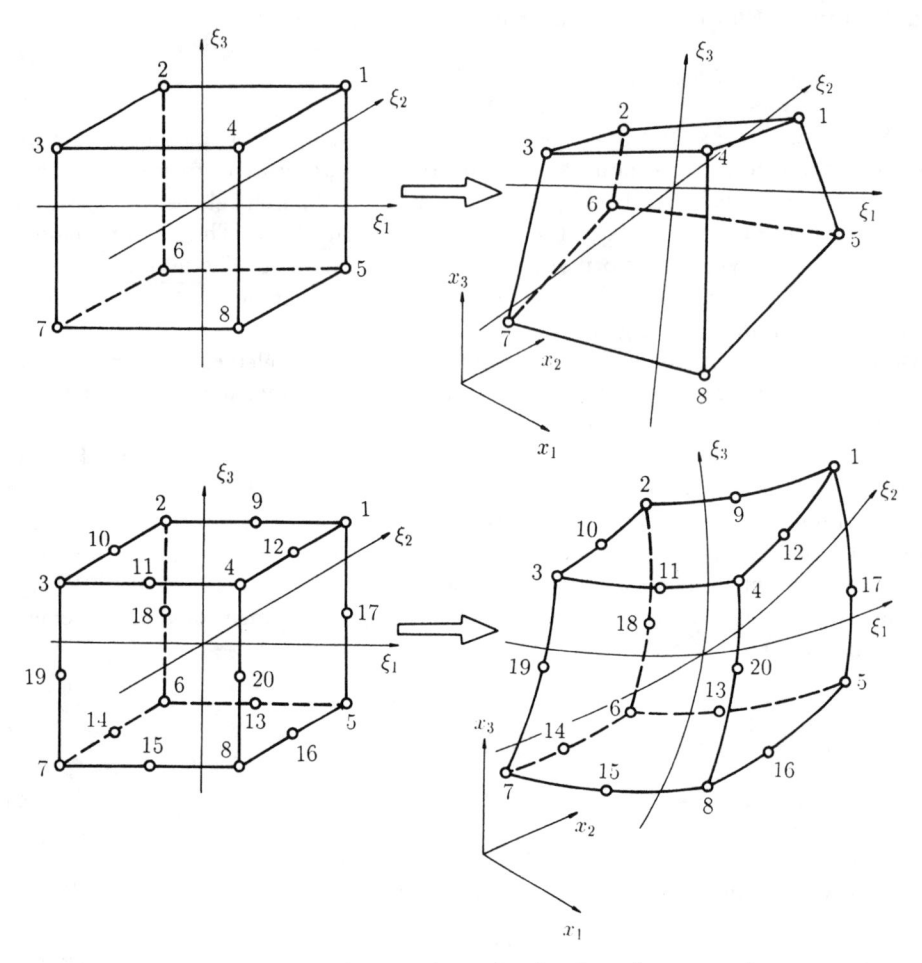

Figure 3.20: Isoparametric mapping of cube-shaped parent elements onto brick elements: (a) linear 8-node element, (b) quadratic 20-node element

$$N^{(i)} = \frac{1}{8}(1+\xi_1^{(i)}\xi_1)(1+\xi_2^{(i)}\xi_2)(1+\xi_3^{(i)}\xi_3)(\xi_1^{(i)}\xi_1 + \xi_2^{(i)}\xi_2 + \xi_3^{(i)}\xi_3 - 2) \,,$$
$$i = 1, 2, \ldots, 8,$$

and for the midside nodes as

$$N^{(i)} = \frac{1}{4}(1+\xi_1^{(i)}\xi_1)(1+\xi_2^{(i)}\xi_2)(1+\xi_3^{(i)}\xi_3)\left\{1 + \left[(\xi_1^{(i)})^2 - 1\right]\xi_1^2 \right.$$
$$\left. + \left[(\xi_2^{(i)})^2 - 1\right]\xi_2^2 + \left[(\xi_3^{(i)})^2 - 1\right]\xi_3^2 \right\} \,, \qquad i = 9, 10, \ldots, 20 \,. \quad (3.215)$$

3.4 Finite Element Modelling of Reinforced and Prestressed Concrete

The shape functions for the linear and the quadratic isoparametric serendipity element can be used to generate elements with any number of nodes between 8 and 20. This feature is attractive for regions of transition from quadratic to linear interpolation of the coordinates and the displacements.

Analogous to the formulation of brick elements, isoparametric tetrahedral elements can be generated. Triangular surfaces of elements can be obtained by mapping neighboring node points of a cubic-shaped parent element onto the same node point. A detailed description is contained, e.g., in [Bathe (1982), Zienkiewicz (1989,1991)].

Commonly, the structural behavior of thick-walled concrete structures is characterized by small displacements. Hence, in what follows, only geometrically linear behavior will be taken into account, i.e., no distinction between the initial and the current configuration will be made. The matrix $\mathbf{B}_0$, appearing, e.g., in the expression for the element stiffness matrix for the geometrically linear theory (see equation (3.49)), contains the derivatives of the shape functions with respect to the global coordinates. The shape functions, however, are formulated in terms of the local coordinates (see equations (3.214) and (3.215)). Hence, the derivatives with respect to the local coordinates must be related to the derivatives with respect to the global coordinates. This relationship is given as

$$\frac{\partial}{\partial \xi} = \mathbf{J} \frac{\partial}{\partial \mathbf{x}} , \qquad (3.216)$$

where

$$\mathbf{J} = \frac{\partial \mathbf{x}}{\partial \xi} = \begin{bmatrix} \frac{\partial x_1}{\partial \xi_1} & \frac{\partial x_2}{\partial \xi_1} & \frac{\partial x_3}{\partial \xi_1} \\ \frac{\partial x_1}{\partial \xi_2} & \frac{\partial x_2}{\partial \xi_2} & \frac{\partial x_3}{\partial \xi_2} \\ \frac{\partial x_1}{\partial \xi_3} & \frac{\partial x_2}{\partial \xi_3} & \frac{\partial x_3}{\partial \xi_3} \end{bmatrix} . \qquad (3.217)$$

is the Jacobian matrix. Thus,

$$\frac{\partial}{\partial \mathbf{x}} = \mathbf{J}^{-1} \frac{\partial}{\partial \xi} . \qquad (3.218)$$

(3.218) is employed for computation of the coefficients of the matrix $\mathbf{B}_0$. The infinitesimal volume element dV_0 in the integral for the element stiffness matrix is given as

$$dV_0 = dx_1 \, dx_2 \, dx_3 = Det(\mathbf{J}) \, d\xi_1 \, d\xi_2 \, d\xi_3 , \qquad (3.219)$$

where $Det(\mathbf{J})$ is the Jacobian determinant. A similar relationship can be derived for surface integrals (see equation (3.20)). The integrations are performed numerically, e.g., by means of Gaussian integration. The integration limits, in terms of the natural coordinates, are -1 and $+1$ in each one of the three coordinate directions.

Shell Elements

On principle, shells (and plates and panels as special cases of shells) could be discretized by means of general three-dimensional finite elements as have been presented in the preceding subsection. However, this approach would be inefficient. Moreover, if the shell thickness is small in comparison with characteristic dimensions of the structure, numerical problems may be encountered [Zienkiewicz (1989,1991)]. Such problems will arise, if the resulting system of algebraic equations becomes ill-conditioned. Thus, instead of using general three-dimensional elements for the discretization of shells, special elements are employed for this purpose.

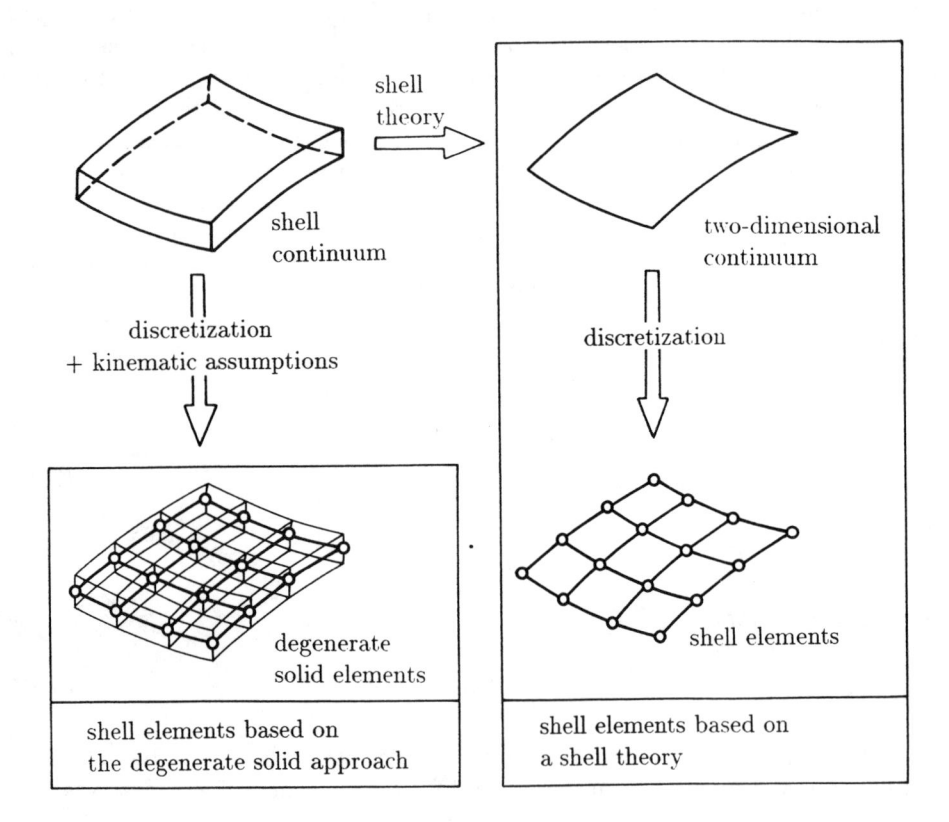

Figure 3.21: Basis of formulation of finite shell elements: shell theory or degenerate solid approach

Departing from the three-dimensional continuum, there are basically two different ways of formulating shell elements (Fig.3.21). One of them consists of replacing the strain-displacement relations (3.12) for the three-dimensional continuum by strain-displacement equations based on a shell theory, representing a two-dimensional

3.4 Finite Element Modelling of Reinforced and Prestressed Concrete

continuum theory. Essential features of such a theory are appropriate kinematic assumptions and admissible simplifications. The other one is characterized by the introduction of the respective kinematic assumptions on the level of the discretized three-dimensional continuum, i.e., on the level of three-dimensional finite elements. This alternative is referred to as the degenerate solid approach. A brief review of both concepts will be given in the following. A detailed comparison of the two approaches can be found in [Buechter (1992)], where the mechanical equivalence of the two formulations, for the same mechanical assumptions, was shown. A review article on shell finite elements with almost 300 references can be found in [Yang (1990)]. It covers both categories of shell elements.

Both approaches are based on the assumption that points on a normal to the middle surface of the undeformed shell remain on a straight line during the deformation process. In addition, the normal stress in the direction of the normal to the middle surface of the shell is neglected. Combined with the restriction to small strains, which is generally justified for concrete shells, this allows neglecting the change of the thickness of the shell during the deformation process. Consequently, the displacements in the direction of the normal to the middle surface are equal for all points on a particular normal.

These assumptions represent a close approximation of the actual structural response of thin and moderately thick shells. They allow description of a three-dimensional problem solely by variables which are referred to the middle surface of the shell. This results in a considerable reduction of the number of degrees of freedom.

In passing, it is mentioned that shells can also be discretized by flat elements which are obtained by superimposing membrane action and bending action. However, such discretizations have serious drawbacks. Firstly, flat elements neither exhibit membrane strains in consequence of transverse displacements nor rotations in consequence of tangential displacements. Secondly, bending moments are "attracted" by the edges of such facet-type discretizations. Since these edges do not exist in reality, the computed concentrations of bending moments in the vicinity of these edges are unrealistic analysis results. Nevertheless, flat elements are appropriate for the analysis of folded plates.

Shell elements based on the degenerate solid approach Degenerate elements, derived directly from three-dimensional finite elements, were originally reported in [Ahmad (1968)]. The degenerate solid approach does not need a shell theory. This is one of the reasons why it it has gained much popularity.

Derivation of a degenerate solid element can depart, e.g., from a three-dimensional isoparametric quadratic serendipity element such as shown in Fig.3.20(b). It is assumed that the curvilinear coordinates ξ_1 and ξ_2 are located in the middle surface of the shell. The two surfaces of the shell are characterized by $\xi_3 = 1$ and $\xi_3 = -1$.

Degeneration of the three-dimensional continuum element to the shell element is accomplished by assuming that in any configuration all shell points with the same values of ξ_1 and ξ_2 are located on a straight line. Its direction is given by the unit vector $\mathbf{v}_3$ denoted as director vector. It follows that the coordinate ξ_3 is a straight line. This kinematic assumption is known from the theory of shear-deformable plates as the Reissner-Mindlin hypothesis. It results in a linear variation of the displacements in the direction of the shell thickness. $\mathbf{v}_3$ can be determined from available information about the geometric form of the middle surface. The concept of parametric elements leads to an approximation of the middle surface. Consequently, $\mathbf{v}_3$ is an approximation of the normal to the exact middle surface. Hence, for a node point which is shared by several elements, slightly different directions of $\mathbf{v}_3$ will be obtained. Alternatively, $\mathbf{v}_3$ can be defined by the user for each node point. This can be done, e.g., by specifying one point on the top surface and one on the bottom surface. In this case, there is only one vector for a particular node point. However, in general, this vector does not represent the exact normal to the middle surface. This is the reason why $\mathbf{v}_3$ is often referred to as pseudo-normal vector.

It has been mentioned previously that the thickness of the shell is assumed to remain unchanged during the deformation process. This implies that the normal strain perpendicular to the middle surface of the shell is neglected. Hence, this assumption is only valid for the case of small strains. It was also mentioned that the normal stress in the direction of the normal to the middle surface is neglected. The assumption of a constant shell thickness allows the displacements of any shell point to be described in terms of the displacements of the middle surface and of two scalar rotations ϕ_1 and ϕ_2 of the director vector $\mathbf{v}_3$ about orthogonal unit vectors $\mathbf{v}_1$ and $\mathbf{v}_2$, perpendicular to $\mathbf{v}_3$. Since there exists an infinite number of possibilities to construct two orthogonal vectors $\mathbf{v}_1$ and $\mathbf{v}_2$, perpendicular to $\mathbf{v}_3$, a particular scheme has to be chosen to ensure a unique definition [Zienkiewicz (1989,1991), Hughes (1987)].

Within the framework of the degenerate solid approach the middle surface of the shell is not described exactly. For any configuration of the shell, the position vector $\mathbf{x}^e$ of an arbitrary point between the top and the bottom surface of a particular shell element e is obtained from an interpolation involving the position vectors $\mathbf{x}_0^{(i)}$ of the node points in the middle surface of the element, the director vectors $\mathbf{v}_3^{(i)}$ and the thicknesses $h^{(i)}$ at the node points as (Fig.3.22)

$$\begin{aligned}\mathbf{x}^e(\xi_1,\xi_2,\xi_3) &= \mathbf{x}_0^e(\xi_1,\xi_2) + \frac{1}{2}h^e(\xi_1,\xi_2)\,\xi_3\mathbf{v}_3^e \\ &\approx \sum_{i=1}^{n} N^{(i)}(\xi_1,\xi_2)\left[\mathbf{x}_0^{(i)} + \frac{1}{2}h^{(i)}\,\xi_3\mathbf{v}_3^{(i)}\right].\end{aligned} \qquad (3.220)$$

In (3.220) $N^{(i)}$ represents the shape function for node i of element e; n is the number of node points per element. The shape functions are formulated in terms of the local coordinates ξ_1 and ξ_2. For a linear isoparametric 4-node serendipity element the shape functions are given as

$$N^{(i)} = \frac{1}{4}(1+\xi_1^{(i)}\xi_1)(1+\xi_2^{(i)}\xi_2)\,, \qquad i=1,2,3,4\,. \qquad (3.221)$$

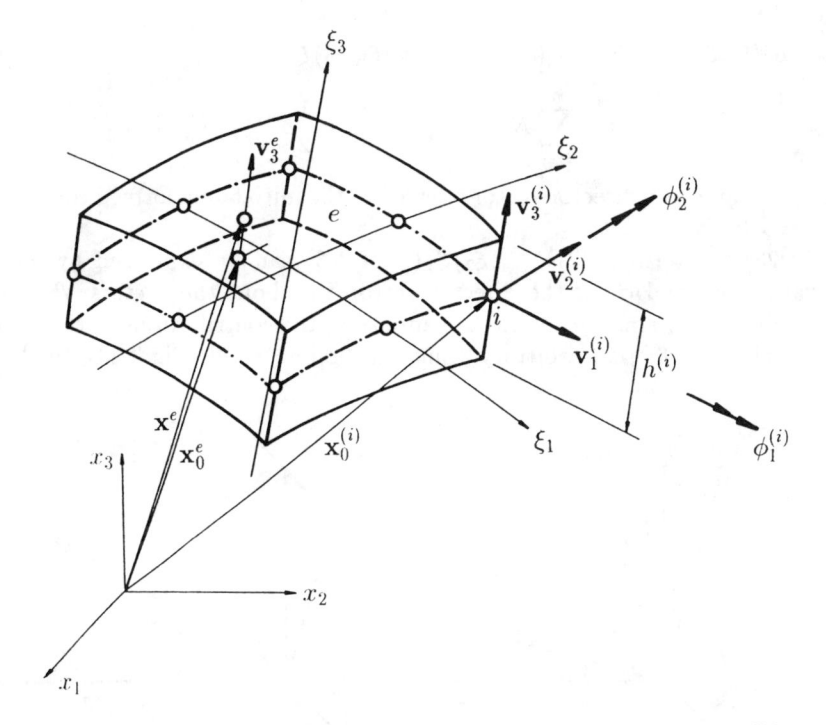

3.4 Finite Element Modelling of Reinforced and Prestressed Concrete

Figure 3.22: Shell element derived on the basis of the degenerate solid approach

For a quadratic isoparametric 8-node serendipity element the shape functions for the corner nodes are given as

$$N^{(i)} = \frac{1}{4}(1 + \xi_1^{(i)}\xi_1)(1 + \xi_2^{(i)}\xi_2)(\xi_1^{(i)}\xi_1 + \xi_2^{(i)}\xi_2 - 1) , \qquad i = 1, 2, 3, 4,$$

and for the midside nodes as

$$N^{(i)} = \frac{1}{2}(1 + \xi_1^{(i)}\xi_1 + \xi_2^{(i)}\xi_2)[1 - (\xi_1^{(i)}\xi_2)^2 - (\xi_2^{(i)}\xi_1)^2] , \qquad i = 5, 6, 7, 8 .$$
(3.222)

Popular degenerate solid elements other than serendipity elements with the node points restricted to the element boundaries are Lagrangian elements. They are characterized by internal node points and by interpolation functions which are based on Lagrange polynomials.

It is noted that (3.220) holds for both the initial and the current configuration. Hence, substitution of (3.220) and of the respective expression for the undeformed configuration into (3.13) yields the displacement of an arbitrary point of the degenerate element e as

$$\mathbf{u}^e(\xi_1,\xi_2,\xi_3) = \mathbf{u}_0^e(\xi_1,\xi_2) + \frac{1}{2}h^e(\xi_1,\xi_2)\,\xi_3\,(\mathbf{v}_3^e - \mathbf{V}_3^e) =$$

$$\approx \sum_{i=1}^{n} N^{(i)}(\xi_1,\xi_2)\left[\mathbf{u}_0^{(i)} + \frac{1}{2}h^{(i)}\xi_3\left(\mathbf{v}_3^{(i)} - \mathbf{V}_3^{(i)}\right)\right], \quad (3.223)$$

where $\mathbf{V}_3$ denotes the director vector referred to the initial undeformed configuration and $\mathbf{u}_0 = \mathbf{x}_0 - \mathbf{X}_0$.

In (3.223) the term $(1/2)h^e(\xi_1,\xi_2)\,\xi_3\,(\mathbf{v}_3^e - \mathbf{V}_3^e)$ can be expressed by means of the rotations ϕ_1 and ϕ_2 of the director vector $\mathbf{V}_3$ about the vectors $\mathbf{V}_1$ and $\mathbf{V}_2$, respectively. If only small rotations are admitted, the contributions of the rotations to the displacement of an arbitrary point can be linearized (Fig.3.23). In this case

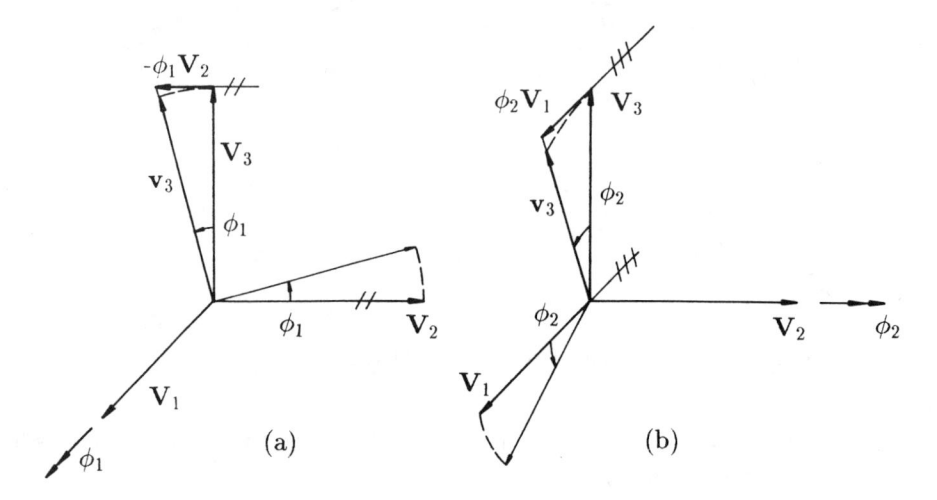

Figure 3.23: Dependence of the displacements on the rotations, for the special case of small rotations: (a) rotation ϕ_1 about $\mathbf{V}_1$, (b) rotation ϕ_2 about $\mathbf{V}_2$

the displacement of an arbitrary point of element e, referred to the global Cartesian coordinate system, is given as

$$\mathbf{u}^e(\xi_1,\xi_2,\xi_3) = \mathbf{u}_0^e(\xi_1,\xi_2) + \frac{1}{2}h^e(\xi_1,\xi_2)\,\xi_3\,(\phi_2^e \mathbf{V}_1^e - \phi_1^e \mathbf{V}_2^e)$$

$$\approx \sum_{i=1}^{n} N^{(i)}(\xi_1,\xi_2)\left[\mathbf{u}_0^{(i)} + \frac{1}{2}h^{(i)}\xi_3\left(\phi_2^{(i)} \mathbf{V}_1^{(i)} - \phi_1^{(i)} \mathbf{V}_2^{(i)}\right)\right].$$

$$(3.224)$$

Independent C_0-interelement-continuous interpolations of mid-surface displacements and rotations are characteristic features of degenerate solid elements. (A function is C_0-interelement continuous, if it is continuous across the element boundaries. In general, the derivative of a C_0-interelement continuous function normal to the element boundaries is discontinuous.)

3.4 Finite Element Modelling of Reinforced and Prestressed Concrete

In the expression for the displacement vector $\mathbf{u}^e$ for an arbitrary point of the degenerate element e (see equation (3.38)) the vector of the nodal displacements is given as

$$\mathbf{q}^e = \lfloor (u_1)_0^{(1)} \ (u_2)_0^{(1)} \ (u_3)_0^{(1)} \ \phi_1^{(1)} \ \phi_2^{(1)} \ (u_1)_0^{(2)} \ \ldots \phi_2^{(n)} \rfloor^T . \qquad (3.225)$$

The matrix of shape functions $\mathbf{N}^e$ can be written formally as

$$\mathbf{N}^e = [\, \mathbf{N}^{(1)} \ \ \mathbf{N}^{(2)} \ \ \ldots \ \ \mathbf{N}^{(n)} \,], \qquad (3.226)$$

where, following from (3.224) and (3.225)

$$\mathbf{N}^{(i)} = \begin{bmatrix} N^{(i)} & 0 & 0 & -c\,(V_2)_{x_1}^{(i)} & c\,(V_1)_{x_1}^{(i)} \\ 0 & N^{(i)} & 0 & -c\,(V_2)_{x_2}^{(i)} & c\,(V_1)_{x_2}^{(i)} \\ 0 & 0 & N^{(i)} & -c\,(V_2)_{x_3}^{(i)} & c\,(V_1)_{x_3}^{(i)} \end{bmatrix}, \quad i=1,\ldots,n, \qquad (3.227)$$

with $c = (1/2)h^{(i)}\xi_3$.

In (3.224) the rotations of the node points were defined with respect to the local base vectors $\mathbf{V}_1$ and $\mathbf{V}_2$. Alternatively, these rotational degrees of freedom can be defined as rotations about the axes of the global Cartesian coordinate system, i.e., $\boldsymbol{\phi} = \lfloor \phi_{x_1}, \phi_{x_2}, \phi_{x_3} \rfloor^T$. This is convenient, if globally aligned rotational boundary conditions are prescribed. For small rotations, the vector $(\mathbf{v}_3^e - \mathbf{V}_3^e)$, appearing in (3.223), is then expressed as

$$\mathbf{v}_3^e - \mathbf{V}_3^e = \boldsymbol{\phi} \times \mathbf{V}_3^e . \qquad (3.228)$$

However, this procedure will fail for nodes on a smooth middle surface since no stiffness will be associated with the rotation about the local base vector $\mathbf{V}_3$. In order to overcome this problem, usually a small fictitious rotational stiffness, corresponding to the rotation about $\mathbf{V}_3$, is introduced. However, this approach is not recommended for nonlinear analyses [Bathe (1985b)].

Linearization of the contribution of the rotations to the displacements, which has led to the second term in (3.224), is also admissible for geometrically nonlinear analyses, provided the load increments are chosen sufficiently small so that the incremental rotations will be small [Bathe (1980), Bathe (1982)]. This linearization is adequate for "moderately large rotations" as occurring in reinforced and prestressed concrete plates and shells. However, if the restriction to small load increments is abandoned, the contributions of the rotations to the displacements will have to be expressed as nonlinear functions of the rotations. A comparison of different options for the mathematical description of finite rotations is contained in [Buechter (1992)]. In [Ramm (1976)] the director vector $\mathbf{v}_3$ is defined, e.g., in terms of the direction cosines of the angles $\psi_k, k = 1, 2, 3$, enclosed by $\mathbf{v}_3$ and the coordinate axes x_k. Thus, in (3.220)

$$\mathbf{v}_3^{(i)} = \begin{Bmatrix} \cos \psi_1^{(i)} \\ \cos \psi_2^{(i)} \\ \cos \psi_3^{(i)} \end{Bmatrix} . \qquad (3.229)$$

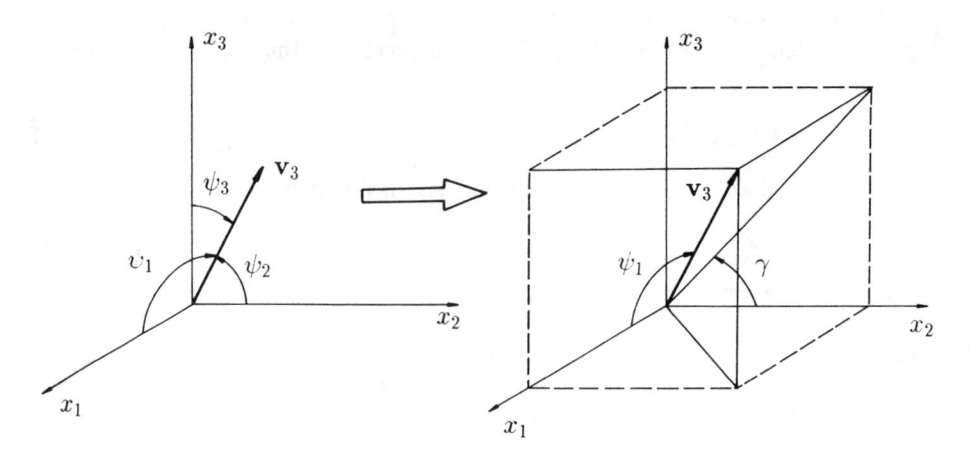

Figure 3.24: Example for the definition of the director vector, taking large rotations into account

The direction cosines in (3.229) are expressed in terms of two independent angles, e.g., in terms of ψ_1 and γ. γ is the angle enclosed by the projection of $\mathbf{v}_3$ onto the x_2-x_3 plane and the x_2-axis (see Fig.3.24). Thus, $\mathbf{v}_3^{(i)}$ can be expressed in terms of $\psi_1^{(i)}$ and $\gamma^{(i)}$ as

$$\mathbf{v}_3^{(i)} = \left\{ \begin{array}{c} \cos \psi_1^{(i)} \\ \sin \psi_1^{(i)} \cos \gamma^{(i)} \\ \sin \psi_1^{(i)} \sin \gamma^{(i)} \end{array} \right\} . \tag{3.230}$$

γ will not be defined for the choice of ψ_1 and γ according to Fig.3.24, if the director vector is parallel to the x_1-axis. In this case, both the x_2-component and the x_3-component of this vector are equal to zero.

For computation of the tangent stiffness matrix within the framework of nonlinear finite element analyses, (3.223) has to be linearized, resulting in

$$\Delta \mathbf{u}^e(\xi_1, \xi_2, \xi_3) \approx \sum_{i=1}^{n} N^{(i)}(\xi_1, \xi_2) \left[\Delta \mathbf{u}_0^{(i)} + \frac{1}{2} h^{(i)} \xi_3 \mathbf{L}^{(i)} \left\{ \begin{array}{c} \Delta \gamma^{(i)} \\ \Delta \psi_1^{(i)} \end{array} \right\} \right], \tag{3.231}$$

where $\Delta \gamma^{(i)}$ and $\Delta \psi_1^{(i)}$ are the unknown linearized nodal rotations and

$$\mathbf{L}^{(i)} = \left[\begin{array}{cc} 0 & -\sin \psi_1^{(i)} \\ -\sin \psi_1^{(i)} \sin \gamma^{(i)} & \cos \psi_1^{(i)} \cos \gamma^{(i)} \\ \sin \psi_1^{(i)} \cos \gamma^{(i)} & \cos \psi_1^{(i)} \sin \gamma^{(i)} \end{array} \right] . \tag{3.232}$$

3.4 Finite Element Modelling of Reinforced and Prestressed Concrete

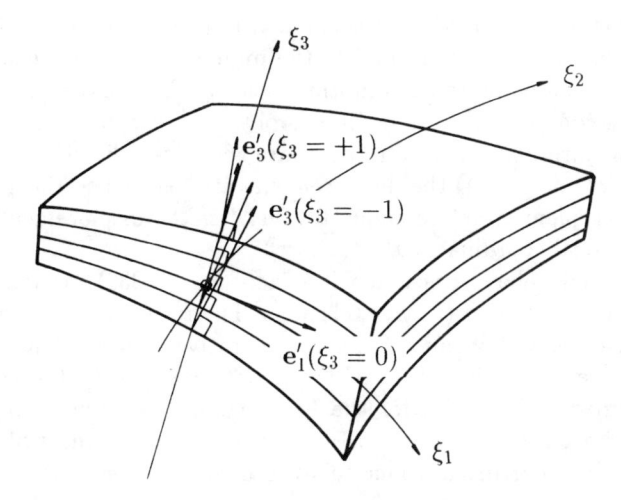

Figure 3.25: Local basis for the formulation of the constitutive relations

The constitutive relations are referred to a local Cartesian $x'y'z'$-coordinate system (Fig.3.25). For an arbitrary point $\mathbf{x}^e(\xi_1, \xi_2, \xi_3)$ of shell element e, the orthonormal basis $\mathbf{e}'_i$, $i = 1, 2, 3$, for this coordinate system is defined such that $\mathbf{e}'_1$ and $\mathbf{e}'_2$ are tangent to the surface $\xi_3 = $ const. and $\mathbf{e}'_3$ is normal to this surface. $\mathbf{e}'_1$ and $\mathbf{e}'_2$ are obtained from the vectors $(\partial \mathbf{x}^e / \partial \xi_1)$ and $(\partial \mathbf{x}^e / \partial \xi_2)$ [Hughes (1987)]. These vectors are tangent to the local curvilinear coordinates ξ_1 and ξ_2. Thus, they define the tangent plane to the surface $\xi_3 = $ const. Formulation of the constitutive relations with respect to local coordinates x' y' z' allows satisfying the requirement $\sigma'_z = 0$. The constitutive relations for linear-elastic material behavior are obtained from generalized Hooke's law by setting $\sigma_{z'} = 0$:

$$\left\{ \begin{array}{c} \sigma_{x'} \\ \sigma_{y'} \\ \tau_{x'y'} \\ \cdots \\ \tau_{y'z'} \\ \tau_{z'x'} \end{array} \right\} = \frac{E}{1-\nu^2} \left[\begin{array}{ccc:cc} 1 & \nu & 0 & 0 & 0 \\ & 1 & 0 & 0 & 0 \\ & & \frac{1-\nu}{2} & 0 & 0 \\ \hdashline \cdots & \cdots & \cdots & \cdots & \cdots \\ & & & \kappa\frac{1-\nu}{2} & 0 \\ \text{sym.} & & & & \kappa\frac{1-\nu}{2} \end{array} \right] \left\{ \begin{array}{c} \varepsilon_{x'} \\ \varepsilon_{y'} \\ \gamma_{x'y'} \\ \cdots \\ \gamma_{y'z'} \\ \gamma_{z'x'} \end{array} \right\}.$$

(3.233)

In (3.233) the first three constitutive equations are equal to the material equations for plane stress conditions. The two remaining constitutive equations relate the transverse shear strains to the corresponding shear stresses. κ denotes the shear correction factor which is well known from the theory of shear-deformable beams.

The condition for determination of κ is the equality of the integral of the true strain energy density along an arbitrary normal to the middle surface of the shell and of the respective integral based on the assumption of constant transverse shear strains. On the basis of this condition the value for κ is obtained as 5/6.

In a displacement formulation the strains ε' are computed directly from the displacements. In (3.223) or (3.224) the displacements are referred to the global coordinate system. Consequently, before computation of ε' the displacements must be transformed to the local coordinates x', y', z'.

If nonlinear material behavior is taken into account, (3.233) must be replaced by appropriate total or rate constitutive relations. The respective incremental constitutive relations, employed in an incremental-iterative solution scheme, are given in (3.31). In case of consideration of geometric nonlinearity the constitutive equations are assumed to hold also for the Lagrangian strain tensor and the 2nd Piola-Kirchhoff stress tensor. Clearly, this assumption is only admissible for small strains, as are usually occurring in concrete structures. The respective incremental constitutive equations are given in (3.29).

Consideration of shear deformations on the basis of independent C_0-interelement-continuous interpolations of the displacements and the rotations yields good results for moderately thick shells, i.e., for shells, for which the Reissner-Mindlin hypothesis serves as a reasonable kinematic assumption. However, in the limiting case of thin shells, frequently so-called locking is observed. It is characterized by a mechanical behavior of the elements, which is by far too stiff. Shear locking results from the inability of the elements to satisfy a fundamental kinematic constraint condition for thin shells. It requires normals to the middle surface of a shell to remain normal to this surface during the deformation process. This constraint condition, known as the Kirchhoff hypothesis, implies the vanishing of transverse shear deformations. However, degenerate elements with interpolation functions of low order cannot approximate states of vanishing transverse shear strains adequately. The reason for this deficiency is that such elements do not contain pure bending modes. Consequently, states of pure bending are frequently accompanied by parasitic transverse shear strains which do not exist in reality (see, e.g., [Andelfinger (1991)]). The thinner the shell, the larger the ratio of the shear stiffness over the bending stiffness. If the thickness of the shell is so small that the shear stiffness exceeds the bending stiffness considerably, errors in the representation of the transverse shear strains will be magnified. In addition to shear locking, membrane locking may occur for thin shells. It is the consequence of the inability of the elements to represent inextensible deformations of their middle surfaces. The thinner the shell, the larger the ratio of the membrane stiffness over the bending stiffness. Consequently, errors in the representation of the membrane strains will be magnified if the thickness of the shell is so small that the membrane stiffness exceeds the bending stiffness considerably. Especially for low order elements severe membrane locking may occur. For coarse meshes shear locking or membrane locking may lead to useless results.

A popular remedy for locking is reduced or selective integration. Reduced integration is characterized by a reduced order of integration for all terms of the tangent stiffness matrix and of the internal force vector. Selective integration is characterized by applying reduced integration only to those terms of the tangent stiffness matrix and of the internal force vector, which are related to the transverse shear strains. However, application of reduced or selective integration may lead to additional zero-energy modes, termed as spurious zero-energy modes.

Another popular remedy for locking is the introduction of assumed fields of transverse shear strains. These shear strains are not computed from the displacements. Rather, they are interpolated from the displacement-based transverse shear strains at selected points of the element. Examples for robust 4-node quadrilateral shell elements based on the degenerate solid approach with assumed fields of transverse shear strains can be found, e.g., in [Dvorkin (1984), Bathe (1985), Bathe (1986), Huang (1986), Andelfinger (1991)]. 8-node and 9-node curved degenerate elements with assumed fields of both transverse shear strains and membrane strains were proposed, e.g., in [Bathe (1986), Huang (1986)]. These elements do not exhibit shear or membrane locking. Moreover, they do not contain spurious zero-energy modes. They are well suited for material nonlinear and geometrically nonlinear analyses of thin and moderately thick shells. A numerical study [Stander (1989)] of several degenerate elements based on assumed strain fields has clearly demonstrated the robustness of the so-called MITC4 element [Dvorkin (1984), Bathe (1985), Bathe (1986)]. Some other elements based on assumed strain fields were found to be significantly less robust. It was shown in [Andelfinger (1991), Onate (1992)] that both reduced integration techniques and assumed strain methods rest on evaluating the transverse shear strains only at such points at which they do not appear in bending modes in the form of spurious transverse shear strains.

Alternatively, the discrete Kirchhoff theory (DKT) can be used as the starting point for the development of thin-shell finite elements. Accordingly, these elements are known as discrete Kirchhoff elements or DKT elements. This approach is characterized by independent C_0-interelement-continuous interpolations of the displacements and the rotations. However, in contrast to degenerate solid elements, the contribution of the transverse shear strains to the strain energy is neglected and the Kirchhoff constraint is enforced exactly at selected discrete points (see, e.g., [Nagtegaal (1981)]).

Shell elements based on a shell theory Shell theories are characterized by describing the geometric form of the middle surface of a shell by means of the concepts of differential geometry for curved surfaces [Flügge (1972), Başar (1985)]. The middle surface of the shell is viewed as a two-dimensional curved subspace embedded in the three-dimensional Euclidean space. Since this subspace is curved, some of the basic concepts of Euclidean geometry such as the concept of parallel directions do not apply [Flügge (1972)].

The position vector $\mathbf{X}_0$ of a point P on a curved surface, representing the middle surface of the undeformed shell, is given as (Fig.3.26)

$$\mathbf{X}_0 = \mathbf{X}_0(\alpha^1, \alpha^2) , \qquad (3.234)$$

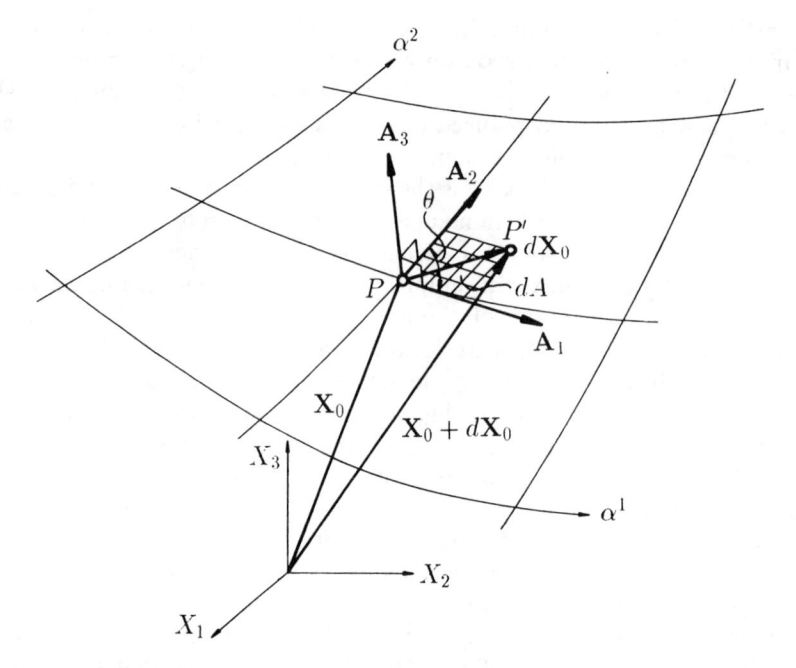

Figure 3.26: Geometric description of a curved surface

where α^1 and α^2 are two independent parameters, denoted as surface coordinates or curvilinear coordinates. The equations

$$\mathbf{X}_0 = \mathbf{X}_0(\alpha^1, \alpha^2 = const.) , \quad \mathbf{X}_0 = \mathbf{X}_0(\alpha^1 = const. , \alpha_2) \tag{3.235}$$

describe parameter lines, also referred to as coordinate lines. The two curvilinear coordinates may have different physical units. E.g., for a cylindrical surface it is convenient to choose one of them as the polar angle and the other one as the length in the direction of the generatrices. In general, the parameter lines are not orthogonal.

The tangent vectors to the parameter lines at a particular point P,

$$\mathbf{A}_i = \frac{\partial \mathbf{X}_0}{\partial \alpha^i} , \quad i = 1, 2, \tag{3.236}$$

determine the tangent plane at this point. They are also referred to as the in-surface base vectors. The third base vector is normal to the tangent plane. Hence, this vector is denoted as normal vector. It is defined as a unit vector. Consequently,

$$\mathbf{A}_3 = \frac{\mathbf{A}_1 \times \mathbf{A}_2}{|\mathbf{A}_1 \times \mathbf{A}_2|} . \tag{3.237}$$

The line element $d\mathbf{X}_0$ on the surface is given as

$$d\mathbf{X}_0 = \mathbf{A}_i d\alpha^i , \quad i = 1, 2 . \tag{3.238}$$

3.4 Finite Element Modelling of Reinforced and Prestressed Concrete

The square of the length of $d\mathbf{X}_0$ is obtained as

$$(dS)^2 = d\mathbf{X}_0 \cdot d\mathbf{X}_0 = (\mathbf{A}_i d\alpha^i) \cdot (\mathbf{A}_j d\alpha^j) = A_{ij} d\alpha^i d\alpha^j, \quad i,j = 1,2. \quad (3.239)$$

In (3.239)

$$A_{ij} = \mathbf{A}_i \cdot \mathbf{A}_j \quad (3.240)$$

denotes the covariant components of the metric tensor of the surface. Knowledge of A_{ij} allows determination of lenghts and areas on curved surfaces. (3.239) is known as the first fundamental form of a curved surface. It follows from (3.238) that the line elements in the direction of α^1 and α^2 are given as

$$d\mathbf{X}_0^{(1)} = \mathbf{A}_1 d\alpha^1, \quad d\mathbf{X}_0^{(2)} = \mathbf{A}_2 d\alpha^2. \quad (3.241)$$

Thus,

$$|d\mathbf{X}_0^{(1)}| = \sqrt{A_{11}} d\alpha^1, \quad |d\mathbf{X}_0^{(2)}| = \sqrt{A_{22}} d\alpha^2, \quad \cos\theta = \frac{A_{12}}{\sqrt{A_{11}A_{22}}}. \quad (3.242)$$

The area dA in Fig.3.26 can be expressed as

$$dA = |d\mathbf{X}_0^{(1)} \times d\mathbf{X}_0^{(2)}| = |\mathbf{A}_1 \times \mathbf{A}_2| d\alpha^1 d\alpha^2. \quad (3.243)$$

Differentiation of the identity

$$\mathbf{A}_i \cdot \mathbf{A}_3 = 0, \quad i = 1,2, \quad (3.244)$$

yields

$$\frac{\partial \mathbf{A}_i}{\partial \alpha^j} \cdot \mathbf{A}_3 = -\mathbf{A}_i \cdot \frac{\partial \mathbf{A}_3}{\partial \alpha^j}. \quad (3.245)$$

The covariant components of the curvaturve tensor are defined as

$$B_{ij} = \frac{\partial \mathbf{A}_i}{\partial \alpha^j} \cdot \mathbf{A}_3 = \frac{\partial^2 \mathbf{X}_0}{\partial \alpha^i \partial \alpha^j} \cdot \mathbf{A}_3. \quad (3.246)$$

Alternatively, these tensor components are defined as

$$B_{ij} = -\mathbf{A}_i \cdot \frac{\partial \mathbf{A}_3}{\partial \alpha^j}. \quad (3.247)$$

(3.247) follows from substitution of (3.245) into (3.246). Occasionally, the covariant components of the curvature tensor are defined without the minus sign on the right-hand side of (3.247). Expansion of (3.247) by $d\alpha^i d\alpha^j$ yields

$$B_{ij} d\alpha^i d\alpha^j = -\mathbf{A}_i d\alpha^i \cdot \frac{\partial \mathbf{A}_3}{\partial \alpha^j} d\alpha^j. \quad (3.248)$$

Substitution of (3.238) into (3.248) results in

$$d\mathbf{X}_0 \cdot d\mathbf{A}_3 = -B_{ij} d\alpha^i d\alpha^j. \quad (3.249)$$

(3.249) is known as the second fundamental form of a curved surface. Knowledge of B_{ij} permits determination of the curvatures and the twist at a particular point of a curved surface.

In the following, the meaning of the curvatures and of the twist of a surface will be explained. Multiplying (3.247) by $d\alpha^j$ and assuming for a moment that only $B_{11} \neq 0$, yields

$$d\mathbf{A}_3 = -\frac{B_{11}}{A_{11}} \mathbf{A}_1 d\alpha^1 . \qquad (3.250)$$

(The assumption of $B_{12} = B_{21} = 0$ implies that α^1 and α^2 are lines of principal curvature. Such parameter lines are orthogonal.)

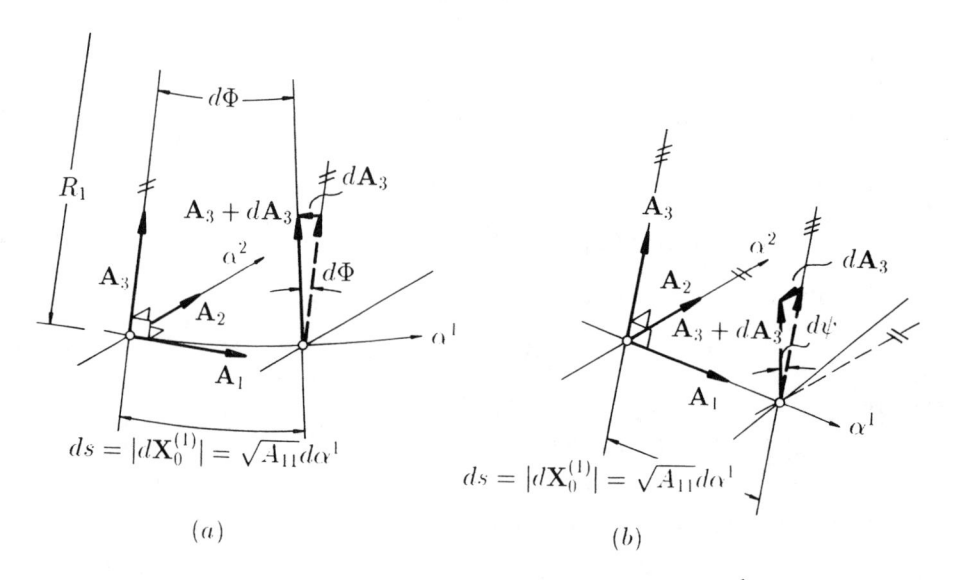

Figure 3.27: Definition of (a) the curvature in the direction of α^1 and (b) the twist

Fig.3.27(a) shows that because of $|d\mathbf{A}_3| = d\Phi$, following from $|\mathbf{A}_3| = 1$, the curvature $1/R_1$ of the surface in the direction of α^1 is given as

$$\frac{1}{R_1} = \frac{d\Phi}{ds} = \frac{|d\mathbf{A}_3|}{\sqrt{A_{11}} d\alpha^1} = \frac{B_{11}}{A_{11}} . \qquad (3.251)$$

Multiplying (3.247) by $d\alpha^j$, assuming for a moment that only $B_{12} = B_{21} \neq 0$ and setting $i = 2$, yields

$$d\mathbf{A}_3 = -\frac{B_{21}}{A_{22}} \mathbf{A}_2 d\alpha^1 . \qquad (3.252)$$

Fig.3.27(b) shows that because of $|d\mathbf{A}_3| = d\Psi$, following from $|\mathbf{A}_3| = 1$, the twist of the surface is given as

3.4 Finite Element Modelling of Reinforced and Prestressed Concrete 227

$$\frac{1}{T} = \frac{d\Psi}{ds} = \frac{|d\mathbf{A}_3|}{\sqrt{A_{11}}d\alpha^1} = \frac{B_{21}}{\sqrt{A_{11}A_{22}}} . \quad (3.253)$$

Within the framework of the total Lagrangian description all quantities referred to an arbitrary point of the shell are expressed in terms of the respective quantities referred to the normal projection of this point onto the middle surface of the undeformed shell and in terms of the distance α^3 between this projection and the shell point. The surface coordinates α^i, $i = 1, 2$, and α^3 are convective coordinates. For a particular material point the values of such coordinates are fixed during the deformation process.

The displacements of an arbitrary point of the shell are formulated in terms of the displacements and the rotations of the normal projection of this point onto the middle surface and in terms of α^3. For moderately thick shells the Reissner-Mindlin kinematics is employed. However, most shells are thin-walled structures for which the Kirchhoff hypothesis (Fig.3.28) is adequate.

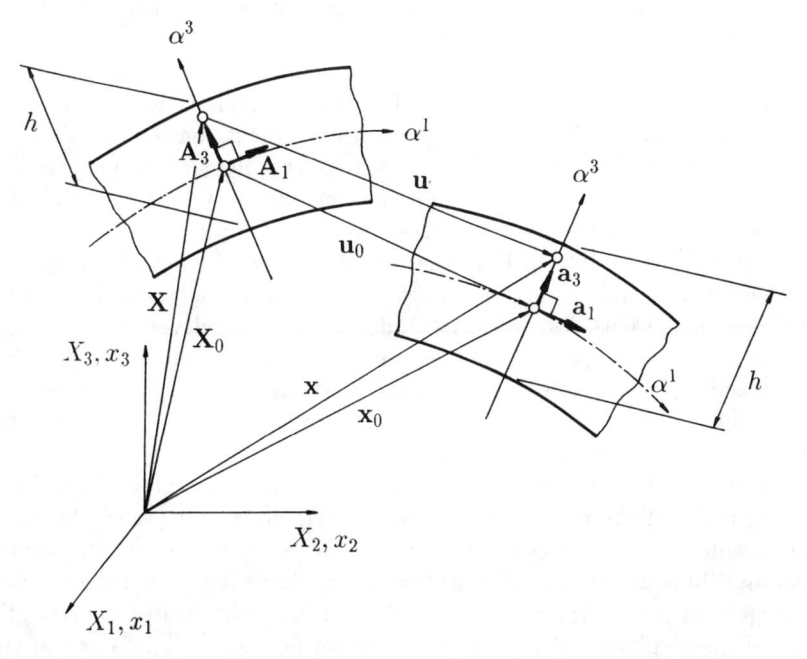

Figure 3.28: Sections $\alpha^2 = 0$ through a thin shell: (a) undeformed configuration $(\mathbf{X}_0, \mathbf{X})$ (b) deformed configuration $(\mathbf{x}_0, \mathbf{x})$ on the basis of the Kirchhoff hypothesis

The strain tensor at an arbitrary point of the shell is formulated in terms of the tensor of the membrane strains and the tensor of the changes of curvature for the normal projection of the shell point onto the middle surface and in terms of α^3. For concrete shells, restriction to small strains is generally justified. Depending on different simplifying assumptions, different shell theories for small strains are obtained.

They are ranging from geometrically nonlinear theories, accounting for large displacements and rotations, to geometrically linear theories. The latter are characterized by small displacements and small rotations. If the displacements do not exceed the order of magnitude of the shell thickness, a geometrically nonlinear shell theory of small displacements and moderately large rotations [Sanders (1963), Koiter (1967)] may be used. Generally, such a shell theory is adequate for concrete shells. Derivation of linear and nonlinear shell theories including evaluation of the errors resulting from the employed simplifying assumptions can be found in [Başar (1985)].

Curved finite shell elements which are based on a shell theory are typically formulated in terms of curvilinear coordinates. Thus, the nodes are not specified in terms of global Cartesian coordinates but rather in terms of curvilinear coordinates. The components of the displacement vector are referred to the base vectors $\mathbf{A}_i$, $i = 1, 2, 3$. If the exact expressions for the components of the metric tensor (see equation (3.240)) and of the curvature tensor (see equation (3.246)) as well as for the first derivatives of these tensor components with respect to α^1 and α^2 are supplied by means of a user-defined subroutine [Floegl (1981)], the geometric properties of the shell to be analyzed can be described exactly.

Derivation of displacement-based finite elements for thin shells is relatively complicated because of the requirement to choose C_1-interelement-continuous functions for the displacements. (A function is C_1-interelement-continuous, if the function and its first derivatives are continuous across the element boundaries.) In order to satisfy this requirement, shape functions of relatively high order must be used for the description of the element displacements [Thomas (1974), Thomas (1975), Floegl (1981), Harte (1986)]. E.g., the list of degrees of freedom for triangular elements with cubic shape functions contains the nodal displacements and their first partial derivatives with respect to α^1 and α^2. The respective list for triangular elements with quintic shape functions, in addition, contains second partial derivatives of the nodal displacements with respect to α^1 and α^2. If the degrees of freedom contain higher-order derivatives of the displacements, it may be difficult to specify proper geometric boundary conditions.

An example for a triangular finite shell element with cubic shape functions is given in Fig.3.29 [Thomas (1974), Floegl (1981), Hofstetter (1987), Walter (1988)]. The element has 33 degrees of freedom. 30 degrees of freedom are displacement-related and three are force-related degrees of freedom. Hence, the element belongs to the category of mixed finite elements. The displacement-related degrees of freedom consist of the displacements at the three corner nodes ($i = 1,2,3$) and at the center node ($i = 4$) and the first partial derivatives of the displacements with respect to α^1 and α^2 at the three corner nodes. For an arbitrary point in the middle surface of element e, the displacement vector is given as (3.38)

$$\mathbf{u}_0^e = \mathbf{N}^e \mathbf{q}^e , \qquad (3.254)$$

with

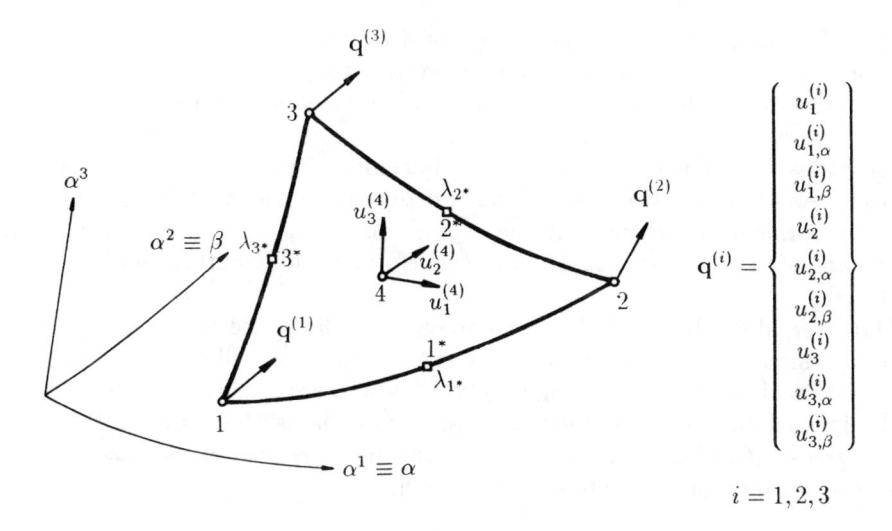

3.4 Finite Element Modelling of Reinforced and Prestressed Concrete

Figure 3.29: Example for a triangular thin-shell finite element

$$\mathbf{q}^e = \lfloor u_1^{(1)}\ u_{1,\alpha}^{(1)}\ u_{1,\beta}^{(1)}\ u_2^{(1)}\ u_{2,\alpha}^{(1)}\ u_{2,\beta}^{(1)}\ u_3^{(1)}\ u_{3,\alpha}^{(1)}\ u_{3,\beta}^{(1)}\ u_1^{(2)}\ u_{1,\alpha}^{(2)} \ldots$$
$$\ldots u_{3,\beta}^{(3)}\ u_1^{(4)}\ u_2^{(4)}\ u_3^{(4)} \rfloor^T \tag{3.255}$$

as the vector of nodal displacements and with $\alpha \equiv \alpha^1$ and $\beta \equiv \alpha^2$. The matrix $\mathbf{N}^e$ in (3.254) contains cubic shape functions, formulated in terms of triangular coordinates. In addition, one Lagrange multiplier λ per element side is employed to enforce C_1-interelement-continuity. For an arbitrary point of the shell element the displacement components in the directions of the base vectors $\mathbf{A}_i$, $i = 1, 2, 3$, are computed as

$$u_i = (u_i)_0 - \alpha^3 \phi_i\ , \quad i = 1, 2, \quad u_3 = (u_3)_0\ , \tag{3.256}$$

where α^3 denotes the distance of this point from the middle surface of the shell and ϕ_1 and ϕ_2 are the rotations of $\mathbf{A}_3$ about $\mathbf{A}_1$ and $\mathbf{A}_2$, respectively. The rotations are defined such that for positive value of α^3 a positive value of ϕ_i yields a contribution to the corresponding displacement component in the direction of $-\mathbf{A}_i$. The rotations are expressed in terms of the tangential displacement components of the middle surface of the shell and of the derivatives of the normal displacement with respect to α^1 and α^2 (see, e.g., [Koiter (1967), Başar (1985)]).

Alternatively, the isoparametric concept can be employed for elements based on shell theory [MARC (1992)]. In this case the geometric shape of the middle surface of the shell is described in an approximate manner, similar to the degenerate solid approach covered in the preceding paragraph.

It is often claimed that the possibility of describing the geometric properties of the middle surface of the shell exactly is an advantage of finite elements based on shell theory. Especially for imperfection-sensitive shells this advantage is considered to be important. However, for such elements consideration of imperfections is a laborous task. In addition to the analytical definition of the geometric shape of the perfect shell, the geometric shape of the imperfect shell must be defined analytically. In general, this results in complicated expressions for the components of the metric tensor and of the curvature tensor for the undeformed middle surface of the imperfect shell.

Moreover, difficulties will be encountered with finite elements based on shell theory, if intersections of shells must be considered or if shell elements have to be connected with beam elements. Special tying options are needed to connect adjacent nodes located on the intersection of two curved surfaces. The tying of the respective degrees of freedom is complicated because the directions of the base vectors at adjacent nodes at such an intersection are different.

The layer concept This concept is characterized by subdividing a finite element across its thickness into several thin layers (Fig.3.30(a)). For each of these layers the appropriate constitutive relations are employed. For thin shells a state of plane stress is assumed in each layer. For moderately thick shells also the transverse shear strains are considered. Across the thickness of a single layer the stresses are assumed to be constant. Thus, the actual distribution of the stresses across the shell thickness is approximated by piecewise constant stresses (Fig.3.30(b)).

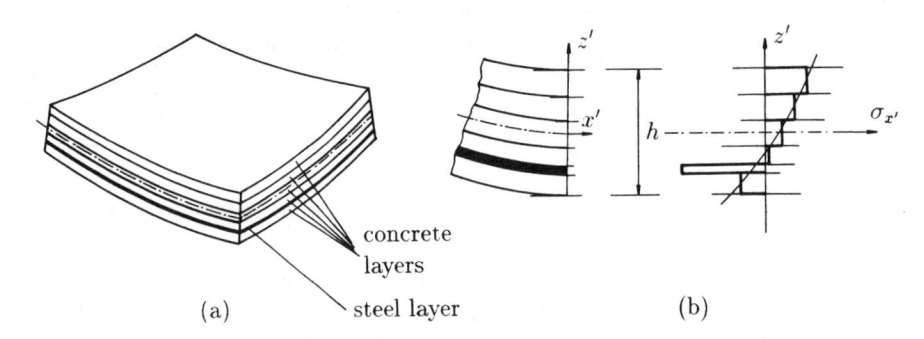

Figure 3.30: Layered finite shell element: (a) subdivision into layers across the thickness, (b) stress distribution across the thickness

The integration in the direction of the thickness, required for the computation of the stiffness matrix and the vector of internal forces, is performed numerically. For each layer one integration point is needed. This point is located in the middle surface of the respective layer. Hence, the total number of integration points per element is equal to the product of the number of integration points in the middle surface of a layer and the number of layers. The layer concept allows consideration

3.4 Finite Element Modelling of Reinforced and Prestressed Concrete

of the nonlinear material behavior of concrete shells. In particular, the nonlinear behavior of concrete under compression and the development of bending cracks, starting at one of the two faces of the shell, can be taken into account. Moreover, the reinforcement can be modelled by means of fictitious steel layers. The thickness of these layers is determined such that the cross-sectional area of the reinforcement is preserved. The distance of a steel layer from the middle surface is equal to the distance of the respective reinforcing bars from the middle surface of the shell. This mode of representation of the reinforcement is known as distributed modelling. It will be described in more detail in subsection 3.4.3.

Plate Elements

A plate may be regarded as a special case of a shell. Hence, shell elements may be specialized for the analysis of plates. Commonly, commercial finite element programs contain both kinds of shell elements, i.e., degenerate solid elements and elements based on shell theory. Both types of elements can be used for the analysis of plates.

Analogous to shell elements, plate elements can either be derived on the basis of the Reissner-Mindlin hypothesis, suitable for moderately thick plates, or on the basis of the Kirchhoff hypothesis, suitable for thin plates. The curved degenerate elements of the previous paragraph, based on the Reissner-Mindlin hypothesis, can be specialized easily to plate elements. Firstly, the middle surface of the plate is assumed to lie in the x_1-x_2 plane of a Cartesian coordinate system. Thus, for a plate element, $\mathbf{u}_0 = \lfloor 0 \; 0 \; (u_3)_0 \rfloor^T$. Secondly, the local base vectors $\mathbf{V}_i$, $i = 1, 2, 3$, in Figs. 3.22 and 3.23 are taken parallel to the global coordinate axes. Thus, $\phi_1 \equiv \phi_{x_1}$, $\phi_2 \equiv \phi_{x_2}$, $\mathbf{V}_1 = \lfloor 1 \; 0 \; 0 \rfloor^T$ and $\mathbf{V}_2 = \lfloor 0 \; 1 \; 0 \rfloor^T$. Substitution of $\mathbf{u}_0$, $\mathbf{V}_1$ and $\mathbf{V}_2$ into (3.224), holding for small rotations, yields

$$\mathbf{u}^e = \left\{ \begin{array}{c} x_3^e \, \phi_2^e(x_1, x_2) \\ -x_3^e \, \phi_1^e(x_1, x_2) \\ (u_3^e)_0(x_1, x_2) \end{array} \right\} \approx \sum_{i=1}^{n} N^{(i)}(\xi_1, \xi_2) \left\{ \begin{array}{c} x_3^{(i)} \phi_2^{(i)} \\ -x_3^{(i)} \phi_1^{(i)} \\ (u_3)_0^{(i)} \end{array} \right\}, \quad (3.257)$$

where $x_3^e = (1/2)\, h^e\, \xi_3$ denotes the distance of the considered point from the middle surface of the plate.

Thus, for a plate element of the form of a degenerate solid element, the vector of the nodal displacements is obtained as

$$\mathbf{q}^e = \lfloor (u_3)_0^{(1)} \; \phi_1^{(1)} \; \phi_2^{(1)} \; (u_3)_0^{(2)} \; \ldots \; \phi_2^{(n)} \rfloor^T. \quad (3.258)$$

Substituting (3.257) into the expression for the linearized Cauchy strain tensor (3.9), yields the in-plane strains as

$$\left\{ \begin{array}{c} \varepsilon_{11}^e \\ \varepsilon_{22}^e \\ \gamma_{12}^e \end{array} \right\} = x_3^e \left\{ \begin{array}{c} \dfrac{\partial \phi_2^e}{\partial x_1} \\ -\dfrac{\partial \phi_1^e}{\partial x_2} \\ \dfrac{\partial \phi_2^e}{\partial x_2} - \dfrac{\partial \phi_1^e}{\partial x_1} \end{array} \right\} \qquad (3.259)$$

and the transverse shear strains as

$$\left\{ \begin{array}{c} \gamma_{13}^e \\ \gamma_{23}^e \end{array} \right\} = \left\{ \begin{array}{c} \dfrac{\partial (u_3^e)_0}{\partial x_1} + \phi_2^e \\ \dfrac{\partial (u_3^e)_0}{\partial x_2} - \phi_1^e \end{array} \right\} . \qquad (3.260)$$

Geometrically nonlinear behavior can be accounted for in the same manner as for shell elements of the form of degenerate solid elements. Shear locking, observed for the limiting case of thin plates, can be prevented in a similar manner as for shell elements by applying reduced or selective integration or by introducing assumed transverse shear strain fields [Bathe (1985)].

If the Kirchhoff constraint, $\gamma_{13} = 0, \gamma_{23} = 0$, is introduced on the analytical level, then the rotations ϕ_1 and ϕ_2 are expressed in terms of the first partial derivatives of $(u_3)_0$ as

$$\phi_1 = \frac{\partial (u_3)_0}{\partial x_2} , \quad \phi_2 = -\frac{\partial (u_3)_0}{\partial x_1} . \qquad (3.261)$$

The main problem with finite plate elements derived from the Kirchhoff plate theory is the requirement to choose a C_1-interelement-continuous function for $(u_3)_0$. This requirement is identical to the C_1-interelement continuity requirement for thin-shell elements based on shell theory and on the Kirchhoff hypothesis. Displacement functions of the form of polynomials with three degrees of freedom per node, i.e., $(u_3)_0$ and its partial derivatives with respect to x_1 and x_2, do not satisfy the requirement of C_1-interelement continuity. The difficulty of satisfying this requirement has stimulated the development of a large number of different types of finite elements for the analysis of thin plates. An extensive review can be found in [Zienkiewicz (1989,1991)].

For consideration of the nonlinear material behavior of concrete plates, the layer concept, described in the previous section, is employed.

Plane Stress Elements

Plane stress elements are suitable for the analysis of thin-walled plane structures loaded such that the middle surface remains plane. A typical example of such a structure is a panel. Assuming the middle surface of the panel to lie in the x_1-x_2 plane of a Cartesian coordinate system, the condition of plane stress requires that

3.4 Finite Element Modelling of Reinforced and Prestressed Concrete

$\sigma_{31} = \sigma_{32} = \sigma_{33} = 0$. Consequently, in the constitutive equations for a general three-dimensional state of stress the respective stress components must be set equal to zero. Popular plane stress elements are the bilinear 4-node and the quadratic 8-node isoparametric serendipity element (Fig.3.31).

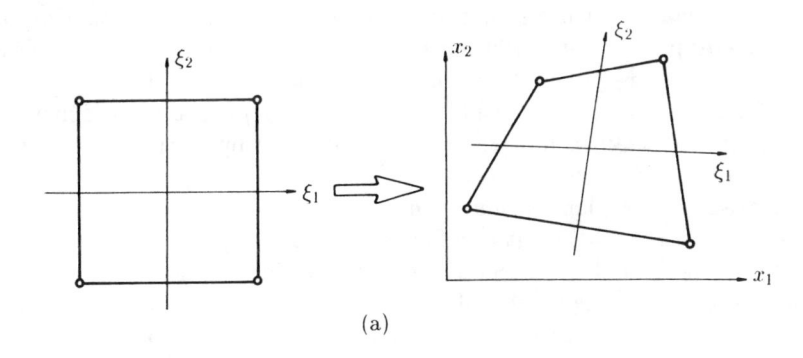

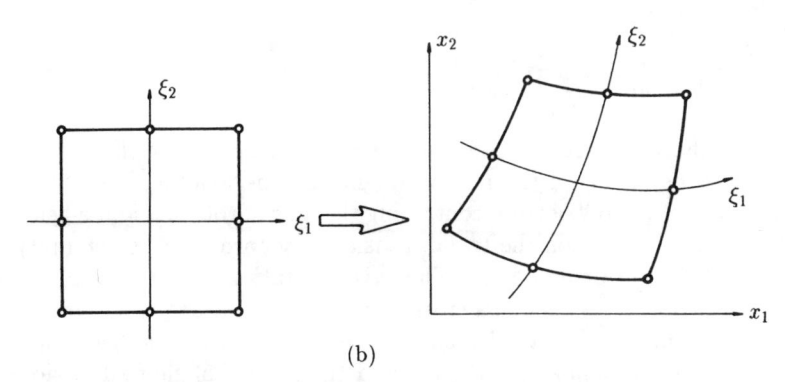

Figure 3.31: Two-dimensional isoparametric serendipity finite elements: (a) bilinear 4-node element, (b) quadratic 8-node element

The displacement vector for an arbitrary point of finite element e is given as

$$\mathbf{u}^e = \lfloor u_1 \ u_2 \rfloor^T . \tag{3.262}$$

The vector of the nodal displacements reads as follows:

$$\mathbf{q}^e = \lfloor u_1^{(1)} \ u_2^{(1)} \ u_1^{(2)} \ \ldots \ u_2^{(n)} \rfloor^T . \tag{3.263}$$

The matrix of the shape functions $\mathbf{N}^e$, employed for both the mapping of parent elements and the description of the element displacements, can be written formally as in (3.213). For plane stress elements, however, $\mathbf{I}$ represents a 2×2 unit matrix. The shape functions $N^{(i)}$ for the 4-node and the 8-node element are given in (3.221) and (3.222).

Plane Strain Elements

If the longitudinal extension of a structure is considerably larger than characteristic dimensions of its cross-section, a state of plane strain will occur, provided the loading and the boundary conditions do not depend on the longitudinal direction. Assuming a typical cross-section of such a structure to lie in the x_1-x_2 plane of a Cartesian coordinate system, plane strain conditions will be encountered, if the deformations are restricted to the x_1-x_2 plane, i.e., if the components $\varepsilon_{31}, \varepsilon_{32}$ and ε_{33} of the strain tensor are equal to zero. In the constitutive equations for a general three-dimensional state of stress, plane strain conditions are accounted for by setting the respective strain components equal to zero.

A typical example for plane strain conditions is a long dam subjected to water pressure. For such conditions it is sufficient to analyze a slice of unit thickness. Hence, the two-dimensional isoparametric serendipity finite elements of Fig.3.31 can be employed. The displacement vector for an arbitrary point of finite element e, $\mathbf{u}^e$, the vector of the nodal displacements, $\mathbf{q}^e$, and the matrix of shape functions, $\mathbf{N}^e$, are identical to the respective vectors and the respective matrix for plane stress elements. Plane strain elements differ from plane stress elements only by the constitutive relations.

Axisymmetric Elements

For a structure characterized by axial symmetry of the geometric shape, the boundary conditions, the layout of the reinforcement and the loading, it is sufficient to discretize one half of a cross-section containing the axis of rotation and a generatrix. Hence, for such a cross-section the two-dimensional isoparametric serendipity finite elements of Fig.3.31 can be employed. The axis of rotation, denoted as x_1, and any axis x_2, perpendicular to x_1, define a cross-sectional plane. The displacement vector $\mathbf{u}^e$ of an arbitrary point of finite element e located in such a cross-sectional plane, the vector of the nodal displacements, $\mathbf{q}^e$, and the matrix of shape functions, $\mathbf{N}^e$, are identical to the respective vectors and the respective matrix for plane stress or plane strain elements. The components of $\mathbf{u}^e$ refer to the axial and the radial displacement. However, in spite of the two-dimensional discretization, axisymmetric finite elements are ringlike three-dimensional elements. In addition, axisymmetric elements differ from plane stress or plane strain elements by the constitutive equations. These relations are obtained from the constitutive equations for a general three-dimensional state of stress by setting the components σ_{31} and σ_{32} of the stress tensor and the components ε_{31} and ε_{32} of the strain tensor equal to zero. The subscripts 1, 2, 3 refer to the axial, the radial and the circumferential coordinate of a cylindrical coordinate system. For axisymmetric shells it is sufficient to discretize a generatrix. The displacement state of axisymmetric finite shell elements is described by means of axial and radial nodal displacements and nodal rotations of the normal [Zienkiewicz (1989,1991), Yuan (1989)].

3.4 Finite Element Modelling of Reinforced and Prestressed Concrete

Beam Elements

Analogous to plate and shell elements, the formulation of beam elements can either be based on the Timoshenko beam theory, suitable for deep beams, or on the Bernoulli hypothesis, suitable for slender beams. The Timoshenko beam theory is based on a kinematic hypothesis which is equivalent to the Reissner-Mindlin hypothesis for moderately thick plates. It requires planes normal to the axis of the undeformed beam to remain plane and undeformed during the deformation process. Consequently, the Timoshenko beam theory yields constant transverse shear deformations. Hence, the displacements of an arbitrary point of the beam can be described in terms of the displacements of the corresponding point on the axis of the beam and of the rotations of the respective cross-section about its principal axes of inertia. The Bernoulli hypothesis requires the cross-sections not only to remain plane and undeformed during the deformation process but also to remain normal to the axis of the deformed beam. Hence, the rotations of the cross-sections can be expressed in terms of the first derivatives of the displacement components of the axis of the beam. Consequently, similar to Kirchhoff plate elements, C_1-interelement-continuous functions must be chosen for the displacements.

In what follows, beam elements based on the Timoshenko beam theory with independent interpolations for the displacements and the rotations will be formulated (Fig.3.32). This formulation is well suited for beams of arbitrary orientation in the three-dimensional Euclidean space as well as for nonlinear analyses. A further advantage of this formulation is that only C_0-interelement continuity is required for the displacements and the rotations. The beam elements are resulting from degeneration of a three-dimensional solid. Hence, they are referred to as degenerate elements.

For simplicity, the cross-sections of the beam elements are assumed to be rectangular. Thus, the shear center coincides with the center of gravity of the cross-section. Moreover, it is assumed that the shape and the area of the cross-section do not change during the deformation. This assumption requires the strains to be small. For each point on the axis of the beam in the current, deformed configuration a local x'_1, x'_2, x'_3 Cartesian coordinate system with the orthonormal base vectors $\mathbf{v}_i$, $i = 1, 2, 3$, is established. The base vector $\mathbf{v}_1$ is tangent to the beam axis and the base vectors $\mathbf{v}_2$ and $\mathbf{v}_3$ are chosen parallel to the principal axes of inertia of the cross-section.

Thus, within the framework of the isoparametric concept, the position vector $\mathbf{x}^e$ of an arbitrary point of the degenerate element in the current, deformed configuration is obtained by means of interpolation of the position vectors $\mathbf{x}_C^{(i)}$ of the node points located on the beam axis, the nodal base vectors $\mathbf{v}_2^{(i)}$ and $\mathbf{v}_3^{(i)}$ and the shape functions $N^{(i)}(\xi_1)$ as

$$\begin{aligned}\mathbf{x}^e(\xi_1,\xi_2,\xi_3) &= \mathbf{x}_C^e(\xi_1) + \frac{1}{2}h_2^e(\xi_1)\,\xi_2\mathbf{v}_2^e + \frac{1}{2}h_3^e(\xi_1)\,\xi_3\mathbf{v}_3^e \\ &\approx \sum_{i=1}^n N^{(i)}(\xi_1)\left[\mathbf{x}_C^{(i)} + \frac{1}{2}h_2^{(i)}\xi_2\mathbf{v}_2^{(i)} + \frac{1}{2}h_3^{(i)}\xi_3\mathbf{v}_3^{(i)}\right],\end{aligned} \quad (3.264)$$

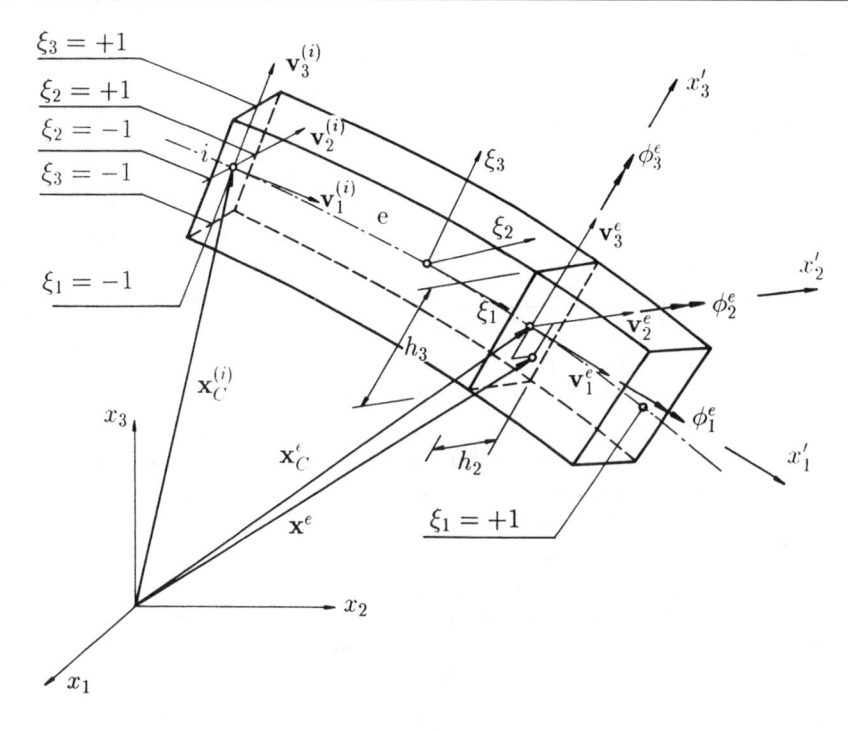

Figure 3.32: Finite beam element derived on the basis of the degenerate solid approach

where $h_2(\xi_1)$ and $h_3(\xi_1)$ are the dimensions of the rectangular cross-section and ξ_i, $i = 1, 2, 3$, are local coordinates ranging from -1 to $+1$. For a linear isoparametric beam element, the shape functions are given as

$$N^{(i)}(\xi_1) = \frac{1}{2}(1 + \xi_1^{(i)}\xi_1), \quad i = 1, 2. \tag{3.265}$$

For a quadratic isoparametric beam element, the shape functions for the corner nodes ($i = 1, 2$) and for the midside node ($i = 3$) are given as

$$N^{(i)}(\xi_1) = \frac{1}{2}\left[\xi_1^{(i)}\xi_1 + (\xi_1)^2\right], \quad i = 1, 2,$$
$$N^{(i)}(\xi_1) = 1 - (\xi_1)^2, \quad i = 3. \tag{3.266}$$

Replacing $\mathbf{x}_C^e, \mathbf{v}_2^e$ and $\mathbf{v}_3^e$ by $\mathbf{X}_C^e, \mathbf{V}_2^e$ and $\mathbf{V}_3^e$ yields the respective relation for the initial, undeformed configuration. Hence, substitution of (3.264) into (3.13) yields the displacements, referred to the global Cartesian coordinate system, as

$$\mathbf{u}^e(\xi_1, \xi_2, \xi_3) = \mathbf{u}_C^e(\xi_1) + \frac{1}{2}h_2^e(\xi_1)\xi_2(\mathbf{v}_2^e - \mathbf{V}_2^e) + \frac{1}{2}h_3^e(\xi_1)\xi_3(\mathbf{v}_3^e - \mathbf{V}_3^e)$$

3.4 Finite Element Modelling of Reinforced and Prestressed Concrete

$$\approx \sum_{i=1}^{n} N^{(i)}(\xi_1) \left[\mathbf{u}_C^{(i)} + \frac{1}{2} h_2^{(i)} \xi_2 (\mathbf{v}_2^{(i)} - \mathbf{V}_2^{(i)}) + \frac{1}{2} h_3^{(i)} \xi_3 (\mathbf{v}_3^{(i)} - \mathbf{V}_3^{(i)}) \right], \tag{3.267}$$

where $\mathbf{u}_C = \mathbf{x}_C - \mathbf{X}_C$.

For the case of small rotations, the second and the third term in (3.267) can be expressed by means of the local rotations ϕ_i, $i = 1, 2, 3$, about the base vectors $\mathbf{V}_i$ as

$$\begin{aligned}
\mathbf{u}^e(\xi_1, \xi_2, \xi_3) &= \mathbf{u}_C^e(\xi_1) + \frac{1}{2}(\phi_2^e h_3^e \xi_3 - \phi_3^e h_2^e \xi_2) \mathbf{V}_1^e - \frac{1}{2} \phi_1^e h_3^e \xi_3 \mathbf{V}_2^e + \frac{1}{2} \phi_1^e h_2^e \xi_2 \mathbf{V}_3^e \\
&\approx \sum_{i=1}^{n} N^{(i)}(\xi_1) \left[\mathbf{u}_C^{(i)} + \frac{1}{2} \left(\phi_2^{(i)} h_3^{(i)} \xi_3 - \phi_3^{(i)} h_2^{(i)} \xi_2 \right) \mathbf{V}_1^{(i)} \right. \\
&\quad \left. - \frac{1}{2} \phi_1^{(i)} h_3^{(i)} \xi_3 \mathbf{V}_2^{(i)} + \frac{1}{2} \phi_1^{(i)} h_2^{(i)} \xi_2 \mathbf{V}_3^{(i)} \right].
\end{aligned} \tag{3.268}$$

Independent C_0-interelement-continuous interpolations of the displacements and the rotations are characteristic features of degenerate solid elements.

The displacement vector of an arbitrary point of the degenerate element e is given according to (3.211). The vector of the nodal displacements for this element reads as follows:

$$\mathbf{q}^e = \lfloor (u_1)_C^{(1)} \; (u_2)_C^{(1)} \; (u_3)_C^{(1)} \; \phi_1^{(1)} \; \phi_2^{(1)} \; \phi_3^{(1)} \; (u_1)_C^{(2)} \ldots \phi_3^{(n)} \rfloor^T. \tag{3.269}$$

Alternatively, for small rotations the vectors $(\mathbf{v}_2 - \mathbf{V}_2)$ and $(\mathbf{v}_3 - \mathbf{V}_3)$ in (3.267) can be expressed, similar to (3.228), in terms of the rotations $\boldsymbol{\phi} = \lfloor \phi_{x_1} \; \phi_{x_2} \; \phi_{x_3} \rfloor^T$ about the axes of the global Cartesian coordinate system:

$$\mathbf{v}_2^e - \mathbf{V}_2^e = \boldsymbol{\phi} \times \mathbf{V}_2^e, \quad \mathbf{v}_3^e - \mathbf{V}_3^e = \boldsymbol{\phi} \times \mathbf{V}_3^e. \tag{3.270}$$

With respect to the local coordinates $x_1' \equiv x'$, $x_2' \equiv y'$, $x_3' \equiv z'$ the non-vanishing components of the stress tensor are the axial stress $\sigma_{x'}$ and the shear stresses $\tau_{x'y'}$ and $\tau_{x'z'}$. Hence, in terms of these coordinates the constitutive equations for a linear-elastic material are given as

$$\left\{ \begin{array}{c} \sigma_{x'} \\ \tau_{x'y'} \\ \tau_{x'z'} \end{array} \right\} = \left[\begin{array}{ccc} E & 0 & 0 \\ 0 & \kappa G & 0 \\ 0 & 0 & \kappa G \end{array} \right] \left\{ \begin{array}{c} \varepsilon_{x'} \\ \gamma_{x'y'} \\ \gamma_{x'z'} \end{array} \right\}. \tag{3.271}$$

For rectangular cross-sections, the value of the shear correction factor κ is equal to $5/6$. The strains $\varepsilon_{x'}$, $\gamma_{x'y'}$ and $\gamma_{x'z'}$ are computed by transforming the displacement components from global to local coordinates and substituting the so-obtained results into the expression for the components of the linearized Cauchy strain tensor (see equation (3.9)).

For the case of nonlinear material behavior, (3.271) must be replaced by appropriate total or rate constitutive equations. Analogous to subdividing shell elements across the thickness into several thin layers, beam elements are subdivided into thin layers, perpendicular either to $\mathbf{V}_2$ or to $\mathbf{V}_3$. For each one of these layers the appropriate constitutive equations are employed. If geometric nonlinearity is taken into account, the linearized contributions of the rotations to the displacements according to (3.268) or (3.270) can still be used, provided the load increments are chosen such that only small incremental rotations occur [Bathe (1982)]. If this restriction is abandoned, the contributions of the rotations to the displacements must be expressed as nonlinear functions of the rotations (see, e.g., [Dvorkin (1988)]).

For the case of very slender beams, degenerate solid elements will exhibit shear locking. Procedures to avoid shear locking are similar to the respective procedures for shell elements derived on the basis of the degenerate solid approach.

Degenerate elements can also be employed to model edge beams and stiffeners of plates and shells. In consequence of the similarity of beam elements to shell elements based on the degenerate solid concept, such elements can easily be combined in a finite element model. Edge beams or stiffeners are assumed to be rigidly attached to the shell. Hence, along the intersection of the beam and the shell the displacements and the rotations of the beam and the shell must be the same. In order to satisfy this requirement, the beam elements must coincide with the boundaries of the shell elements. The order of the interpolation polynomials for the displacements as well as for the rotations must be the same for both types of elements. Finally, the nodal rotations of the shell elements and the beam elements must be referred to a common basis.

If the axis of the stiffening beam does not lie in the middle surface of the shell, then, for the case of small rotations, the displacement $\mathbf{u}_C$ of the center of gravity of a typical cross-section of the beam can be expressed in terms of the displacement and the rotations of the corresponding point on the middle surface of the shell (Fig.3.33):

$$\mathbf{u}_C = \mathbf{u}_0 + \boldsymbol{\phi} \times e\mathbf{V}_3 . \qquad (3.272)$$

In (3.272) $\mathbf{u}_0$ denotes the displacement vector of this point, $\boldsymbol{\phi}$ is the vector of rotations about the axes of the global Cartesian coordinate system at this point and e is the distance between the center of gravity of the cross-section of the beam and its projection onto the middle surface of the shell. This distance is a positive quantity if the mentioned center of gravity is located on the same side of the middle surface of the shell as the shell director vector $\mathbf{V}_3$.

3.4.3 Representation of the Reinforcing Steel

Discrete Modelling

Discrete representation of the reinforcement is based on modelling the reinforcing bars as separate elements. Commonly, truss or cable elements are used for this

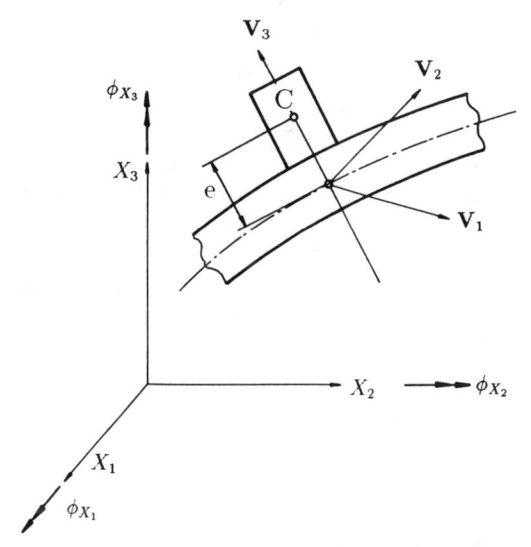

3.4 Finite Element Modelling of Reinforced and Prestressed Concrete

Figure 3.33: Connexion of beam element with shell element (both elements are based on the degenerate solid concept)

purpose (Fig.3.34). However, for the investigation of structural details, occasionally two-dimensional or even three-dimensional elements are used for modelling of the reinforcement. Truss and cable elements do not have rotational degrees of freedom because they can only carry axial forces. The displacement vector $\mathbf{u}^e$ of an arbitrary point of the axis of a truss or cable element e can be computed with the help of (3.38). For three-dimensional problems, the vectors $\mathbf{u}^e$, $\mathbf{q}^e$ and the matrix $\mathbf{N}^e$ which is employed for both the mapping of parent elements and the description of the element displacements, are given according to (3.211), (3.212) and (3.213). For two-dimensional problems, $\mathbf{u}^e$ and $\mathbf{q}^e$ are defined according to (3.262) and (3.263). The 3×3 unit matrix $\mathbf{I}$ in (3.213) is replaced by a 2×2 unit matrix. For a 2-node truss element with linear interpolation of the displacements and, consequently, constant axial force, the shape functions $N^{(i)}$ are given in (3.265). For a 3-node cable element with quadratic interpolation of the displacements, the shape functions are specified in (3.266).

The material behavior of truss and cable elements is described by means of the one-dimensional constitutive relations presented in section 2.5. In order to guarantee compatibility of the displacements of the concrete and the reinforcement, truss and cable elements must coincide with the boundaries of the concrete elements. The node points of both types of elements must also coincide. Hence, the shape functions for the concrete elements and the truss or cable elements must be of the same order. E.g., three-dimensional isoparametric trilinear 8-node elements and two-dimensional isoparametric bilinear 4-node elements for the representation of concrete are compatible with linear 2-node truss elements for the reinforcing steel. Three-dimensional isoparametric quadratic 20-node elements and two-dimensional

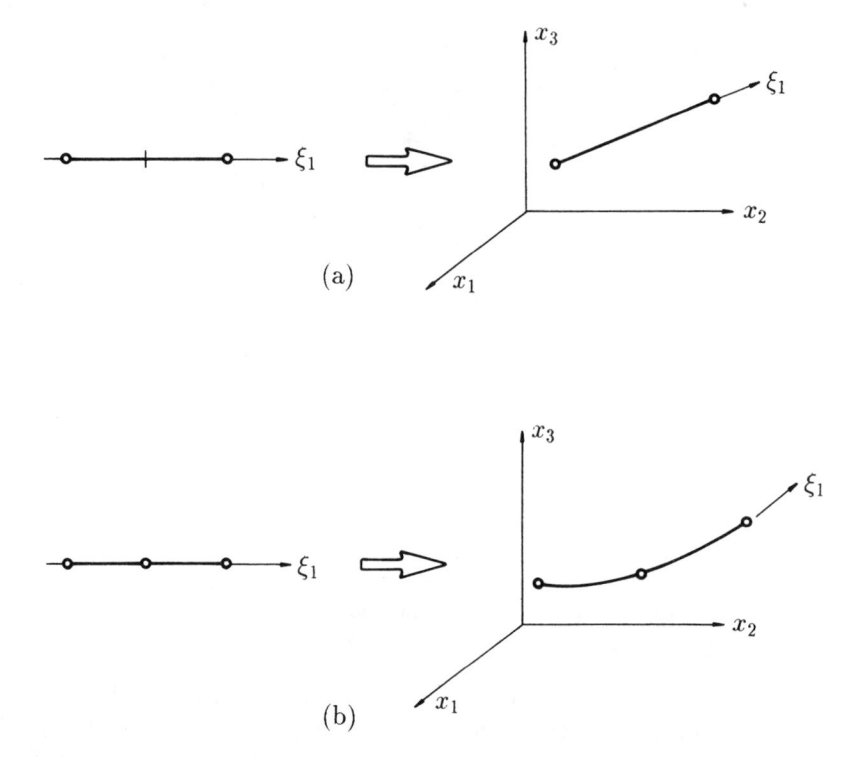

Figure 3.34: One-dimensional isoparametric finite elements: (a) linear 2-node truss element, (b) quadratic 3-node cable element

isoparametric quadratic 8-node elements for the representation of concrete are compatible with quadratic 3-node cable elements for the reinforcing bars. For axisymmetric analyses, reinforcing bars in the circumferential direction are represented by single node axisymmetric ring elements.

The location of the steel elements is determined by the layout of the reinforcement. Consequently, the boundaries of the concrete elements must follow the reinforcing bars. Thus, the layout of the reinforcement has a strong influence on the generation of the finite element mesh for a concrete structure. The discrete representation of the reinforcing steel by means of truss elements will be employed in subsection 4.2.1 for the analysis of a short reinforced concrete cantilever beam.

Commonly, when the overall structural behavior is investigated, coinciding nodes of concrete and steel elements are assigned the same degrees of freedom. Bond slip and dowel action are either disregarded or considered implicitly by modifying the constitutive relations of concrete or steel. However, especially for the investigation of the behavior of structural details such as, e.g., the anchorage of prestressing tendons, it may be necessary to model bond slip and dowel action more accurately. For this purpose, different degrees of freedom are assigned to coinciding nodes of concrete and steel elements. Special interface elements, referred to as bond or contact elements, are employed to connect the different degrees of freedom of coinciding nodes of steel and concrete elements. Simple interface elements connect a single node of a concrete element with a single node of a steel element [Ngo (1967)]. Such elements are springs with a nonlinear relationship between nodal forces and respective relative nodal displacements, i.e., differences of displacements of the concrete and the reinforcing steel at a particular node. An alternative to nodal interface elements are continuous interface elements [Mehlhorn (1985), Keuser (1987), Mehlhorn (1990)]. Such elements are characterized by a continuous concrete-steel interface along the entire length of the reinforcing bars. Compared with nodal interface elements, their performance is better [Keuser (1987)]. Discrete steel elements and continuous interface elements can be combined to steel-interface elements [Miguel (1990)], Such elements allow modelling of the behavior of both the reinforcing bar and the interface. Interface elements will be described in subsection 3.4.5.

The main advantage of modelling reinforced concrete by superposition of concrete and steel elements is the relatively accurate representation of the mechanical behavior of the reinforcement and the interface. The discrete representation of the reinforcement is the only way of accounting for bond slip and dowel action directly. Disadvantages of this approach are the great effort required for the discretization of a structure and the significant increase of the number of degrees of freedom. These disadvantages are the consequence of having to consider each reinforcing bar in the finite element mesh. Therefore, discrete modelling of the reinforcement is restricted to the analysis of structural details.

Embedded Modelling

Separate elements for the concrete and the reinforcing steel are also used for the embedded representation of the reinforcement. However, for this representation of the reinforcement, the same type of elements with the same number of nodes and degrees of freedom and, consequently, the same shape functions are used for the concrete and the reinforcement. Hence, the embedded approach is characterized by incorporating the one-dimensional reinforcing bars into two- or three-dimensional elements. Fig.3.35 shows a reinforcing bar embedded in a bilinear isoparametric serendipity element. The stiffness matrix and the internal force vector of embedded reinforcement elements only contain the contributions of the reinforcing bars. They are computed by integration along the curves representing the segments of the reinforcing bars within the respective element. The embedded reinforcement elements are then superimposed on the respective concrete elements. The reinforcing bars do not have to follow the boundaries of the concrete elements. Hence, the embedded representation of the reinforcement allows generating a finite element mesh without taking much care about the layout of the reinforcement. Rather, the reinforcing bars may pass through the concrete elements in an arbitrary manner. Since the reinforcement elements and the concrete elements must be assigned the same degrees of freedom, perfect bond between concrete and steel is obtained. Hence, bond slip and dowel action can only be modelled implicitly by modifying the constitutive

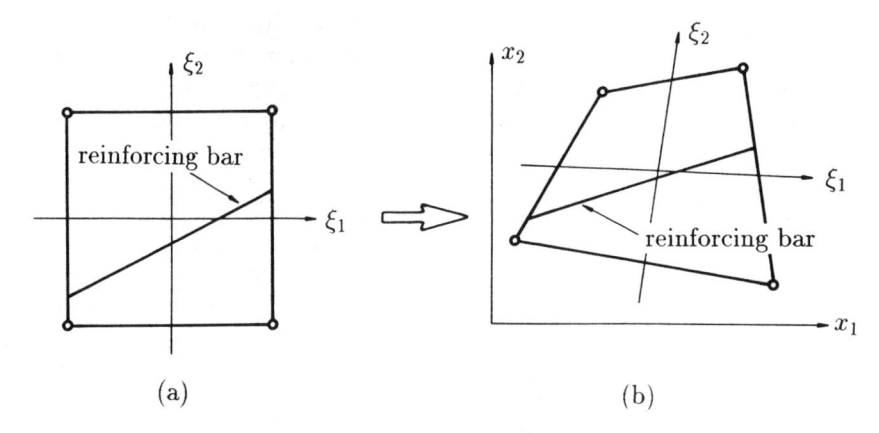

Figure 3.35: Embedded steel element: (a) in the local coordinate system, (b) in the global Cartesian coordinate system

relations for concrete or steel. A disadvantage of this type of representation of the reinforcement is that special reinforcement elements are required. Such elements may not exist in the available finite element program. Moreover, similar to the discrete approach, each reinforcing bar must be considered when preparing the input data for a finite element analysis.

The embedded approach has also been used for modelling of the tendons of prestressed concrete structures. Subsection 3.4.4 contains a detailed description of the embedded representation of tendons.

Distributed Modelling

The distributed representation of the reinforcement is characterized by smearing reinforcing bars to thin layers of mechanically equivalent thickness within a particular concrete element. This approach is suited for regions of uniformly or nearly uniformly distributed reinforcing bars. Such conditions are mainly encountered in reinforced concrete surface structures. Reinforcing bars, smeared to a layer of mechanically equivalent thickness, must have the same direction and the same distance from the middle surface of a surface structure. E.g., concrete barrel vaults are typically reinforced with one reinforcing mesh close to the upper face of the shell and one close to the lower face. The reinforcing bars are parallel and normal to the longitudinal direction. Diagonal reinforcement is supplemented in the edge regions. The reinforcing meshes are parallel to the middle surface.

Such a reinforced concrete shell is conveniently modelled by composite concrete - steel elements based on the layer concept which was introduced in subsection 3.4.2. This concept will be employed in chapter 4 for the representation of the reinforcing steel in the finite element models for the analysis of reinforced and prestressed concrete surface structures. In contrast to the discrete and the embedded representation

3.4 Finite Element Modelling of Reinforced and Prestressed Concrete

of the reinforcement, the concrete and the reinforcing bars are combined to composite finite elements. Composite plate or shell elements, e.g., consist of several layers of concrete and of reinforcing steel, parallel to the middle surface. The constitutive equations for a typical layer of smeared unidirectional reinforcement are referred to the local directions 1 and 2, which are parallel and normal to the reinforcing bars, respectively. In rate form, these material relations are given as:

$$\left\{\begin{array}{c} \dot{\sigma}_1 \\ \dot{\sigma}_2 \\ \dot{\tau}_{12} \end{array}\right\} = \left[\begin{array}{ccc} E_T^S & 0 & 0 \\ 0 & 0 & 0 \\ 0 & 0 & 0 \end{array}\right] \left\{\begin{array}{c} \dot{\varepsilon}_1 \\ \dot{\varepsilon}_2 \\ \dot{\gamma}_{12} \end{array}\right\}, \qquad (3.273)$$

where E_T^S is the tangent material modulus of the reinforcing steel. To obtain the contribution of the steel layer to the tangent stiffness matrix of the composite element, the tangent material stiffness matrix in (3.273) is transformed to the global coordinate system, using (2.44). The incremental nodal displacements and the incremental strains are referred to the global coordinate system. Thus, computation of the stress increments in the reinforcing bars must be preceded by transformation of the actual strains to the direction parallel to the reinforcing bars, i.e., to the local direction 1.

A combination of the distributed and the embedded representation of the reinforcement is obtained by smearing the reinforcement to thin layers, embedding the smeared layers into elements of the same type as the concrete elements and superimposing these elements on the concrete elements. This approach is convenient, e.g., for three-dimensional concrete structures with arbitrarily oriented layers of reinforcement.

Combining concrete and steel within an element requires the assumption of perfect bond between the concrete and the steel layers. Hence, bond slip and dowel action can only be modelled implicitly by modifying the constitutive relations of concrete or steel.

The advantages and disadvantages of the distributed approach are similar to the ones of the embedded representation. However, smearing of the reinforcement to thin layers of mechanically equivalent thickness only makes sense for uniformly distributed reinforcing bars.

3.4.4 Representation of Tendons

A popular method to account for the tendons in prestressed concrete structures is the load-balancing concept [Leonhardt (1980), Naaman (1982)]. It is characterized by treating the reinforced concrete structure and each one of the tendons as a free body. Thus, application of the principle of virtual displacements to the reinforced concrete structure requires consideration of the forces exerted from the tendons on the adjacent concrete (Fig.3.36).

For the special case of only one tendon, the mathematical formulation of the principle of virtual displacements, referred to the initial configuration of the reinforced concrete structure, follows from (3.36) as

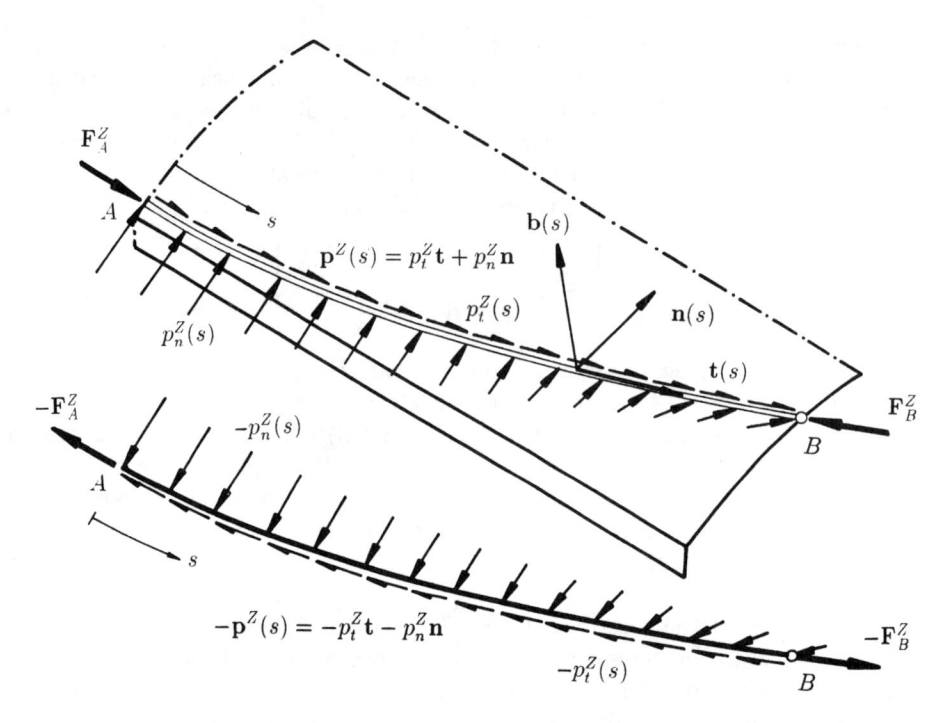

Figure 3.36: Concrete structure and tendon treated as free bodies within the framework of the load-balancing concept for consideration of prestressing

$$-\int_{V_0}(\delta\mathbf{E})^T\mathbf{S}\,dV_0 + \int_{V_0}(\delta\mathbf{u})^T\bar{\mathbf{b}}\,dV_0 + \int_{S_0^\sigma}(\delta\mathbf{u})^T\bar{\mathbf{t}}\,dS_0^\sigma$$

$$+ \int_{l_0}(\delta\mathbf{u}(s))^T\mathbf{p}^Z(s)ds + (\delta\mathbf{u}_A)^T\mathbf{F}_A^Z + (\delta\mathbf{u}_B)^T\mathbf{F}_B^Z = 0\,. \quad (3.274)$$

In (3.274) the first term refers to the virtual work of the internal forces of the concrete and the reinforcing steel. The second and the third term represent the virtual work done by the volume forces and the surface tractions acting on the reinforced concrete structure. The three remaining terms refer to the virtual work done by the forces exerted from the tendon on the reinforced concrete structure. $\mathbf{p}^Z(s)$, $\mathbf{F}_A^Z$ and $\mathbf{F}_B^Z$ denote the vector of line loads and the vectors of the anchor forces, respectively. The parameter s is the arc length of the tendon measured from point A; l_0 denotes the total length of the tendon.

The mathematical formulation of the principle of virtual displacements for the tendon, treated as a free body, requires consideration of the forces $-\mathbf{p}^Z(s)$, $-\mathbf{F}_A^Z$ and $-\mathbf{F}_B^Z$ exerted from the concrete on the tendon, yielding

3.4 Finite Element Modelling of Reinforced and Prestressed Concrete

$$- A^Z \int_{l_0} \delta E^Z(s) S^Z(s) \ ds - \left[\int_{l_0} (\delta \mathbf{u}(s))^T \mathbf{p}^Z(s) ds + (\delta \mathbf{u}_A)^T \mathbf{F}_A^Z + (\delta \mathbf{u}_B)^T \mathbf{F}_B^Z \right] = 0 \ .$$
(3.275)

In (3.275) δE^Z and S^Z denote the virtual axial strain and the axial stress in the tendon; A^Z is the area of the cross-section of the tendon.

In the sum of (3.274) and (3.275) the virtual work done by the forces $\mathbf{p}^Z(s)$, $\mathbf{F}_A^Z$ and $\mathbf{F}_B^Z$ cancels out. Hence, the virtual work done by the internal and the external forces acting on the prestressed concrete structure is obtained as

$$- \int_{V_0} (\delta \mathbf{E})^T \mathbf{S} \ dV - A^Z \int_{l_0} \delta E^Z(s) S^Z(s) \ ds + \int_{V_0} (\delta \mathbf{u})^T \bar{\mathbf{b}} \ dV_0 + \int_{S_0^\sigma} (\delta \mathbf{u})^T \bar{\mathbf{t}} \ dS_0^\sigma = 0 \ .$$
(3.276)

Thus, alternatively to (3.274), the effect of prestressing can be taken into account on the basis of (3.276). This was done in [Povoas (1989)].

In any case, the crucial point of modelling of the tendons is the description of their geometric properties, because both the forces exerted from the tendons on the concrete and the internal forces acting in the tendons depend on the layout of the tendons.

Therefore, out of the three different approaches for the representation of reinforcing bars, the embedded representation is best suited for the modelling of tendons. This approach allows the accurate description of the geometric properties of a tendon without having to care much about the tendon layout during mesh generation.

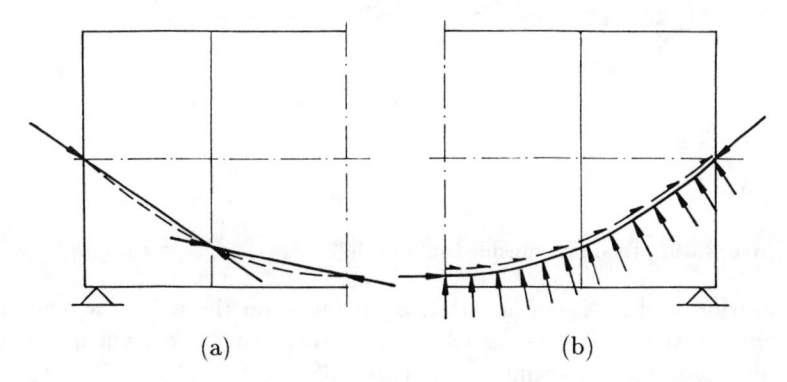

Figure 3.37: Modelling of a tendon: (a) approximation by straight segments, (b) representation as a curve

Early models for consideration of tendons are characterized by approximating the geometric shape of the tendon axes by a series of straight segments. Their end points are located on the interfaces of finite elements representing the concrete [Kang (1980), Van Greunen (1983)] (Fig.3.37(a)). In this approach the line loads resulting from the curvature of the tendons are approximated by point loads acting at the ends of the tendon segments. Thus, for coarse meshes the quality of the approximation of the influence of prestressing on the structural response is poor. Moreover, the preparation of the input data is a tedious task. For each tendon, all segments must be specified along with the numbers of the respective finite concrete elements.

Analytical Description of the Tendon Geometry

Improved models for consideration of tendons in prestressed concrete panels, plates and shells are characterized by a more accurate description of the geometric shape of the tendon axes (Fig.3.37(b)) [Hofstetter (1986a), Hofstetter (1986b), Hofstetter (1987), Roca (1988), Povoas (1989), Roca (1993a), Roca (1993b)].

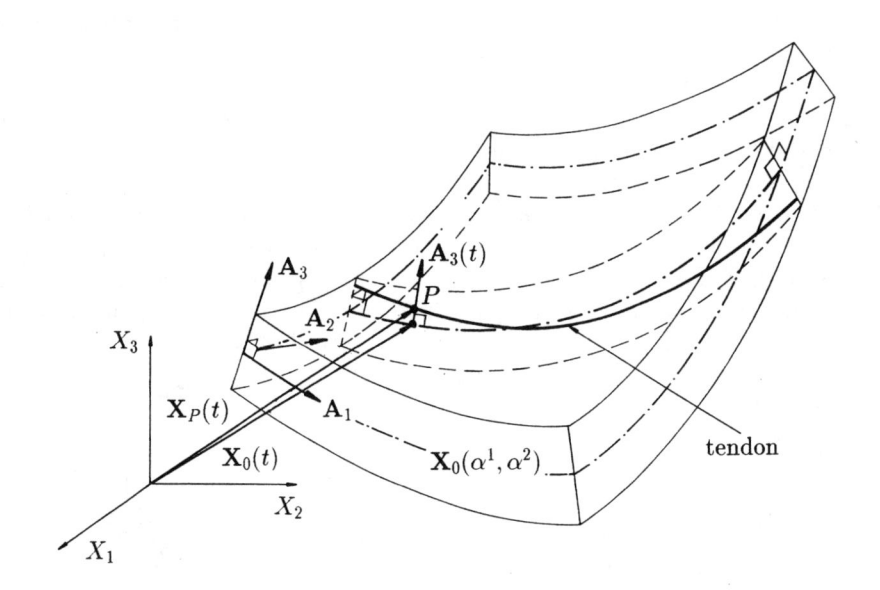

Figure 3.38: Tendon embedded in a shell

The position vector $\mathbf{X}_P$ of an arbitrary point P on the axis of a tendon in the initial configuration of a prestressed concrete shell can be written in terms of an arbitrarily chosen curve parameter t as (Fig.3.38)

$$\mathbf{X}_P(t) = \mathbf{X}_0(t) + \alpha^3(t)\mathbf{A}_3(t) , \qquad (3.277)$$

where

$$\mathbf{X}_0(t) = \mathbf{X}_0(\alpha^1(t), \alpha^2(t)) \qquad (3.278)$$

represents a surface curve, located in the middle surface $\mathbf{X}_0(\alpha^1, \alpha^2)$ of the prestressed concrete shell.

3.4 Finite Element Modelling of Reinforced and Prestressed Concrete

In contrast to (3.277), in (3.274) - (3.276) the curve describing the course of a tendon is given in terms of the arc length s. Taking an arbitrarily chosen curve parameter t instead of s, the integrations over the arc length in (3.274) - (3.276) require formulation of the relationship between a differential of the arc length, ds, and a differential of the curve parameter, dt. This relation is given as

$$ds = \sqrt{\frac{d\mathbf{X}_P(t)}{dt} \cdot \frac{d\mathbf{X}_P(t)}{dt}}\, dt \ . \qquad (3.279)$$

Choosing the curve parameter t to be equal to one of the two surface coordinates, the explicit form of the analytic representation of a surface curve is obtained as

$$\alpha^2 = \alpha^2(\alpha^1) \quad \text{or} \quad \alpha^1 = \alpha^1(\alpha^2) \ . \qquad (3.280)$$

The first (second) one of these two alternative representations allows consideration of the special case $\alpha^2 = \text{const.}$ ($\alpha^1 = \text{const.}$).

The absolute value of the second term on the right-hand side of (3.277) is equal to the distance $\alpha^3(t)$ of the tendon axis from the middle surface of the shell, measured in the direction of the vector $\mathbf{A}_3$ (see equation (3.237)) normal to the tangent plane of the middle surface of the shell. Obviously, the inequality $\alpha^3(t) < h(t)/2$ must hold, where $h(t) = h(\alpha^1(t), \alpha^2(t))$ denotes the thickness of the shell at the respective point of the tendon axis. In prestressed concrete shells and panels the axes of the tendons are usually located in the middle surface, i.e., $\alpha^3(t) = 0$. In concrete slabs, however, the location of the tendon axes is eccentric with respect to the middle surface.

In order to compute work-equivalent node forces of the forces exerted from the tendons on the reinforced concrete structure, the numbers of the elements passed through by the individual tendons must be determined. Each tendon is subdivided into a number of segments such that each segment represents the section of the tendon within a particular finite concrete element. Moreover, the points of intersection of the tendon axes with the boundaries of the finite concrete elements passed through by the tendon must be determined.

Geometric Modelling of Tendons Embedded into Shell Elements Based on Shell Theory

The starting point for the derivation of such elements is the mathematical description of the middle surface of the shell. Usually, this description is available in analytical form. Hence, the analytical description of the geometric form of the tendon axis is ideally suited for geometric modelling of tendons embedded into elements based on shell theory. The node points of such elements are defined in terms of the surface coordinates α^1 and α^2 (see Fig.3.29). The parametric mapping of an edge of an element, with corner node points only, onto the α^1-α^2 plane is a straight line. The surface curve according to (3.280) represents the parametric mapping of the projection of the tendon axis to the middle surface of the shell, i.e., of $\mathbf{X}_0(t)$, $t = \alpha^1 \vee \alpha^2$, onto the α^1-α^2 plane. Hence, the coordinates α^1 and α^2 of the points of intersection of the tendon axes with the edges of the element can be determined in the α^1-α^2 plane. Once these coordinates are known, the eccentricity $\alpha^3(t)$, with $t = \alpha^1 \vee \alpha^2$, of the tendon axis at these points of intersection can be computed.

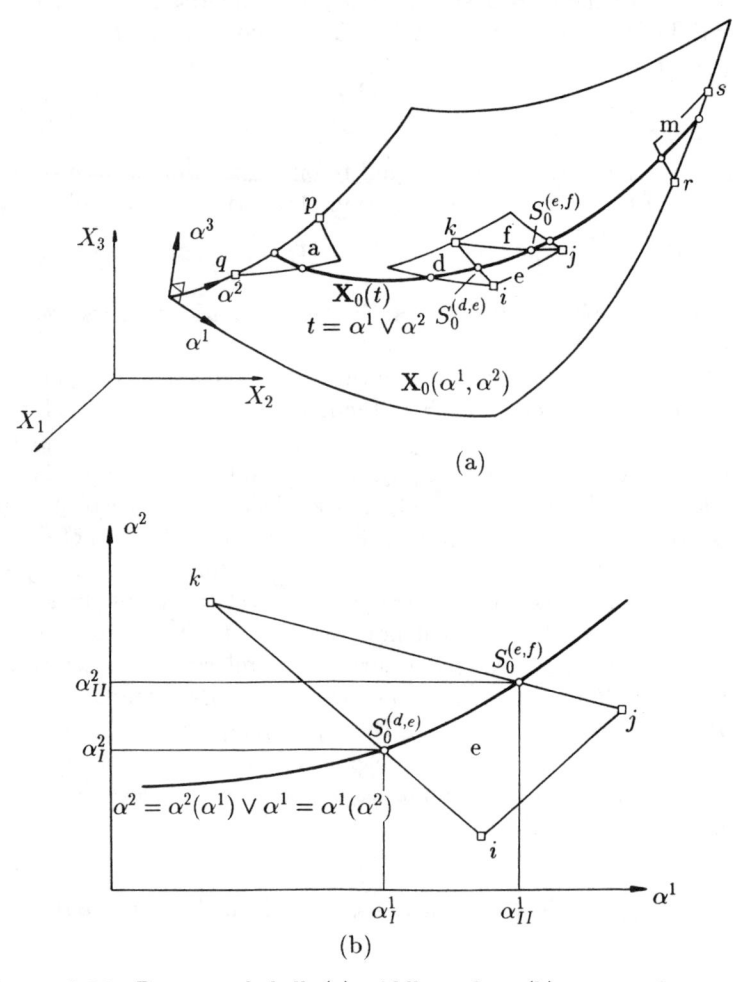

Figure 3.39: Prestressed shell: (a) middle surface, (b) parametric mapping of element e onto the α^1-α^2 plane

Let the numbers of the finite elements passed through by the tendon in Fig.3.39(a) be known up to element d. Furthermore, let the surface coordinates of the points of intersection of the tendon axis with the edges of these elements be known. The last one of these points, $S^{(d,e)}$, is located on the interface of elements d and e. The corresponding point on the surface curve $\mathbf{X}_0(t)$ is denoted as $S_0^{(d,e)}$ (Fig.3.39). The number of the next element passed through by the tendon, i.e., element e (see Fig.3.39), follows from the incidence table containing the numbers of all elements

3.4 Finite Element Modelling of Reinforced and Prestressed Concrete

and of their node points. The next point of intersection of the tendon axis with an element interface is point $S^{(e,f)}$. The projection of this point to the middle surface of the shell is denoted as $S_0^{(e,f)}$. Its surface coordinates are determined as the coordinates of the second point of intersection of the parametric mappings of the surface curve $\mathbf{X}_0(t)$, $t = \alpha^1 \vee \alpha^2$, and of the edges of element e onto the α^1-α^2 plane (see Fig.3.39(b)). The equation of the parametric mapping of the edge of element e containing the node points j and k can be written as

$$\alpha^2 = k^{(jk)} \alpha^1 + d^{(jk)} , \qquad (3.281)$$

where

$$k^{(jk)} = \frac{(\alpha^2)^{(k)} - (\alpha^2)^{(j)}}{(\alpha^1)^{(k)} - (\alpha^1)^{(j)}} , \quad d^{(jk)} = \frac{(\alpha^2)^{(j)}(\alpha^1)^{(k)} - (\alpha^2)^{(k)}(\alpha^1)^{(j)}}{(\alpha^1)^{(k)} - (\alpha^1)^{(j)}} . \qquad (3.282)$$

The mentioned point of intersection is obtained as a root of the nonlinear equation resulting from substitution of (3.281) into the left-hand side of (3.280$_1$). This root is determined in an iterative manner, using the regula falsi in the interval $[(\alpha^1)^{(j)}, (\alpha^1)^{(k)}]$. If the second point of intersection of the tendon axis with the boundaries of element e is not located on the interface of elements e and f, then no root will exist in the interval $[(\alpha^1)^{(j)}, (\alpha^1)^{(k)}]$.

For initiation and termination of the process of automatic determination of all points of intersection of the tendon axis with the edges of the finite elements passed through by the tendon, the number of the first and of the last element (elements a and m in Fig.3.39(a)) needs to be known. Moreover, the numbers of the node points defining those element edges, on which the two ends of the tendon axis are located (node points p, q and r, s, respectively, in Fig.3.39(a)), are required. If, instead of the first of the two equations (3.280), the second of these relations is used for specification of the parametric mapping of $\mathbf{X}_0(t)$ onto the α^1-α^2 plane, then the superscripts 1 and 2 in (3.281) will have to be exchanged.

Geometric Modelling of Tendons Embedded into Shell Elements of the Form of Degenerate Solid Elements

For the case of shell elements of the form of degenerate solid elements, the geometric properties of the shell are considered in an approximate manner by means of the parametric concept. For this case it has been proposed to employ the analytical representation of the tendon axes (see equation (3.277)) only for determination of the points of intersection of these axes with the interfaces of the degenerate elements. Within a particular shell element, however, the geometric form of the axis of a tendon is approximated by a parabola [Roca (1988), Roca (1993a), Roca (1993b)]. The surface coordinates α^1 and α^2 of the points of intersection of a tendon axis with the edges of the degenerate solid elements passed through by the tendon are determined in the α^1-α^2 plane. Hence, the middle surfaces of the degenerate solid elements must be defined in the α^1-α^2 plane. The projections of these points of intersection are determined in the same way as has been described for elements based on shell theory.

However, with regards to degenerate solid elements, it is not sufficient to know the surface coordinates α^1 and α^2 of these points of intersection. Since the middle surfaces of the degenerate elements are defined in terms of the natural coordinates ξ_1 and ξ_2 (see, e.g., the 4-node or the 8-node serendipity degenerate solid elements described in subsection 3.4.2), the mentioned projections of points of intersection also must be specified in terms of ξ_1 and ξ_2. For any point on the middle surface of the degenerate solid element e, an implicit relation for ξ_1 and ξ_2 is given by

$$\mathbf{X}_0(\alpha^1, \alpha^2) = \mathbf{X}_0^e(\xi^1, \xi^2) . \tag{3.283}$$

The left-hand side of (3.283) represents the analytical description of the middle surface of the shell (see equation (3.234)). The right-hand side of this equation represents the approximation of the middle surface of the part of the shell which is modelled by element e. This approximation is based on the parametric concept (see equation (3.220)). Consequently,

$$\mathbf{X}_0^e(\xi^1, \xi^2) = \sum_{i=1}^{n} N^{(i)}(\xi_1, \xi_2) \mathbf{X}_0^{(i)} . \tag{3.284}$$

Use of (3.283) for determination of ξ_1 and ξ_2 for given α^1 and α^2 is computationally expensive. In order to reduce this expense, it was proposed in Roca [Roca (1993b)] to use the approximate transformation rule

$$\left\{ \begin{array}{c} \alpha^1 \\ \alpha^2 \end{array} \right\} \approx \sum_{i=1}^{n} N^{(i)}(\xi_1, \xi_2) \left\{ \begin{array}{c} (\alpha^1)^{(i)} \\ (\alpha^2)^{(i)} \end{array} \right\} \tag{3.285}$$

for the mentioned purpose. In (3.285) $(\alpha^1)^{(i)}$ and $(\alpha^2)^{(i)}$ denote the surface coordinates of node point i of the degenerate solid element. For the sake of simplicity, in (3.285) the shape functions $N^{(i)}(\xi_1, \xi_2)$ of the 4-node serendipity element (3.221) were taken [Roca (1993b)].

For a shell element of the form of a degenerate solid element, the projection of a tendon segment to the middle surface of the element is approximated as [Roca (1993b)]

$$\xi_1 = \xi_1(\tau) \quad \text{and} \quad \xi_2 = \xi_2(\tau) , \tag{3.286}$$

where ξ_1 and ξ_2 are chosen as parabolic functions of the curve parameter τ. Within the respective shell element, $-1 \leq \tau \leq 1$. The six constants required for the definition of the two parabolic functions of (3.286) are determined from known data of the tendon geometry at the element boundaries such that C_1-interelement-continuity is guaranteed. Similarly, the eccentricity of a tendon with respect to the middle surface, now denoted as $e_3(t)$, is approximated in terms of τ from known data at the element boundaries such that C_1-interelement-continuity of the eccentricity is achieved at the element interfaces. Hence, the approximation of the geometric form of the tendon axis within a single degenerate solid element follows from the representation of the position vector $\mathbf{x}^e$ for an arbitrary point within a degenerate solid element according to (3.220) as

3.4 Finite Element Modelling of Reinforced and Prestressed Concrete

$$\begin{aligned}\mathbf{x}_P^e &= \mathbf{x}_0^e + e_3(t)\mathbf{v}_3^e \\ &\approx \sum_{i=1}^{n} N^{(i)}(\xi_1(\tau),\xi_2(\tau))\left[\mathbf{x}_0^{(i)} + e_3(\tau)\mathbf{v}_3^{(i)}\right] .\end{aligned} \qquad (3.287)$$

Geometric Modelling of Tendons Embedded into Beam Elements

(3.287) can be modified to describe the position vector $\mathbf{x}_P^e$ of an arbitrary point of a tendon located in a spatially curved beam. In this case, the eccentricity of the tendon with respect to the axis of the beam is described by the functions $e_2(t)$ and $e_3(t)$ measured in the directions of the local base vectors $\mathbf{v}_2$ and $\mathbf{v}_3$, which are parallel to the principal directions of the cross-section of the beam (see Fig.3.32). Hence, following from (3.264), $\mathbf{x}_P^e$ is given as

$$\mathbf{x}_P^e = \mathbf{x}_C^e + e_2(t)\mathbf{v}_2^e + e_3(t)\mathbf{v}_3^e . \qquad (3.288)$$

Geometrically and materially nonlinear time-dependent analyses of reinforced and prestressed space frames are described in [Mari (1984)].

Determination of Tendon Forces

The mechanical behavior of prestressed concrete structures is strongly influenced by the prestressing forces in the tendons. These forces depend on the jacking forces applied at one or both ends of the tendons and on instantaneous as well as time-dependent prestress losses.

Losses of prestressing forces depend on the method of prestressing. Because of relaxation of the prestressing steel and of shrinkage of concrete in pretensioned members, there are prestress losses already before the transfer of the prestress forces from the fixed bulkheads to the concrete. In addition, at this transfer instantaneous prestress losses occur. They are caused by the deformations of the concrete. In posttensioned members, the instantaneous prestress losses during the tensioning operations are the consequence of friction between the tendon and the duct and of slip of the tendon at the anchorage resulting from locking the tendon at the completion of jacking. Additional prestress losses will occur if the tensioning operations are carried out sequentially. These losses result from the shortening of the concrete in consequence of the prestressing of yet untensioned tendons.

For any method of prestressing, time-dependent prestress losses occur. They are caused by the relaxation of the prestressing steel and by creep and shrinkage of concrete. In addition, the tendon forces are influenced by the application or the removal of external loads after jacking. In any case, losses of prestress after completion of jacking are caused by changes of the strain state in the concrete adjacent to the tendon.

Tendon forces resulting from prestressing Prestress losses caused by friction between the tendon and the duct are determined from the conditions of equilibrium formulated for a differential of the arc length, ds, of the axis of a spatially curved tendon.

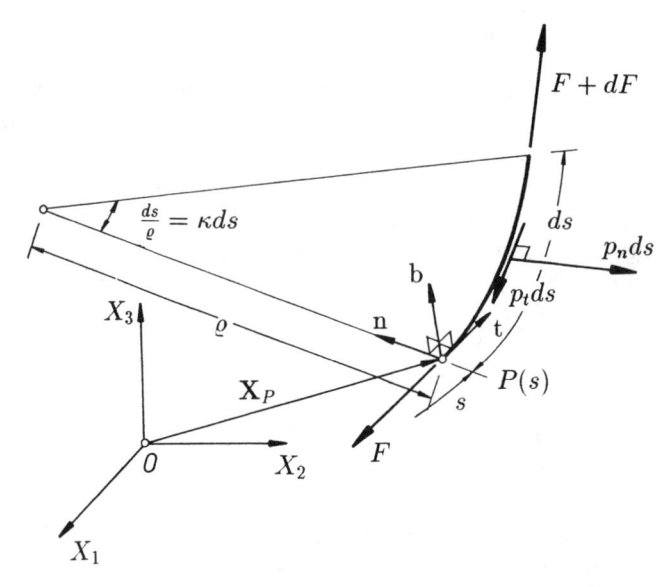

Figure 3.40: Forces acting on a differential of the arc element, ds, of a tendon

Fig.3.40 shows the forces acting on ds. It also contains the moving trihedral of the tendon axis at point $P(s)$, consisting of the tangent vector $\mathbf{t}$, the principal normal vector $\mathbf{n}$ and the binormal vector $\mathbf{b}$. These vectors are computed from the analytical representation of the tendon curve (see equation (3.277)) as

$$\begin{aligned}
\mathbf{t}(s) &= \frac{d\mathbf{X}_P(s)}{ds}, \\
\mathbf{n}(s) &= \frac{1}{\kappa(s)} \frac{d^2\mathbf{X}_P(s)}{ds^2}, \\
\mathbf{b}(s) &= \mathbf{t}(s) \times \mathbf{n}(s).
\end{aligned} \qquad (3.289)$$

The vectors $\mathbf{t}(s), \mathbf{n}(s)$ and $\mathbf{b}(s)$ are unit vectors; $\kappa(s)$ denotes the curvature of the tendon at point $P(s)$, defined as

$$\kappa(s) = \left| \frac{d^2\mathbf{X}_P(s)}{ds^2} \right|. \qquad (3.290)$$

If, instead of the arc length s, an arbitrary curve parameter t is used for determination of the moving trihedral, then $\mathbf{t}$ and $\mathbf{n}$ must be computed by means of the chain rule making use of the relationship between ds and dt according to (3.279).

3.4 Finite Element Modelling of Reinforced and Prestressed Concrete

The equations of equilibrium in the directions of **n** and **t** are given as

$$F(s)\,\kappa(s)\,ds - p_n(s)\,ds = 0\;,\quad dF(s) - p_t(s)\,ds = 0\;, \tag{3.291}$$

where $F(s)$ denotes the tendon force at point $P(s)$, $p_n(s)$ is the force per unit length, exerted from the concrete on the tendon in the direction of $\mathbf{n}(s)$, and $p_t(s)$ is the frictional force per unit length, acting in the direction of the tangent to the tendon curve. This frictional force is assumed as

$$p_t(s) = \pm\mu p_n(s) \tag{3.292}$$

with μ as the coefficient of sliding friction. The sign on the right-hand side of (3.292) must be chosen such that p_t is acting in the opposite direction of the sliding motion of the tendon. Following from (3.291_1),

$$p_n(s) = \kappa(s)F(s)\;. \tag{3.293}$$

Combination of the equations (3.291) and (3.292) and integration of the resulting relationship in the interval $[s_P, s_Q]$ where $s_Q > s_P$, assuming $F(s)$ to be a smooth function in this interval, yields

$$F(s_Q) = F(s_P)\exp(\pm\mu\int_{s_P}^{s_Q}\kappa(s)ds)\;. \tag{3.294}$$

Let P and Q be the points on the tendon axis at the common edge of the finite elements d and e and e and f, respectively (see Fig.3.39). The corresponding tendon forces are denoted as $F(s_P)$ and $F(s_Q)$. Then, (3.294) allows computation of $F(s_Q)$, provided $F(s_P)$ is known. Starting at an anchor point of the tendon with a given jacking force, the tendon forces at the points of intersection of the tendon axis with the edges of the finite elements passed through by the tendon can be computed sequentially. Within the individual finite elements it is usually sufficient to assume a linear variation of the tendon force. In this case, (3.291_2) results in constant frictional forces within the elements. For element e, the frictional force is obtained as

$$p_t = \pm\frac{F(s_Q) - F(s_P)}{l_e}\;, \tag{3.295}$$

where l_e denotes the length of the tendon segment in this element. (3.295) is also employed for bonded tendons to determine the tangential forces resulting from bond between the concrete and the prestressing steel.

Algorithms for (a) computation of tendon forces for the case that the jacking forces are applied at both anchors, (b) determination of prestress losses resulting from anchorage slip and (c) quantification of the compensation of prestress losses in consequence of friction by overstressing and stress-release operations can be found in [Hofstetter (1986a), Hofstetter (1987), Roca (1993b)]. Wobble of the tendon in the duct is accounted for by adding an empirical constant to the curvature $\kappa(s)$ in (3.294).

Changes of the tendon forces Basically, there are two different reasons for changes of the prestressing forces after completion of the tensioning operation. One of them are the changes of the strain state in the concrete, caused by applying or removing external loads, by temperature changes and by creep and shrinkage of the concrete. Algorithms for the computation of strains induced by creep and shrinkage of concrete were presented in section 3.3. The other reason for changes of the prestressing forces is relaxation of the prestressing steel. The magnitude of these changes depends on the stress level in the prestressing steel. An algorithm for consideration of time-dependent prestress losses resulting from relaxation of the prestressing steel was presented in subsection 3.3.3.

In the following, the algorithmic treatment of changes of the tendon forces, induced by changes of the strain state in the concrete, will be described. Let the equilibrium configuration be known up to load step n. Consequently, the equations (3.274) - (3.276) are satisfied up to this load step. Thus, for an arbitrary point on the axis of a tendon, defined by the arc length s, the longitudinal strain and stress, $\varepsilon_n^Z(s)$ and $\sigma_n^Z(s)$, are known.

In order to determine the equilibrium configuration for load step $n+1$ within the framework of an incremental-iterative finite element analysis, (3.274), (3.275) or, alternatively, (3.276) must be linearized at the known equilibrium configuration n (see subsection 3.1.3). Linearization of equation (3.274) at this equilibrium configuration yields

$$-\int_{V_0} \left((\delta \mathbf{E}_n)^T \Delta \mathbf{S}_{n+1} + (\Delta \delta \mathbf{E}_{n+1})^T \mathbf{S}_n\right) dV_0$$

$$+ \int_{V_0} (\delta \mathbf{u})^T \Delta \bar{\mathbf{b}}_{n+1} \, dV_0 + \int_{S_0^\sigma} (\delta \mathbf{u})^T \Delta \bar{\mathbf{t}}_{n+1} \, dS_0^\sigma$$

$$+ \int_{l_0} (\delta \mathbf{u}(s))^T \Delta \mathbf{p}_{n+1}^Z(s) ds + (\delta \mathbf{u}_A)^T (\Delta \mathbf{F}_A^Z)_{n+1} + (\delta \mathbf{u}_B)^T (\Delta \mathbf{F}_B^Z)_{n+1} = 0 \; .$$

(3.296)

Similarly, linearization of (3.275) at the known equilibrium configuration n gives

$$-A^Z \int_{l_0} \left(\delta E_n^Z(s) \Delta S_{n+1}^Z(s) + \Delta \delta E_{n+1}^Z(s) S_n^Z(s)\right) ds$$

$$- \left[\int_{l_0} (\delta \mathbf{u}(s))^T \Delta \mathbf{p}_{n+1}^Z(s) ds + (\delta \mathbf{u}_A)^T (\Delta \mathbf{F}_A^Z)_{n+1} + (\delta \mathbf{u}_B)^T (\Delta \mathbf{F}_B^Z)_{n+1}\right] = 0 \; .$$

(3.297)

Substitution of (3.297) into (3.296) yields

3.4 Finite Element Modelling of Reinforced and Prestressed Concrete

$$-\int_{V_0} \left((\delta \mathbf{E}_n)^T \Delta \mathbf{S}_{n+1} + (\Delta \delta \mathbf{E}_{n+1})^T \mathbf{S}_n\right) dV_0$$

$$-A^Z \int_{l_0} \left(\delta E_n^Z(s) \Delta S_{n+1}^Z(s) + \Delta \delta E_{n+1}^Z(s) S_n^Z(s)\right) ds \qquad (3.298)$$

$$+\int_{V_0} (\delta \mathbf{u})^T \Delta \bar{\mathbf{b}}_{n+1} \, dV_0 + \int_{S_0^\sigma} (\delta \mathbf{u})^T \Delta \bar{\mathbf{t}}_{n+1} \, dS_0^\sigma = 0 \ .$$

Alternatively, (3.298) can be obtained directly through linearization of (3.276).

(3.298) is the starting point for determination of the global vector of incremental nodal displacements $\Delta \mathbf{q}_{n+1}$ within the framework of an incremental-iterative finite element analysis. Details of this mode of finite element analysis were presented in subsection 3.1.5. Once $\Delta \mathbf{q}_{n+1}$ is known, the incremental displacement $\Delta \mathbf{u}_{n+1}^e$ of an arbitrary point of a particular element e is obtained, analogous to (3.38), as

$$\Delta \mathbf{u}_{n+1}^e = \mathbf{N}^e \Delta \mathbf{q}_{n+1}^e \ , \qquad (3.299)$$

where $\Delta \mathbf{q}_{n+1}^e$ is the local vector of incremental nodal displacements. Since all incremental quantities refer to load step $n+1$, the subscript indicating the load step will be omitted in the following. For the update of the tendon forces it is irrelevant whether the incremental displacements $\Delta \mathbf{u}^e$ have been produced by time-independent or time-dependent sources. From the displacements $\mathbf{u}_n^e$ and $\mathbf{u}_{n+1}^e = \mathbf{u}_n^e + \Delta \mathbf{u}_{n+1}^e$ the incremental concrete strains $\Delta \mathbf{E}^C$ can be computed [Bathe (1982)]. From the incremental strains $\Delta \mathbf{E}^C(s)$ in the concrete adjacent to the tendon the change of the respective concrete strain $\Delta E_t^C(s)$ in the direction of the tangent $\mathbf{t}(s)$ to the axis of the tendon can be computed. With the help of the transformation law (2.47) for second order tensors $\Delta E_t^C(s)$ is obtained as

$$\Delta E_t^C(s) = n_{ti}(s) \, n_{tj}(s) \, \Delta E_{ij}^C(s) \ , \quad i,j = 1,2,3 \ . \qquad (3.300)$$

For the case of finite elements based on shell theory the strains are referred to the local base vectors $\mathbf{A}_i$, whereas for the case of degenerate solid elements the strains are related to a local $x'y'z'$ coordinate system. The subscripts i and j in (3.300) refer to the respective base vectors. For moderately thick shells the square of the cosine n_{t3} of the angle enclosed by $\mathbf{t}$ and the normal to the middle surface of the shell is negligibly small. Consequently, the contribution of the incremental strain component in the direction of the normal to the middle surface of the shell to the change of the axial deformation of the tendon is negligible. For thin shells, according to the Kirchhoff hypothesis, also the transverse shear strains can be neglected. Hence, in this case only the contribution of the incremental in-plane strains to the change of the axial deformation of the tendon needs to be considered.

For the case of bonded tendons, assuming perfect bond between the concrete and the prestressing steel, the incremental displacements of the prestressing steel and of the adjacent concrete are identical. Thus, the change of strain, $\Delta E_t^C(s)$, at a particular point in the concrete adjacent to the tendon, measured in the direction of the tangent $\mathbf{t}(s)$ to the axis of the tendon, results in an identical change of the longitudinal strain $\Delta E^Z(s)$ at the respective point of the tendon. Hence, for an arbitrary point of the tendon the condition

$$\Delta E_t^C(s) = \Delta E^Z(s) \tag{3.301}$$

holds. Because of identical incremental displacement fields for the concrete and the prestressing steel, the superscript Z of the terms δE_n^Z and $\Delta \delta E_{n+1}^Z$ in (3.298) can be omitted. Thus, the second integral in (3.298) can be expressed in terms of the vector of the incremental nodal displacements of the finite concrete elements. This integral contains the contribution of the bonded tendons to the tangential stiffness of the prestressed concrete structure.

In contrast to bonded tendons, for unbonded tendons the changes of the strains in the concrete adjacent to the tendons are not directly related to the changes of the longitudinal strains in the tendons. The reason for this is the possibility of relative movements between the tendons and the adjacent concrete. Nevertheless, in this case the total change of length of each unbonded tendon must be equal to the total change of length of the respective duct bonded with the adjacent concrete. For small strains this condition is given as

$$\int_{l_0} \Delta E_t^C(s) ds = \int_{l_0} \Delta E^Z(s) ds , \tag{3.302}$$

where l_0 denotes the length of the tendon. The left-hand side of (3.302) can be evaluated, provided the incremental concrete strain $\Delta E_t^C(s)$ is known. The incremental tendon strain $\Delta E^Z(s)$ depends on the incremental tendon force $\Delta F(s)$. If friction between the tendon and the duct is taken into account, $\Delta F(s)$ will be a function of s. However, for greased unbonded tendons, frictional effects frequently can be neglected because of the small value of the coefficient of friction. This approximation results in a uniform incremental strain in the tendon, i.e., $\Delta E^Z = $ const. Specializing (3.302) for $\Delta E^Z = $ const. yields

$$\Delta E^Z = \frac{1}{l_0} \int_{l_0} \Delta E_t^C(s) ds . \tag{3.303}$$

Analysis techniques accounting for frictional effects between unbonded tendons and the respective ducts have been reported in [Hofstetter (1987), Roca (1988)]. The evaluation of the contribution of unbonded tendons to the tangential stiffness of a prestressed concrete structure is less straightforward than for bonded tendons. This contribution is considered in the second integral in (3.298). Since relative movements between unbonded tendons and the adjacent concrete are permitted, the incremental strains and the virtual strains in the prestressing steel and in the adjacent concrete are different. Consequently, the contribution of unbonded tendons to

3.4 Finite Element Modelling of Reinforced and Prestressed Concrete

the tangential stiffness of a prestressed concrete structure cannot be expressed in terms of the incremental nodal displacements of the finite concrete elements. However, in general, the contribution of unbonded tendons to the tangential stiffness is by far less important than for bonded tendons. Therefore, this contribution may be neglected in the analysis model. Neglect of the contribution of unbonded tendons to the tangential stiffness of a prestressed concrete structure only affects the rate of convergence of the Newton iteration. It does not affect the results as such.

The change of stress ΔS^Z at a particular point of the tendon, produced by the change of strain ΔE^Z, is obtained from the uniaxial constitutive relation for prestressing steel, i.e., from relations such as equations (2.249) and (2.250), as

$$\Delta S^Z = S^Z(E^Z + \Delta E^Z) - S^Z(E^Z) \ . \tag{3.304}$$

For geometrically nonlinear analyses it is assumed that the constitutive relations formulated in (2.249) and (2.250) for ε and σ also hold for E and S. Clearly, this assumption is only valid for small strains.

Work-equivalent nodal forces from prestress   The line loads $\mathbf{p}^Z(s)$ exerted from the tendons on the concrete are given as

$$\mathbf{p}^Z(s) = p_t(s)\mathbf{t}(s) + p_n(s)\mathbf{n}(s) \ . \tag{3.305}$$

The tangential force p_t resulting from friction or bond is obtained from (3.295). The forces p_n, acting in the direction of $\mathbf{n}$, are determined by means of (3.293). The virtual work δW^Z of the forces exerted from a particular tendon on the adjacent concrete is given as

$$\delta W^Z = \int_{l_0} (\delta \mathbf{u}(s))^T \mathbf{p}^Z(s) ds + (\delta \mathbf{u}_A)^T \mathbf{F}_A^Z + (\delta \mathbf{u}_B)^T \mathbf{F}_B^Z \ , \tag{3.306}$$

i.e., as the sum of the last three terms on the left-hand side of (3.274). Discretization of (3.306) by subdividing the tendon into segments such that each segment represents the section of the tendon within a particular finite concrete element allows formulation of the virtual work of the tendon segment embedded in the concrete element e (Fig.3.41). The respective expression is given as

$$\delta W^{Z,e} = \int_{l_e} (\delta \mathbf{u}^e(s))^T \mathbf{p}^Z(s) ds \ . \tag{3.307}$$

It is assumed that the anchors of the tendon are not located on the boundaries of element e. Consequently, the expression for the virtual work of the point loads acting at the two ends of the tendon are not included in (3.307).

For the degenerate solid element e, the work-equivalent nodal forces $\mathbf{f}_{ex}^{Z,e}$ from prestress are computed by substituting the virtual displacements $\delta \mathbf{u}^e = \mathbf{N}^e \delta \mathbf{q}^e$, with $\mathbf{N}^e$, e.g., according to (3.226), and $\delta \mathbf{q}^e$ analogous to (3.225), into (3.307). This substitution leads to [Roca (1993b)]

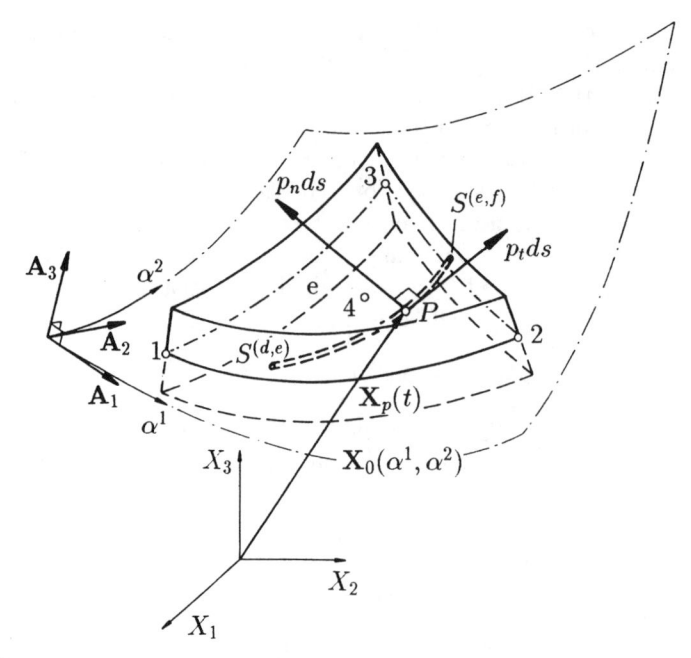

Figure 3.41: Forces exerted from a tendon on a finite element

$$\mathbf{f}_{ex}^{Z,e} = \int_{l_e} \mathbf{N}^{e^T} \mathbf{p}^{Z,e}(s)\, ds \ . \tag{3.308}$$

For the case of finite elements based on shell theory, firstly, in (3.256) the rotations ϕ_i must be expressed in terms of the tangential displacements of the middle surface of the shell and of the partial derivatives of the transverse displacement of the middle surface with respect to α^1 and α^2, respectively. Secondly, on the basis of these expressions, the respective virtual displacements must be determined. Thirdly, the virtual displacements of the middle surface of the shell and their partial derivatives must be expressed in terms of virtual nodal displacements. This is achieved with the help of the variation of $\mathbf{u}_0^e = \mathbf{N}^e \mathbf{q}^e$. Finally, the so-obtained result for the virtual displacements is used for the evaluation of (3.307). This leads to the vector of work-equivalent nodal forces from prestress. A detailed derivation of work-equivalent nodal forces from prestress for finite elements based on shell theory can be found in [Hofstetter (1986a), Hofstetter (1987)].

3.4.5 Models for Consideration of the Interface Behavior

In subsection 3.4.3 it was emphasized that the discrete representation of the reinforcing bars allows explicit consideration of bond slip and dowel action by means of special interface elements, referred to as bond or contact elements. Moreover, if a discrete crack model is used (see subsection 3.2.5), then the concrete-to-concrete

3.4 Finite Element Modelling of Reinforced and Prestressed Concrete

interface behavior at cracks, governed by aggregate interlock (see subsection 1.4.2), can be modelled by interface elements.

If, on the other hand, the embedded or distributed representation is chosen for the reinforcement, then the interface behavior can only be modelled implicitly by means of appropriate empirical modifications of the constitutive relations for concrete or steel.

Explicit Representation of the Interface Behavior

Usually, the displacements along the common boundary of two adjacent finite elements are required to be compatible. However, if slips may occur at the interfaces of finite elements, this requirement will not reflect the physical reality. Such interfaces are encountered between concrete and steel elements as well as between concrete elements. The mechanical behavior at an interface between concrete and steel is governed by the relationship between the bond stress and bond slip. The interface behavior at discrete cracks in the concrete is dominated by aggregate interlock. In order to model slips at interfaces explicitly, special elements must be used.

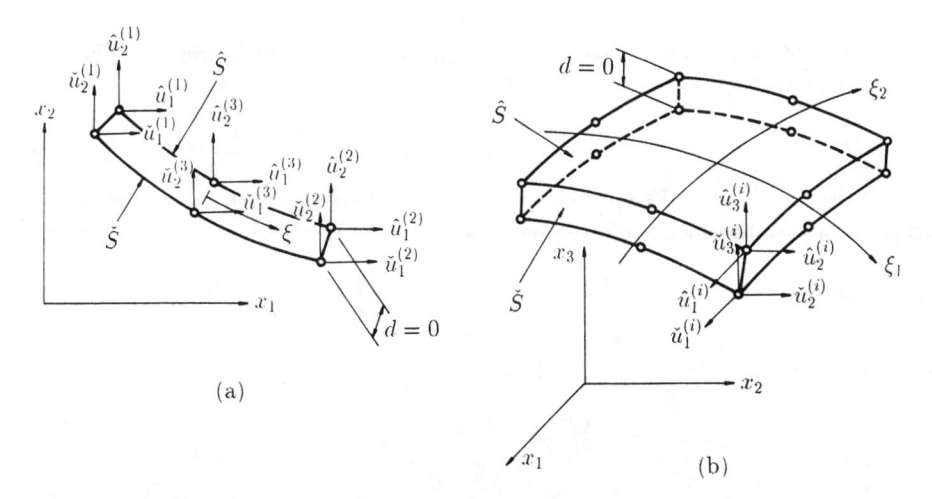

Figure 3.42: Interface elements: (a) one-dimensional, (b) two-dimensional

Continuous interface elements for two-dimensional and three-dimensional problems are shown in Fig.3.42. They provide continuous relations for the relative displacements of common edges of adjacent two-dimensional elements and of common surfaces of adjacent three-dimensional elements [Schäfer (1975), Mehlhorn (1985), Keuser (1987), Feenstra (1991a), Feenstra (1991b)]. The thickness d of the interface elements shown in Fig.3.42 is equal to zero. Hence, these interface elements are one-dimensional and two-dimensional elements. The displacements on each one of the two sides of the interface, i.e., the displacements on $\hat{S}$ and $\check{S}$ in Fig.3.42, are interpolated using the isoparametric concept. In what follows, for the sake of simplicity of the formulation, the basic equations will be derived for the one-dimensional quadratic interface element shown in Fig.3.42(a). The derivation is restricted to geometric linearity. The displacements of the interface element e are obtained from the relations

$$\hat{\mathbf{u}}^e(\xi) = \sum_{i=1}^{3} N^{(i)}(\xi)\hat{\mathbf{u}}^{(i)}, \qquad \check{\mathbf{u}}^e(\xi) = \sum_{i=1}^{3} N^{(i)}(\xi)\check{\mathbf{u}}^{(i)}, \qquad (3.309)$$

where $N^i(\xi)$ is a one-dimensional shape function according to (3.266), related to the double node i. The superscripts $\hat{}$ and $\check{}$ denote the respective side of the interface.

The two equations (3.309) can be written as

$$\left\{ \begin{array}{c} \hat{\mathbf{u}}^e \\ \check{\mathbf{u}}^e \end{array} \right\} = \mathbf{M}^e \left\{ \begin{array}{c} \hat{\mathbf{q}}^e \\ \check{\mathbf{q}}^e \end{array} \right\}, \qquad (3.310)$$

where

$$\mathbf{M}^e = \left[\begin{array}{cc} \mathbf{N}^e & 0 \\ 0 & \mathbf{N}^e \end{array} \right]. \qquad (3.311)$$

$\mathbf{N}^e$ is obtained from (3.213), where $n = 3$ and $\mathbf{I}$ is a 2×2 unit matrix. The nodal displacement vectors are given as

$$\hat{\mathbf{q}}^e = \lfloor \hat{u}_1^{(1)} \; \hat{u}_2^{(1)} \; \ldots \; \hat{u}_2^{(3)} \rfloor^T, \qquad \check{\mathbf{q}}^e = \lfloor \check{u}_1^{(1)} \; \check{u}_2^{(1)} \; \ldots \; \check{u}_2^{(3)} \rfloor^T. \qquad (3.312)$$

For an arbitrary point of the interface element e, the relative displacements of the two sides of the interface are given as

$$\mathbf{u}^{e,rel} = \hat{\mathbf{u}}^e - \check{\mathbf{u}}^e = \mathbf{L} \left\{ \begin{array}{c} \hat{\mathbf{u}}^e \\ \check{\mathbf{u}}^e \end{array} \right\}, \qquad \mathbf{L} = \left[\begin{array}{cccc} 1 & 0 & -1 & 0 \\ 0 & 1 & 0 & -1 \end{array} \right]. \qquad (3.313)$$

Substitution of (3.310) into (3.313$_1$) yields

$$\mathbf{u}^{e,rel} = \mathbf{B}^e \mathbf{q}^e, \qquad (3.314)$$

where

$$\mathbf{B}^e = \mathbf{L}\mathbf{M}^e \qquad \text{and} \qquad \mathbf{q}^e = \left\{ \begin{array}{c} \hat{\mathbf{q}}^e \\ \check{\mathbf{q}}^e \end{array} \right\}. \qquad (3.315)$$

The matrix $\mathbf{B}^e$ relates the nodal displacements to the relative displacements of the two sides of the interface. The designation of this matrix is identical to the designation of the matrix relating the nodal displacements to the strains (see equation (3.40)).

The constitutive relation between the relative displacements $\mathbf{u}^{e,rel}$ and the corresponding tractions $\mathbf{t}^e$ acting on both sides of the interface element e (Fig.3.43) is given either as a total relationship or in rate form, i.e.,

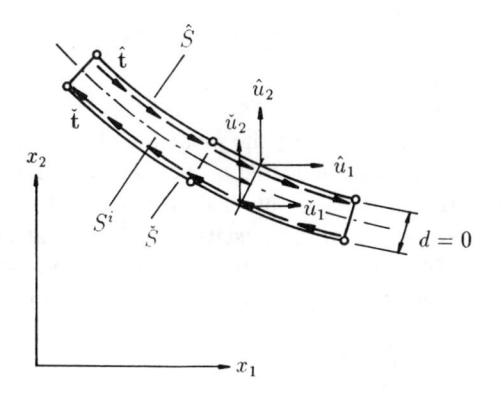

3.4 Finite Element Modelling of Reinforced and Prestressed Concrete

Figure 3.43: Displacements and tractions acting on an interface element

$$\mathbf{t}^e = \mathbf{D}_S^e \mathbf{u}^{e,rel} \quad \text{or} \quad \dot{\mathbf{t}}^e = \mathbf{D}_T^e \dot{\mathbf{u}}^{e,rel}, \tag{3.316}$$

where $\mathbf{D}_S^e$ denotes a $[2 \times 2]$ secant material stiffness matrix and $\mathbf{D}_T^e$ stands for a $[2 \times 2]$ tangent material stiffness matrix. Depending on the type of interface to be modelled, the constitutive relations (3.316) are either obtained from bond tests or from experiments concerning dowel action or from tests referring to aggregate interlock at discrete cracks. For the explicit representation of the interface behavior, use of the constitutive models described in section 2.6 may be made. The constitutive relations obtained from the aforementioned experiments are usually formulated in terms of local coordinates. Typically, the axes of such coordinates are parallel and perpendicular to the interface. Hence, the relative displacements must be transformed to the local coordinate system. With the help of the constitutive relations the secant material stiffness matrix or the tangent material stiffness matrix is determined in the local coordinate system. Finally, the respective constitutive matrix is transformed back to the global coordinate system, yielding (3.316).

Considering the interface as the part S^i of the boundary S^σ of the investigated structure (Fig.3.43), the term

$$\int_{S^i} \left(\hat{\mathbf{t}}_n : \delta\hat{\mathbf{u}} + \check{\mathbf{t}}_n : \delta\check{\mathbf{u}} \right) dS^i \tag{3.317}$$

must be added to the mathematical formulation of the principle of virtual displacements for the load increment n (see equation (3.22). (For the special case of geometric linearity, in (3.22) $\mathbf{S}_n$ and $\delta \mathbf{E}_n$ are replaced by σ_n and $\delta\varepsilon$, respectively.) Noting that $\hat{\mathbf{t}} \equiv \mathbf{t} = -\check{\mathbf{t}}$, (3.317) can be rewritten as

$$\int_{S^i} \mathbf{t}_n : \delta\mathbf{u}^{rel} dS^i. \tag{3.318}$$

(3.318) represents the virtual work done by the tractions $\mathbf{t}_n$ on the virtual relative displacements $\delta\mathbf{u}^{rel}$ of the two sides of the interface. Consequently, the term

$$-\int_{S^i} \left(\frac{\partial \mathbf{t}_n}{\partial \mathbf{u}_n^{rel}} : \Delta \mathbf{u}_{n+1}^{rel} \right) : \delta \mathbf{u}^{rel} dS^i \qquad (3.319)$$

must be added to the left-hand side of (3.33). This relation represents the linearization of the mathematical formulation of the principle of virtual displacements for determination of the incremental displacements for load step $n+1$.

Replacing S^i, $\mathbf{t}_n$ and $\delta \mathbf{u}^{rel}$ in (3.318) by $S^{e,i}$ which denotes the surface of the interface element e, $\mathbf{t}_n^e$ and $\delta \mathbf{u}^{e,rel}$, and substituting the variation of $\mathbf{u}^{e,rel}$ which follows from (3.314) as $\delta \mathbf{u}^{e,rel} = \mathbf{B}^e \delta \mathbf{q}^e$ into the so-obtained relation, yields the contribution of interface element e to the vector of the nodal forces as

$$\int_{S^{e,i}} \mathbf{B}^{e^T} \mathbf{t}_n^e dS^{e,i} . \qquad (3.320)$$

The contribution of this interface element e to the tangent stiffness matrix is obtained as

$$-\int_{S^{e,i}} \mathbf{B}^{e^T} \mathbf{D}_T^e \mathbf{B}^e dS^{e,i} . \qquad (3.321)$$

This expression follows from substitution of the expression for $\delta \mathbf{u}^{e,rel}$ together with an analogous expression for $\Delta \mathbf{u}^{e,rel}$ and with $\partial \mathbf{t}_n^e / \partial \mathbf{u}_n^{e,rel} = \mathbf{D}_T^e$, obtained from (3.316$_2$), into an element-related term analogous to (3.319).

The numerical integration in (3.320) and (3.321) is usually performed by means of Gaussian integration. However, especially if some of the coefficients of the constitutive matrix $\mathbf{D}_T^e$ are large in comparison to the coefficients of the constitutive matrix for the adjacent material, then oscillations of the tractions may occur. In such a case nodal lumping schemes may yield better results [Rots (1990)]. It is noted that, in general, the constitutive tangent matrix $\mathbf{D}_T^e$ is not symmetric. Hence, the tangent stiffness matrix (3.321) is not symmetric.

In contrast to continuous interface elements, lumped interface elements only connect the two nodes which are part of a double node. Hence, for such interface elements the constitutive relations (3.316) are evaluated at the individual double nodes. Consequently, for lumped interface elements there is no coupling between individual double nodes. Originally, such interface elements were proposed in [Ngo (1967)]. These elements do not have geometric dimensions. Rather, they represent springs. The constitutive properties of these springs govern the relationship between the relative displacements of the double nodes and the respective node forces.

For a particular double node i a lumped interface element is obtained by specialization of the equations for continuous interface elements. In (3.313$_1$) the superscript e is replaced by the superscript (i), yielding

$$\mathbf{u}^{(i),rel} = \hat{\mathbf{u}}^{(i)} - \check{\mathbf{u}}^{(i)} = \mathbf{L} \left\{ \begin{array}{c} \hat{\mathbf{u}}^{(i)} \\ \check{\mathbf{u}}^{(i)} \end{array} \right\} = \mathbf{L} \left\{ \begin{array}{c} \hat{\mathbf{q}}^{(i)} \\ \check{\mathbf{q}}^{(i)} \end{array} \right\} , \qquad (3.322)$$

where

$$\hat{\mathbf{q}}^{(i)} = \lfloor \hat{u}_1^{(i)} \; \hat{u}_2^{(i)} \rfloor^T , \qquad \check{\mathbf{q}}^{(i)} = \lfloor \check{u}_1^{(i)} \; \check{u}_2^{(i)} \rfloor^T . \qquad (3.323)$$

Consequently, the contribution of interface element e to the vector of nodal forces (see equation (3.320)) is replaced by the contribution

$$\mathbf{L}^T \mathbf{t}_n^{(i)} A^{(i)} \qquad (3.324)$$

of double node i at the interface to the vector of nodal forces. In (3.324) $A^{(i)}$ denotes the area of the part of the interface which is represented by the double node i. The contribution of this double node to the tangent stiffness is obtained by replacing (3.321) by

$$-\mathbf{L}^T \mathbf{D}_T \mathbf{L} A^{(i)} \ . \qquad (3.325)$$

In [Feenstra (1991b)] interface elements were used for the investigation of five different constitutive models for consideration of aggregate interlock. However, the results from nonlinear finite element analysis of a moderately deep, shear-critical beam, presented in this paper, indicate that the influence of aggregate interlock on both the stress distribution and the load-carrying capacity of the beam is insignificant.

Implicit Representation of the Interface Behavior

The implicit representation of the interface behavior is characterized by an appropriate empirical modification of the constitutive laws for the concrete and/or the steel. Especially for the analysis of relatively large structures, where the reinforcement is modelled by the embedded or the distributed approach and cracking is taken into account by a smeared crack model, the implicit approach is the only possibility to model the interface behavior.

Aggregate interlock at cracks is considered implicitly by introducing a modified shear modulus into the constitutive relations for concrete. This approach was described in section 3.2.5.

The interface behavior at concrete-to-steel interfaces, caused by bond slip, is modelled implicitly by relating the tension stiffening effect (see section 1.4.1) either to the concrete or to the steel. Hence, either the constitutive law for the concrete or the one for the steel is modified appropriately.

Concrete-related models for consideration of tension stiffening are more popular than steel-related models. In concrete-related constitutive models, tension stiffening is accounted for by replacing the softening branch of the tensile stress - strain diagram for plain concrete, shown in Fig.1.4(c), by the respective average stress - average strain diagram for the concrete component of reinforced concrete, plotted in Fig.1.28(c). The difference between the two softening branches is given by the magnitude of the ultimate strain. The values for the ultimate tensile strain of reinforced concrete reported in the literature are characterized by a large scatter. However, as a rule of thumb, the ultimate tensile strain of reinforced concrete is about one order of magnitude larger than the ultimate strain of plain concrete. Strictly speaking, according to Fig.1.28(b) also the constitutive relations for the reinforcing steel should be modified. In particular, the yield strength of the steel should be reduced slightly. A suitable combination of modified constitutive relations for concrete and for the reinforcing steel should allow simulation of the mechanical behavior of the reinforced concrete specimen of Fig.1.27(a), reflected by the diagrams of Fig.1.28.

Modified constitutive relations for the concrete were obtained from a concrete specimen, reinforced only in the longitudinal direction and subjected to uniaxial tension. Hence, the cracks are normal to the reinforcement. Consequently, the constitutive model must be extended to biaxial conditions, characterized by reinforcement in two directions and cracks which are not necessarily normal to the reinforcement. Often, such an extension only consists of applying the modified uniaxial tensile post-peak constitutive relations for the concrete to the principal directions of the concrete strains without consideration of the layout of the reinforcement. However, since tension stiffening is caused by bond stresses between the concrete and the reinforcing bars, it is preferable that concrete-related tension stiffening models are referred to the directions of the reinforcement. This can be done by considering tension stiffening as a function of the concrete strains in the directions of the reinforcing bars.

Alternatively, a reinforcement-related tension stiffening model can be employed. In this approach the residual tensile load-carrying capacity of the cracked concrete is accounted for by a modified stress-strain curve of the reinforcing steel. Consequently, the steel-related tension stiffening model is *a priori* referred to the directions of the reinforcing bars.

Figs.3.44(a) and (b) refer to the concrete-related and the steel-related tension stiffening model proposed in [Kollegger (1988), Mehlhorn (1990), Kollegger (1990)]. These models were derived from an experimental force - average strain diagram of a reinforced concrete specimen subjected to uniaxial tension, such as the one shown in Fig.1.28(a). In Fig.3.44 $\tilde{\varepsilon}$ denotes the average tensile strain, σ_{ts}^C is the average residual tensile stress carried by the cracked concrete and ρ denotes the reinforcement ratio. If the residual tensile load-carrying capacity of the cracked concrete is related to the reinforcing steel, the additional stress in the reinforcement $\Delta\sigma_{ts}^S$ is computed from σ_{ts}^C and ρ, i.e., $\Delta\sigma_{ts}^S = \sigma_{ts}^C/\rho$. Hence, the material parameters for the tension stiffening models shown in Fig.3.44 are σ_{ts}^C, $\tilde{\varepsilon}_{ts,A}$, $\tilde{\varepsilon}_{ts,B}$ and $\tilde{\varepsilon}_{ts,C}$. The modified constitutive relations are based on experimental results indicating that tension stiffening mainly depends on the reinforcement ratio and that it is practically independent of the angle enclosed by the reinforcement and the cracks (provided tension stiffening is evaluated in the direction of the reinforcement).

The tension stiffening models illustrated in Fig.3.44 are based on simplifying assumptions. Firstly, the average concrete strain in the direction of the reinforcement at the initiation of cracking is set equal to zero. This is the reason why the stress at $\tilde{\varepsilon} = 0$, shown in Fig.3.44, is not equal to zero. Secondly, the tension stiffening effect is assumed to vanish shortly before yielding of the reinforcement is initiated. The first assumption is motivated by the fact that, in contrast to experiments on uniaxial specimens, for two-dimensional or three-dimensional problems, the concrete strain in the direction of the reinforcement at the initiation of cracking may be positive,

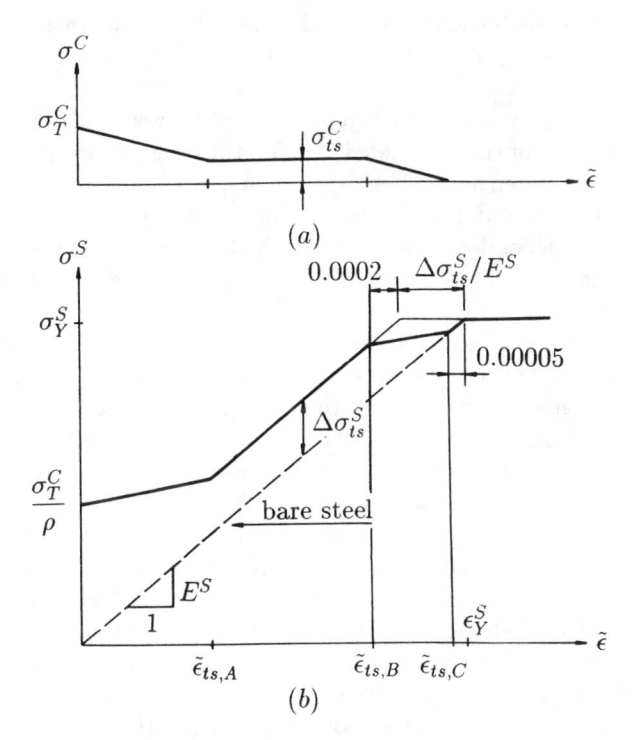

3.4 Finite Element Modelling of Reinforced and Prestressed Concrete

Figure 3.44: Modelling of tension stiffening by modifying the constitutive relations (a) for the concrete and (b) for the reinforcing steel

zero or even negative. The reason for this is the possibility of an arbitrary orientation of the cracks with respect to the reinforcement. The mentioned concrete strain is zero, e.g., in a quadratic concrete panel with reinforcing bars parallel to the edges, subjected to pure shear along the edges. Application of the tension stiffening models of Fig.3.44 to commonly encountered cases, characterized by a positive value of the concrete strain in the direction of the reinforcement at the initiation of cracking, results in a discontinuity of the stress at the initiation of cracking.

It is shown in [Kollegger (1988), Mehlhorn (1990), Kollegger (1990)] that the concrete-related model of Fig.3.44(a) yields almost the same structural response as the reinforcement-related model of Fig.3.44(b), provided it is formulated in terms of the concrete strain in the direction of the reinforcement. However, if the concrete-related model is formulated in terms of the principal tensile strains of the concrete, the stiffening effect will vanish too early. It is also noted that in reinforcement-related tension stiffening models the compressive stresses in the concrete are somewhat overestimated. The reason for this is the neglect of the (tensile) stress-carrying capacity of the cracked concrete.

A different reinforcement-related tension stiffening model was proposed in [Floegl (1982)]. This model is based on a simplified distribution of the bond stresses between the cracks. The actual distribution of the bond stresses between two adjacent cracks, shown in Fig.1.27(e), was approximated by piecewise constant average bond stresses. In addition, the residual tensile stresses of the cracked concrete (Fig.1.27(d)) and the steel stresses (Fig.1.27(c)) were assumed to be constant between two adjacent cracks. On the basis of these assumptions so-called tension stiffening factors, greater or equal to one, were derived. The tension stiffening effect was accounted for by multiplying the steel stresses and the stiffness of the reinforcement by the tension stiffening factors. These factors depend on the average bond stress, the crack spacing, the angle enclosed by the normal to the crack and the reinforcement, the average steel stress and the diameter of the reinforcing bars.

Constitutive models for reinforced concrete beams and panels, considering the composite material as a continuum on the macrolevel, i.e., following approach (b) in subsection 3.4.1, with special emphasis on the post-cracking behavior of the composite material, can be found in [Cervenka (1990)] and [Gupta (1989a), Gupta (1989b)].

Advantages of the implicit representation of the interface behavior are the relative simplicity of the discretization procedure as well as the computational simplicity. These advantages result from avoiding bond or contact elements by incorporating the effects of the interface behavior into the constitutive relations for concrete or steel. Consequently, it is possible to assign the same degrees of freedom to coinciding nodes of concrete and steel elements in the embedded approach or to use composite concrete - steel elements in the distributed representation of the reinforcement. A disadvantage of the implicit representation of the interface behavior is the greater inaccuracy of the steel and concrete stresses compared with the stresses obtained with bond elements.

4 Application to Engineering Problems

4.1 Introduction

This chapter is devoted to the application of the mathematical models for reinforced and prestressed concrete, presented in chapter 2, to the analysis of reinforced and prestressed concrete structures by means of finite element methods described in chapter 3. The numerical analyses reported in this chapter have been carried out at the Institute for Strength of Materials of the University of Technology of Vienna during the last fifteen years.

This report is organized as follows: In the next section finite element ultimate load analyses of thick-walled three-dimensional reinforced concrete structures, restricted to consideration of material nonlinearity, will be described. These analyses were conducted by Eberhardsteiner [Eberhardsteiner (1991)] and Meschke [Meschke (1991)] on the basis of several well-known nonlinear-elastic and elastic-plastic constitutive models for three-dimensional stress-states, which have been implemented into the commercial finite element program MARC [MARC (1992)].

The next section deals with finite element analyses of reinforced concrete surface structures, including both material and geometric nonlinearity. Most of these examples were performed with the help of the computer program FESIA for ultimate load analyses of reinforced concrete surface structures, initially developed by Floegl [Floegl (1981)]. FESIA is based on Koiter's nonlinear shell theory of moderately large rotations [Koiter (1967)]. Material nonlinear behavior is taken into account either by means of the nonlinear-elastic biaxial stress-strain relations proposed in [Liu (1972)] or on the basis of a biaxial elastic-plastic constitutive model [Walter (1988)]. Fracture of concrete is treated within the framework of the smeared crack concept. In addition, the capability of reinforced concrete to carry tensile forces between neighboring cracks through stress transfer from the reinforcement to the surrounding concrete by bond slip is taken into account. These investigations include one of the first finite element ultimate load analyses of reinforced concrete cooling towers subjected to quasi-static wind loading [Mang (1983)]. The other examples presented in this section were carried out with the finite element program MARC [MARC (1992)] employing shell elements based on the degenerate solid approach.

The last section deals with ultimate load analyses of prestressed concrete surface structures including time-dependent effects. Prestressing was implemented into FESIA by Hofstetter [Hofstetter (1987)]. Constitutive models accounting for the time-dependent behavior were incorporated by Walter [Walter (1988)].

4.2 Nonlinear FE-Analyses of 3D Reinforced Concrete Structures

Reports on nonlinear finite element (FE) analyses of general three-dimensional (3D) and axisymmetric reinforced concrete structures are contained, e.g., in [Buyukozturk (1985), Bathe (1989), Eberhardsteiner (1991), Meschke (1991), Mang (1991), Vidosa (1991), Pagnoni (1992), Ashour (1993)]. In what follows, a detailed report on three-dimensional finite element analysis of a short reinforced concrete cantilever [Eberhardsteiner (1991), Meschke (1991)] and on axisymmetric finite element analysis of a reinforced concrete cylinder subjected to a concentrated compressive force [Meschke (1991)] will be given.

4.2.1 Short Reinforced Concrete Cantilever

Description of the Structure

[Mehmel (1967)] contains a report on an experimental program conducted on short reinforced concrete cantilevers with different geometric shapes, varying layouts of the reinforcement and two types of loading. The short cantilever, denoted as I.2 in [Mehmel (1967)] and shown in Fig.4.1, was chosen for comparing the computed ultimate load-carrying capacity to the experimental result. The ratio of the length to the height and the width of the cantilever of 0.60 m : 0.65 m : 0.40 m is such that the cantilever can be viewed as a thick-walled three-dimensional structure. The layout of the reinforcement is also shown in Fig.4.1. The line load is acting on steel plates of 40 mm thickness, parallel to the front face of the cantilever, at a distance of 0.50 m from the vertical column. In the experiments the load was increased incrementally. The magnitude of the increments was 0.10 MN. The corresponding crack patterns are contained in [Mehmel (1967)]. In Fig.4.2 these crack patterns are shown for three typical load levels. However, information on the deformation of the cantilever as a function of the load intensity is missing.

The ultimate load in the experiment was obtained as 0.933 MN. Hence, the crack pattern in Fig.4.2(c) refers to the ultimate load. Failure was initiated by the formation of a crack extending from beneath the acting load to the lower horizontal edge between the cantilever and the column. Subsequently, the stress in the diagonal reinforcing bars attained the yield strength. This was accompanied by an increase of the plastic strains in the diagonal reinforcing bars at the intersection of the mentioned crack, resulting in a rapid increase of the crack width. Failure of the concrete in the compressive region finally caused the structural failure.

The modulus of elasticity, Poisson's ratio, the uniaxial compressive cubic strength and the tensile strength of concrete are given as $E^C = 21\,870$ MPa, $\nu^C = 0.20$, $\sigma_P^C = 22.6$ MPa and $\sigma_T^C = 2.3$ MPa, respectively. For the reinforcing steel the modulus of elasticity and the yield strength are given as $E^S = 206\,000$ MPa and $\sigma_y^S = 430$ MPa, respectively.

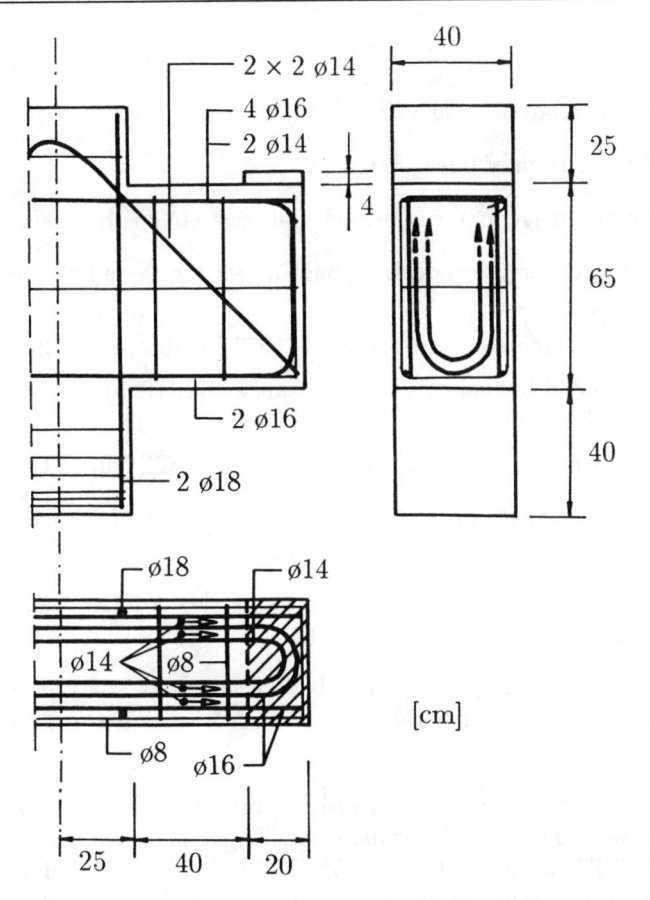

4.2 Nonlinear FE-Analyses of 3D Reinforced Concrete Structures

Figure 4.1: Short reinforced concrete cantilever

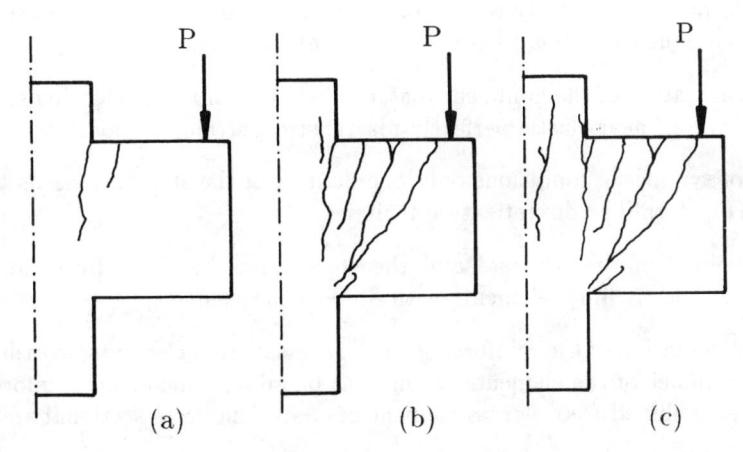

Figure 4.2: Crack patterns at typical load levels: (a) P = 0.300 MN, (b) P = 0.700 MN, (c) P = 0.933 MN

Analysis Model

The main features of the analysis model are

- assumption of geometrically linear behavior,
- consideration of nonlinear material behavior of concrete on the basis of
    (a) the linear-elastic fracture model, accounting for tensile failure only (subsection 2.3.1),
    (b) the hypoelastic orthotropic model [Elwi (1979)] (subsection 2.3.2),
    (c) the invariant-based hypoelastic model [Stankowski (1985)] (subsection 2.3.2),
    (d) the elastic-plastic model by Han and Chen [Han (1987)] (subsection 2.3.3),
- consideration of cracking of concrete within the framework of the fixed orthogonal smeared crack concept (subsection 3.2.5),
- modelling of failure of concrete on the basis of the criterion reported in [Hsieh (1982)] (subsection 2.2.3, equations (2.23) and (2.24)),
- consideration of tension softening by linearly descending stress-strain relations corresponding to constant tension-softening moduli, within the range from $-1\,000$ MPa to $-\infty$,
- assumption of the shear stiffness of cracked concrete, resulting from aggregate interlock, as either constant (the value of the constant is within the range from zero up to 50% of the elastic shear stiffness) or depending linearly on a fictitious crack strain normal to the crack (subsection 2.6.1, equation (2.254)),
- disregard of the ability of reinforced concrete to carry forces between cracks in consequence of bond between concrete and steel,
- consideration of the nonlinear material behavior of the reinforcing steel on the basis of a linear-elastic perfectly-plastic stress-strain relationship.

Because of symmetry conditions only one quarter of the structure needs to be discretized (Fig.4.3). The discretization includes

- modelling of the concrete and the steel plates by three-dimensional linear isoparametric finite elements with 8 nodes per element,
- representation of the reinforcing steel by linear truss elements; in order to keep the number of the elements as small as possible, some of the reinforcing bars were combined to one truss element of equivalent cross-sectional area.

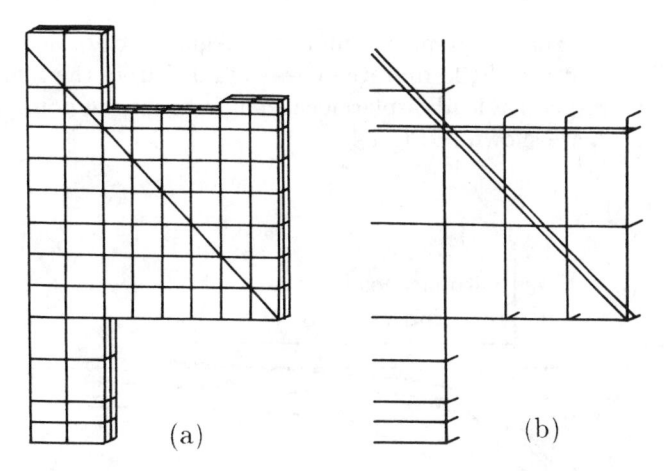

4.2 Nonlinear FE-Analyses of 3D Reinforced Concrete Structures

Figure 4.3: Finite element discretization: (a) modelling of the concrete and the steel plates by three-dimensional finite elements, (b) representation of the reinforcement by truss elements

According to subsection 3.4.3, the truss elements must coincide with the boundaries of the concrete elements. This restriction guarantees compatibility of the displacements of the reinforcement and the adjacent concrete. This is the reason why the shapes of some of the three-dimensional concrete elements are those of triangular prisms. The triangular surfaces of these elements were obtained by mapping neighboring node points of the cube-shaped parent elements onto the same node point. In order to avoid a refinement of the finite element mesh, in the analysis model the concrete cover was only taken into account for the horizontal and the diagonal loop reinforcement.

This mode of discretization requires the assumption of perfect bond between the concrete and the reinforcement. According to [Van Mier (1987)], the modelling of bond slip by bond-slip elements connecting common nodes of three-dimensional elements and truss elements does not have a significant influence on the computed structural behavior of short cantilevers.

The analyses were performed by applying displacement increments of 0.15 mm each at the location of the acting load.

Results

In what follows, the main findings of several analyses including parameter studies [Eberhardsteiner (1991), Meschke (1991)] will be summarized.

- A good approximation of the ultimate load can be obtained even by treating concrete as a linear-elastic material, accounting only for tensile failure of concrete and assuming a linearly-elastic, perfectly-plastic stress-strain relationship for the reinforcing steel. Constitutive models taking the nonlinear material behavior in the pre-peak compressive region into account yield results which differ relatively little from the ones obtained from the linear material models. The respective load-displacement diagrams for the point directly beneath the load are shown in Fig.4.4.

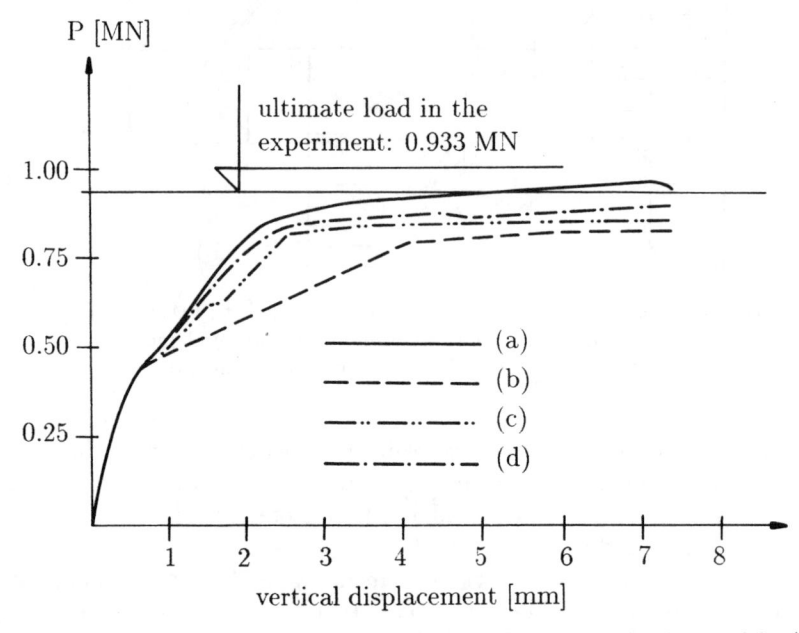

Figure 4.4: Load-displacement diagrams for different constitutive models: (a) linear-elastic fracture model, (b) hypoelastic orthotropic model, (c) invariant-based hypoelastic model, (d) elastic-plastic model

	ultimate load in MN
experiment	0.933
linear-elastic fracture model	0.960
hypoelastic orthotropic model [Elwi (1979)]	0.820
invariant-based hypoelastic model [Stankowski (1985)]	0.850
elastic-plastic model by Han and Chen [Han (1987)]	0.910

Table 4.1 Values of the ultimate load of the cantilever

Table 4.1 contains the experimentally obtained value of the ultimate load of the cantilever and the computed values of this ultimate load. The diagrams in Fig.4.4 and the values in Table 4.1 are based on a shear stiffness G_{AI} of the cracked concrete, decreasing linearly as a function of the increase of the crack

4.2 Nonlinear FE-Analyses of 3D Reinforced Concrete Structures

width according to (2.254), with $F = 0.25$ and $\bar{\varepsilon}_{11} = 0.008$, and on a softening modulus of $-2\,500$ MPa. At a deflection of the load point of about 0.5 mm, the response of the cantilever deviates from linearly-elastic structural behavior (see Fig.4.4). The onset of pronounced nonlinear structural behavior is caused by the formation of a crack emanating from the upper horizontal edge at the connexion of the cantilever with the column and extending vertically over two thirds of the height of the cantilever (Fig.4.2(a)). In consequence of the formation of this crack, the structural stiffness is reduced. It remains almost constant at the reduced level until, at a deflection of the load point of about 2 to 2.5 mm, yielding of the upper horizontal reinforcement is initiated, which is followed by yielding of the diagonal bars.

Because of the small influence of the nonlinear material behavior of concrete in the pre-peak compressive region on the structural response, the subsequent parameter study was conducted by means of the linear-elastic fracture model, considering tensile failure only.

- According to (2.254), the shear stiffness of cracked concrete is linearly decreasing with increasing crack width. Replacing (2.254) in the sense of a simplifying assumption by a constant value for the shear stiffness of cracked concrete, has a considerable influence on the computed structural behavior. The ratio of the shear stiffness of cracked concrete G_{AI} over the elastic shear stiffness G, i.e., the shear-retention factor r, was chosen as 0, 0.2 and 0.5, respectively. Fig.4.5 contains the load-displacement diagrams for these three values of r.

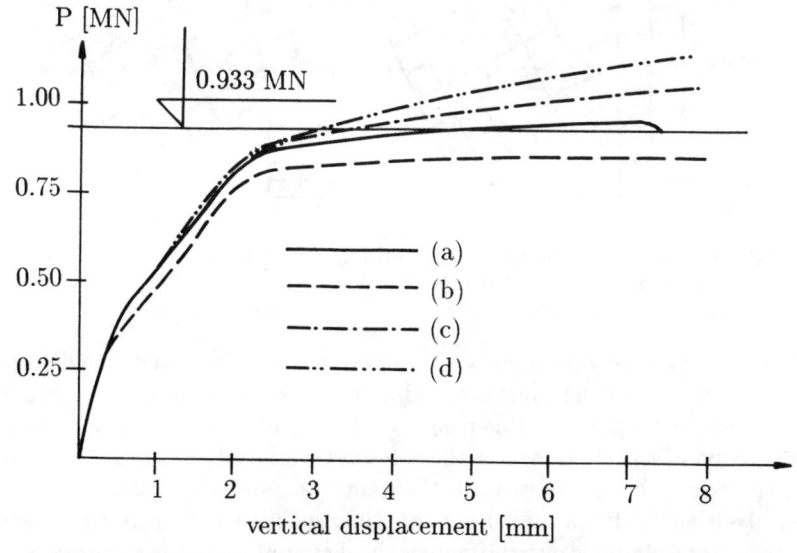

Figure 4.5: Load-displacement diagrams for different assumptions of the shear retention factor r: (a) value of r linearly decreasing with increasing crack width, (b) $r = 0$, (c) $r = 0.2$, (d) $r = 0.5$

It can be seen that disregard of aggregate interlock ($r = 0$) underestimates the residual stiffness of cracked concrete. The frequently used values for r in the range of 0.2 to 0.5, however, result in an erroneous stiffness of cracked concrete. It is the consequence of not accounting for the pronounced deterioration of aggregate interlock with increasing crack width, which is observed in the experiments. For the case of a constant shear-retention factor no horizontal plateau of the load-displacement diagram is obtained. The cantilever rather fails because of rupture of the reinforcing bars. The assumption of a constant shear stiffness of cracked concrete in combination with the fixed orthogonal smeared crack concept results in the accumulation of incremental shear stresses $\Delta\tau = rG\Delta\gamma$, transferred across the crack planes. Transformation of these shear stresses to principal directions may yield principal stresses which are considerably larger than the tensile strength of concrete. However, within the framework of the fixed orthogonal crack model only the stresses acting in directions normal to already existing cracks are checked for cracking. In Fig. 4.6 the deformed cantilever is plotted for the two limiting values of the shear retention factor, i.e., for $r = 0$ and $r = 0.5$, at a deflection of the load point of 6 mm.

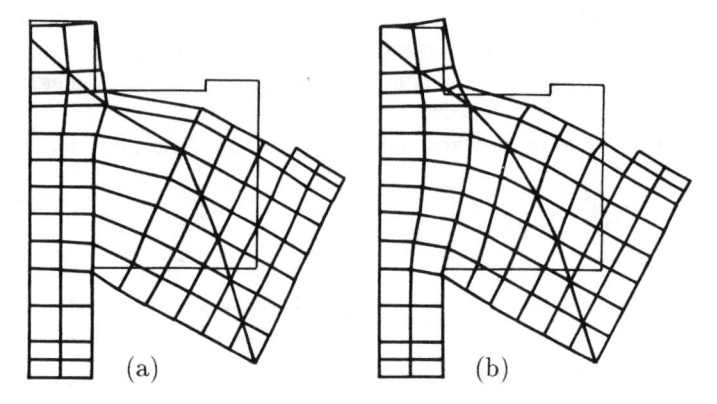

Figure 4.6: Deformed structure for limiting values of the shear retention factor; (a) $r = 0$, (b) $r = 0.5$

Disregard of aggregate interlock results in a concentration of the deformations in the elements of the cantilever, adjacent to the column. In the context of the smeared crack approach this means that the width of the vertical crack at the connexion of the cantilever to the column is relatively large. The deformations of the remaining parts of both the cantilever and the column, however, are relatively small. If, on the other hand, the capability of transferring shear forces across crack planes is overestimated, the distribution of the deformations of the cantilever will be relatively uniform. Significant deformations of the column will be obtained in this case. They are caused by large stresses at the cracks. Such erroneous predictions are the consequence of neglecting the dependence

4.2 Nonlinear FE-Analyses of 3D Reinforced Concrete Structures

of aggregate interlock on the crack width.

- The choice of the softening modulus only affects the part of the load-displacement curve following the initiation of cracking. The influence of this choice on the ultimate load is small. The load-displacement curves shown in Fig.4.7 for two different values of the softening modulus were computed on the basis of the linear-elastic material model, considering tensile failure only, and of a shear stiffness of the cracked concrete which is linearly decreasing with increasing crack width. Fig.4.8 shows the respective crack patterns at a deflection of the

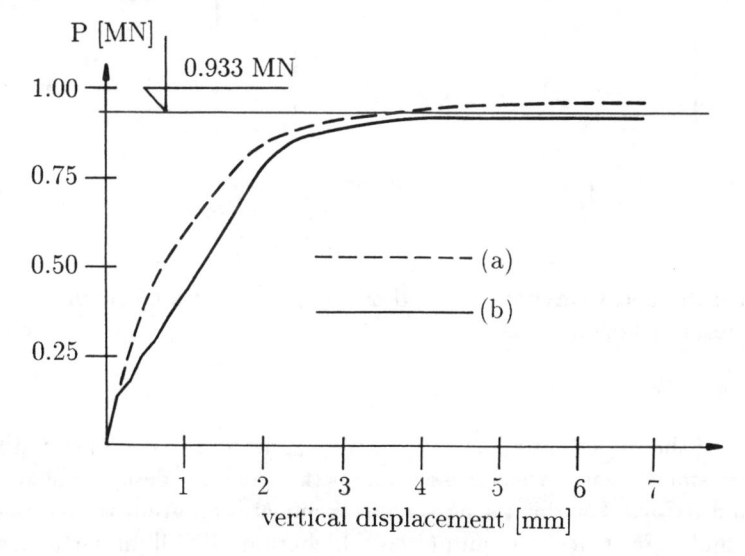

Figure 4.7: Load-displacement diagrams for different values of the softening modulus: (a) $E^C_{soft} = -1\,000$ MPa, (b) $E^C_{soft} = -\infty$

load point of 0.6 mm, computed on the basis of the smeared crack concept. Fig.4.8(a) is based on the assumption of a linearly decreasing stress-strain relationship in the softening regime, characterized by a softening modulus of $-1\,000$ MPa. Fig.4.8(b) refers to the limiting case of no transfer of normal stresses across cracks after the onset of cracking. Expectedly, disregard of tension softening yields a structural response which is considerably softer after the onset of cracking than is the case if tension softening is taken into account. This is caused by the more rapid crack propagation in the case of disregard of tension softening. It is noted that use of a constant softening modulus yields mesh-dependent results. For this example, however, the influence of tension softening on the ultimate load is small. The reason for this is that shortly before structural failure most of the cracks have opened completely. There is no transfer of normal stresses across such cracks.

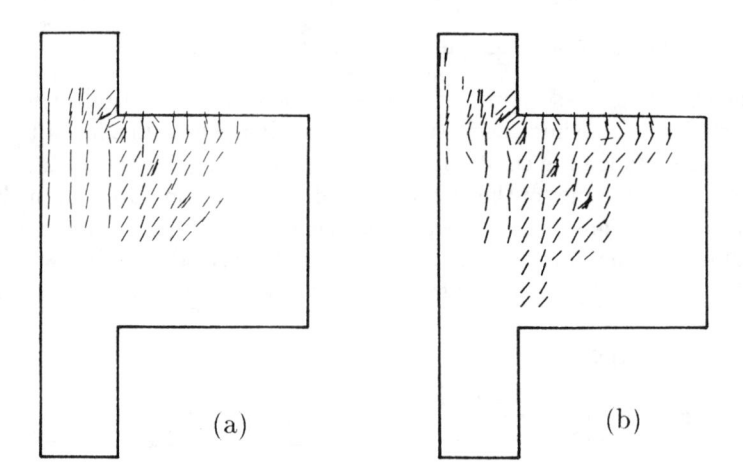

Figure 4.8: Crack patterns at a deflection of the load point of 0.6 mm:
(a) $E_{soft}^C = -1\,000$ MPa, (b) $E_{soft}^C = -\infty$

4.2.2 Reinforced Concrete Cylinder Subjected to a Concentrated Compressive Force

Description of the Structure

Knowledge of the material behavior of concrete subjected to concentrated loads acting on very small bearing areas is essential for the economic design of heavily loaded anchors and fixings. For this purpose an experimental program was carried out by Lieberum and Reinhardt [Lieberum (1987), Lieberum (1989)]. In particular, the penetration of a concentrated load and the mechanism of failure of concrete under such a load were studied. Cylindrical concrete specimens were used in the experiments. Their diameter was 0.40 m and their height was 0.45 m. They were subjected to concentrated compressive forces by steel punches with diameters ranging from 0.013 m to 0.032 m. The tests were conducted by controlling the displacements. Because of the relatively large scatter of the experimental results for the steel punches with the smaller diameters, an experiment conducted with a steel punch of 0.032 m diameter, denoted as A/17 in [Lieberum (1987)], is chosen for the comparison of experimental and numerical results. The part of the experimental setup which is relevant for the analysis is shown in Fig.4.9. In order to prevent the premature splitting of concrete, six reinforcing steel bars of 8 mm diameter each were placed in the circumferential direction close to the outer surface of the concrete cylinder.

In [Lieberum (1987)] the modulus of elasticity of concrete, Poisson's ratio, the uniaxial compressive cubic strength and the tensile strength are given, in this order, as $E^C = 30\,000$ MPa, $\nu^C = 0.18$, $\sigma_P^C = 39$ MPa and $\sigma_T^C = 3.0$ MPa. The modulus of elasticity and the yield strength of the reinforcing steel are given as $E^S = 193\,740$ MPa and $\sigma_Y^S = 374$ MPa, respectively. The modulus of elasticity for the steel punch

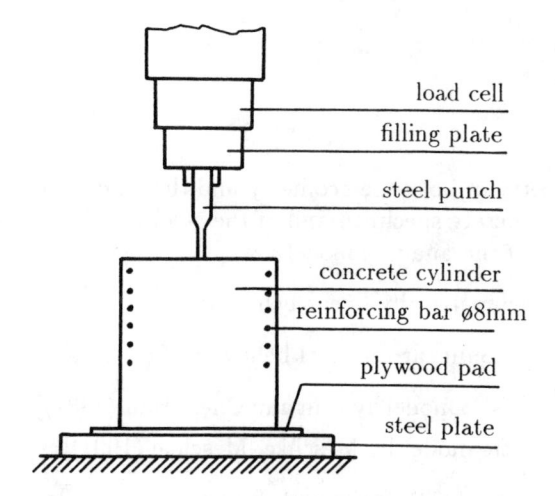

4.2 Nonlinear FE-Analyses of 3D Reinforced Concrete Structures

Figure 4.9: Experimental setup for a reinforced concrete cylinder subjected to a concentrated compressive force

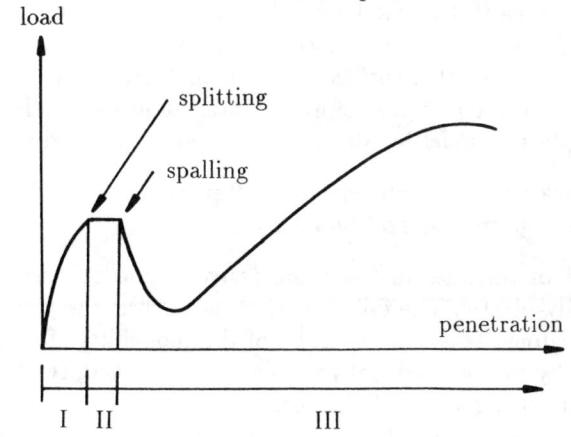

Figure 4.10: Qualitative shape of the load-penetration diagram

is taken as 206 000 MPa.

The qualitative shape of the experimentally obtained load-penetration diagram is shown in Fig.4.10. This diagram consists of three regions. Region I contains the initial, ascending part of the load-penetration diagram, characterized by local compaction of the concrete under the steel punch. Region II coincides with the crack plateau indicating local failure which is characterized by radial cracking and spalling of concrete around the steel punch. Upon further increase of the penetration, region III is reached. A feature of this region is the sudden drop of the load. It is followed by the final increase of the load until the beginning of the formation of splitting cracks. This ascending part of the load-penetration diagram depends on the reinforcement of the concrete specimen. Consideration of this part of the load-penetration diagram was beyond the scope of the experiments.

Analysis Model

Because of axisymmetry of both the geometry and the loading, only one half of the cross-section of the concrete specimen and of the steel punch were discretized.

The main features of the analysis model are:

- Assumption of geometrically linear behavior.

- Consideration of nonlinear material behavior of concrete on the basis of

    (a) the elastic-plastic model by Han and Chen [Han (1987)] (subsection 2.3.3),
    (b) the hypoplastic model by Meschke [Meschke (1991)].

    The geometric shape of the failure surface of the hypoplastic material model by Meschke differs significantly from the geometric shape of the failure surface of the elastic-plastic constitutive model by Han and Chen. In the model by Han and Chen the failure surface proposed in [Willam (1975)] is used. The mathematical description of this surface is given in (2.15) and (2.16). The failure surface is open in the direction of hydrostatic compression. The failure surface of the hypoplastic model by Meschke, however, is a closed surface.

- Consideration of cracking of concrete within the framework of the fixed orthogonal smeared crack concept (subsection 3.2.5).

- Modelling of failure of concrete in the compressive region by assuming the material to be ideally-plastic. The failure surface is used as the yield surface for the post-peak regime. This simplification of the modelling of crushing of concrete is justified by the confinement of concrete in the vicinity of the steel punch, resulting in ductile material behavior.

- Idealization of tension softening by a linearly descending stress-strain relationship, corresponding to a constant tension-softening modulus of $-2\,500$ MPa.

- Assumption of a constant shear retention factor for cracked concrete, given as $r = 0.20$ (subsection 2.6.1).

- Disregard of tension stiffening. This is justified by the fact that no reinforcement has been provided in the vicinity of the steel punch where local failure occurs in the concrete specimen. Hence, the ability of reinforced concrete to carry forces between cracks in consequence of bond between the reinforcing bars and the surrounding concrete is not relevant for the analysis.

- Consideration of nonlinear material behavior of the reinforcing steel on the basis of linear elasticity and perfect plasticity.

The discretization includes (Fig.4.11):

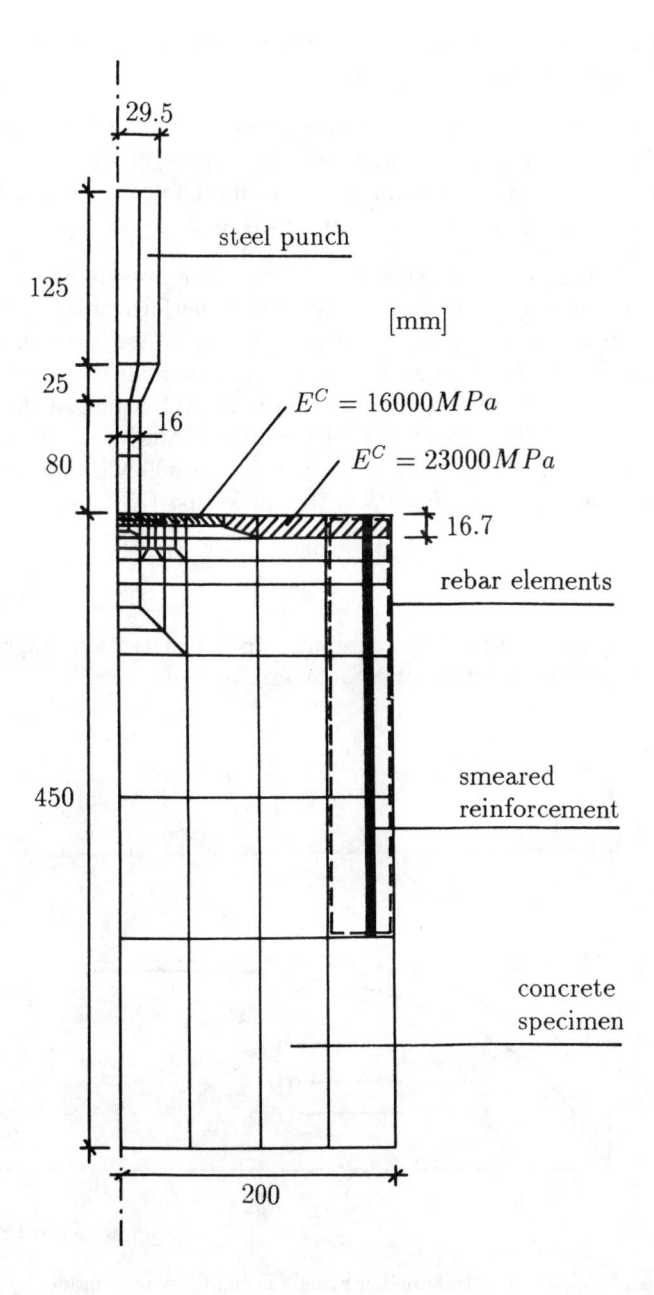

4.2 Nonlinear FE-Analyses of 3D Reinforced Concrete Structures

Figure 4.11: Finite element model of the concrete specimen and the steel punch

- Subdivision of the concrete cylinder and the steel punch into quadratic isoparametric axisymmetric finite elements.

- Modelling of the circumferential reinforcement by means of rebar elements [MARC (1992)]. These are finite elements representing only the reinforcing bars which are smeared to thin layers of mechanically equivalent thickness. They are superimposed on the finite concrete elements.

The numerical analyses were restricted to the initial, ascending part of the load-penetration diagram (region I). The analyses were performed by applying vertical displacement increments of 0.015 mm each at the top of the steel punch.

In the vicinity of the top surface of the concrete cylinder there is a thin layer of concrete with a considerably larger content of mortar. Because of the considerably lower stiffness of mortar as compared with concrete, the value of the modulus of elasticity of this concrete was taken as only $E^C = 16\,000$ MPa in the thinner part of this layer and as $E^C = 23\,000$ MPa in the thicker part (Fig.4.11).

Results

Fig.4.12 allows a comparison of the computed load-penetration diagrams with the experimentally obtained load-penetration diagram.

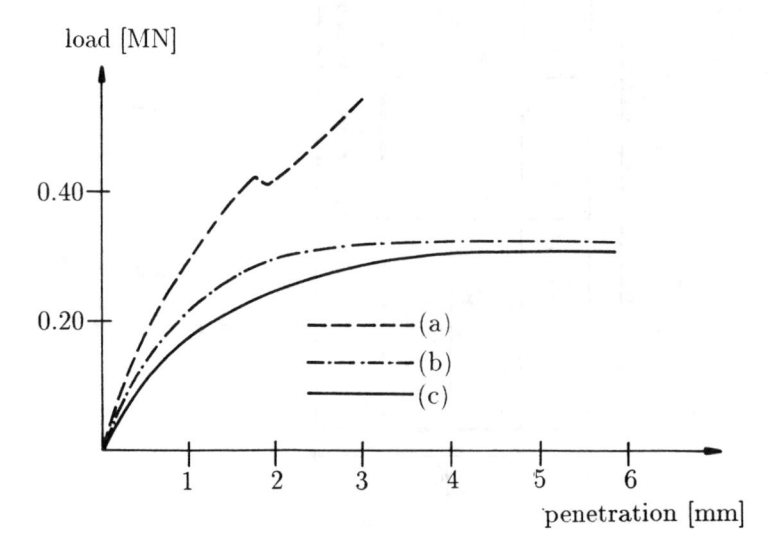

Figure 4.12: Load-penetration diagrams: (a) elastic-plastic model by Han and Chen, (b) hypoplastic model by Meschke, (c) experiment

The figure shows that the material model has a considerable influence on the results. Apart from overestimating the structural stiffness at the onset of the test,

4.2 Nonlinear FE-Analyses of 3D Reinforced Concrete Structures

the hypoplastic model by Meschke yields a good approximation of the experimentally obtained load-penetration diagram. In contrast to this constitutive model, the material model by Han and Chen does not result in a good approximation of the experimentally obtained load-penetration diagram. The reason for this is that the failure surface is open in the direction of hydrostatic compression. The stress states for points in the vicinity of the steel punch are characterized by the dominance of the hydrostatic compressive stress. In comparison to this stress the deviatoric octahedral stress is small. Hence, in case of an open failure surface the respective stress paths do not reach this surface. Consequently, the stress continues to increase. Concrete specimens do not fail under hydrostatic compression. However, there is no experimental evidence of such an unlimited load-carrying capacity for stress states characterized by small deviations from hydrostatic compression. Fig.4.13 shows the stress paths for a material point of concrete, located beneath the steel punch close to the axis of rotation, computed by means of the two applied material models. It can be seen that the stress path resulting from the material model by Han and Chen does not approach the failure surface, which indicates an overly stiff response. On the other hand, the stress path for the same material point, computed on the basis of the constitutive model by Meschke, gradually approaches the failure surface, which indicates a decrease of the stiffness.

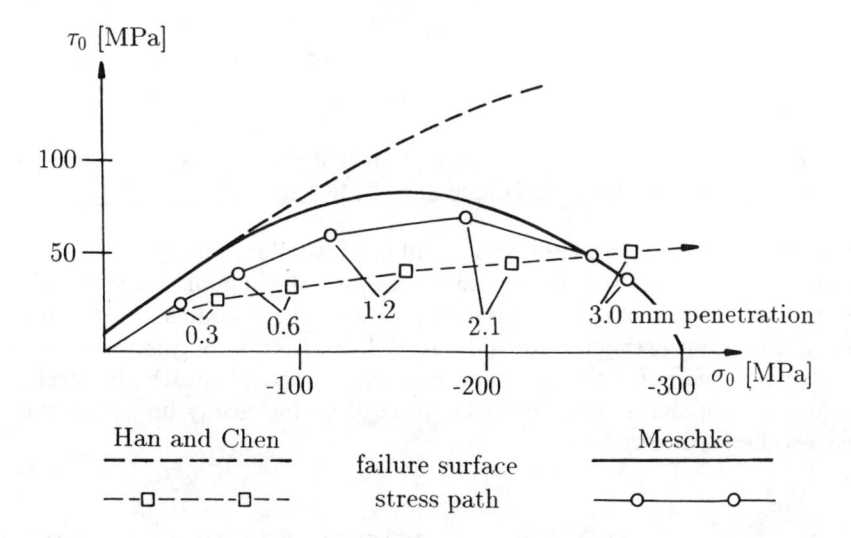

Figure 4.13: Stress paths for a material point of concrete beneath the steel punch, computed by means of (a) the elastic-plastic model by Han and Chen and (b) the hypoplastic model by Meschke

Hence, the stiffness and the load-carrying capacity of structures characterized by the dominance of hydrostatic compression may be overestimated considerably if a material model with an open failure surface is used.

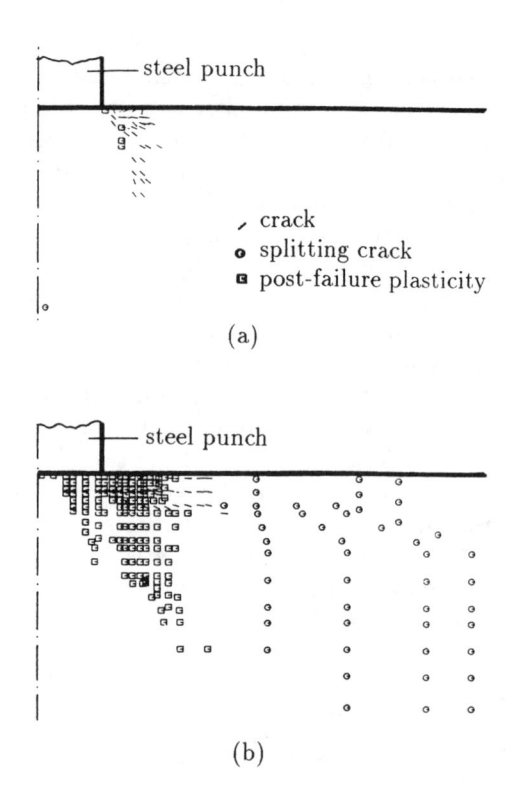

Figure 4.14: Regions of cracked and plasticized concrete: (a) at 0.3 mm penetration; (b) at 3 mm penetration

Fig.4.14 shows the location of cracks and of post-failure plastic zones of concrete for two levels of penetration, computed by means of the hypoplastic material model. At a penetration of 2 mm, the maximum plastic concrete strains are within a range of up to 20%. However, the upper limit of this range is only reached in a very small region extending over two finite elements directly beneath the steel punch. Therefore, in a global sense, the assumption of geometrically linear behavior still appears to be justified.

4.3 Nonlinear FE-Analyses of RC Panels, Slabs and Shells

A large number of papers on nonlinear finite element analyses of reinforced concrete panels, slabs and shells has been published during the last two decades (see, e.g., [Darwin (1976), Kabir (1977), Floegl (1981), Mang (1983), Owen (1983), Kollegger (1988), May (1988), Walter (1988), Crisfield (1989), Scordelis (1989), Kollegger (1990), Mang (1990), Wang (1990), Meschke (1991b), Abbasi (1992), Grote (1992), Kraetzig (1992), Balkaya (1993), Di (1993), Min (1993), Polak (1993), Shahrooz

(1994)].

The vast majority of nonlinear finite element analyses of reinforced concrete surface structures reported in the literature refers to structures, for which documentations of experimentally obtained results were available at the time of the analyses. Truly predictive nonlinear finite element analyses, however, were carried out, e.g., in the context of a round robin analysis, organized jointly by the ASCE-ACI-Committee 334 and the ACI-Committee 444 in 1991/92. The organizers prepared data on three reinforced concrete shell structures which were tested in the service-level and the strength-level regime. The experimental program was carried out for a cylindrical shell with edge beams, a folded plate without edge beams and a hyperboloid of revolution. The participants were not given any information on the observed behavior or on the data collected during the tests. The writers participated in this round robin analysis by providing predictions for the cylindrical shell and the folded plate. A summary of the round robin analysis will be published [Krauthammer (1995)]. The degree of agreement between the structural response, predicted by the writers and the experimentally observed structural behavior is good.

As an example for nonlinear finite element analyses of reinforced concrete surface structures, the analysis of a cooling tower [Meschke (1991b)] has been chosen. In the following, a report on this analysis will be presented.

4.3.1 Reinforced Concrete Cooling Tower

Description of the Structure

The cooling tower is part of a power station in Ptolemaïs, Greece. It was built in the mid-sixties. The present condition of the structure is characterized by a large number of long, meridional membrane cracks. The purpose of the analysis [Meschke (1991b)] was to examine the structural safety of the cracked structure.

The height of the hyperbolic reinforced concrete shell is 82 m. The diameter of the base and of the crown of the shell is 54.8 m and 39.0 m, respectively (Fig.4.15(a)). The diameter of the throat is 35.4 m. The throat is located 59.0 m above the base. Stiffening rings were provided at the crown and at the base. At the base, the thickness of the shell is 0.50 m. It is gradually reduced to 0.10 m at a height of 8.3 m above the base. Only within this relatively small region an inner and an outer layer of reinforcing steel, each in the circumferential and the meridional direction, was provided. The remaining part of the shell contains only one layer of reinforcing steel in the circumferential and the meridional direction, placed in the middle between the two faces of the shell. Details of the reinforcement and of the stiffening rings are shown in Fig.4.15(b).

The cylindrical compressive strength, the tensile strength and the modulus of elasticity of concrete are given as $\sigma_P^C = 22.5$ MPa, $\sigma_T^C = 1.80$ MPa and $E^C = 26\,000$ MPa, respectively. The value taken for σ_T^C represents a conservative assumption. This was confirmed by experiments yielding a tensile strength of 2.60 MPa. The

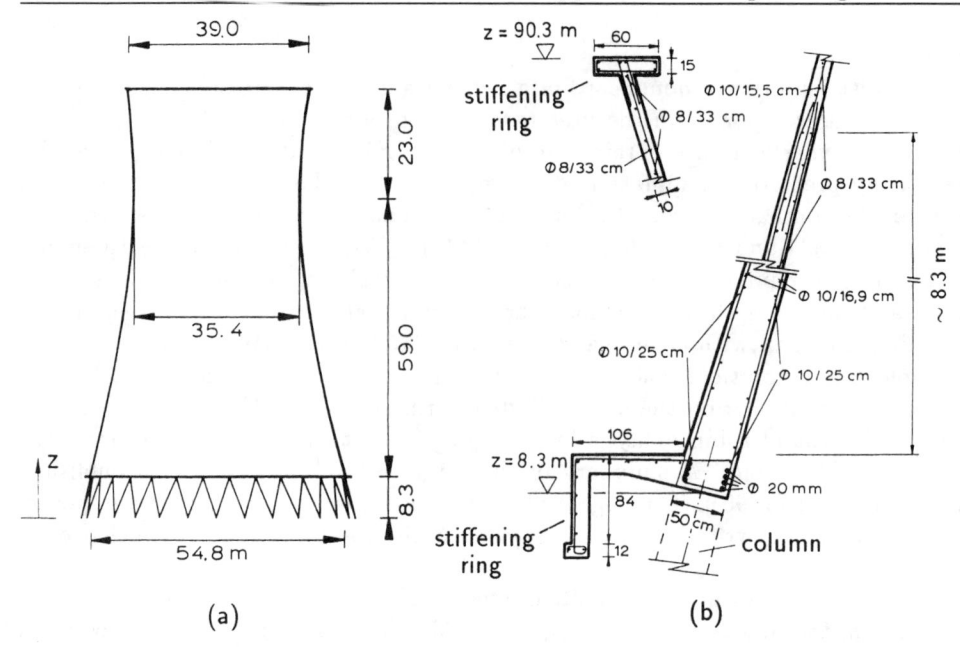

Figure 4.15: Hyperbolic cooling tower: (a) geometric dimensions, (b) cross-sections

yield strength and the modulus of elasticity of the reinforcing steel are given as 400 MPa and 206 000 MPa, respectively.

The load history consists of dead load, wind load corresponding to a wind velocity of 140 km/h, and thermal loading. The latter is an important load case, because probably the temperature difference between the inner and the outer surface of the shell has caused the aforementioned meridional cracks. The maximum temperature difference occurs in winter. It was estimated conservatively as 45°C.

Analysis Model

In what follows, the main features of the analysis model will be summarized:

- Geometrically nonlinear behavior of the shell is taken into account on the basis of an updated Lagrangian formulation.

- Consideration of nonlinear material behavior of concrete is restricted to cracking. Biaxial stress states in the shell are predominantly in the tension-compression material domain. Failure is governed by cracking of concrete and yielding of the reinforcement. Hence, the simplifying assumption of linear-elastic material behavior of concrete in the compressive region is admissible.

- Cracking of concrete is modelled within the framework of the fixed orthogonal smeared crack concept (subsection 3.2.5).

4.3 Nonlinear FE-Analyses of RC Panels, Slabs and Shells

- The shear stiffness of cracked concrete, resulting from aggregate interlock, is assumed to depend linearly on a fictitious crack strain, normal to the crack (subsection 2.6.1, equation (2.254)).

- Tension softening is taken into account on the basis of a linearly descending stress-strain diagram in the post-peak region.

- Tension-stiffening is neglected.

- For the reinforcing steel a linearly-elastic perfectly-plastic constitutive law is adopted.

Because of symmetry, only one half of the cooling tower needs to be discretized. The discretization of the shell consists of

- subdivision of one half of the shell into 180 bilinear (4 node) thick-shell elements, formulated on the basis of the degenerate solid approach,

- subdivision of the concrete shell through the thickness into 13 layers,

- distribution of the reinforcement to thin layers of mechanically equivalent thickness,

- subdivision of one half of the stiffening ring at the crown of the shell into 15 shell elements on each side of the shell,

- subdivision of one half of the stiffening ring at the base of the shell into 15 shell elements,

- modelling of the supporting columns with the help of beam elements with two nodes each, formulated on the basis of the degenerate solid approach.

Hence, in total, the (coarse) finite element mesh consists of 225 shell elements and 30 beam elements (Fig.4.16(a)). A fine mesh (Fig.4.16(b)) was generated by a consistent refinement of the coarse mesh. The discretization of the two stiffening rings by beam elements is not visible in Fig.4.16.

The softening modulus of concrete, E^C_{soft}, was determined from experimentally obtained values as reported, e.g., in [Milford (1984)]. A conservative assumption for this modulus is $E^C_{soft} = -20\,000$ MPa. Use of a constant softening modulus yields mesh-dependent results [Bažant (1979c)]. Comparative analyses were performed on the basis of the fictitious crack model within the fixed orthogonal smeared crack approach [Ottosen (1986)]. Both meshes were used. However, no significant mesh-dependence was observed.

Closing and re-opening of cracks was considered by means of a secant unloading and reloading branch in the tension domain of the stress-strain diagram. If the crack is closed, the axial stiffness of concrete will be fully recovered for subsequent compressive loading. The shear stiffness, however, does not regain the original value for uncracked concrete.

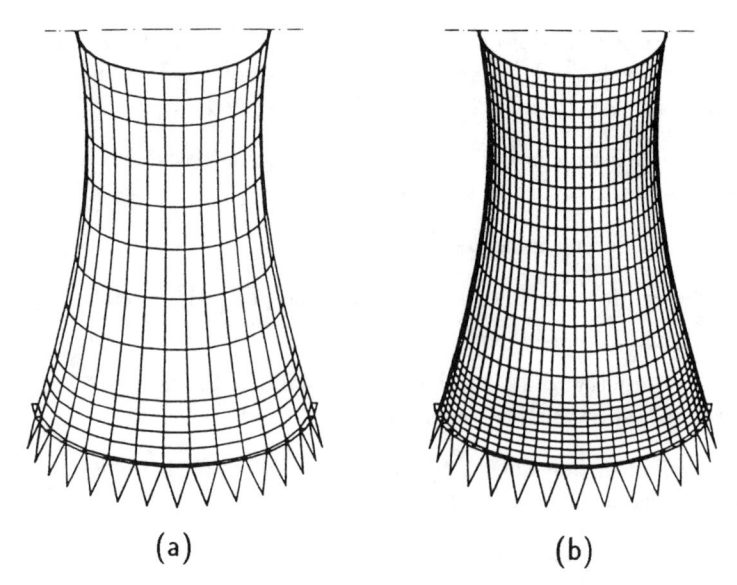

Figure 4.16: Finite element meshes of the cooling tower: (a) coarse, (b) fine

The specific weight of concrete was taken as $\gamma = 0.025$ MN/m^3. The maximum quasi-static wind load w was determined according to [VGB-Rules (1980)], resulting in

$$w(z, \theta) = c_p(\theta) \, q_D(z) \,, \tag{4.1}$$

where $q_D(z)$ denotes the wind pressure distribution in the vertical direction and $c_p(\theta)$ represents the modification of $q_D(z)$ to account for the variation of the wind load in the circumferential direction (z is the distance from the ground; $\vartheta = 0$ defines the windward meridian). For the design wind pressure, corresponding to wind zone III [VGB-Rules (1980)] characterized by a wind velocity of 140 km/h, the vertical pressure distribution is given as [VGB-Rules (1980)]

$$q_D(z) = 0.75 \left(\frac{z}{10}\right)^{0.22} ; \tag{4.2}$$

the dimension of z are meters; the dimension of $q_D(z)$ is kN/m^2. Plots of $q_D(z)$ and $c_p(\theta)$ are shown in Fig.4.17.

Only two load cases out of those investigated in [Meschke (1991b)] are described in the following. Load case I consists of dead load followed by the application of wind load. The second load case, denoted as load case III in [Meschke (1991b)], consists of dead load, consideration of a temperature difference between the inner and the outer surface of the shell of $\Delta T = 45°$C, followed by the removal of ΔT and, subsequently, by application of the wind load. The temperature difference causes cracking of the shell. Hence, load case III is characterized by applying the wind load to the pre-damaged shell. The analysis was performed by means of the nonlinear finite element program MARC [MARC (1992)].

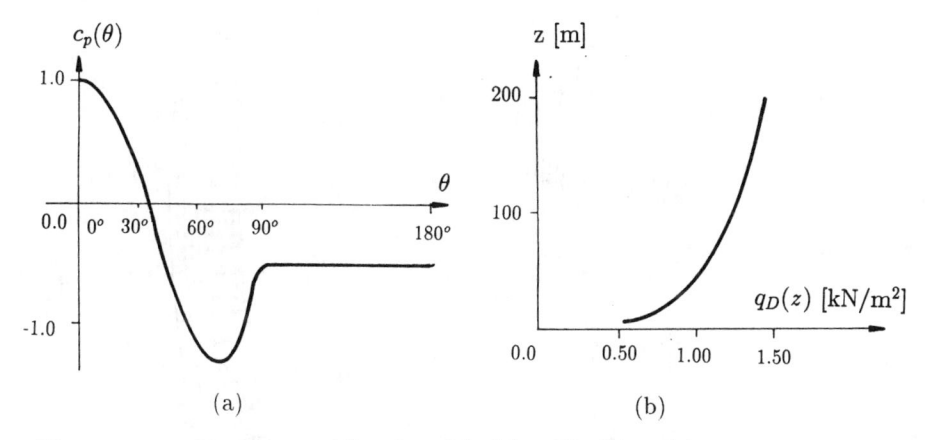

4.3 Nonlinear FE-Analyses of RC Panels, Slabs and Shells

Figure 4.17: Functions $c_p(\theta)$ and $q_D(z)$: (a) $c_p(\theta)$, (b) $q_D(z)$

Results

Prior to the numerical simulation of the structural response resulting from the described load history, an axisymmetric analysis was performed. The purpose of this analysis was (a) to verify the assumption that the observed cracks have been caused by a temperature difference ΔT between the inner and the outer surface of the shell during winter time and (b) to determine the value of ΔT at which cracks are opening at the outer surface of the shell. After application of dead load, the temperature difference ΔT was increased incrementally.

Fig.4.18 shows the crack distribution in a meridional cross-section of the shell. (In the figure, the thickness of the shell is magnified by a factor of 100.) At $\Delta T = 12.5°C$, meridional cracks occur at the outer surface (Fig.4.18(a)). They can be interpreted as microcracks, because the strains normal to the cracks are still within the softening region. Upon further increase of ΔT the cracks gradually penetrate into the shell. At $\Delta T = 22.5°C$, the meridional microcracks are opening at the outer surface (Fig.4.18(b)). Macroscopically open cracks are characterized by a complete loss of the capacity to transfer tensile stresses.

The numerical simulation of the structural response of the cooling tower subjected to dead load and wind load yields load-deflection diagrams shown in Fig.4.19. One of these two diagrams (solid curve) is characterized by disregard of the pre-damage resulting from a temperature history, the other one (dashed curve) by consideration of such a pre-damage. The diagrams illustrate the horizontal displacement of the windward meridian at the crown of the shell, plotted as a function of the dimensionless factor λ, by which the wind load w has been multiplied. It follows from Fig.4.19 that the pre-damage induced by thermal loading has a significant influence on the load factor at the onset of cracking, denoted as λ_C. However, this pre-damage has only little influence on the load factor at the initiation of yielding of the reinforcement, denoted as λ_Y, and next to no influence on the load factor at failure,

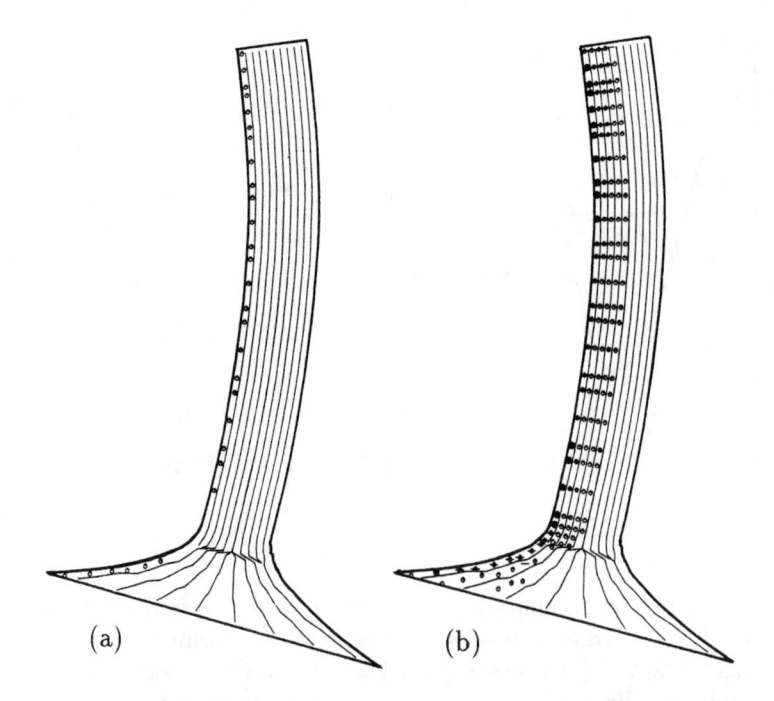

○ meridional cracks within the softening region
● meridional cracks beyond the softening region
— circumferential cracks within the softening region

Figure 4.18: Distribution of cracks in consequence of a temperature difference ΔT: (a) $\Delta T = 12.5°C$, (b) $\Delta T = 22.5°C$

——— load case I
− − − − load case III

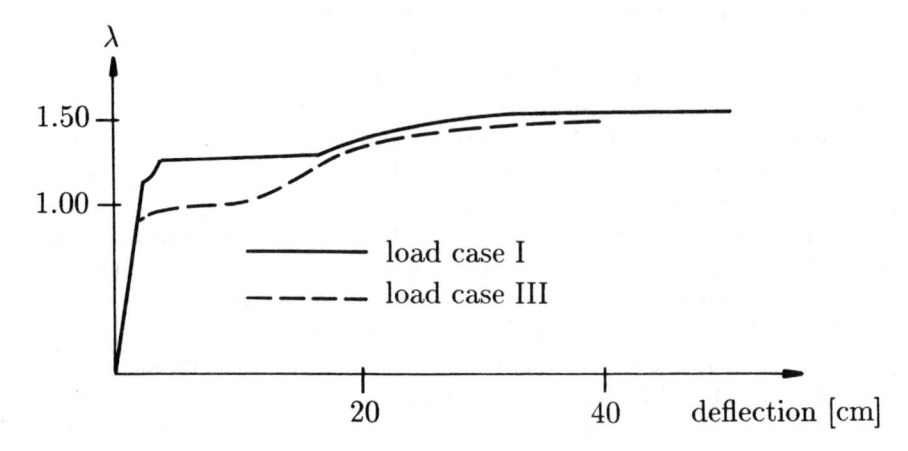

Figure 4.19: Load-deflection diagrams

4.3 Nonlinear FE-Analyses of RC Panels, Slabs and Shells

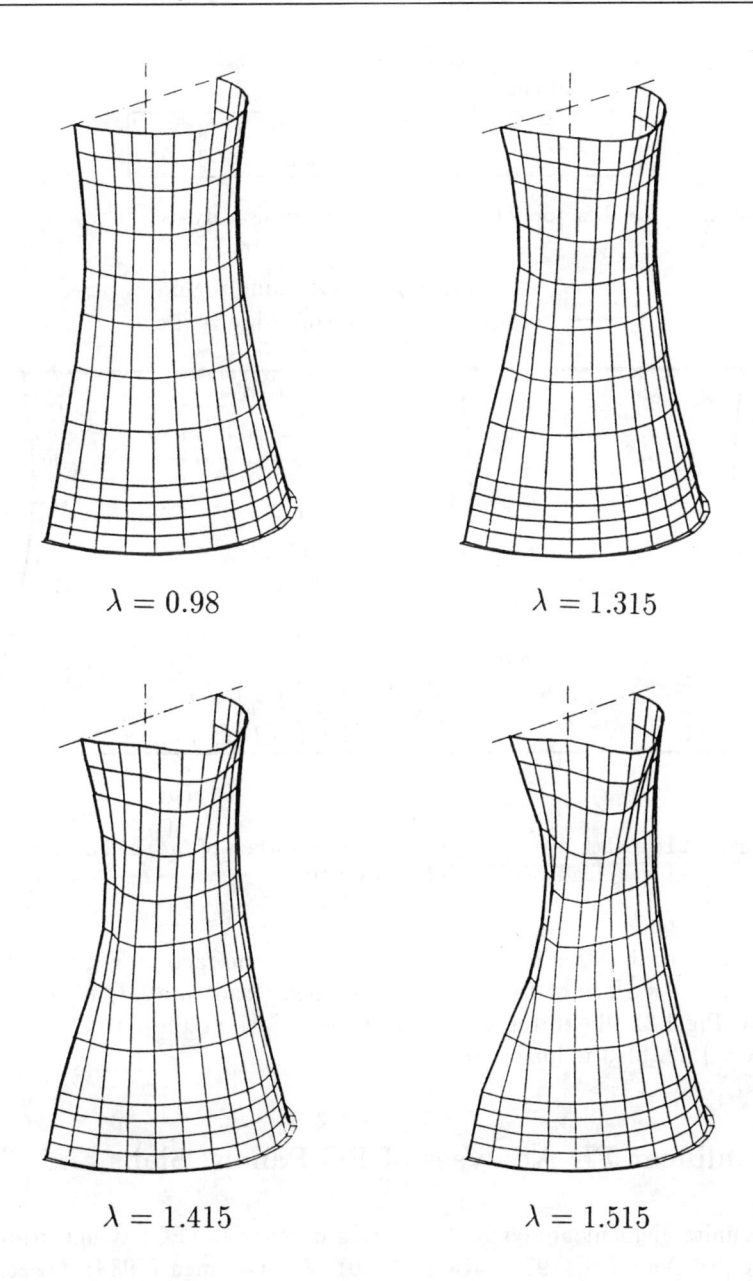

$\lambda = 0.98$

$\lambda = 1.315$

$\lambda = 1.415$

$\lambda = 1.515$

Figure 4.20: Deformed middle surface of the pre-damaged shell at four different levels of the wind load (deflections magnified by a factor of 100)

load case	λ_C	λ_Y	λ_U
I	1.26	1.29	1.56
III	0.98	1.24	1.52

Table 4.2   Load factors obtained from numerical analyses

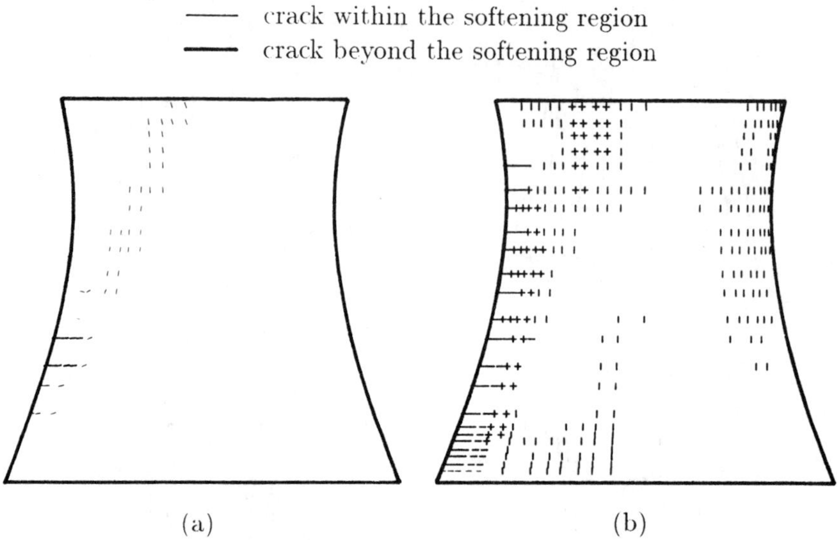

—— crack within the softening region
—— crack beyond the softening region

Figure 4.21: Crack distributions at the outer surface of the shell at $\lambda = 1.22$: (a) load case I, (b) load case III

denoted as λ_U. Table 4.2 contains the values of the load factors. Fig.4.20 shows the deformed middle surface of the pre-damaged shell (load case III) at four different levels of λ. Fig.4.21 illustrates the crack distributions at the outer surface of the shell, at $\lambda = 1.22$, for the load cases I and III.

4.4   Nonlinear FE-Analyses of PC Panels, Slabs and Shells

Nonlinear finite element analyses of prestressed concrete (PC) structures are contained, e.g., in [Van Zyl (1978), Kang (1980), Van Greunen (1983), Owen (1983), Mari (1984), Hofstetter (1986b), Hofstetter (1987), Roca (1988), Walter (1988), Hofstetter (1989), Povoas (1989), Scordelis (1989), Mang (1991), Roca (1993a), Roca (1993b)]. In what follows, nonlinear finite element analyses of a prestressed concrete slab, a prestressed concrete cylindrical shell with edge beams, and a prestressed concrete secondary containment structure will be described in detail.

4.4 Nonlinear FE-Analyses of PC Panels, Slabs and Shells

4.4.1 Prestressed Concrete Slab

Description of the Structure

Fig.4.22(a) shows a prestressed concrete slab tested by Ritz et al. [Ritz (1975)]. The length and the span of the rectangular slab are 3.80 m and 3.60 m, respectively. Its thickness is 0.18 m. The slab is simply supported along its edges.

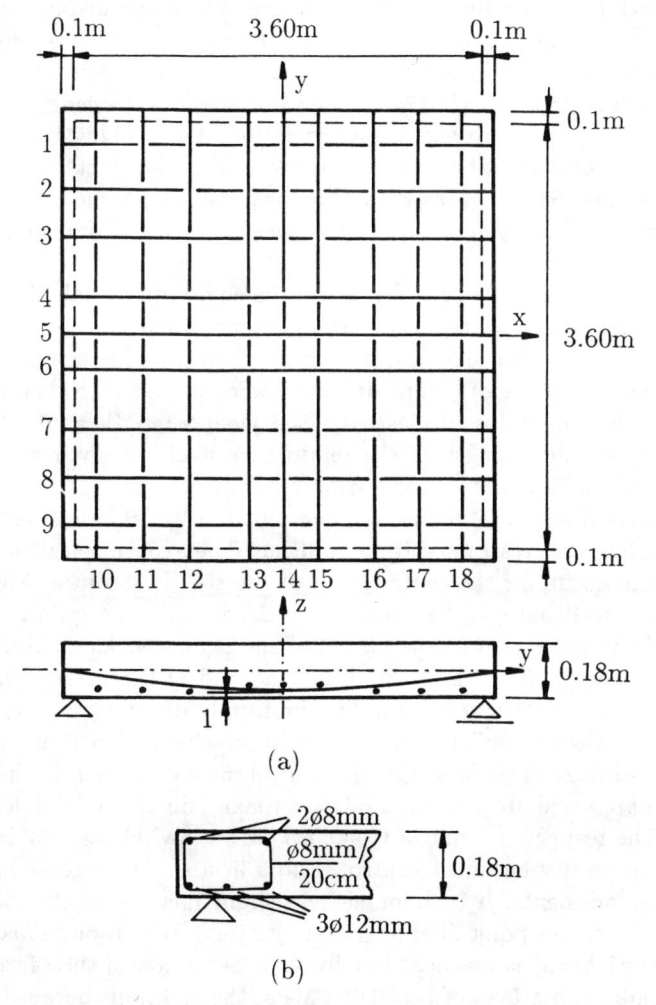

Figure 4.22: Prestressed concrete slab: (a) layout of the tendons, (b) reinforcement along the edges

The modulus of elasticity, Poisson's ratio and the specific weight of concrete are given as $E^C = 35\,800$ MPa, $\nu^C = 0.16$ and $\rho^C = 2470$ kg/m^3 [Ritz (1975)]. The uniaxial prism strength and the modulus of rupture of concrete are specified as

$\sigma_P^C = 33$ MPa and $\sigma_{T,ru}^C = 6.0$ MPa, respectively.

The geometric form of the axes of the 18 tendons consisting of 1/2" strand wires is one of quadratic parabolae located in planes perpendicular to the x-axis and the y-axis. The cross-sectional area of the tendons is given as $A^Z = 0.93 \cdot 10^{-4}$ m^2, the modulus of elasticity as $E^Z = 190\,000$ MPa, the yield strength at 0.2% permanent strain as $\sigma_{0.2}^Z = 1\,703$ MPa, the tensile strength as $\sigma_T^Z = 1\,803$ MPa, and the ultimate strain at failure as 2.66%. The ends of the tendon axes are located at the edges of the middle surface of the slab. In the symmetry plane the distance of the axes of the tendons No. 13, 14 and 15 from the lower face of the slab is about 3.2 cm. The respective distance of the remaining tendon axes is about 1.8 cm. Each tendon was prestressed with a force of 0.1176 MN. One week after prestressing, at the beginning of the ultimate load test, average prestressing forces of 0.1125 MN were measured. In order to reduce friction between the tendons and the ducts, the tendons were greased. As pointed out in [Ritz (1975)], it was possible to remove the tendons after the test from the slab by hand. Hence, it was justified to neglect friction in the analyses.

Reinforcing steel was only provided along the edges of the slab. This was done in order to reinforce the anchor zones of the tendons. 2 bars of 8 mm diameter each were placed on the upper side and 3 bars of 12 mm diameter each on the lower side of the slab. Stirrups of 8 mm diameter were placed at a distance of 0.20 m (Fig.4.22(b)). The modulus of elasticity, the yield strength at 0.2 % permanent strain and the tensile strength of the reinforcing steel are given as $E^S = 210\,000$ MPa, $\sigma_{0.2}^S = 520$ MPa and $\sigma_T^S = 650$ MPa, respectively.

The uniformly distributed load was approximated by 16 point loads. The corner points of the slab were restrained from uplifting. The experimentally obtained load-displacement diagram for the center point of the slab is shown in Fig.4.23. In this figure, g, p and w denote, in this order, the dead load, the uniformly distributed live load and the deflection of the center of the slab, measured from the position of this point after application of the dead load and of the prestressing forces. Up to a load intensity which corresponds to a uniformly distributed load of $p = 0.0579$ MPa, the mechanical behavior of the slab is linear-elastic. Upon further increase of the load, the stiffness is reduced gradually until, at a deflection of the center point of the slab of approximately 8 cm, a relative maximum of the load-deflection curve is attained. The respective value of p is 0.0955 MPa. Switching from load control in the experiment to displacement control results in a slight decrease of the load for increasing displacements. It is followed by a slight increase of the load until, at a deflection of the center point of approximately 21 cm, the absolute maximum of the load ($p = 0.0967$ MPa) is reached. Finally, at a deflection of the center point of 26 cm, corresponding to a load of $p = 0.093$ MPa, the slab fails because of rupture of tendons.

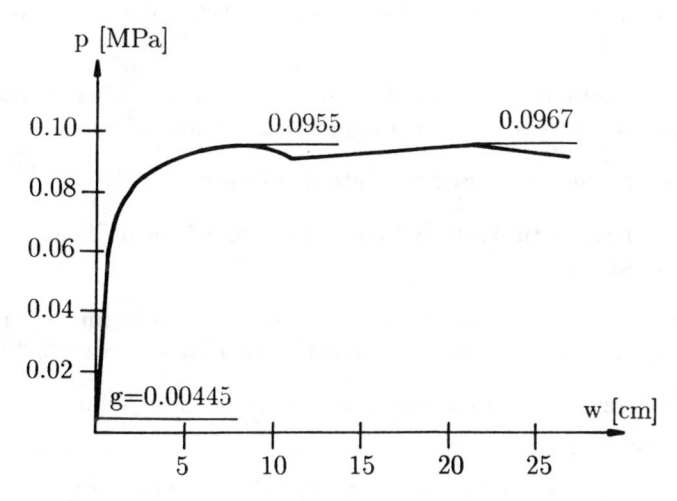

4.4 Nonlinear FE-Analyses of PC Panels, Slabs and Shells

Figure 4.23: Experimentally obtained load-deflection diagram

Analysis Model

The main features of the analysis model are:

- consideration of geometrically nonlinear behavior on the basis of Koiter's nonlinear shell theory for moderately large rotations [Koiter (1967)],

- consideration of nonlinear material behavior of concrete, based on the nonlinear elastic constitutive law proposed by Liu, Nilson and Slate [Liu (1972)] (see subsection 2.3.2, equation (2.35)),

- consideration of cracking of concrete within the framework of the fixed orthogonal smeared crack concept (subsection 3.2.5),

- disregard of tension softening by setting the tensile stress equal to zero when the tensile strength is exceeded,

- assumption of a linear dependence of the shear stiffness of cracked concrete, resulting from aggregate interlock, on a fictitious crack strain normal to the crack (subsection 2.6.1, equation (2.254)),

- consideration of the ability of reinforced concrete to carry forces between neighboring cracks in consequence of bond between the reinforcement and the surrounding concrete on the basis of the steel-related tension-stiffening model reported in [Floegl (1982)],

- consideration of nonlinear material behavior of the reinforcing steel by means of a bilinear stress-strain relation (see subsection 2.5),

- description of the tendons as arbitrary spatial curves embedded in the shell (see subsection 3.4.4),

- consideration of nonlinear material behavior of the prestressing steel on the basis of a nonlinear stress-strain relation (see subsection 2.5).

The discretization of the prestressed concrete slab consists of:

- subdivision of the structure into triangular C_1-conforming shell elements (see Fig.3.29 [Thomas (1975)]),

- subdivision of the concrete slab through the thickness into thin layers such that a state of plane stress may be assumed in each layer (see Fig.3.30),

- smearing of the reinforcement to thin layers of mechanically equivalent thickness (see Fig.3.30),

- application of the embedded representation of the tendons in order to determine element-related forces exerted from the tendons on the concrete (see subsection 3.4.4).

Because of symmetry, only one quarter of the slab needs to be discretized. The part of the slab extending beyond the supports is neglected. Three different meshes, consisting of 8, 32 and 128 elements, respectively, were employed for the finite element analyses. The coarse and the medium mesh are shown in Fig.4.24. The reason for the irregularity of the finite element meshes is the reinforcement which is restricted to the vicinity of the edges of the slab. The fine mesh is a systematic refinement of the medium mesh.

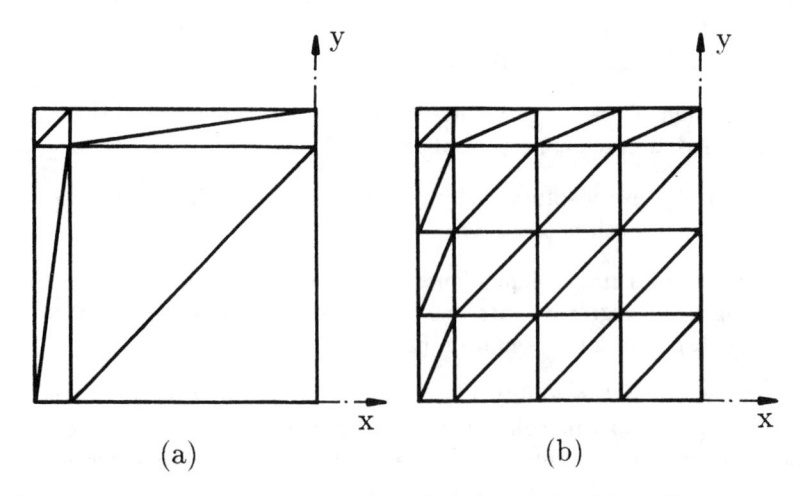

Figure 4.24: Finite element meshes: (a) coarse mesh, (b) medium mesh

4.4 Nonlinear FE-Analyses of PC Panels, Slabs and Shells

Determination of the failure envelope for biaxially stressed concrete requires specification of the tensile strength of concrete, σ_T^C. According to [Leonhardt (1980)], σ_T^C is related to the modulus of rupture $\sigma_{T,ru}^C$ as follows:

$$\sigma_T^C = 0.52\, \sigma_{T,ru}^C . \tag{4.3}$$

Hence, $\sigma_T^C = 3.12$ MPa. To account for tension stiffening, the bond strength and the initial distance of cracks are assumed as 5.5 MPa and 0.20 m, respectively. However, because of the restriction of the reinforcement to the vicinity of the edges of the slab, the influence of tension stiffening on the structural behavior is small. On the basis of a bilinear stress-strain law and an assumed ultimate strain of 10 %, the hardening modulus of the reinforcing steel is obtained as 1 330 MPa.

The tendon axes are spatial curves. The geometric description of the axes of the tendons parallel to the x-axis and parallel to the y-axis, respectively, is given as

$$\mathbf{x} = \left\{ \begin{array}{c} x \\ const. \\ z = a_i x^2 + b_i \end{array} \right\},\ i = 1,\ldots,5,\quad \mathbf{x} = \left\{ \begin{array}{c} const. \\ y \\ z = a_i y^2 + b_i \end{array} \right\},\ i = 10,\ldots,14, \tag{4.4}$$

where a_i and b_i denote constants. These constants are determined such that the curvature of the tendon axes is the same as in the experiment and that there is no eccentricity of these axes at the boundaries of the slab. The values of a_i and b_i are summarized in Table 4.3. With the help of these constants the direction of the tangent vector and of the principal normal vector and the curvature of the tendon axis can be te determined for each point of this axis.

tendon #	a_i [m^{-1}]	b_i [m]
1 – 5	0.0201	−0.065
10 – 12	0.0201	−0.065
13 – 14	0.0154	−0.050

Table 4.3 Coefficients for the geometric description of the tendon axes

Results

Table 4.4 contains the values for the deflection of the center of the prestressed slab subjected to dead load, obtained from linear analyses of the three finite element models of the slab.

These values can be compared with the result of an approximate calculation by hand, which is based on the analytic solution for the center deflection w of a quadratic slab of span l, subjected to a uniform load p. This solution is given as [Girkmann (1963)]

mesh	upward deflection at center in [cm]
8 elements	0.0433
32 elements	0.0664
128 elements	0.0674

Table 4.4 Deflection of the center of a PC slab subjected to dead load

$$w = 0.00406\,\frac{p\,l^4}{K} \quad \text{with} \quad K = \frac{E\,h^3}{12(1-\nu^2)}, \tag{4.5}$$

where h denotes the thickness of the slab. An approximation of the line load exerted from a tendon on the slab in the direction of the principal normal vector is obtained by multiplying the mean curvature of the tendon with the prestressing force. For the tendons # 1 to 5 and # 10 to 12 the mean curvature is 0.04 m^{-1}; for the tendons # 13 and 14 it is 0.031 m^{-1}. Thus, the line loads in the direction of the principal normal vector are obtained as $0.04 \cdot 0.1125 \cdot 10^6 = 4\,500$ N/m and $0.031 \cdot 0.1125 \cdot 10^6 = 3\,488$ N/m, respectively. Conversion of these line loads to a uniformly distributed load per unit area, p^Z, gives

$$p^Z = (4\,500 \cdot 15 + 3\,488 \cdot 3) \cdot 3.6/3.6^2 \approx 21\,657 \text{ N/m}^2. \tag{4.6}$$

Superposition of p^Z and dead load ($g = 4450$ N/m^2), noting that p^Z is acting in the opposite direction of gravity, yields a uniformly distributed load of 0.0172 MN/m^2. Substitution of this value into (4.5) yields an upward deflection of the center of 0.066 cm. The experimental result is reported as 0.06 cm.

Fig.4.25 contains the experimentally obtained load-deflection diagram for the center of the slab and the respective load-deflection diagrams resulting from finite element analyses. The load-deflection diagrams for the medium and the fine mesh agree well with the experimentally obtained diagram. In the analyses the load was increased incrementally. Hence, it was only possible to determine the structural response up to the first maximum value of the load. In Table 4.5 the values for the ultimate load, as obtained from such load-controlled finite element analyses, are summarized.

mesh	ultimate load in [MN/m^2]
8 elements	0.107
32 elements	0.093
128 elements	0.084

Table 4.5 Results from ultimate load analyses of a PC slab

It follows from Table 4.5 and from Fig.4.23 that the maximum difference between the experimentally obtained and the computed value of the ultimate load is

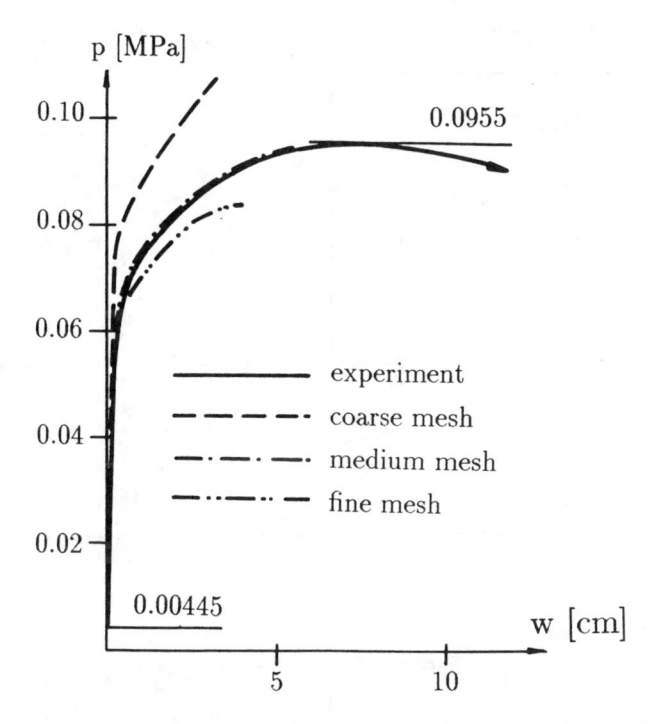

4.4 Nonlinear FE-Analyses of PC Panels, Slabs and Shells

Figure 4.25: Experimentally obtained and computed load-deflection diagrams

12%. The rate of convergence of the computed results for the ultimate load is relatively poor. The difference between the values of the ultimate load obtained by the coarse and the medium mesh, respectively, is 13 %. The difference of the respective values between the medium and the fine mesh is not much smaller, namely 10 %. Such a situation indicates that the results from finite element analyses are mesh-dependent. This is the consequence of using the strength criterion for the determination of cracking in combination with the assumption of an instantaneous drop of the tensile stress to zero at the onset of cracking [Bažant (1979c)]. Since, apart from the boundary region, the investigated prestressed slab does not contain reinforcing bars, the deficiency of this approach becomes apparent. Hence, the constitutive model should be improved by replacing the employed crack model by an objective crack model.

In Fig.4.26 the cracks on the top and the bottom surface of the slab are plotted for different load levels. Because of symmetry, only one quarter of the slab is shown. The diagonal cracks in the corner region of the top surface are caused by restraining the upward deflection of the corner points of the slab.

The maximum value of the prestressing forces in the tendons was obtained for tendon # 5. The respective value of 0.1435 MN is associated with the ultimate load. Hence, the increase of the initial prestressing force in this tendon is 28 %. Disregard

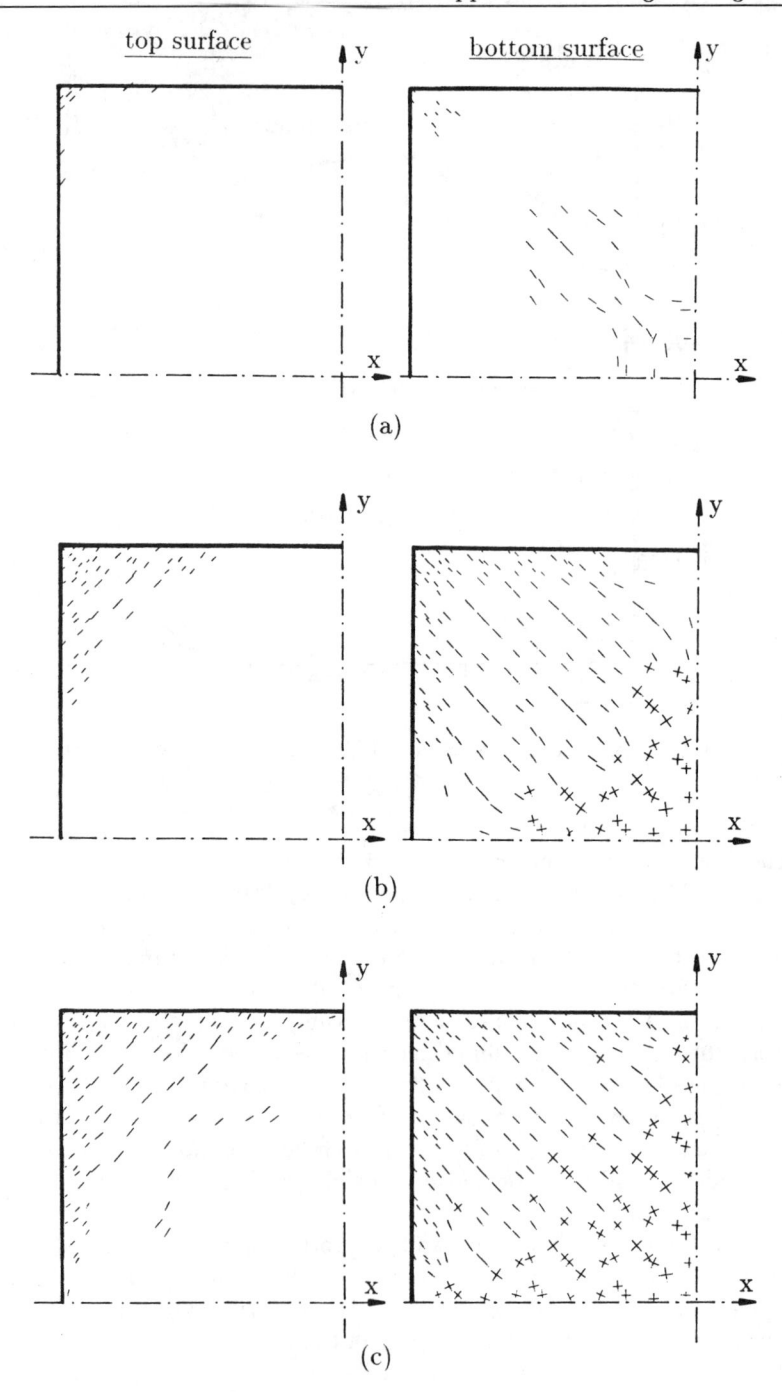

Figure 4.26: Development of cracks with increasing load: (a) $p = 0.0654$ MPa, (b) $p = 0.070$ MPa, (c) $p = 0.090$ MPa

of the changes of the prestressing forces with increasing deflections of the slab results in a 13 % decrease of the value computed for the ultimate load.

Remarkably, neglect of the relatively small amount of reinforcing steel in the boundary region of the slab yields a 22 % decrease of the value computed for the ultimate load. The respective deflections are significantly smaller than the ones obtained for the reinforced slab. If, however, an orthogonal mesh of reinforcement of 0.3 % of the cross-sectional area is provided throughout the slab, then the value computed for the ultimate load will increase by 27 %.

4.4.2 Prestressed Concrete Shell with Edge Beams

Description of the Structure

Fig.4.27 shows the model of a prestressed circular cylindrical barrel vault with edge beams and diaphragms, tested by Bouma et al. [Bouma (1961)]. The dimensions are given in centimeters. The scale factor (model dimensions : dimensions of the actual shell) is 1:8.

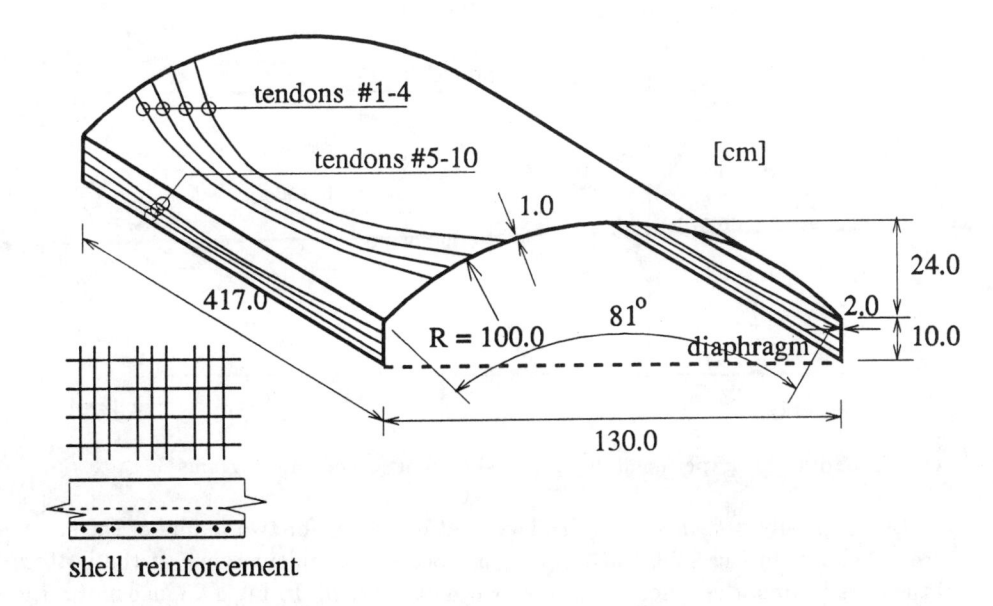

Figure 4.27: Circular cylindrical barrel vault with edge beams and diaphragms

The shell only has a bottom reinforcement consisting of 1 mm diameter wires at 2 cm distance in the longitudinal as well as in the transverse direction. In addition, 1 mm wires at 4 cm distance were placed in the transverse direction. The area of the steel reinforcement in the cross-sections of the edge beams is 0.4% of the area of the cross-sections. Both the shell and the edge beams were prestressed by means of posttensioned bonded tendons. Eight tendons were placed in the middle surface of the shell and six tendons each in the two edge beams. The shell is supported at the four corner points.

In [Bouma (1961)] the cube strength and the modulus of rupture of the concrete are reported as 35 MPa and 6.5 MPa, respectively. The yield stress and the tensile strength of the reinforcement are specified as 300 MPa and 390 MPa, respectively. The modulus of elasticity and the tensile strength of the prestressing steel are given as 200 000 MPa and 1 700 MPa, respectively. The tendons were prestressed from both ends. The prestressing forces were chosen such that the stress in the prestressing wires at midspan was equal to 1 000 MPa.

According to [Bouma (1961)], the design load p for the shell membrane and the edge beams was chosen as $10\,g_s$ and $5\,g_e$, respectively, where g_s and g_e denote the dead load given as $g_s = \gamma h_s$ and $g_e = \gamma h_e$, respectively. γ is the specific weight of concrete, h_s is the thickness of the shell membrane and h_e is the thickness of the edge beams. With $\gamma = 0.025$ MN/m^3, $h_s = 0.01$ m and $h_e = 0.02$ m, $g_s = g_e/2$ is obtained as 0.00025 MPa. The load was increased up to failure of the shell.

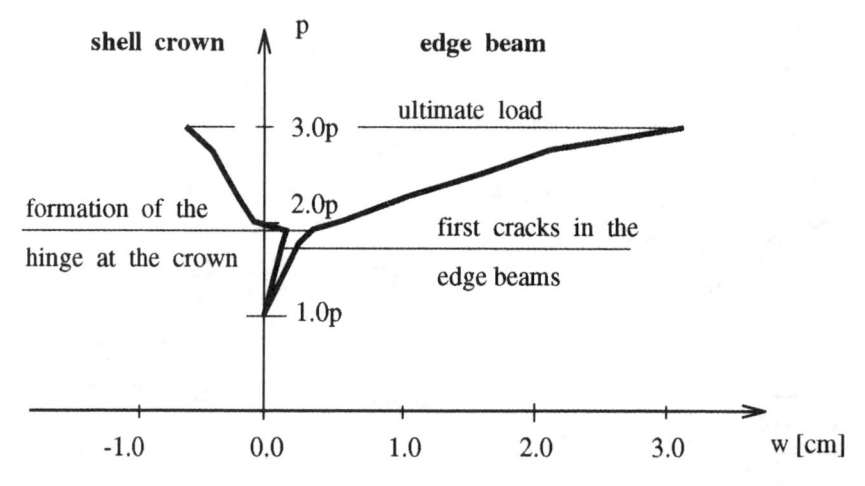

Figure 4.28: Experimentally obtained load-displacement diagrams

Experimentally obtained load-displacement diagrams for two points of the structure are shown in Fig.4.28. Both points are located at midspan, one at the bottom of the edge beams, the other one at the crown of the shell. In Fig.4.28, w denotes the vertical displacement of the respective points. In the experiment, at a load level of $1.85\,p$, the tensile strength of concrete was reached at midspan of the edge beams in the bottom edge fibers, resulting in the initiation of cracking. Upon further increase of the load, cracking also began at the crown of the shell, in the longitudinal direction. Since no top layer of shell reinforcement was provided, a large longitudinal crack developed at the crown of the shell, forming a hinge. In consequence of this hinge, upon further increase of the load, the vertical displacements of the crown of

4.4 Nonlinear FE-Analyses of PC Panels, Slabs and Shells

the shell changed their sign, i.e., the crown of the shell started moving upwards. Because of the reduction of the stiffness of the shell in consequence of cracking, the reinforcement at the crown of the shell began yielding. Subsequently, the reinforcement of the edge beams started yielding. Finally, the prestressed shell failed at a load level of $3.0\ p$.

The mechanical behavior of the shell in the nonlinear range is influenced considerably by the missing top layer of reinforcement in the shell membrane. Nevertheless, it is interesting to check, whether a nonlinear finite element analysis is able to provide a close approximation of the experimentally observed nonlinear structural behavior.

Analysis Model

Because of symmetry, only one quarter of the shell needs to be discretized. The main features of the analysis model are identical to the ones of the analysis model for the prestressed concrete slab described in the preceding subsection. Thus, the structure was discretized by means of curved, triangular C_1-conforming shell elements (Fig.3.29), formulated in terms of surface coordinates α and β. In Fig.4.29 these coordinates are shown for both the shell and the edge beam.

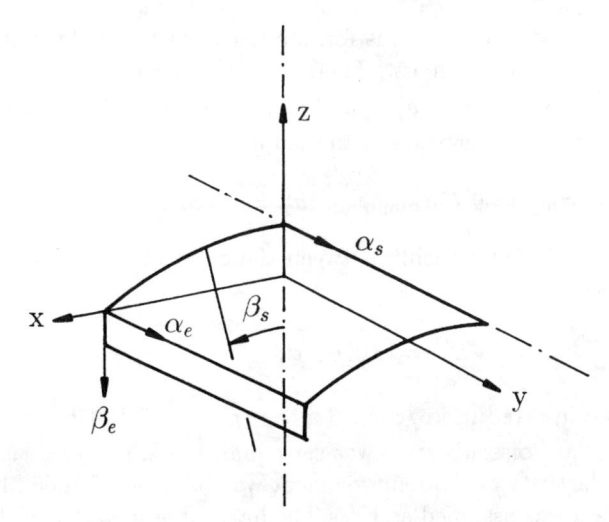

Figure 4.29: Surface coordinates of the shell (α_s, β_s) and the edge beam (α_e, β_e)

For both the middle surface of the shell and the vertical plane of symmetry of the edge beam, α was chosen parallel to the longitudinal direction of the structure. For the middle surface of the shell, β was chosen as the angle enclosed by the z-axis and the vector normal to the middle surface of the shell. As far as the vertical plane of symmetry of the edge beam is concerned, β is parallel to the z-axis.

The diaphragms at the longitudinal ends of the shell were assumed to be infinitely stiff in their planes and completely flexible normal to these planes. On the basis of these assumptions the diaphragms were omitted in the finite element discretizations. Three meshes consisting of 8, 24 and 96 shell elements, respectively, were considered in the analysis. The coarse and the medium mesh are shown in Fig.4.30. The fine mesh was obtained by a consistent refinement of the medium mesh.

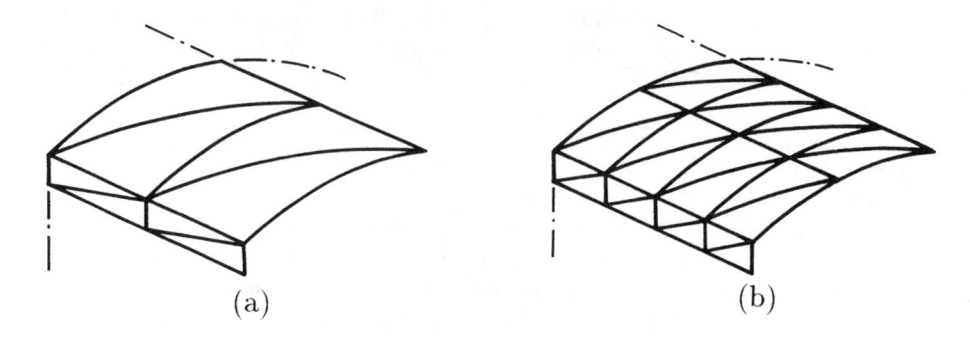

Figure 4.30: Finite element meshes: (a) coarse, (b) medium

The failure envelope for concrete was formulated in terms of the prism strength $\sigma^C_{P,pr}$ and of the tensile strength σ^C_T. Following [Leonhardt (1980)], the relation between the (known) cube strength $\sigma^C_{P,cu}$ and $\sigma^C_{P,pr}$ and the one between the (known) modulus of rupture $\sigma^C_{T,ru}$ and σ^C_T were assumed as

$$\sigma^C_{P,pr} = 0.79 \sigma^C_{P,cu}, \qquad \sigma^C_T = 0.52 \sigma^C_{T,ru}. \tag{4.7}$$

In [Leonhardt (1980)] the relationship between the cube strength and the modulus of elasticity is proposed as

$$E^C = 18\,000 \sqrt{\sigma^C_{P,cu}}, \tag{4.8}$$

where $\sigma^C_{P,cu}$ must be inserted in kp/cm². Hence, $\sigma^C_{P,pr} = 27.7$ MPa, $\sigma^C_T = 3.4$ MPa and $E^C = 33\,700$ MPa. Poisson's ratio was set equal to 0.2.

The modulus of elasticity of the reinforcement was taken as 210 000 MPa and the ultimate tensile strain was assumed as 10%. The hardening modulus of the assumed bilinear stress-strain diagram was obtained as 910 MPa. The horizontal reinforcing bars in the edge beams were modelled as mechanically equivalent vertical steel layers. To account for tension stiffening, the bond strength and the initial distance of the cracks were assumed to be 5.5 MPa and 0.2 m, respectively.

The layout of the tendons was reconstructed from a photograph in [Bouma (1961)]. In terms of the surface coordinates α and β (see Fig.4.29) the tendon axes can be described mathematically as

$$\beta = a_i \alpha^2 + b_i. \tag{4.9}$$

4.4 Nonlinear FE-Analyses of PC Panels, Slabs and Shells

The coefficients a_i and b_i are listed in Table 4.6. The tendons were placed in the middle surface of the shell. In the edge beams they were arranged symmetrically to the vertical plane of symmetry. However, in order to simplify the analysis, they were assumed to be located in this plane of symmetry.

	Tendon #			
	1	2	3	4
a_i [m^{-2}]	−0.09896	−0.08285	−0.06674	−0.05063
b_i [rad]	0.66707	0.67702	0.68697	0.69692

	Tendon #					
	5	6	7	8	9	10
a_i [m^{-1}]	−0.01572	−0.01336	−0.00904	−0.00668	−0.00236	0.0
b_i [m]	0.07138	0.07988	0.07988	0.08838	0.08838	0.09688

Table 4.6 Coefficients a_i and b_i for the geometric description of the tendon axes

Prestress losses because of friction between the tendons and the ducts were taken into account by assuming the coefficient of friction to be equal to 0.2 and the wobble friction coefficient to be equal to $0.04°$/m. In order to obtain a tendon stress of 1 000 MPa at midspan, the applied tendon forces must vary from 3 170 N to 3 450 N, depending on the length and the curvature of the respective tendon.

Results

Fig.4.31 shows experimentally obtained load-deflection diagrams and corresponding diagrams resulting from the finite element analysis based on the medium mesh. The diagrams refer to the deflection of the crown of the shell membrane and to the deflections of the edge beams at midspan. The deflections are caused by the load λp where λ is the load factor. Table 4.7 contains values of λ, for the three considered finite element meshes, at initial cracking in the edge beams and at failure of the structure. The values for the load factor λ at initial cracking in the edge beams, as

mesh	λ at initial cracking in the edge beams	λ at failure
coarse	2.1	3.40
medium	1.8	2.54
fine	1.7	2.33

Table 4.7 Load factor λ for three different finite element meshes

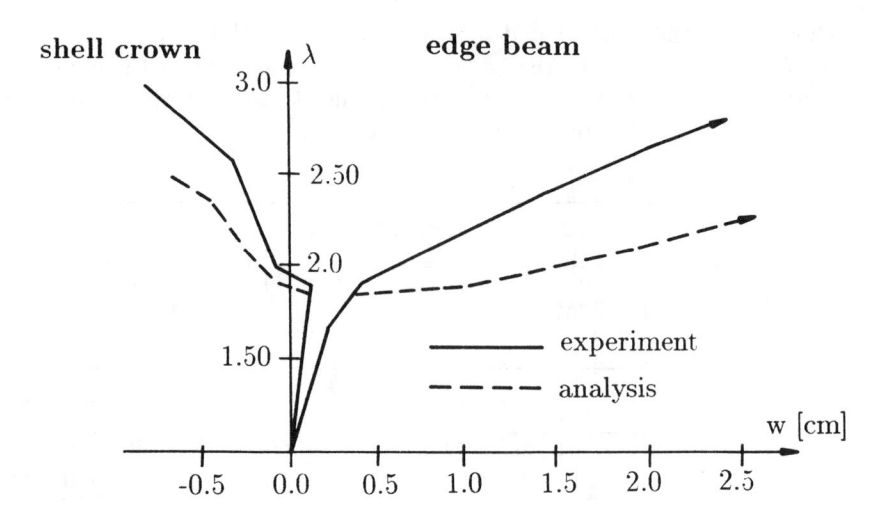

Figure 4.31: Load-deflection diagrams

obtained for the medium and the fine mesh, agree well with the respective experimentally observed value of $\lambda = 1.85$. Upon a slight increase of the load, the analysis correctly signals the development of longitudinal cracks in the crown domain of the shell membrane, causing the formation of the mentioned hinge. Upon further increase of the load, for the case of the medium and the fine mesh the analyses indicate the change in direction of the deflection of the crown. However, within this range of loading, especially for the edge beams, the computed displacements are larger than the experimentally obtained displacements. Nevertheless, the difference between the computed and the experimentally obtained value for the ultimate load of about 20% is acceptable. Compared with the previous example, the rate of convergence of the numerical results for the ultimate load, based on the three meshes, is better. The difference between the values for the ultimate load computed on the basis of the coarse and the medium mesh is 25 %. The difference between respective values computed by means of the medium and the fine mesh is 8 %. Hence, the mesh-dependence of the results for the prestressed shell is smaller than the one for the prestressed slab. This mesh-dependence is caused by the application of the strength criterion for the determination of crack initiation in combination with the assumption of an instantaneous drop of the tensile stress to zero at the onset of cracking. The reason for the relatively small mesh-dependence of the results for the prestressed shell is the existence of the reinforcement.

Fig.4.32(a) shows the undeformed structure. Figs.4.32(b) - (d) illustrate the deformed structure at different load levels. In Figs.4.32(b) - (d) the deflections are magnified by a factor of 200, 100 and 10, respectively. Distributions of cracks on the outer and the inner surface of the shell at two different load levels are shown in Figs.4.33 and 4.34. These crack distributions were computed on the basis of the medium mesh.

4.4 Nonlinear FE-Analyses of PC Panels, Slabs and Shells

(a)

(b)

(c)

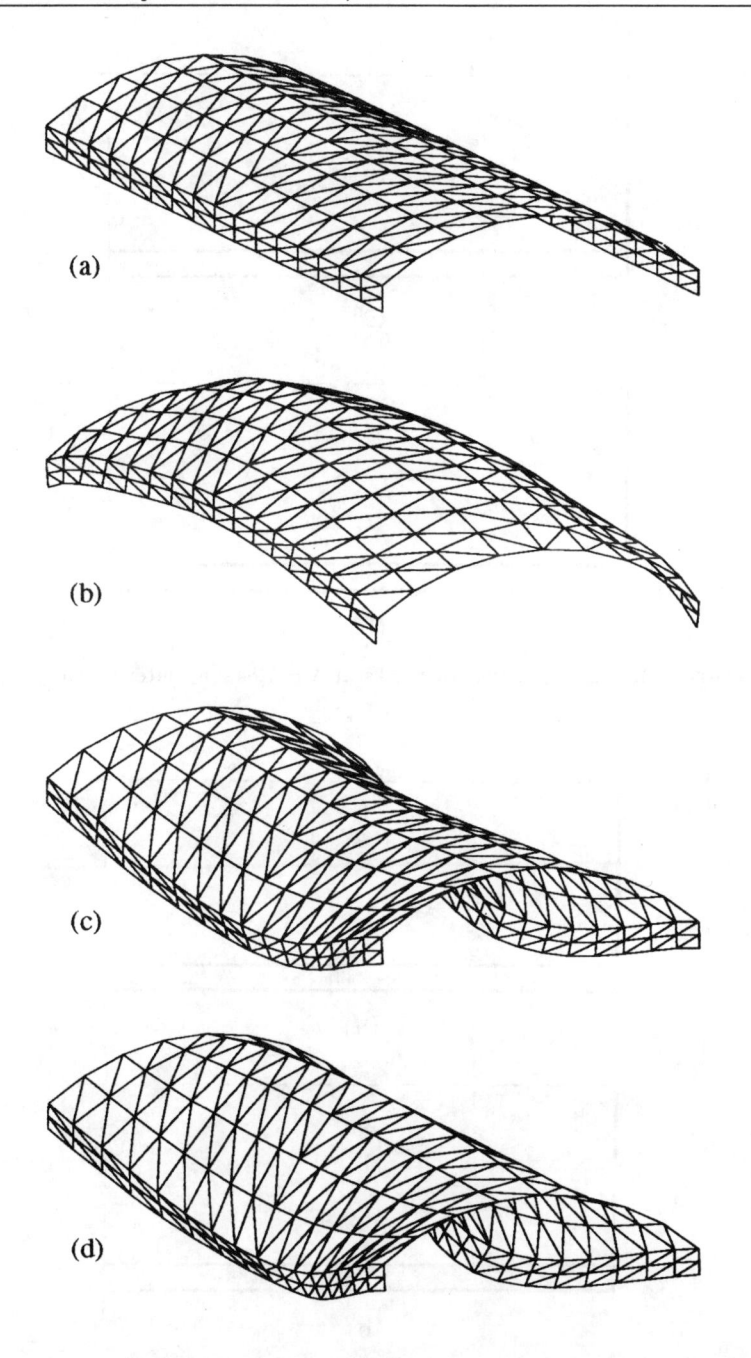

(d)

Figure 4.32: Undeformed and deformed shell: (a) undeformed, (b) $\lambda = 0.80$ (magnification factor: 200), (c) $\lambda = 1.7$ (magnification factor: 100), (d) $\lambda = 2.2$ (magnification factor: 10)

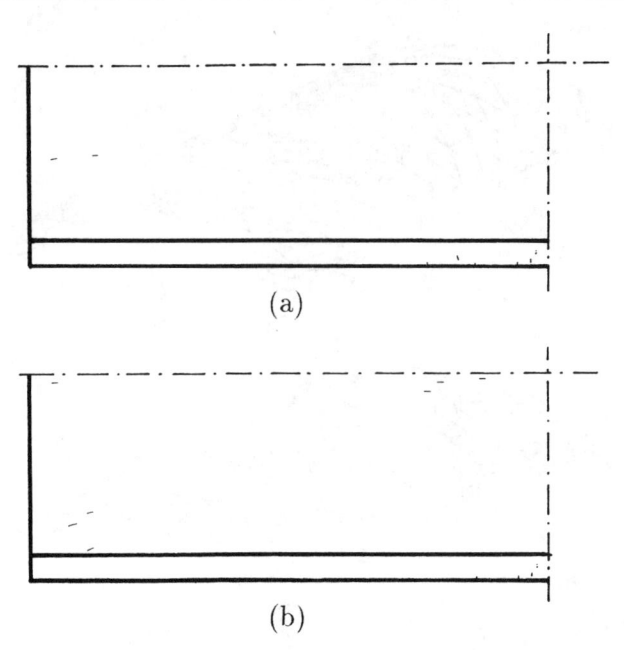

Figure 4.33: Distributions of cracks at $\lambda = 1.84$: (a) outer surface, (b) inner surface

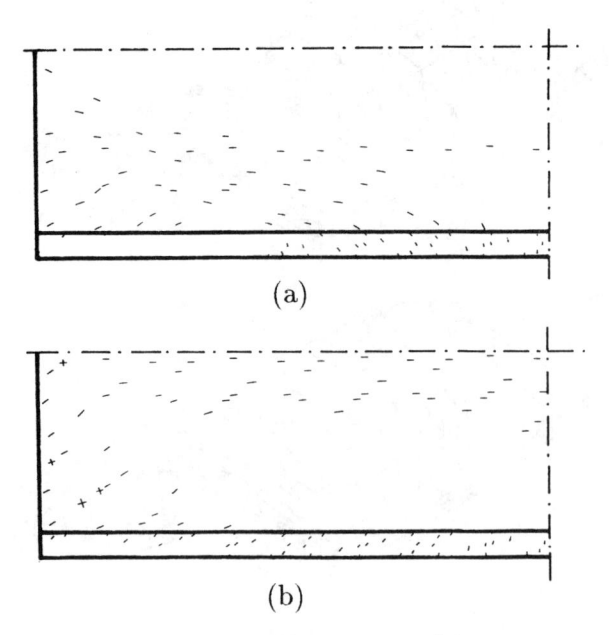

Figure 4.34: Distributions of cracks at $\lambda = 2.0$: (a) outer surface, (b) inner surface

Consideration of geometric nonlinearity is indispensable for the analytical reproduction of the change in direction of the deflection of the crown of the shell after the formation of the aforementioned hinge.

If the reinforcement of the shell was subdivided into two equal layers and rearranged such that one layer was close to the top surface and the other one close to the bottom surface, then the finite element analysis would not yield the aforementioned pronounced upward movement of the crown of the shell.

Disregard of the increase of the tendon forces with increasing external loads results in values for the ultimate load which are considerably smaller than the values of Table 4.7. For the medium mesh this value was obtained as $1.95\,p$.

4.4.3 Prestressed Concrete Secondary Containment Structure

Description of the Structure

Fig.4.35 shows a vertical and a horizontal cross-section of the model of a prestressed concrete secondary containment structure. The scale of the model is 1:14. The model was tested by McGregor et al. [MacGregor (1980), Rizkalla (1984)]. It consists of a cylindrical wall, prestressed in the circumferential and the axial direction, and a dome, prestressed in two orthogonal directions. The containment rests on a rigid base. The horizontal tendons in the cylindrical wall are anchored in four vertical buttresses. The tendons in the dome are anchored in a prestressed ring connecting the dome and the wall. The height of the model (without the base) is 3.81 m, the outer diameter is 3.2 m. The thickness of the wall and of the dome is 0.127 m and 0.102 m, respectively. The cross-section of the ring is 0.419×0.254 m. Construction details concerning the prestressed ring and the connexion between the cylindrical wall and the base are shown in Fig.4.36. Construction details of a buttress and of the dome are illustrated in Fig.4.37.

Material data of the concrete used for the lower half of the cylindrical wall (up to 1.81 m above base), of the concrete taken for the ring and the dome and of the shotcrete used for the upper half of the cylindrical wall are summarized in Table 4.8. These data were taken from [MacGregor (1980)]. They refer to the uniaxial cylindrical strength $\sigma^C_{P,cy}$, the splitting tensile strength $\sigma^C_{T,sp}$ and the secant modulus of elasticity E^C_{sec}, measured at $0.4\sigma^C_{P,cy}$. No creep and shrinkage data were reported in [MacGregor (1980)].

Table 4.9 contains values of the yield strength σ^S_Y, the tensile strength σ^S_T and the modulus of elasticity E^S of the reinforcing steel. This steel is characterized by a pronounced yield plateau, ranging up to a strain of 2%. In the cylindrical wall an inner and an outer layer of reinforcing steel was placed, consisting of 6 mm diameter deformed bars in the vertical direction and #3 bars in the circumferential direction, both at a distance of 76 mm. In the dome an inner and an outer layer of reinforcing steel, consisting of orthogonal meshes of 6 mm diameter deformed bars, spaced at 102 mm, was provided. Generally, the concrete cover was 13 mm. Details of the reinforcement can be found in [MacGregor (1980)].

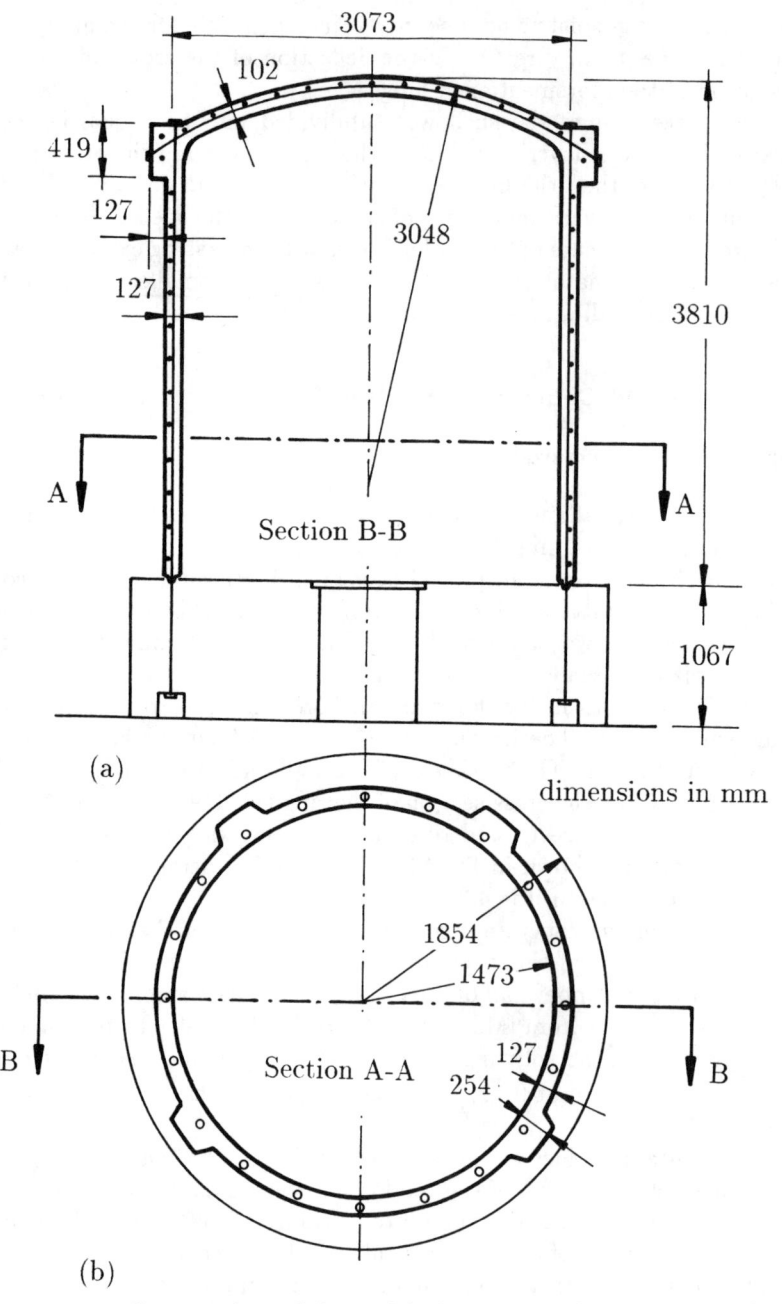

Figure 4.35: Cross-sections of the model of a prestressed concrete secondary containment structure: (a) vertical, (b) horizontal

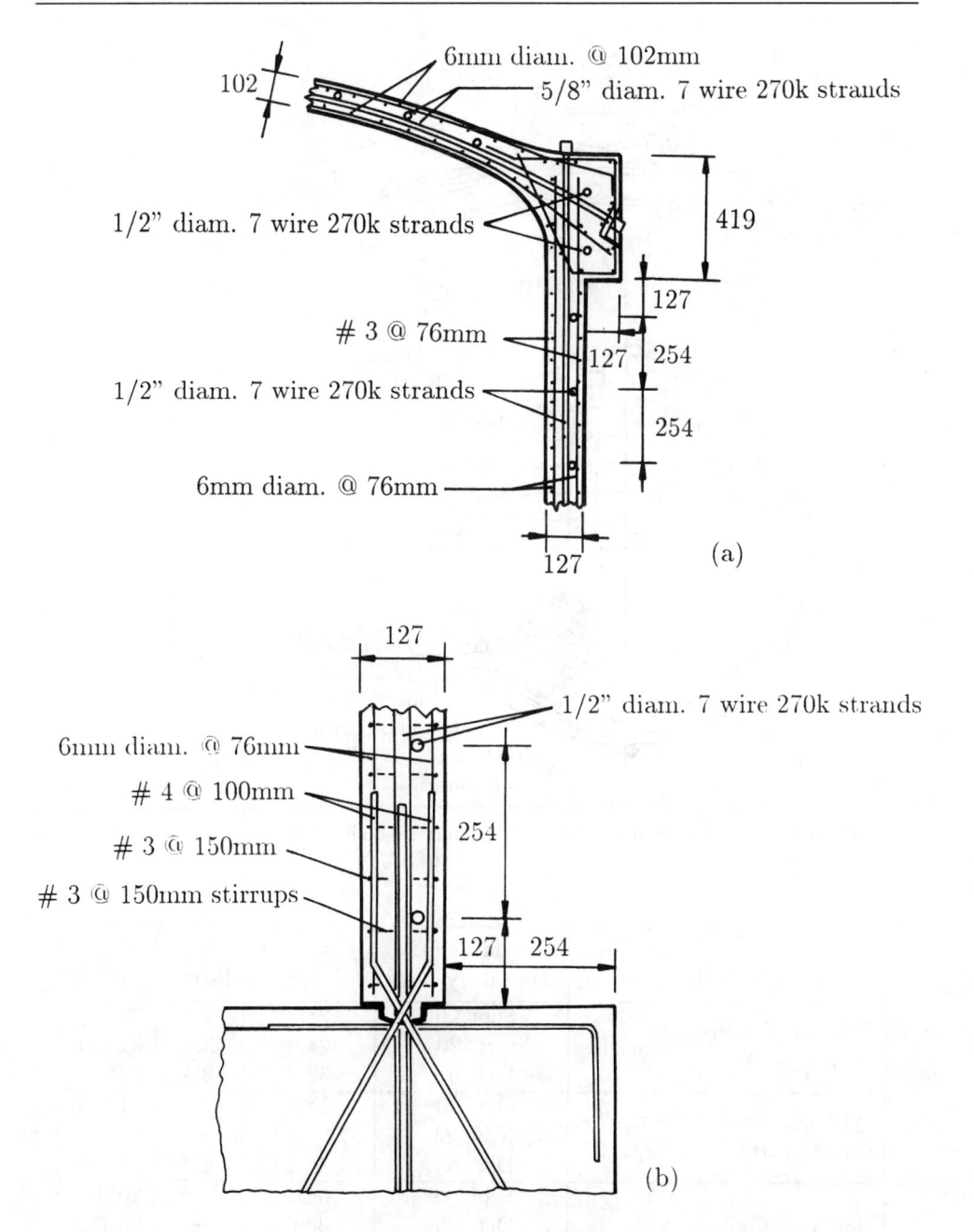

4.4 Nonlinear FE-Analyses of PC Panels, Slabs and Shells

Figure 4.36: Construction details: (a) prestressed ring, (b) connexion between the cylindrical wall and the base

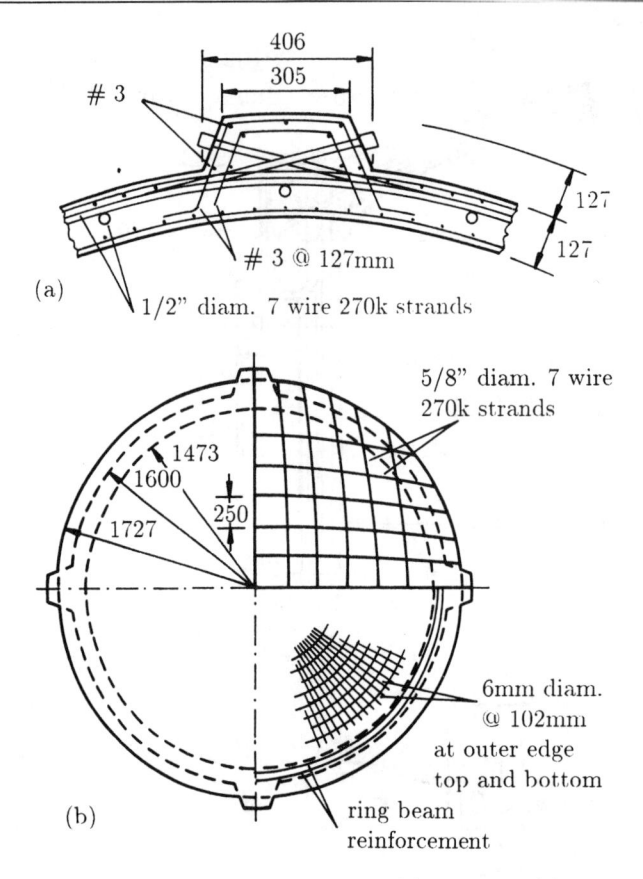

Figure 4.37: Construction details: (a) buttress, (b) dome

	date of placement	date of laboratory tests	$\sigma^C_{P,cy}$ [MPa]	$\sigma^C_{T,sp}$ [MPa]	E^C_{sec} [MPa]
cylindrical wall (lower part)	June 7	Sept. 29	28.2	—	20 700.
		Oct. 20	28.6	3.6	21 400.
		Dec. 5	32.2	2.3	22 800.
cylindrical wall (upper part)	Sept. 19 - - Sept. 21	Oct. 20	18.9	—	11 200.
		Nov. 23	27.4	—	—
		Dec. 5	25.3 , 18.2	4.2	11 700.
ring and dome	Sept. 1	Sept. 29	34.2	—	20 000.
		Oct. 20	30.0	—	20 300.
		Dec. 5	25.2	2.7	18 300.

Table 4.8 Experimentally obtained material data of concrete

4.4 Nonlinear FE-Analyses of PC Panels, Slabs and Shells

	σ_Y^S [MPa]	σ_T^S [MPa]	E^S [MPa]
6 mm diam. deformed bars	498	675	202 700
#3 ($\frac{3}{8}$" diam.) outer reinforcement	357	531	197 900
#3 ($\frac{3}{8}$" diam.) inner reinforcement	346	505	197 900
#4 ($\frac{4}{8}$" diam.)	369	557	204 800
#5 ($\frac{5}{8}$" diam.)	365	552	196 500

Table 4.9 Material properties of reinforcing steel

1/2" diameter 7 wire 270 k strands were used for prestressing of the cylindrical wall and the ring and 5/8" diameter 7 wire 270 k strands were placed in the dome. The cylindrical wall contains 20 tendons in the axial direction and 24 tendons in the circumferential direction. 4 tendons in the circumferential direction were placed in the ring. All horizontal tendons only extend over one half of the circumference. They are staggered by $90°$. Hence, four buttresses were provided. The orthogonal mesh of tendons in the dome consists of 11 tendons each in both directions. The horizontal tendons and the tendons in the dome were tensioned from both ends. All tendons were grouted after the tensioning operations.

	A^Z [mm^2]	$\sigma_{0.1}^Z$ [MPa]	σ_T^Z [MPa]	E^Z [MPa]
1/2" diam. 7 wire 270 k strand	98.1	1 703.	1 958.	206 842.
5/8" diam. 7 wire 270 k strand	150.0	1 531.	1 730.	206 842.

Table 4.10 Material properties of prestressing steel

Values for the cross-sectional area A^Z, the yield strength at 0.1% permanent strain, $\sigma_{0.1}^Z$, the tensile strength σ_T^Z and the modulus of elasticity E^Z are given in Table 4.10. No information about friction between the tendons and the ducts was provided in [MacGregor (1980)]. However, it was mentioned that the tendons were retensioned to maintain a uniform stress distribution in the tendons. The prestressing forces are summarized in Table 4.11.

The model was loaded by a uniformly distributed internal pressure. The loading history is summarized in Table 4.12. Fig.4.38 shows the load-deflection diagrams obtained from test F and from the initial part of test G for the point of the cylindrical wall located 1.81 m above the base midway between two consecutive buttresses and for the apex of the dome.

The load-deflection diagrams obtained from the complete test G are illustrated in Fig.4.39. In the experiment the first crack developed at the inner face of the dome at an internal pressure p of 0.2 MPa. Axial and circumferential cracking in the wall was noticed at 0.28 MPa. At this level of the internal pressure, cracking was also

tendon location	prestressing force [MN]
cylindrical wall	
horizontal	0.0649
vertical	0.0924
ring	0.0932
dome	0.1305

Table 4.11 Prestressing forces

Date	
Sept. 1	placement of concrete for the ring and the dome
Sept. 21 - Sept. 25	tensioning of the tendons
Nov. 1 - Nov. 15	five test runs (A to E) up to an internal pressure of $p = 0.27$ MPa which is below the initial cracking pressure
Nov. 20	test F up to $p = 0.55$ MPa; unloading
Dec. 11	test G up to $p = 1.0$ MPa
Dec. 12	test G up to failure at $p = 1.1$ MPa

Table 4.12 Loading history

observed in the dome. For p exceeding 0.56 MPa, outward bulging of the cylindrical walls became apparent. At 0.76 MPa a significant increase of bulging was observed. Widespread yielding of the reinforcement was the reason for this increase. At 0.90 MPa the crack patterns in the wall and in the dome were fully developed. Only the part of the dome, extending about 1.22 m from the outer edge of the ring beam into the dome, was free of visible cracks, even at the end of the test. At 0.97 MPa concrete was spalling at the top of the dome. Finally, the containment failed at a pressure of 1.10 MPa. The reason for the failure was rupture of a circumferential and an axial tendon in the cylindrical wall.

Analysis Model

Except for the constitutive laws, the main features of the analysis model are identical to the ones of the analysis model for the prestressed concrete slab described in subsection 4.4.1. For the concrete, a rate-type constitutive law was used [Walter (1988)]. It allows to distinguish between inelastic loading and linear-elastic unloading and reloading. This material law is an extension of the nonlinear-elastic (total) stress-strain law proposed by Liu, Nilson and Slate [Liu (1972)] (see subsection 2.3.2, equation (2.35)). It is similar to constitutive models for consideration of plasticity

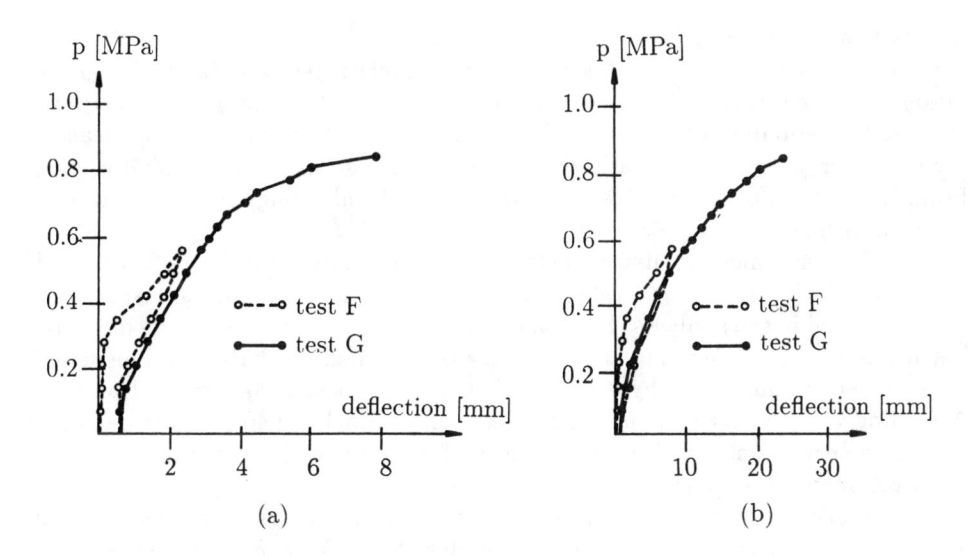

4.4 Nonlinear FE-Analyses of PC Panels, Slabs and Shells 313

Figure 4.38: Load-deflection diagrams obtained from test F and from the initial part of test G: (a) for the point of the cylindrical wall located 1.81 m above the base midway between two consecutive buttresses, (b) for the apex of the dome

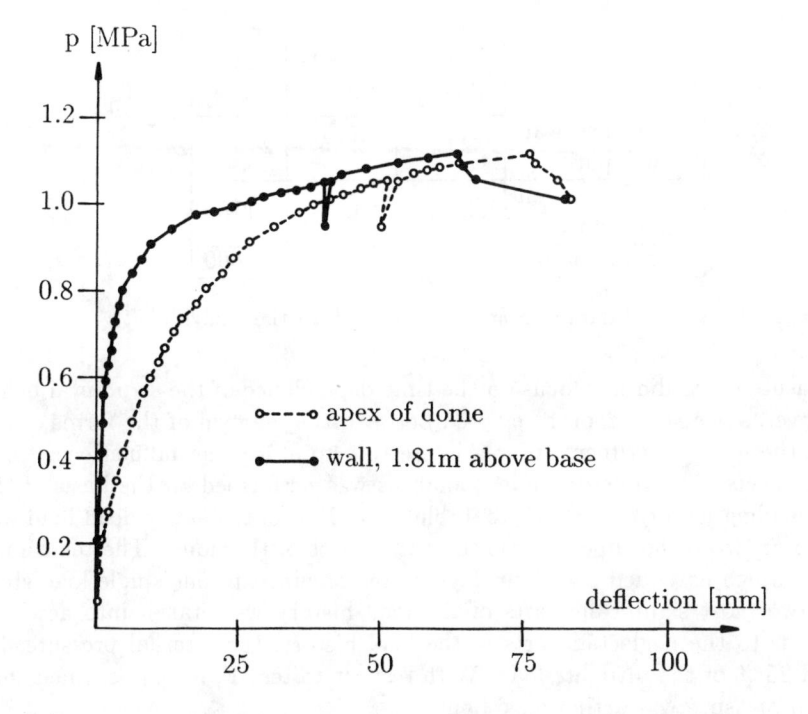

Figure 4.39: Load-deflection diagrams from the complete test G

with isotropic hardening.

For the reinforcing steel, a bilinear stress-strain relationship, suitable for consideration of plasticity with kinematic hardening (see subsection 2.5, Fig. 2.31(a)), was employed. The material behavior of the prestressing steel was modelled by means of the nonlinear stress-strain relation according to the equations (2.249) and (2.250). Both for the reinforcing and the prestressing steel unloading and reloading were assumed as linear-elastic.

Because of symmetry, only one octant of the structure needs to be discretized. Two different finite element meshes, consisting of 37 and 76 elements, respectively, were employed in the analyses (Fig.4.40(a) and (b)). In contrast to the fine mesh, the buttresses were not taken into account in the coarse mesh. In the analysis model the tendons were accounted for by the embedded approach described in subsection 3.4.4. Accordingly, the courses of the tendon axes, shown in Fig.4.40(c), were described as surface curves embedded in the middle surface of the cylindrical wall and of the spherical dome, respectively.

For the analysis the material properties given in Table 4.8 were referred to an age of concrete of 28 days. The secant modulus of elasticity, E_{sec}^C, was substituted by the tangent modulus E_{28}^C. The resulting material properties were determined on the basis of the CEB/FIP model code [CEB-FIP (1978)]. They are summarized in Table 4.13 [Walter (1988)].

	$\sigma_{P,cy,28}^C$ [MPa]	$\sigma_{T,sp,28}^C$ [MPa]	E_{28}^C [MPa]
cylindrical wall (lower part)	29.7	2.95	24 100
cylindrical wall (upper part)	22.4	4.18	10 700
ring and dome	29.8	2.68	22 600

Table 4.13   Material data of concrete, employed for the analysis

The measurements did not focus on the time-dependence of the structural behavior. However, since three months elapsed between the removal of the forms of the dome and the final experiment, it was interesting to study the influence of time-dependent effects. The time-dependent analysis was performed on the basis of the simplified loading history according to Table 4.14. Hence, the entire dead load was assumed to act from the time of removal of the forms of the dome. The tensioning operations, which extended over four days, were combined to one single load step. Only the two most significant parts of the load history were taken into account. With regards to the neglected parts of the load history, the internal pressure did not exceed 25 % of the ultimate load. With respect to test F, it was assumed that the internal pressure was acting for 6 hours.

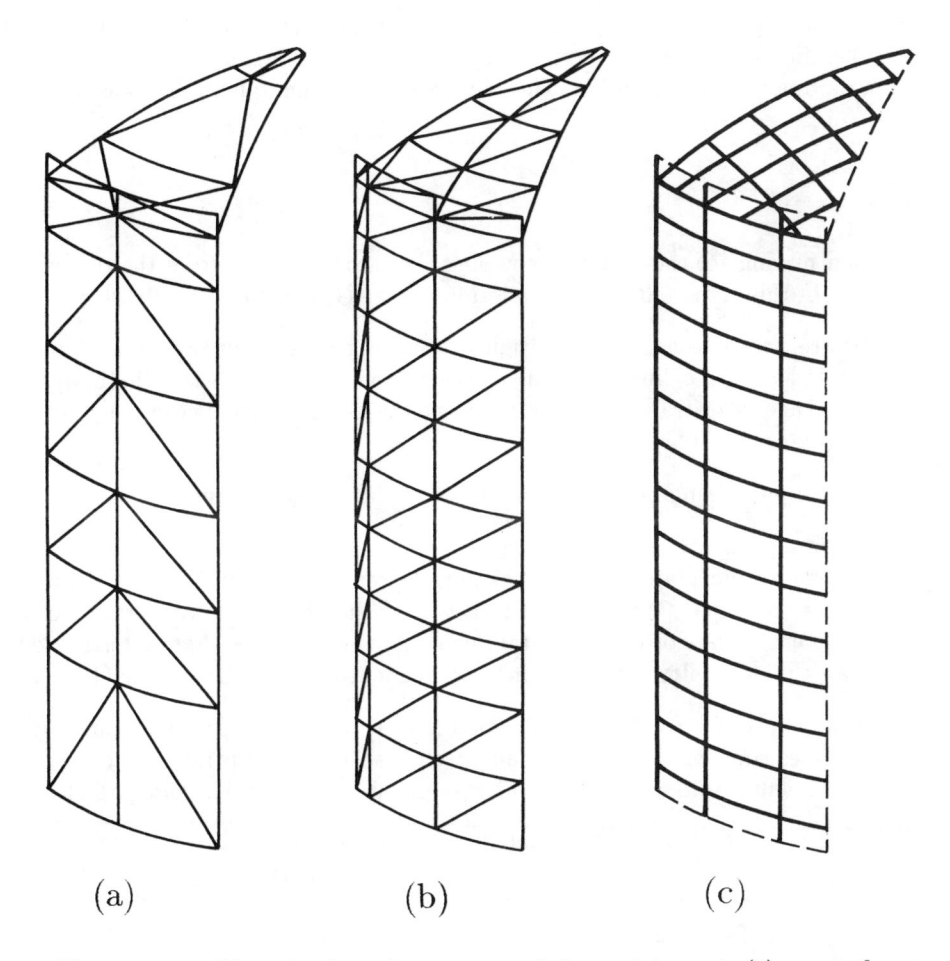

4.4 Nonlinear FE-Analyses of PC Panels, Slabs and Shells

Figure 4.40: Discretization of one octant of the containment: (a) coarse finite element mesh, (b) fine finite element mesh, (c) tendons, as considered for both meshes

t = 0 (Sept. 1)	
t = 7 days	removal of the forms of the dome
	→ dead load is acting on the model
t = 22 days	prestressing
t = 80 days	test F up to $p = 0.55$ MPa; unloading
t = 101 days	test G up to failure

Table 4.14 Loading history considered in the finite element analyses

Since no creep and shrinkage data from experiments were available, the CEB/FIP creep and shrinkage laws were employed (see section 2.4). Standard values were used for the parameters in these laws. The relative environmental humidity was assumed to be 70 %.

Results

The main findings of the analysis, covering the period of time from the removal of the forms of the dome until the completion of test G, are [Walter (1988)]:

- Before prestressing, when only dead load is acting, shrinkage cracks are indicated in the cylindrical wall in the region above the base. In this region shrinkage is restrained by the hinge at the connexion between the wall and the base.

- After prestressing of the tendons some of these cracks close.

- In consequence of creep, the magnitudes of the computed deflections for the situation immediately after tensioning increase by about 50 % until the beginning of test G. With respect to this test, the numerical investigation of the structural behavior of the containment indicated a somewhat softer response than the one following from time-independent analysis [Hofstetter (1987)].

- However, as far as the ultimate load is concerned, the results obtained from analyses considering time-dependent effects differ very little from corresponding results from analyses in which these effects have been disregarded.

Fig.4.41 refers to test F. It shows the experimentally obtained load-deflection diagram for a point on the cylindrical wall, 1.81 m above the base, midway between two adjacent buttresses, and the corresponding diagram resulting from finite element analysis. Fig.4.42 illustrates the respective load-deflection diagrams for the apex of the dome.

For this point the computed load-deflection diagram agrees relatively well with the experimentally obtained diagram. For the aforementioned point of the cylindrical wall, however, most of the computed values for the deflection are considerably smaller than the respective experimentally obtained values. It is noted in [MacGregor (1980)] that this difference is probably caused by an unwanted imperfection, resulting in a value measured for the deflection of this point of the wall which is larger than that for all other points of the wall.

The finite element analyses essentially reproduce the experimentally observed structural behavior. At 0.30 MPa, cracks in the dome and in the lower part of the wall are indicated by the analyses. At 0.50 MPa, the entire lower part of the cylindrical wall is cracked. At 0.70 MPa, yielding of the reinforcement is signalled by the analysis. At 0.80 MPa, the vertical reinforcement in the cylindrical wall is yielding. At 0.85 MPa, yielding of the circumferential reinforcement in the cylindrical wall is indicated. The computed diagnosis of the reason for the failure of the structure

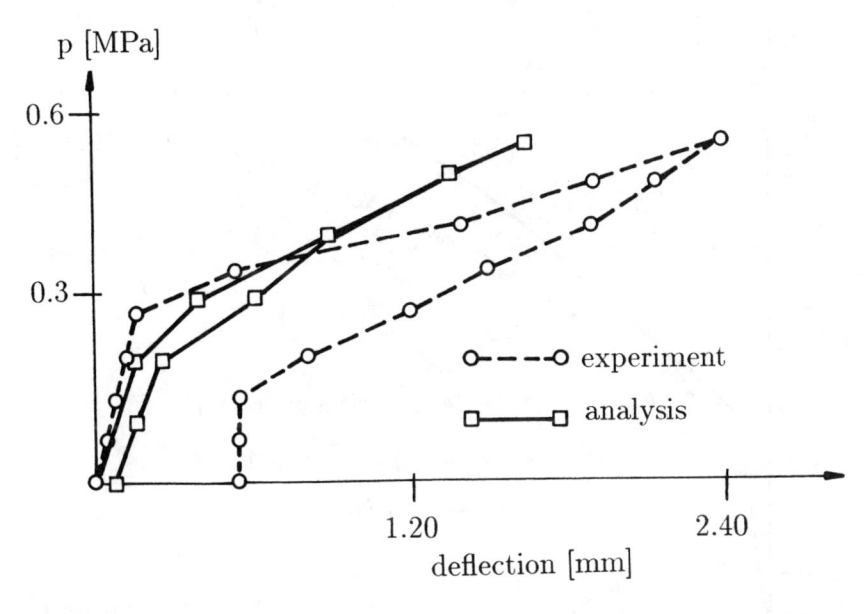

4.4 Nonlinear FE-Analyses of PC Panels, Slabs and Shells

Figure 4.41: Test F: load-deflection diagrams for a point of the cylindrical wall

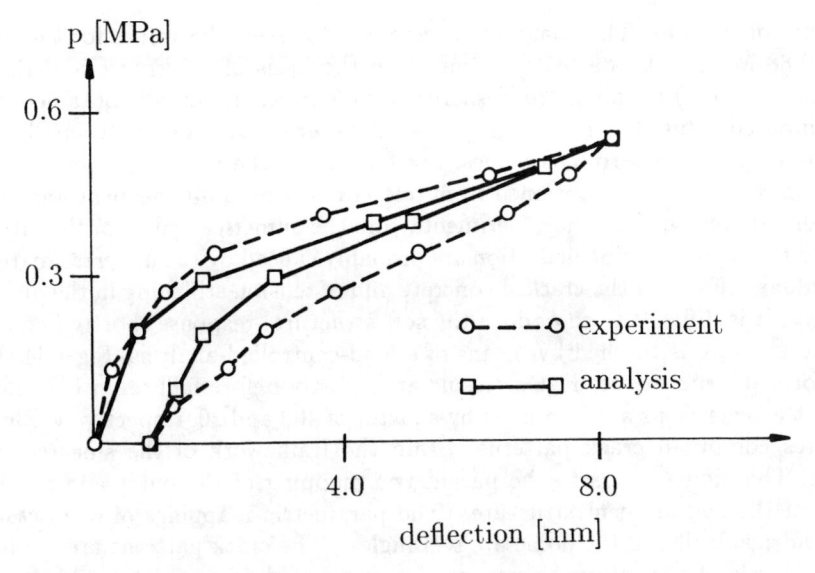

Figure 4.42: Test F: load-deflection diagrams for the apex of the dome

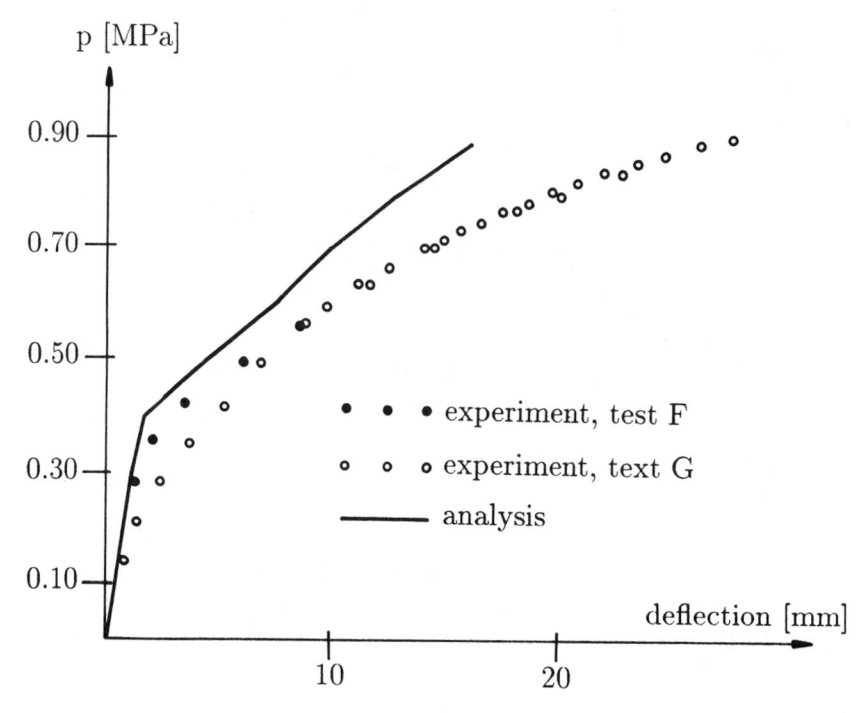

Figure 4.43: Test G: load-deflection diagrams for the apex of the dome

is rupture of tendons. This diagnosis is correct. However, the values for the failure loads (0.88 MPa and 0.89 MPa, obtained on the basis of the coarse and the fine mesh, respectively) are about 20 % smaller than the experimentally obtained value. The computed values for the deflections before failure are considerably smaller than the experimentally determined values. Fig.4.43 shows the load-deflection diagrams for the apex of the dome, obtained from test G and from finite element analysis.

The differences between the experimental and the numerical value for the ultimate load and the corresponding deflection are probably caused by the underestimation of the residual stiffness of the cracked concrete and of tension stiffening in the analysis. Moreover, it is difficult to reproduce the soft structural response shortly before the ultimate pressure is reached by means of a load-controlled analysis. Fig.4.44 shows the deformed structure after prestressing and shortly before failure. In Figs.4.44(a) and (b) the deflections are magnified by a factor of 400 and 20, respectively. Fig.4.45 illustrates computed crack patterns within the framework of the smeared crack concept. This figure refers to the parametric mapping of the outer and the inner surface of the containment structure. (The parametric mappings of one octant of the cylindrical wall and the dome are rectangles.) The crack patterns are related to different levels of the internal pressure. The computed advance of cracking agrees well with the experimentally observed advance.

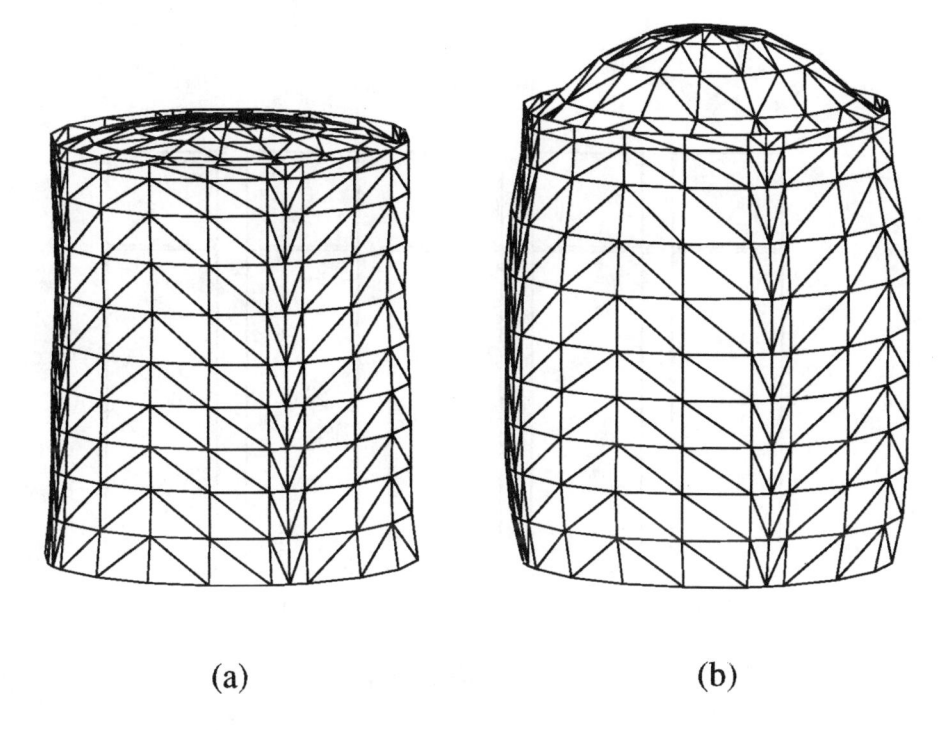

4.4 Nonlinear FE-Analyses of PC Panels, Slabs and Shells

(a) (b)

Figure 4.44: Deformed containment structure: (a) after prestressing (magnification factor for the deflections: 400), (b) shortly before failure (magnification factor for the deflections: 20)

The relative environmental humidity has a considerable influence on the deformations. The value for the ultimate load, however, is practically unaffected by this parameter, because this value is controlled by the load-carrying capacity of the reinforcing and the prestressing steel. At this stage, the concrete has lost all of its load-carrying capacity. This is also the reason why no mesh-dependence of the ultimate load is observed in this case.

outer surface

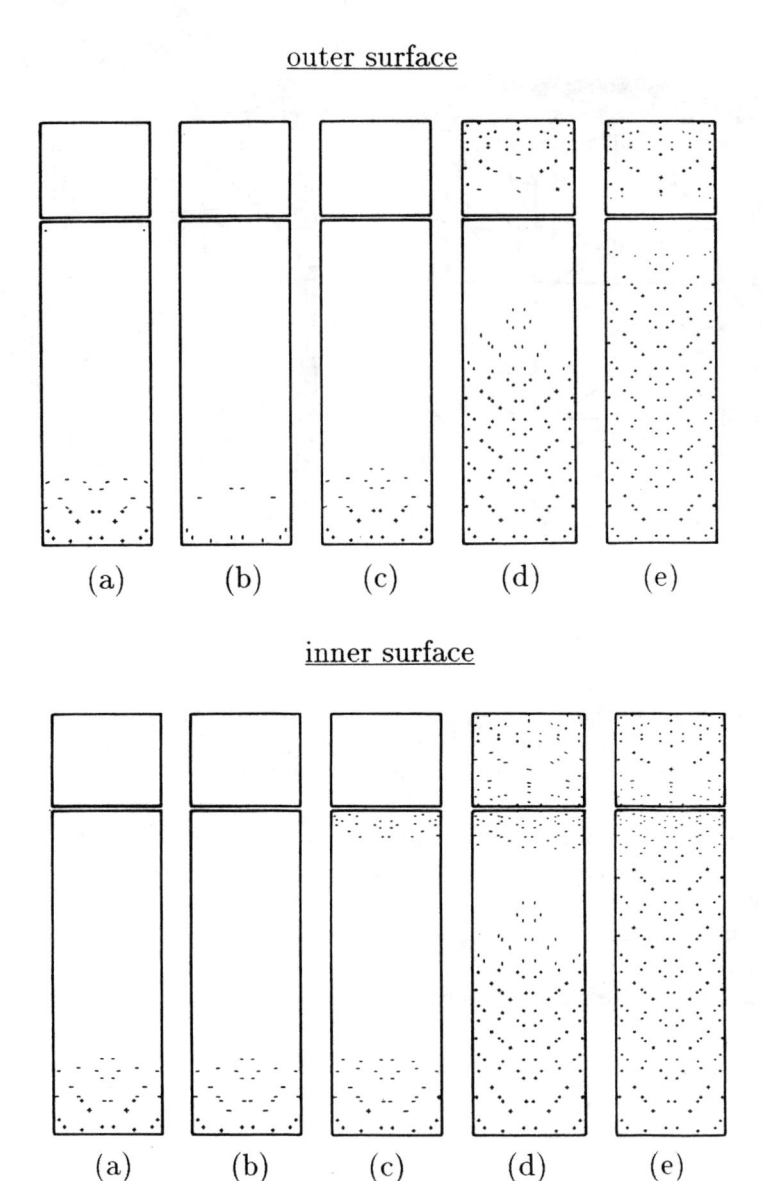

(a) (b) (c) (d) (e)

inner surface

(a) (b) (c) (d) (e)

Figure 4.45: Crack patterns on the inner and outer surface of the model of the containment structure for a few selected levels of the pressure (p): (a) dead load (dl), (b) dl + prestress (ps), (c) dl + ps + p (= 0.2 MPa), (d) dl + ps + p (= 0.3 MPa), (e) dl + ps + p (= 0.85 MPa)

A Notation

In general, superscripts of material parameters denote the type of material whereas subscripts indicate the specific meaning of these parameters. E.g., σ^C_{Pi} denotes the stresses in the concrete (indicated by the superscript "C") at maximum strength (indicated by the subscript "P" for "peak"), referred to the principal directions (indicated by the subscript i, i=1,2,3).

A.1 Displacements

$\mathbf{u}$	displacement vector
$u_1 \equiv u, u_2 \equiv v, u_3 \equiv w$	components of the displacement vector, referred to a local or global orthogonal system of reference
$\mathbf{u}^e$	displacement vector for an arbitrary point within element e
$\Delta\mathbf{u}$	vector of incremental displacements
$\delta\mathbf{u}$	vector of virtual displacements
$\mathbf{q}$	vector of nodal displacements
$\Delta\mathbf{q}$	vector of incremental nodal displacements
$\mathbf{q}^e$	vector of nodal displacements for a particular element e

A.2 Strains

ε	Cauchy strain tensor
e	deviatoric strain tensor
E	Lagrangian (Green) strain tensor
I'_1, I'_2, I'_3	invariants of the Cauchy strain tensor ε
J'_2, J'_3	invariants of the deviatoric strain tensor e
ε	strain in the direction of the uniaxial stress σ
ε^C_P	concrete strain corresponding to the uniaxial compressive strength (peak stress) of concrete
ε^{cr}	creep strain
ε^{in}	instantaneous strain
ε^{sh}	shrinkage strain
ε^T	temperature-induced strain
ε^0	stress-independent strain

ε^σ	stress-induced strain
ε_V	volumetric strain
ε_0	octahedral normal strain
$\bar{\varepsilon}_p$	equivalent plastic strain
$\varepsilon_i,\ i=1,2,(3)$	principal strains
$\varepsilon_x,\varepsilon_y,\varepsilon_z,(\varepsilon_{xy},\varepsilon_{yz},\varepsilon_{zx})$	normal strains (shear strains) with respect to the axes x, y, z of a Cartesian coordinate system
$\varepsilon_{11},\varepsilon_{22},\varepsilon_{33},(\varepsilon_{12},\varepsilon_{23},\varepsilon_{31})$	normal strains (shear strains) with respect to the principal axes of orthotropy
$\varepsilon^C_{Pi},\ i=1,2,(3)$	principal strains corresponding to σ^C_{Pi}
$\varepsilon^C_{Si},\ i=1,2,(3)$	concrete strains corresponding to σ^C_{Si}
$\varepsilon^{(j)}_N$	fictitious strain perpendicular to a crack
ε^{eq}_i	fictitious equivalent-uniaxial strains referred to principal directions
ε^{eq}_{Pi}	fictitious equivalent-uniaxial strains corresponding to σ_{Pi}
ε^{eq}_{Si}	fictitious equivalent-uniaxial strains corresponding to σ_{Si}
$\varepsilon^S_U(\varepsilon^Z_U)$	ultimate strain of reinforcing (prestressing) steel
$\tilde{\varepsilon}$	average strain
γ_0	octahedral engineering shear strain
$\gamma_{xy},\gamma_{yz},\gamma_{zx}$	engineering shear strains $\gamma_{ij}=2\varepsilon_{ij}, i\neq j$
$\Delta\varepsilon$	tensor of incremental strains
$\delta\varepsilon$	tensor of virtual strains
ε^a	tensor of inelastic strains
ε^e	tensor of elastic strains
ε^p	tensor of plastic strains
ε^{vp}	tensor of viscoplastic strains

A.3 Stresses

σ	Cauchy stress tensor
s	deviatoric stress tensor
S	2nd Piola-Kirchhoff stress tensor
I_1, I_2, I_3	invariants of the Cauchy stress tensor σ
J_2, J_3	invariants of the deviatoric stress tensor s
k	yield stress as part of the yield function
s	$= \sigma^C/\sigma^C_P$, stress intensity for concrete
σ	uniaxial stress
σ^C_P	uniaxial compressive strength (peak stress) of concrete (cubic, prismatic or cylindrical strength)
σ^C_T	uniaxial tensile strength of concrete

σ_0	hydrostatic stress
$\sigma_Y^S(\sigma_Y^Z)$	yield stress of reinforcing (prestressing) steel
$\sigma_{0.01}^S(\sigma_{0.01}^Z)$	proportional limit of reinforcing (prestressing) steel
$\sigma_{0.1}^S(\sigma_{0.1}^Z), \sigma_{0.2}^S(\sigma_{0.2}^Z)$	yield stress of reinforcing (prestressing) steel without a well-defined yield plateau, defined as the stress at 0.1 % or 0.2 % permanent strain
$\sigma_T^S(\sigma_T^Z)$	ultimate tensile strength of reinforcing (prestressing) steel
$\sigma_{Yc}^C(\sigma_{Yt}^C)$	initial yield stress in uniaxial compression (tension)
σ_M	mean stress within a single cycle of cyclic loading
$\sigma_i, i=1,2,(3)$	principal stresses
$\sigma_x, \sigma_y, \sigma_z, (\tau_{xy}, \tau_{yz}, \tau_{zx})$	normal stresses (shear stresses) with respect to the axes x, y, z of a Cartesian coordinate system
$\sigma_{11}, \sigma_{22}, \sigma_{33}, (\tau_{12}, \tau_{23}, \tau_{31})$	normal stresses (shear stresses) with respect to the principal axes of orthotropy
$\sigma_{Pi}^C, i=1,2,(3),$	peak stresses of concrete under multiaxial loading, referred to principal directions
$\sigma_{Si}^C, i=1,2,(3),$	concrete stresses in the softening region, referred to principal directions
$\sigma_i^{(f)}$	fictitious uniaxial stresses referred to principal directions
$\tilde{\sigma}$	average stress
τ^C	shear stress in the concrete
τ_0	octahedral shear stress
τ^B	bond stress
τ_P^B	ultimate bond strength
$\Delta\boldsymbol{\sigma}$	tensor of incremental stresses
ρ	back stress

A.4 Moduli

E	modulus of elasticity (Young's modulus) for a linear-elastic isotropic material
E_0^C	initial tangent modulus of concrete determined from a uniaxial compression test
E_{28}^C	initial tangent modulus of concrete referred to the age of concrete of 28 days
E_{soft}^C	softening modulus of concrete
$E_0^S(E_0^Z)$	initial tangent modulus of reinforcing (prestressing) steel determined from a uniaxial tension test

E_1, E_2, E_3	moduli of elasticity for a linear-elastic orthotropic material, referred to principal axes of orthotropy
E_{T1}, E_{T2}, E_{T3}	tangent stiffness moduli for a nonlinear orthotropic material, referred to principal axes of orthotropy
G	shear modulus for a linear-elastic isotropic material
G_{AI}	modified shear modulus to account for aggregate interlock of cracked concrete
$H_i(H_k)$	isotropic (kinematic) hardening modulus
$\mathbf{C}\ (C_{ijkl})$	tensor (component) of the elastic material moduli
$\mathbf{C}_T$	tensor of the tangent material moduli
$\mathbf{D}\ (D_{ijkl})$	tensor (component) of the material compliance moduli
ν	Poisson's ratio for a linear-elastic isotropic material
ν_0	initial Poisson's ratio determined from a uniaxial compression test
ν_1, ν_2	Poisson's ratios for a linear-elastic orthotropic material, referred to principal axes of orthotropy (2D)
$\nu_{12}, \nu_{13}, \nu_{21}, \nu_{23}, \nu_{31}, \nu_{32}$	Poisson's ratios for a linear-elastic orthotropic material, referred to principal axes of orthotropy (3D)

A.5 Miscellaneous

$\cdot$	derivative with respect to time
f	yield function
F	part of f depending on the stress
$\mathbf{f}$	tensor function
g	plastic potential
$\mathbf{g}$	tensor function to describe the evolution of the inelastic strain
h_k^*, h_k	functions to describe the evolution of the internal variable ξ_k
H	plastic modulus
r	distance of a stress point in the principal stress space from the hydrostatic axis
t	discrete point of time
t_0	age of concrete at loading
Δt	time interval

A.5 Miscellaneous

T	temperature
W^p	plastic work
α	dilatancy factor
α_i	$= \sigma_j/\sigma_i$, ratio of principal stresses (biaxial state of stress)
θ	angle of similarity
$\delta_T^C(\delta_N^C)$	relative displacement of concrete parallel (normal) to a crack surface
$\dot{\lambda}$	consistency parameter
κ	hardening parameter
$\phi(f)$	scaling function for the evolution of the internal variables
$\rho_C(\rho_T)$	distance of a stress point on the compressive (tensile) meridian from the hydrostatic axis
ξ	coordinate denoting the hydrostatic axis in the principal stress space
$\boldsymbol{\xi}$	array of internal variables

Bibliography

[ACI-Committee 209] American Concrete Institute Committee 209, Subcommittee II, "Prediction of Creep, Shrinkage and Temperature Effects in Concrete Structures", in *Designing for Effects of Creep, Shrinkage and Temperature*, ACI, Special Publication No. 27, 1971, 51 - 93.

[ACI-Committee 224] "Cracking of Concrete Members in Direct Tension", Report of ACI Committee 224, *ACI Journal*, **83**, 1986, 3-13.

[ACI-Committee 439] "Steel Reinforcement - Physical Properties and U.S. Availability", Report of ACI Committee 439, *ACI Materials Journal*, **86**, 1989, 63-76.

[ASCE (1982)] "Finite Element Analysis of Reinforced Concrete Structures", Task Committee on Finite Element Analysis of Reinforced Concrete Structures of the Structural Division Committee on Concrete and Masonry Structures, ASCE, New York, 1982.

[Abbasi (1992)] Abbasi M.S.A., Baluch M.H., Azad A.K. and Abdel Rahman H.H., "Nonlinear Finite Element Modelling of Failure Modes in RC Slabs", *Computers & Structures*, **42**, 1992, 815-823.

[Acker (1989)] Acker P., "Shrinkage Stresses", in *Fracture Mechanics of Concrete Structures, From Theory to Applications*, in Elfgren L. (ed.), RILEM Report, Chapman and Hall, London, 1989, 155-161.

[Ahmad (1968)] Ahmad S., Irons B.M. and Zienkiewicz O.C., "Analysis of Thick and Thin Shell Structures by Curved Finite Elements", *International Journal for Numerical Methods in Engineering*, **2**, 1970, 419-451.

[Ahmad (1982)] Ahmad S.H. and Shah S.P., "Complete Triaxial Stress-Strain Curves for Concrete", *Journal of Structural Engineering*, **108**, 1982, 728-742.

[Ahmad (1986)] Ahmad S.H., Shah S.P. and Khaloo A.R., "Orthotropic Model of Concrete for Triaxial Stresses", *Journal of Structural Engineering*, **112**, 1986, 165-181.

[Almudaiheem (1987)] Almudaiheem J.A. and Hansen W., "Effect of Specimen Size and Shape on Drying Shrinkage", *ACI Materials Journal*, **84**, 1987, 130-135.

[Andelfinger (1991)] Andelfinger U., "Untersuchungen zur Zuverlässigkeit hybrid-gemischter Finiter Elemente für Flächentragwerke", *Doctoral Dissertation*, Institut für Baustatik, Universität Stuttgart, Stuttgart, 1991.

[Andenaes (1977)] Andenaes E., Gerstle K.H. and H.Y. Ko, "Response of Mortar and Concrete to Biaxial Compression", *Journal of the Engineering Mechanics Division*, ASCE, **103**, 1977, 515-526.

[Ansari (1987)] Ansari F., "Stress-Strain Response of Microcracked Concrete in Direct Tension", *ACI Materials Journal*, **84**, 1987, 481-490.

[Ashour (1993)] Ashour A.F. and Morley C.T., "Three-Dimensional Nonlinear Finite Element Modelling of Reinforced Concrete Structures", *Finite Element in Analysis and Design*, **15**, 1993, 43-55.

[Başar (1985)] Başar Y., and Krätzig W.B., *Mechanik der Flächentragwerke*, Vieweg, Braunschweig, 1985.

[Bažant (1972)] Bažant Z.P., "Prediction of Concrete Creep Effects Using Age-Adjusted Effective Modulus Method", *ACI Journal*, **69**, 1972, 212-217.

[Bažant (1973)] Bažant Z.P. and Wu S.T., "Dirichlet Series Creep Function for Aging Concrete", *Journal of the Engineering Mechanics Division*, ASCE, **99**, 1973, 367-387.

[Bažant (1975)] Bažant Z.P., "Theory of Creep and Shrinkage in Concrete Structures: a Précis of Recent Developments", *Mechanics Today*, **2**, Pergamon Press, New York, 1975, 1-93.

[Bažant (1976)] Bažant Z.P. and Bhat P., "Endochronic Theory of Inelasticity and Failure of Concrete", *Journal of the Engineering Mechanics Division*, ASCE, **102**, 1976, 701-722.

[Bažant (1976b)] Bažant Z.P. and Osman E., "Double Power Law for Basic Creep of Concrete", *Matériaux et Constructions*, RILEM, **9**, 1976, 3-11.

[Bažant (1976c)] Bažant Z.P., "Instability, Ductility, and Size Effect in Strain-Softening Concrete", *Journal of the Engineering Mechanics Division*, ASCE, **102**, 1976, 331-344.

[Bažant (1978)] Bažant Z.P. and Panula L., "Practical Prediction of Creep and Shrinkage of Concrete", *Materials and Structures*, Parts I and II: **11**, 1978, 307-328, Parts III and IV: **11**, 1978, 415-434, Parts V and VI: **12**, 1979, 169-183.

[Bažant (1978b)] Bažant Z.P. and Shieh C.L., "Endochronic Model for Nonlinear Triaxial Behavior of Concrete", *Nuclear Engineering and Design*, **47**, 1978, 305-315.

[Bažant (1978c)] Bažant Z.P., "Endochronic Inelasticity and Incremental Plasticity", *International Journal of Solids and Structures*, **14**, 1978, 691-714.

[Bažant (1979)] Bažant Z.P. and Kim S.S., "Plastic-Fracturing Theory for Concrete", *Journal of the Engineering Mechanics Division*, ASCE, **105**, 1979, 407-428, with Errata in **106**.

[Bažant (1979b)] Bažant Z.P., "Thermodynamics of Solidifying or Melting Viscoelastic Material", *Journal of the Engineering Mechanics Division*, ASCE, **105**, 1979, 933-952.

[Bažant (1979c)] Bažant Z.P. and Cedolin L., "Blunt Crack Band Propagation in Finite Element Analysis", *Journal of the Engineering Mechanics Division*, ASCE, **105**, 1979, 297-315.

[Bažant (1979d)] Bažant Z.P. and Kim S.S., "Approximate Relaxation Function for Concrete", *Journal of the Structural Division*, ASCE, **105**, 1979, 2695-2705.

[Bažant (1980)] Bažant Z.P. and Gambarova P., "Rough Cracks in Reinforced Concrete", *Journal of the Structural Division*, ASCE, **106**, 1980, 819-842.

[Bažant (1980b)] Bažant Z.P. and Panula L., "Creep and Shrinkage Characterization for Analyzing Prestressed Concrete Structures, *PCI Journal*, **25**, 1980, 86-122.

[Bažant (1980c)] Bažant Z.P. and Shieh C.L., "Hysteretic Fracturing Endochronic Theory for Concrete", *Journal of the Engineering Mechanics Division*, ASCE, **106**, 1980, 929-950.

[Bažant (1980d)] Bažant Z.P., "Work Inequalities for Plastic-Fracturing Materials", *International Journal of Solids and Structures*, **16**, 1980, 873-901.

[Bažant (1982)] Bažant Z.P. and Oh B.H., "Strain-Rate in Rapid Triaxial Loading of Concrete", *Journal of the Engineering Mechanics Division*, ASCE, **108**, 1982, 764-782.

[Bažant (1982b)] Bažant Z.P. and Wittmann F.H. (eds.), "Creep and Shrinkage in Concrete Structures", John Wiley & Sons, New York, 1982.

[Bažant (1983)] Bažant Z.P., "Comment on Orthotropic Models for Concrete and Geomaterials", *Journal of Engineering Mechanics*, **109**, 1982, 849-865.

[Bažant (1983b)] Bažant Z.P. and Oh B.H., "Crack Band Theory for Fracture of Concrete", *Materials & Structures*, **16**, 1983, 155-177.

[Bažant (1984)] Bažant Z.P. and Panula L., "Practical Prediction of Creep and Shrinkage of High Strength Concrete", *Matériaux et Constructions*, RILEM, **101**, 1984, 137-151.

[Bažant (1985)] Bažant Z.P. and Chern J.C., "Log Double Power Law for Concrete Creep", *ACI Journal*, **82**, 1985, 665-675.

[Bažant (1985b)] Bažant Z.P. and Chern J.C., "Triple Power Law for Concrete Creep", *Journal of Engineering Mechanics*, **111**, 1985, 63-83.

[Bažant (1985c)] Bažant Z.P. and Wang T.S., "Practical Prediction of Cyclic Humidity Effect in Creep and Shrinkage of Concrete", *Materials and Structures*, RILEM, **18**, 1985, 247-252.

[Bažant (1986)] Bažant Z.P., Bishop F.C. and Chang T.P., "Confined Compression Tests of Cement Paste and Concrete up to 300 ksi" *ACI Journal*, **83**, 1986, 553-560.

[Bažant (1987)] Bažant Z.P., Wittmann F.H., Kim J.K. and Alou F., "Statistical Extrapolation of Shrinkage Data - Part I: Regression", *ACI Materials Journal*, **84**, 1987, 20-34.

[Bažant (1988)] Bažant Z.P. and Sener S., "Size Effect in Pullout Tests", *ACI Materials Journal*, **85**, 1988, 347-351.

[Bažant (1988b)] Bažant Z.P. and Prat P.C., "Microplane Model for Brittle-Plastic Material - I. Theory and II. Verification", *Journal of Engineering Mechanics*, **114**, 1988, 1672-1702.

[Bažant (1989)] Bažant Z.P. and Prasannan S., "Solidification Theory for Concrete Creep; Part I: Formulation; Part II: Verification and Application; *Journal of Engineering Mechanics*, **115**, 1989, 1691-1725.

[Bažant (1990)] Bažant Z.P. and Ožbolt J., "Nonlocal Microplane Model for Fracture, Damage and Size Effect in Structures", *Journal of Engineering Mechanics*, **116**, 1990, 2485-2505.

[Bakht (1989)] Bakht B., Jaeger L.G. and Mufti A.A., "Elastic Modulus from Compression Tests", *ACI Materials Journal*, **86**, 1989, 220-224.

[Balakrishnan (1988)] Balakrishnan S. and Murray D.W., "Concrete Constitutive Model for NLFE Analysis of Structures", *Journal of Structural Engineering*, **114**, 1988, 1449-1466.

[Balkaya (1993)] Balkaya C. and Schnobrich W.C., "Nonlinear 3-D Behavior of Shear-Wall Dominant RC Building Structures", *Structural Engineering and Mechanics*, **1**, 1993, 1-16.

[Barzegar (1989)] Barzegar F., "Analysis of RC Membrane Elements with Anisotropic Reinforcement", *Journal of Structural Engineering*, **115**, 1989, 647-665.

[Bathe (1975)] Bathe K.J., Ramm E. and Wilson E.L., "Finite Element Formulations for Large Deformation Dynamic Analysis", *International Journal for Numerical Methods in Engineering*, **9**, 1975, 353-386.

[Bathe (1980)] Bathe K.J. and Bolourchi S., "A Geometric and Material Nonlinear Plate and Shell Element", *Computers & Structures*, **11**, 1980, 23-48.

[Bathe (1982)] Bathe K.J., *Finite Element Procedures in Engineering Analysis*, Prentice-Hall, Englewood Cliffs, 1982. German Translation: Bathe K.J., *Finite-Elemente-Methoden*, Springer, Berlin, 1986.

[Bathe (1985)] Bathe K.J. and Dvorkin E.N., "A Four-Node Plate Bending Element Based on Reissner-Mindlin Plate Theory and Mixed Interpolation", *International Journal for Numerical Methods in Engineering*, **21**, 1985, 367-383.

[Bathe (1985b)] Bathe K.J. and Dvorkin E.N., "A Formulation of General Shell Elements - The Use of Mixed Interpolation of Tensorial Components", in Middleton J. and Pande G.N. (eds.), *Proceedings of the NUMETA'85 Conference*, Vol. 2, Balkema, Rotterdam, 1985, 551-563.

[Bathe (1986)] Bathe K.J. and Dvorkin E.N., "A Formulation of General Shell Elements - The Use of Mixed Interpolation of Tensorial Components", *International Journal for Numerical Methods in Engineering*, **22**, 1986, 697-722.

[Bathe (1989)] Bathe K.J., Walczak J., Welch A. and Mistry N., "Nonlinear Analysis of Concrete Structures", *Computers & Structures*, **32**, 1989, 563-590.

[Bellamy (1961)] Bellamy C.J., "Strength of Concrete Under Combined Stresses", *ACI Journal*, **58**, 1961, 367-382.

[Bhide (1987)] Bhide S.B. and Collins M.P., "Reinforced Concrete Elements in Shear and Tension", *Publication No. 87-02*, Department of Civil Engineering, University of Toronto, Toronto, 1987.

[Bonzel (1959)] Bonzel J., "Zur Gestaltabhängigkeit der Betondruckfestigkeit", *Beton- und Stahlbetonbau*, **54**, 1959, 223-228.

[Bonzel (1963)] Bonzel J., "Über die Biegezugfestigkeit des Betons", *Beton*, **13**, 1963, 179-182, and **13**, 1963, 227-232.

[Bonzel (1964)] Bonzel J., "Über die Spaltzugfestigkeit des Betons", *Beton*, **14**, 1964, 108-114 and **14**, 1964, 150-157.

[Bortolotti (1988)] Bortolotti L., "Double-Punch Test for Tensile and Compressive Strengths in Concrete", *ACI Materials Journal*, **85**, 1988, 26-32.

[Bouma (1961)] Bouma A.L., Van Riel A.C., Van Koten H. and Beranek W.J., "Investigations on Models of Eleven Cylindrical Shells Made of Reinforced and Prestressed Concrete", *Proceedings of the Symposium on Shell Research*, Delft, North-Holland, Amsterdam, 1961, 79-101.

[Bryant (1987)] Bryant A.H. and Vadhanavikkit C., "Creep, Shrinkage-Size, and Age at Loading Effects", *ACI Materials Journal*, **84**, 1987, 117-123.

[Buechter (1992)] Buechter N., and Ramm E., "Shell Theory Versus Degeneration - A Comparison in Large Roatation Finite Element Analysis", *International Journal for Numerical Methods in Engineering*, **34**, 1992, 39-59.

[Buyukozturk (1984)] Buyukozturk O., Einstein H.H., Soon K.A. and Shrestinian D., "Experimental Study of the Behavior of Pressure Confined Concrete in Cyclic Loading", *Annual Progress Report*, Dept. of Civil Eng., Massachusetts Institute of Technology, Cambridge, 1984.

[Buyukozturk (1985)] Buyukozturk O. and Shareef S.S., "Constitutive Modeling of Concrete in Finite Element Analysis", *Computers & Structures*, **21**, 1985, 581-610.

[CEB-FIP (1978)] CEB-FIP, "Model Code for Concrete Structures", Vol. 2, App. E, Comité Euro-International du Béton, Paris, 1978.

[CEB (1991)] "Behavior and Analysis of Reinforced Concrete Structures under Alternate Actions Including Inelastic Response", *Bulletin d'Information No. 210*, Comité Euro-International du Béton, Lausanne, 1991.

[Carrasquillo (1981)] Carrasquillo R.L., Slate F.O. and Nilson A.H., "Microcracking and Behavior of High Strength Concrete Subject to Short-Term Loading", *ACI Journal*, **78**, 1981, 179-186.

[Carreira (1985)] Carreira D.J. and Chu K.H., "Stress-Strain Relationship for Plain Concrete in Compression", *ACI Journal*, **82**, 1985, 797-804.

[Carreira (1986)] Carreira D.J. and Chu K.H., "Stress-Strain Relationship for Reinforced Concrete in Tension", *ACI Journal*, **83**, 1986, 21-28.

[Casey (1981)] Casey J. and Naghdi P.M., "On the Characterization of Strain-Hardening in Plasticity", *Journal of Applied Mechanics*, **48**, 1981, 285-296.

[Casey (1983)] Casey J. and Naghdi P.M., "On the Nonequivalence of the Stress Space and Strain Space Formulations of Plasticity Theory", *Journal of Applied Mechanics*, **50**, 1983, 350-354.

[Cedolin (1976)] Cedolin L. and Dei Poli S., "Finite Element Nonlinear Plane Stress Analysis of Reinforced Concrete", Estratto da costruzioni in Cemento Armato, *Studie e Rendiconti*, No. 13, 1976.

[Cedolin (1977)] Cedolin L. and Dei Poli S., "Finite Element Studies of Shear-Critical R/C Beams", *Journal of the Engineering Mechanics Division*, ASCE, **103**, 1977, 395-410.

[Cedolin (1977b)] Cedolin L., Crutzen Y.R.J. and Dei Poli S., "Triaxial Stress-Strain Relationship for Concrete", *Journal of the Engineering Mechanics Division*, ASCE, **103**, 1977, 423-439.

[Cedolin (1987)] Cedolin L., Dei Poli S. and Iori I., "Tensile Behavior of Concrete", *Journal of Engineering Mechanics*, **113**, 1987, 431-449.

[Cervenka (1985)] Cervenka V., "Constitutive Model for Cracked Reinforced Concrete", *ACI Journal*, **82**, 1985, 877-882.

[Cervenka (1990)] Cervenka V., Pukl H. and Eligehausen R., "Computer Simulation of Anchoring Technique in Reinforced Concrete Beams", in Bićanić N. and Mang H.A. (eds.), *Proceedings of the Second International Conference on Computer Aided Analysis and Design of Concrete Structures*, Pineridge Press, Swansea, 1990, 1-19.

[Chapman (1987)] Chapman R.A. and Shah S.P., "Early-Age Bond Strength in Reinforced Concrete", *ACI Materials Journal*, **84**, 1987, 501-510.

[Chen (1975)] Chen A.C.T. and Chen W.F., "Constitutive Relations for Concrete", *Journal of the Engineering Mechanics Division*, ASCE, **101**, 1975, 465-481.

[Chen (1980)] Chen W.F. and Yuan R.L., "Tensile Strength of Concrete: Double-Punch Test", *Journal of the Structural Division*, ASCE, **106**, 1980, 1673-1693.

[Chen (1982)] Chen W.F., *Plasticity in Reinforced Concrete*, McGraw-Hill, New York, 1982.

[Chiorino (1984)] Chiorino M.A., Napoli P., Mola F. and Koprna M. (eds.), "Structural Effects of Time-Dependent Behavior of Concrete", *CEB-Design Manual*, Georgi Publishing Company, Saint-Saphorin, 1984.

[Collins (1985)] Collins M.P., Vecchio F.J. and Mehlhorn G., "An International Competition to Predict the Response of Reinforced Concrete Panels", *Canadian Journal of Civil Engineering*, **12**, 1985, 626-644.

[Cope (1980)] Cope R.J., Rao P.V., Clark L.A. and Norris P., "Modelling of Reinforced Concrete Behavior for Finite Element Analysis of Bridge Slabs", in Taylor C., Hinton E. and Owen D.J.R. (eds.), *Numerical Methods for Non-Linear Problems*, Pineridge Press, Swansea, 1980, 457-470.

[Crisfield (1981)] Crisfield M.A., "A Fast Incemental/Iterative Solution Procedure that Handles Snap-Through", *Computers & Structures*, **13**, 1981, 55-62.

[Crisfield (1983)] Crisfield M.A., "An Arc-Length Method Including Line Searches and Accelerations", *International Journal for Numerical Methods in Engineering*, **19**, 1983, 1269-1289.

[Crisfield (1987)] Crisfield M.A. and Wills J., "Numerical Comparisons Involving Different Concrete Models", Colloquium on Computational Mechanics of Concrete Structures, *IABSE Report Vol.54*, International Association for Bridge and Structural Engineering, 1987, 177-187.

[Crisfield (1989)] Crisfield M.A. and Wills J., "Analysis of R/C Panels Using Different Concrete Models", *Journal of Engineering Mechanics*, **115**, 1989, 578-597.

[Crisfield (1991)] Crisfield M.A., *Non-linear Finite Element Analysis of Solids and Structures*, J. Wiley, Chichester, 1991.

[Dafalias (1975)] Dafalias Y.F. and Popov E.P., "A Model of Nonlinearly Hardening Materials for Complex Loading", *Acta Mechanica*, **21**, 1975, 173-192.

[Dahlblom (1990)] Dahlblom O. and Ottosen N.S., "Smeared Crack Analysis Using Generalized Fictitious Crack Model", *Journal of Engineering Mechanics*, **116**, 1990, 55-76.

[Darwin (1976)] Darwin D. and Pecknold D.A., "Analysis of RC Shear Panels Under Cyclic Loading", *Journal of the Structural Division*, ASCE, **102**, 1976, 355-369.

[Darwin (1977)] Darwin D. and Pecknold D.A., "Nonlinear Biaxial Stress-Strain Law for Concrete", *Journal of the Engineering Mechanics Division*, ASCE, **103**, 1977, 229-241.

[Daschner (1982)] Daschner F. und Kupfer H., "Versuche zur Schubübertragung in Rissen von Normal- und Leichtbeton", *Bauingenieur*, **57**, 1982, 57-60.

[de Borst (1985)] de Borst R. and Nauta P., "Non-Orthogonal Cracks in a Smeared Finite Element Model", *Engineering Computations*, **2**, 1985, 35-46.

[de Borst (1987)] de Borst R., "Smeared Cracking, Plasticity, Creep and Thermal Loading - A Unified Approach", *Computational Methods in Applied Mechanics and Engineering*, **62**, 1987, 89-110.

[Dilger (1984)] Dilger W.H., Koch R. and Kowalczyk R., "Ductility of Plain and Confined Concrete Under Different Strain Rates", *ACI Journal*, **81**, 1984, 73-81.

[Divakar (1987)] Divakar M.P., Fafitis A. and Shah S.P., "Constitutive Model for Shear Transfer in Cracked Concrete", *Journal of Structural Engineering*, **113**, 1987, 1046-1062.

[Di (1993)] Di S. and Cheung Y.K., "Nonlinear Analysis of RC Shell Structures Using Laminated Element", *Journal of Structural Engineering*, **119**, 1993, part I: 2059-2073, part II: 2074-2094.

[Drucker (1950)] Drucker D.C., *Quarterly Journal of Applied Mathematics*, **7**, 1950, 411.

[Duda (1991)] Duda H., "Bruchmechanisches Verhalten von Beton unter monotoner und zyklischer Zugbeanspruchung", *Deutscher Ausschuß für Stahlbeton*, Heft 419, Beuth Verlag, Berlin, 1991.

[Dvorkin (1984)] Dvorkin E.N. and Bathe K.J., "A Continuum Mechanics Based Four-Node Shell Element for General Nonlinear Analysis", *Engineering Computations*, **1**, 1984, 77-88.

[Dvorkin (1988)] Dvorkin E.N., Onate E. and Oliver J., "A Non-Linear Formulation for Curved Timoshenko Beam Elements Considering Large Displacement/Rotation Increments", *International Journal for Numerical Methods in Engineering*, **26**, 1988, 1597-1613.

[Eberhardsteiner (1991)] Eberhardsteiner J., "Synthese aus konstitutivem Modellieren von Beton mittels dreiaxialer, nichtlinear-elastischer Werkstoffgesetze und Finite-Elemente-Analysen dickwandiger Stahlbetonkonstruktionen", *Doctoral Dissertation*, Dissertationen der Technischen Universität Wien, No. **48**, Verband der wissenschaftlichen Gesellschaften Österreichs (VWGÖ), Vienna, 1991.

[Elwi (1979)] Elwi A.A. and Murray D.W., "A 3D Hypoelastic Concrete Constitutive Relationship", *Journal of the Engineering Mechanics Division*, ASCE, **105**, 1979, 623-641.

[Elwi (1980)] Elwi A.A. and Murray D.W., "Nonlinear Analysis of Axisymmetric Concrete Structures", *Structural Engineering Report No.87*, Dept. of Civil Eng., University of Alberta, Edmonton, 1980.

[Epstein (1978)] Epstein M. and Murray D.W., "A Biaxial Constitutive Law for Concrete Incorporated in BOSOR code", *Computers and Structures*, **9**, 1978, 57-63.

[Fafard (1993)] Fafard M. and Massicotte B., "Geometrical Interpretation of the Arc-Length Method", *Computers & Structures*, **46**, 1993, 603-615.

[Fardis (1979)] Fardis M.N. and Buyukozturk O., "Shear Transfer Model for Reinforced Concrete", *Journal of the Engineering Mechanics Division*, ASCE, **105**, 1979, 255-275.

[Feenstra (1991a)] Feenstra P.H., de Borst R. and Rots J.G., "Numerical Study on Crack Dilatancy. I: Models and Stability Analysis", *Journal of Engineering Mechanics*, **117**, 1991, 733-753.

[Feenstra (1991b)] Feenstra P.H., de Borst R. and Rots J.G., "Numerical Study on Crack Dilatancy. II: Applications", *Journal of Engineering Mechanics*, **117**, 1991, 754-769.

[Feenstra (1993a)] Feenstra P.H. and de Borst R., "Aspects of Robust Computational Modeling for Plain and Reinforced Concrete", *Heron*, **38**, No.4, Delft, 1993.

[Feenstra (1993b)] Feenstra P.H., "Computational Aspects of Biaxial Stress in Plain and Reinforced Concrete", *Doctoral Dissertation*, Technical University Delft, Delft University Press, 1993.

[Ferguson (1966)] Ferguson P.M., "Bond Stress - The State of the Art", *ACI Journal*, **63**, 1966, 1161-1190.

[Flügge (1972)] *Tensor Analysis and Continuum Mechanics*, Springer, Berlin, 1972.

[Floegl (1981)] Floegl H., "Traglastermittlung dünner Stahlbetonschalen mittels der Methode der Finiten Elemente unter Berücksichtigung wirklichkeitsnahen Werkstoffverhaltens sowie geometrischer Nichtlinearität", *Doctoral Dissertation*, Technical University of Vienna, Vienna, 1981.

[Floegl (1982)] Floegl H. and Mang H.A., "Tension Stiffening Concept Based on Bond Slip", *Journal of the Structural Division*, ASCE, **108**, 1982, 2681-2701.

[Foster (1992)] Foster S., "An Application of the Arc-Length Method Involving Concrete Cracking", *International Journal for Numerical Methods in Engineering*, **33**, 1992, 269-285.

[Fumagalli (1965)] Fumagalli E., Gunasekaran M., Linger D.A., Gillespie H.A., Vile G.W.D. and Sigvaldason O.T., Discussion of the paper by Iyengar K.T.S.R., Chandrashekara K. and Krishnaswamy K.T., "Strength of Concrete Under Biaxial Compression" *ACI Journal*, **62**, 1965, 1187-1198.

[Gerstle (1980)] Gerstle K.H., Aschl H., Bellotti R., Bertacchi P., Kotsovos M.D., Ko H.Y., Linse D., Newman J.B., Rossi P., Schickert G., Taylor M.A., Traina L.A., Winkler H. and Zimmerman R.M., "Behavior of Concrete under Multiaxial Stress States", *Journal of the Engineering Mechanics Division*, ASCE, **106**, 1980, 1383-1403.

[Gerstle (1981a)] Gerstle K.H., "Simple Formulation of Biaxial Concrete Behavior", *ACI-Journal*, **78**, 1981, 62-68.

[Gerstle (1981b)] Gerstle K.H., "Simple Formulation of Triaxial Concrete Behavior", *ACI-Journal*, **78**, 1981, 382-387.

[Girkmann (1963)] Girkmann K., *Flächentragwerke*, Springer, Wien, 1963.

[Glemberg (1986)] Glemberg R., Oldenburg M., Nilsson L. and Samuelsson A., "A General Constitutive Model for Concrete Structures", in Hinton E. and Owen R. (eds.), *Computational Modelling of Reinforced Concrete Structures*, Pineridge Press, Swansea, 1986.

[Gopalaratnam (1985)] Gopalaratnam V.S. and Shah S.P., "Softening Response of Plain Concrete in Direct Tension" *ACI Journal*, **82**, 1985, 310-323.

[Goto (1971)] Goto Y., "Cracks Formed in Concrete around Deformed Tension Bars", *ACI Journal*, **68**, 1971, 1491-1505.

[Grootenboer (1981)] Grootenboer H.J., Leijten S.F.C.H. and Blaauwendraad J., "Numerical Models for Reinforced Concrete Structures in Plane Stress", *Heron*, **26**, No.1c, Delft, 1981.

[Grote (1992)] Grote K., "Theorie und Anwendung geometrisch und physikalisch nichtlinearer Algorithmen auf Flächentragwerke aus Stahlbeton", *Bericht 1*, Fachgebiet Baustatik, Universität Kaiserslautern, Kaiserslautern, 1992.

[Gupta (1984)] Gupta A.K. and Akbar H., "Cracking in Reinforced Concrete Analysis", *Journal of Structural Engineering*, **110**, 1984, 1735-1746.

[Gupta (1989a)] Gupta A.K. and Maestrini S.R., "Post-Cracking Behavior of Membrane Reinforced Concrete Elements Including Tension-Stiffening", *Journal of Structural Engineering*, **115**, 1989, 957-976.

[Gupta (1989b)] Gupta A.K. and Maestrini S.R., "Unified Approach to Modelling Post-Cracking Membrane Behavior of Reinforced Concrete", *Journal of Structural Engineering*, **115**, 1989, 977-993.

[Hansen (1987)] Hansen W. and Almudaiheem J.A., "Ultimate Drying Shrinkage of Concrete - Influence of Major Parameters", *ACI Materials Journal*, **84**, 1987, 217-223.

[Han (1984)] Han D.J., "Constitutive Modelling in Analysis of Concrete Structures", *Doctoral Dissertation*, Purdue University, West Lafayette, 1984.

[Han (1985)] Han D.J. and Chen W.F., "A Nonuniform Hardening Plasticity Model for Concrete Materials", *Journal of Mechanics of Materials*, **4**, 1985, 1-20.

[Han (1987)] Han D.J. and Chen W.F., "Constitutive Modelling in Analysis of Concrete Structures", *Journal of Engineering Mechanics*, **113**, 1987, 577-593.

[Harte (1986)] Harte R. and Eckstein U., "Derivation of Geometrically Nonlinear Finite Shell Elements via Tensor Notation", *International Journal for Numerical Methods in Engineering*, **23**, 1986, 367-384.

[Heilmann (1969)] Heilmann H.G., Hilsdorf H. and Finsterwalder K., "Festigkeit und Verformung von Beton unter Zugspannungen", *Deutscher Ausschuß für Stahlbeton*, Heft 203, W. Ernst u. Sohn, Berlin, 1969.

[Hillerborg (1976)] Hillerborg A., Modéer M. and Peterson P.E., "Analysis of Crack Formation and Crack Growth in Concrete by Means of Fracture Mechanics and Finite Elements", *Cement and Concrete Research*, **6**, 1976, 773-782.

[Hinton (1992)] Hinton, E. (ed.), *NAFEMS Introduction to Nonlinear Finite Element Analysis*, NAFEMS Birniehill, Glasgow, 1992.

[Hofstetter (1986a)] Hofstetter G. and Mang H.A., "Work Equivalent Node Forces from Prestress of Concrete Shells", in Hughes T.J.R. and Hinton E. (eds.), *Finite Element Methods for Plate and Shell Structures, Vol. 2: Formulations and Algorithms*, Pineridge Press International, Swansea, 1986, 312-347.

[Hofstetter (1986b)] Hofstetter G. and Mang H.A., "Nonlinear Finite Element Analysis of Prestressed Concrete Shells", in Taylor C., Owen D.R.J., Hinton E. and Damjanić F.B. (eds.), *Numerical Methods for Non-Linear Problems*, Vol. 3, Pineridge Press, Swansea, 1986, 344-362.

[Hofstetter (1987)] Hofstetter G., "Physikalisch und geometrisch nichtlineare Traglastanalysen von Spannbetonscheiben, -platten und -schalen mittels der Methode der Finiten Elemente", *Doctoral Dissertation*, Dissertationen der Technischen Universität Wien, No. **47**, Verband der wissenschaftlichen Gesellschaften Österreichs (VWGÖ), Vienna, 1990.

[Hofstetter (1989)] Hofstetter G., Walter H. und Mang H.A., "Finite-Elemente Berechnungen von Flächentragwerken aus Spannbeton unter Berücksichtigung von Langzeitverformungen und Zustand II", *Bauingenieur*, **64**, 1989, 449-461.

[Hofstetter (1993)] Hofstetter G., Simo J.C. and Taylor R.L., "A Modified Cap Model: Closest Point Solution Algorithms", *Computers & Structures*, **46**, 1993, 203-214.

[Hognestad (1955)] Hognestad E., Hanson N.W. and McHenry D., "Concrete Stress Distribution in Ultimate Strength Design", *ACI Journal*, **52**, 1955, 455-480.

[Hsieh (1982)] Hsieh S.S., Ting E.C. and Chen W.F., "A Plastic Fracture Model for Concrete", *International Journal of Solids and Structures*, **18**, 1982, 181-197.

[Hsu (1963)] Hsu T.T.C., Slate F.O., Sturman G.M. and Winter G., "Microcracking of Plain Concrete and the Shape of the Stress-Strain Curve", *ACI Journal*, **60**, 1963, 209-224.

[Hsu (1987)] Hsu T.T.C., Mau S.T. and Chen B., "Theory of Shear Transfer Strength of Reinforced Concrete", *ACI Structural Journal*, **84**, 1987, 149-160.

[Huang (1986)] Huang H.C. and Hinton E., "A New Nine Node Degenerated Shell Element with Enhanced Membrane and Shear Interpolation", *International Journal for Numerical Methods in Engineering*, **22**, 1986, 73-92.

[Hughes (1987)] Hughes T.J.R., *The Finite Element Method*, Prentice-Hall, Englewood Cliffs, 1987.

[Iyengar (1965)] Iyengar K.T.S.R., Chandrashekara K. and Krishnaswamy K.T., "Strength of Concrete Under Biaxial Compression" *ACI Journal*, **62**, 1965, 239-249.

[Jiang (1984)] Jiang D.H., Shah S.P. and Andonian A.T., "Study of the Transfer of Tensile Forces by Bond", *ACI Journal*, **81**, 1984, 251-259.

[Kabaila (1964)] Kabaila A., Saenz L.P., Tulin L.G. and Gerstle K.H., "Equation for the Stress-Strain Curve of Concrete", *ACI Journal*, **61**, 1964, 1227-1239.

[Kabir (1977)] Kabir A.F., "Nonlinear Analysis of Reinforced Concrete Panels, Slabs and Shells for Time-Dependent Effects", *Doctoral Dissertation*, University of California at Berkeley, Berkeley, 1977.

[Kang (1980)] Kang Y.J. and Scordelis A.C., "Nonlinear Analysis of Prestressed Concrete Frames", *Journal of the Structural Division*, ASCE, **106**, 1980, 445-462.

[Karsan (1969)] Karsan I.D. and Jirsa J.O., "Behavior of Concrete under Compressive Loadings", *Journal of the Structural Division*, ASCE, **95**, 1969, 2543-2563.

[Keuser (1987)] Keuser M. and Mehlhorn G., "Finite Element Models for Bond Problems", *Journal of Structural Engineering*, **113**, 1987, 2160-2173.

[Khaloo (1988)] Khaloo A.R. and Ahmad S.H., "Behavior of Normal and High-Strength Concrete under Combined Compression-Shear Loading", *ACI Materials Journal*, **85**, 1988, 551-559.

[Koiter (1953)] Koiter W.T., "Stress Strain Relations, Uniqueness and Variational Theorems for Elastic-Plastic Materials with a Singular Yield Surface", *Quarterly Journal of Applied Mathematics*, **11**, 1953, 350-354.

[Koiter (1967)] Koiter W.T., "General Equations of Elastic Stability for Thin Shells", *Proceedings of the Symposium on the Theory of Shells to Honor Lloyd Hamiltion Donell*, University of Houston, Houston, 1967.

[Kollegger (1988)] Kollegger J., "Ein Materialmodell für die Berechnung von Stahlbetonflächentragwerken", *Doctoral Dissertation*, Fachbereich Bauingenieurwesen der Gesamthochschule Kassel, Kassel, 1988.

[Kollegger (1990)] Kollegger J. and Mehlhorn G., "Material Model for the Analysis of Reinforced Concrete Surface Structures", *Computational Mechanics*, **6**, 1990, 341-357.

[Kotsovos (1977)] Kotsovos M.D. and Newman J.B., "Behavior of Concrete under Multiaxial Stress", *ACI Journal*, **74**, 1977, 443-446.

[Kotsovos (1978)] Kotsovos M.D. and Newman J.B., "Generalized Stress-Strain Relations for Concrete", *Journal of the Engineering Mechanics Division*, ASCE, **104**, 1978, 845-856.

[Kotsovos (1979)] Kotsovos M.D. and Newman J.B., "Effect of Stress Path on the Behavior of Concrete under Triaxial Stress States", *ACI-Journal*, **76**, 1979, 213-223.

[Kotsovos (1984a)] Kotsovos M.D., "Deformation and Failure of Concrete in a Structure", *Proceedings of the International Conference on Concrete under Multiaxial Conditions*, RILEM - CEB - CNRS, Presses de'l Université Paul Sabatier, Toulouse, 1984, 104-113.

[Kotsovos (1984b)] Kotsovos M.D. and Cheong H.K., "Applicability of Test Specimen Results for the Description of the Behavior of Concrete in a Structure", *ACI Journal*, **81**, 1984, 358-363.

[Krauthammer (1995)] Krauthammer T. and Swartz S.E., "Finite Element Simulation of Three Concrete Shell Structures: Capabilities and Limitations", *ACI Structural Journal*, 1995, to be published.

[Krätzig (1992)] Krätzig W.B. and Y. Zhuang, "Collapse Simulation of Reinforced Concrete Natural Draught Cooling Towers", *Engineering Structures*, **14**, 1992, 291-299.

[Kupfer (1969)] Kupfer H., Hilsdorf H.K. and Rüsch H., "Behavior of Concrete under Biaxial Stresses", *ACI Journal*, **66**, 1969, 656-666.

[Kupfer (1973)] Kupfer H., "Das Verhalten des Betons unter mehrachsiger Kurzzeitbelastung unter besonderer Berücksichtigung der zweiachsigen Beanspruchung", *Deutscher Ausschuß für Stahlbeton*, Heft 229, W. Ernst u. Sohn, Berlin, 1973.

[Lahnert (1986)] Lahnert B.J., Houde J. and Gerstle K.H., "Direct Measurement of Slip Between Steel and Concrete", Concrete", *ACI Journal*, **83**, 1986, 974-982.

[Lam (1992)] Lam W.F. and Morley C.T., "Arc-Length Method for Passing Limit Points in Structural Calculation", *Journal of Structural Engineering*, **118**, 1992, 169-185.

[Lanig (1991a)] Lanig N., Stöckl S. and Kupfer H., "Versuche zum Kriechen und zur Restfestigkeit von Beton bei mehrachsiger Beanspruchung", *Deutscher Ausschuß für Stahlbeton*, Heft 420, Beuth Verlag, Berlin, 1991, 1-81.

[Lanig (1991b)] Lanig N. and Stöckl S., "Kriechen von Beton nach langer Lasteinwirkung", *Deutscher Ausschuß für Stahlbeton*, Heft 420, Beuth Verlag, Berlin, 1991, 83-110.

[Leonhardt (1980)] Leonhardt F., *Vorlesungen über Massivbau*, Vol. 1 - 6 , Springer, Berlin, 1980.

[Lieberum (1987)] Lieberum K.H., "Das Tragverhalten von Beton bei extremer Teilflächenbelastung", *Doctoral Dissertation*, Technical University of Darmstadt, Darmstadt, 1987.

[Lieberum (1989)] Lieberum K.H. and Reinhardt H.W., "Strength of Concrete on an Extremely Small Bearing Area", *ACI Structural Journal*, **86**, 1989, 67-76.

[Linse (1976a)] Linse D. und Stegbauer A., "Festigkeit und Verformungsverhalten von Beton unter hohen zweiachsigen konstanten Dauerbelastungen und Dauerschwellbelastungen", *Deutscher Ausschuß für Stahlbeton*, Heft 254, W. Ernst u. Sohn, Berlin, 1976.

[Liu (1972)] Liu T.C., Nilson A.H. and Slate F.O., "Stress Strain Response and Fracture of Concrete in Uniaxial and Biaxial Compression", *ACI Journal*, **69**, 1972, 291-295.

[Liu (1972b)] Liu T.C.Y., Nilson A.H. and Slate F.O., "Biaxial Stress-Strain Relations for Concrete", *Journal of the Structural Division*, ASCE, **98**, 1972, 1025-1034.

[Lubliner (1989)] Lubliner J., Oliver J., Oller S. and Onate E., "A Plastic-Damage Model for Concrete", *International Journal of Solids and Structures*, **25**, 1989, 299-326.

[Lubliner (1990)] Lubliner J., *Plasticity Theory*, Macmillan, New York, 1990.

[Luenberger (1984)] Luenberger D.G., *Linear and Nonlinear Programming*, 2nd ed., Addison-Wesley, Reading, 1984.

[MARC (1992)] MARC General Purpose Finite Element Program, *User Information Manual*, Rev.K5.2, MARC Analysis Research Corporation, Palo Alto, 1992.

[MacGregor (1980)] MacGregor J.G., Simmonds S.H. and Rizkalla S.H., "Test of a Prestressed Concrete Secondary Containment Structure", *Structural Engineering Report No. 85*, Dept. of Civil Eng., University of Alberta, Edmonton, 1980.

[Magura (1964)] Magura D.D., Sozen M.A. and Siess C.P.," A Study of Stress Relaxation in Prestressing Reinforcement", *PCI Journal*, **9**, 1964, 13-28.

[Malvern (1969)] Malvern L.E., *Introduction to the Mechanics of a Continuous Medium*, Prentice-Hall, Englewood Cliffs, 1969.

[Mang (1983)] Mang H.A., Floegl H. and Walter H., "Wind-Loaded Reinforced-Concrete Cooling Towers: Buckling or Ultimate Load?", *Engineering Structures*, **5**, 1983, 163-180.

[Mang (1990)] Mang H.A., "Fulfilled and Unfulfilled Expectations: A Review of a Decade of Nonlinear FE Analysis of Reinforced Concrete and Prestressed Concrete Structures at Technical University of Vienna", in Bićanić N. and Mang H.A. (eds.), *Proceedings of the Second International Conference on Computer Aided Analysis and Design of Concrete Structures*, Pineridge Press, Swansea, 1990, 1283-1309.

[Mang (1991)] Mang H.A. and Meschke G., "Nonlinear Finite Element Analysis of Reinforced and Prestressed Concrete Structures", *Engineering Structures*, **13**, 1991, 211-226.

[Mari (1984)] Mari A.R., "Nonlinear Geometric, Material and Time Dependent Analysis of Three Dimensional Reinforced and Prestressed Concrete Frames", *Report No. UCB/SESM-84/12*, Department of Civil Engineering, University of California at Berkeley, Berkeley, 1984.

[Marsden (1988)] Marsden J.E. and Tromba A.J., *Vector Calculus*, 3rd ed., Freeman, New York, 1988.

[Marti (1989)] Marti P., "Size Effect in Double-Punch Tests on Concrete Cylinders", *ACI Materials Journal*, **86**, 1989, 597-601.

[May (1988)] May I.M. and Ganaba T.H., "A Full Range Analysis of Reinforced Concrete Slabs Using Finite Elements", *International Journal for Numerical Methods in Engineering*, **26**, 1988, 973-985.

[Mehlhorn (1985)] Mehlhorn G., Kollegger J., Keuser M. and Kolmar W., "Nonlinear Contact Problems - A Finite Element Approach Implemented in ADINA", *Computers & Structures*, **21**, 1985, 69-80.

[Mehlhorn (1990)] Mehlhorn G., "Some Developments for Finite Element Analyses of Reinforced Concrete Structures", in Bićanić N. and Mang H.A. (eds.), *Proceedings of the Second International Conference on Computer Aided Analysis and Design of Concrete Structures*, Pineridge Press, Swansea, 1990, 1319-1336.

[Mehmel (1967)] Mehmel A. and Freitag W., "Tragfähigkeitsversuche an Stahlbetonkonsolen", *Der Bauingenieur*, **42**, 1967, 362-369.

[Meschke (1991)] Meschke G., "Synthese aus konstitutivem Modellieren von Beton mittels dreiaxialer, elasto-plastischer Werkstoffmodelle und Finite Elemente Analysen dickwandiger Stahlbetonkonstruktionen", *Doctoral Dissertation*, Dissertationen der Technischen Universität Wien, No. **49**, Verband der wissenschaftlichen Gesellschaften Österreichs (VWGÖ), Vienna, 1991.

[Meschke (1991b)] Meschke G., Mang H.A. and Kosza P., "Finite Element Analysis of Cracked Cooling Tower Shell", *Journal of Structural Engineering*, **117**, 1991, 2620-2639.

[Meschke (1994)] Meschke G., "A Multisurface Viscoplastic Model for Shotcrete. Algorithmic Aspects and Finite Element Analyses of Tunnel Linings", in Mang H., Bićanić N. and de Borst R. (eds.), *Computer Modelling of Concrete Structures*, Proceedings of EURO-C 1994, Cromwell Press, Melksham, 1994, 243-253.

[Miguel (1990)] Miguel P.F., Jawad M.A. and Fernandez M.A., "A Discrete-Crack Model for the Analysis of Concrete Structures", in Bićanić N. and Mang H. (eds.), *Proceedings of the 2nd International Conference on Computer Aided Analysis and Design of Concrete Structures*, Pineridge Press, Swansea, 1990, 847-908.

[Milford (1984)] Milford R.V. and Schnobrich W.C., "The Effect of Cracking on the Ultimate Load of Reinforced Concrete Cooling Towers", in Gould P.L., Krätzig W.B., Mungan I. and Wittek U. (eds.), *Proceedings of the Second International Symposium on Natural Draught Cooling Towers*, Springer, Berlin, 1984, 319-332.

[Milford (1985)] Milford R.V. and Schnobrich W.C., "Numerical Model for Cracked Reinforced Concrete", in Damjanić F., Hinton E., Owen D.R.J., Bićanić N. and Simović V. (eds.), *Proceedings of the International Conference on Computer Aided Analysis and Design of Concrete Structures*, Pineridge Press, Swansea, 1985, 71-84.

[Mills (1970)] Mills L.L. and Zimmermann R.M., "Compressive Strength under Multiaxial Loading Conditions", *ACI Journal*, **67**, 1970, 802-807.

[Min (1993)] Min C.S. and Gupta A.K., "Inelastic Behavior of Hyperbolic Cooling Tower", *Journal of Structural Engineering*, **119**, 1993, 2235-2255.

[Mirza (1979)] Mirza S.A. and MacGregor J.G., "Variability of Mechanical Properties of Reinforcing Bars", *Journal of the Structural Division*, ASCE, **105**, 1979, 921-937.

[Mitchell (1988)] Mitchell G.P. and Owen D.R.J., "Numerical Solutions for Elastic-Plastic Problems", *Engineering Computations*, **5**, 1988, 274-284.

[Mizuno (1992)] Mizuno E. and Hatanaka S., "Compressive Softening Model for Concrete", *Journal of Engineering Mechanics*, **118**, 1992, 1546-1563.

[Mlakar (1985)] Mlakar P.F., Vitaya-Udom K.P. and Cole R.A., "Dynamic Tensile-Behavior of Concrete" *ACI Journal*, **82**, 1985, 484-491.

[Murray (1979)] Murray D.W., Chitnuyanondh L., Rijub-Agha K.Y. and Wong C., "Concrete Plasticity Theory for Biaxial Stress Analysis", *Journal of the Engineering Mechanics Division*, ASCE, **105**, 1979, 989-1005.

[Naaman (1982)] Naaman A.E., *Prestressed Concrete, Analysis and Design*, McGraw-Hill, New York, 1982.

[Nagtegaal (1981)] Nagtegaal J.C. and Slater J.G., "A Simple Noncompatible Thin Shell Element Based on Discrete Kirchhoff Theory", in Hughes T.J.R., Pifko A. and Jay A. (eds.), *Nonlinear Finite Element Analysis of Plates and Shells*, ASME, 1981.

[Nasser (1965)] Nasser K.W. and Neville A.M., "Creep of Concrete at Elevated Temperatures", *ACI Journal*, **62**, 1965, 1567-1579.

[Nelson (1988)] Nelson E.L., Carrasquillo R.L. and Fowler D.W., "Behavior and Failure of High-Strength Concrete Subjected to Biaxial-Cyclic Compression Loading", *ACI Materials Journal*, **85**, 1988, 248-253.

[Neville (1981)] Neville A.M., *Properties of Concrete*, 3rd ed., Pitman, London, 1981.

[Ngo (1967)] Ngo D. and Scordelis A.C., "Finite Element Analysis of Reinforced Concrete Beams", *ACI Journal*, **64**, 1967, 152-163.

[Nianxiang (1989)] Nianxiang X. and Wenyan L., "Determining Tensile Properties of Mass Concrete by Direct Tensile Test", *ACI Materials Journal*, **86**, 1989, 214-219.

[Onate (1992)] Onate E., Zienkiewicz O.C., Suarez B. and Taylor R.L., "A General Methodology for Deriving Shear Constrained Reissner-Mindlin Plate Elements", *International Journal for Numerical Methods in Engineering*, **33**, 1992, 345-367.

[Ohtani (1989)] Ohtani Y. and Chen W.F., "A Plastic-Softening Model for Concrete Materials", *Computers & Structures*, **33**, 1989, 1047-1055.

[Ortiz (1985)] Ortiz M. and Popov E.P., "Accuracy and Stability of Integration Algorithms for Elastoplastic Constitutive Relations", *International Journal for Numerical Methods in Engineering*, **21**, 1985, 1561-1576.

[Ottosen (1977)] Ottosen N.S., "A Failure Criterion for Concrete", *Journal of the Engineering Mechanics Division*, ASCE, **103**, 1977, 527-535.

[Ottosen (1986)] Ottosen N.S., "Thermodynamic Consequences of Strain Softening in Tension", *Journal of Engineering Mechanics*, **112**, 1986, 1152-1164.

[Owen (1983)] Owen D.R.J., Figueiras J.A. and Damjanić F., "Finite Element Analysis of Reinforced and Prestressed Concrete Structures Including Thermal Loading", *Computer Methods in Applied Mechanics and Engineering*, **41**, 1983, 323-366.

[Pagnoni (1992)] Pagnoni T., Slater J., Ameur-Moussa R. and Buyukozturk O., "A Nonlinear Three-Dimensional Analysis of Reinforced Concrete Based on a Bounding Surface Model", *Computers & Structures*, **43**, 1992, 1-12.

[Palaniswamy (1974)] Palaniswamy R. and Shah S.P., "Fracture and Stress-Strain Relationship of Concrete under Triaxial Compression", *Journal of the Structural Division*, ASCE, **100**, 1974, 901-916.

[Pekau (1992)] Pekau O.A., Zhang Z.X. and Liu G.T., "Constitutive Model for Concrete in Strain Space", *Journal of Engineering Mechanics*, **118**, 1992, 1907-1927.

[Pochanart (1989)] Pochanart S. and Harmon T., "Bond-Slip Model for Generalized Excitations Including Fatigue", *ACI Materials Journal*, **86**, 1989, 465-474.

[Polak (1993)] Polak M.A. and Vecchio F.J., "Nonlinear Analysis of Reinforced-Concrete Shells", *Journal of Structural Engineering*, **119**, 1993, 3439-3462.

[Povoas (1989)] Povoas R.H.C.F. and Figueiras J.A., "Nonlinear Analysis of Prestressed Concrete Shells. Tendon Formulation", in Owen D.R.J., Hinton E. and Onate E. (eds.), *Proceedings of the 2nd International Conference on Computational Plasticity*, Pineridge Press, 1989, 1379-1394.

[Probst (1978)] Probst P. und Stöckl S., "Kriechen und Rückkriechen von Beton nach langer Lasteinwirkung", *Deutscher Ausschuß für Stahlbeton*, Heft 295, W. Ernst u. Sohn, Berlin, 1978.

[RILEM (1986)] Preprints of the Fourth RILEM International Symposium on "Creep and Shrinkage of Concrete: Mathematical Modeling", Bažant Z.P. (ed.), Dept. of Civil Eng. and Center for Concrete and Geomaterials, Northwestern University, Evanston, 1986.

[RILEM (1989)] Elfgren L. (ed.), *Fracture Mechanics of Concrete Structures*, RILEM-Report, Chapman and Hall, London, 1989.

[Rüsch (1960)] Rüsch H., "Researches Toward a General Flexural Theory for Structural Concrete", *ACI Journal*, **57**, 1960, 1-28.

[Rüsch (1968)] Rüsch H., Sell R., Rasch C., Grasser E., Hummel A., Wesche K. and Flatten H., "Festigkeit und Verformung von unbewehrtem Beton unter konstanter Dauerlast", *Deutscher Ausschuß für Stahlbeton*, Heft 198, W. Ernst u. Sohn, Berlin, 1968.

[Rüsch (1983)] Rüsch H., Jungwirth D. and Hilsdorf H.K., *Creep and Shrinkage - Their Effect on the Behavior of Concrete Structures*, Springer Verlag, New York, 1983.

[Ramm (1976)] Ramm E., "Geometrisch nichtlineare Elastostatik und Finite Elemente", *Bericht Nr. 76-2*, Institut für Baustatik der Universität Stuttgart, Stuttgart, 1976.

[Ramm (1981)] Ramm E., "Strategies for Tracing the Nonlinear Response Near Limit Points", in Wunderlich W., Stein E. and Bathe K.J. (eds.), *Nonlinear Finite Element Analysis in Structural Mechanics*, Springer, Berlin, 1981, 63-89.

[Raphael (1984)] Raphael J.M., "Tensile Strength of Concrete", *ACI Journal*, **81**, 1984, 158-165.

[Rasch (1962)] Rasch Ch., "Spannungs-Dehnungslinien des Betons und Spannungsverteilung in der Biegedruckzone bei konstanter Dehngeschwindigkeit", *Deutscher Ausschuß für Stahlbeton*, Heft 154, W. Ernst u. Sohn, Berlin, 1962.

[Rehm (1961)] Rehm G., "Über die Grundlagen des Verbundes zwischen Stahl und Beton", *Deutscher Ausschuß für Stahlbeton*, Heft 138, W. Ernst u. Sohn, Berlin, 1961.

[Reinhardt (1984)] Reinhardt H.W., "Fracture Mechanics of an Elastic Softening Material like Concrete", *Heron*, **29**, No. 2, Delft, 1984, 1-42.

[Reinhardt (1986)] Reinhardt H.W., Cornelissen H.A.W. and Hordijk D.A., "Tensile Tests and Fracture Analysis of Concrete", *Journal of Structural Engineering*, **112**, 1986, 2462-2477.

[Ritz (1975)] Ritz P., Marti P. and Thürlimann B., "Versuche über das Biegeverhalten von vorgespannten Platten ohne Verbund", *Bericht Nr. 7305-1*, Institut für Baustatik und Konstruktion ETH Zürich, Zürich, 1975.

[Rizkalla (1984)] Rizkalla S.H., Simmonds S.H. and MacGregor J.G., "Prestressed Concrete Containment Model", *Journal of Structural Engineering*, ASCE, **104**, 1984, 730-743.

[Roca (1988)] Roca P., "A Numerical Model for the Nonlinear Analysis of Prestressed Concrete Shells", *Doctoral Dissertation*, Technical University of Catalonia, Barcelona, 1988 (in Spanish).

[Roca (1993a)] Roca P. and Mari A.R., "Numerical Treatment of Prestressing Tendons in the Nonlinear Analysis of Prestressed Concrete Structures", *Computers & Structures*, **46**, 1993, 905-916.

[Roca (1993b)] Roca P. and Mari A.R., "Nonlinear Geometric and Material Analysis of Prestressed Concrete General Shell Structures", *Computers & Structures*, **46**, 1993, 917-929.

[Rots (1989)] Rots J.G. and Blaauwendraad J., "Crack Models for Concrete: Discrete or Smeared? Fixed, Multi-Directional or Rotating?, Heron, **34**, No.1, Delft, 1989.

[Rots (1990)] Rots J.G. and Schellekens J.C.J., "Interface Elements in Concrete Mechanics", in Bićanić N. and Mang H.A. (eds.), *Proceedings of the Second International Conference on Computer Aided Analysis and Design of Concrete Structures*, Pineridge Press, Swansea, 1990, 909-918.

[Saleeb (1982)] Chen W.F. and Saleeb A.F., *Constitutive Equations for Engineering Materials*, John Wiley, New York, 1982.

[Sanders (1963)] Sanders J.R., "Nonlinear Theories for Thin Shells", *Quarterly Journal of Applied Mathematics*, **21**, 1963, 21-36.

[Scavuzzo (1983)] Scavuzzo R., Stankowski T., Gerstle K.H. and Ko H.-Y., "Stress-Strain Curves for Concrete under Multiaxial Load Histories", *Report*, Dept. of Civil, Environmental and Architectural Eng., University of Colorado, Boulder, 1983.

[Schäfer (1975)] Schäfer H., "A Contribution to the Solution of Contact Problems with the Aid of Bond Elements", *Computer Methods in Applied Mechanics and Engineering*, **6**, 1975, 335-354.

[Schickert (1980)] Schickert G. "Schwellenwerte beim Betondruckversuch", *Deutscher Ausschuß für Stahlbeton*, Heft 312, W. Ernst u. Sohn, Berlin, 1980.

[Schickert (1977)] Schickert G. and Winkler H., "Versuchsergebnisse zur Festigkeit und Verformung von Beton bei mehraxialer Druckbeanspruchung", *Deutscher Ausschuß für Stahlbeton*, Heft 277, W. Ernst u. Sohn, Berlin, 1977.

[Schnobrich (1991)] "Preliminary Proceedings of the International Workshop on Finite Element Analysis of Reinforced Concrete", Chapter 1: Introduction by W.C. Schnobrich, Columbia University, New York, 1991.

[Scordelis (1985)] Scordelis A.C., "Past, Present and Future Analysis of Reinforced Concrete Structures", *Proceedings of the US-Japan Joint Seminar on Finite Element Analysis of Reinforced Concrete Structures*, Tokyo, 1985.

[Scordelis (1989)] Scordelis A.C., "Nonlinear Material, Geometric and Time-Dependent Analysis of Reinforced and Prestressed Concrete Shells", *Proceedings of the IASS-Congress on Shell and Spatial Structures*, Madrid, 1989, 57-70.

[Shafer (1985)] Shafer G.S. and Ottosen N.S., "An Invariant-Based Constitutive Model", *Structural Research Series 8506*, Dept. of Civil, Environmental, and Architectural Engineering, University of Colorado, Boulder, 1985.

[Shahrooz (1994)] Shahrooz B.M., Ho I.K., Aktan A.E., de Borst R., Blaauwendraad J., van der Veen C., Iding R.H. and Miller R.A., "Nonlinear Finite Element Analysis of Deteriorated RC Slab Bridge", *Journal of Structural Engineering*, **120**, 1994, 422-440.

[Shah (1966)] Shah S.P. and Winter G., "Inelastic Behavior and Fracture of Concrete", *ACI Journal*, **63**, 1966, 925-930.

[Shah (1968)] Shah S.P. and Chandra S., "Critical Stress, Volume Change, and Microcracking of Concrete", *ACI Journal*, **65**, 1968, 770-781.

[Shah (1987)] Shah S.P. and Sankar R., "Internal Cracking and Strain-Softening Response of Concrete under Uniaxial Compression", *ACI Materials Journal*, **84**, 1987, 200-212.

[Shima (1990)] Shima H., Shin H.M. and Okamura H., "A Constitutive Model for Tension Stiffness of Reinforced Concrete under Reversed Loading", in Bićanić N. and Mang H.A. (eds.), *Proceedings of the Second International Conference on Computer Aided Analysis and Design of Concrete Structures*, Pineridge Press, Swansea, 1990, 1079-1090.

[Simo (1985)] Simo J.C. and Taylor R.L., "Consistent Tangent Operators for Rate Independent Elasto-Plasticity", *Computational Methods in Applied Mechanics and Engineering*, **48**, 1985, 101-118.

[Simo (1986)] Simo J.C. and Taylor R.L., "A Return Mapping Algorithm for Plane Stress Elastoplasticity", *International Journal for Numerical Methods in Engineering*, **22**, 1986, 649-670.

[Simo (1988)] Simo J.C. and Hughes T.J.R., *Elastoplasticity and Viscoplasticity - Computational Aspects*, Springer, in press.

[Simo (1988b)] Simo J.C., Kennedy J.G. and Govindjee S., "Unconditionally Stable Return Mapping Algorithms for Non-Smooth Multi-Surface Plasticity Amenable to Exact Linearization", *International Journal for Numerical Methods in Engineering*, **26**, 1988, 2161-2185.

[Simo (1989)] Simo J.C., Kennedy J.G. and Taylor R.L., "Complementary Mixed Finite Element Formulations for Elastoplasticity", *Computer Methods in Applied Mechanics and Engineering*, **74**, 1989, 177-206.

[Simo (1993)] Simo J.C. and Meschke G., "A New Class of Algorithms for Classical Plasticity Extended to Finite Strains. Application to Geomaterials", *Computational Mechanics*, **11**, 1993, 253-278.

[Sinha (1964)] Sinha B.P., Gerstle K.H. and Tulin L.G., "Stress-Strain Relations for Concrete under Cyclic Loading", *ACI Journal*, **61**, 1964, 195-211.

[Slate (1963)] Slate F.O. and Olsefski S., "X-Rays for Study of Internal Structure and Microcracking of Concrete", *ACI Journal*, **60**, 1963, 575-588.

[Sloan (1987)] Sloan S.W., "Substepping Schemes for the Numerical Integration of Elastoplastic Stress-Strain Relations", *International Journal for Numerical Methods in Engineering*, **24**, 1987, 893-911.

[Smadi (1985)] Smadi M.M., Slate F.O. and Nilson A.H., "High-, Medium-, and Low-Strength Concretes Subject to Sustained Overloads - Strains, Strengths, and Failure Mechanisms", *ACI Journal*, **82**, 1985, 657-664.

[Smadi (1987)] Smadi M.M., Slate F.O. and Nilson A.H., "Shrinkage and Creep of High-, Medium- and Low-Strength Concretes, Including Overloads", *ACI Materials Journal*, **84**, 1987, 224-234.

[Smadi (1989)] Smadi M.M. and Slate F.O., "Microcracking of High and Normal Strength Concretes under Short- and Long-Term Loadings", *ACI Materials Journal*, **86**, 1989, 117-127.

[Smith (1989)] Smith S.S., Willam K.J., Gerstle K.H. and Sture S., "Concrete over the Top, or: Is There Life after Peak?", *ACI Materials Journal*, **86**, 1989, 491-497.

[Soroushian (1986)] Soroushian P., Choi K.B. and Alhamad A., "Dynamic Constitutive Behavior of Concrete", *ACI Journal*, **83**, 1986, 251-259.

[Soroushian (1986b)] Soroushian P., Obaseki K., Rojas M.C. and Sim J., "Analysis of Dowel Bars Acting Against Concrete Core", *ACI Journal*, **83**, 1986, 642-649.

[Soroushian (1987a)] Soroushian P., Obaseki K., Rojas M. and Najm H.S., "Behavior of Bars in Dowel Action against Concrete Cover", *ACI Structural Journal*, **84**, 1987, 170-176.

[Soroushian (1987b)] Soroushian P., Obaseki K. and Rojas M.C., "Bearing Strength and Stiffness of Concrete under Reinforcing Bars", *ACI Materials Journal*, **84**, 1987, 179-184.

[Soroushian (1988)] Soroushian P., Obaseki K., Baiyasi M.I., El-Sweidan B. and Choi K.B., "Inelastic Cyclic Behavior of Dowel Bars", *ACI Structural Journal*, **85**, 1988, 23-29.

[Stander (1989)] Stander N., Matzenmiller A. and Ramm E., "An Assessment of Assumed Strain Methods in Finite Rotation Shell Analysis", *Engineering Computations*, **6**, 1989, 58-66.

[Stankowski (1983)] Stankowski T., "Concrete under Multiaxial Load Histories", *Report*, Dept. of Civil, Environmental, and Architectural Engineering, University of Colorado, Boulder, 1983.

[Stankowski (1985)] Stankowski T. and Gerstle K.H., "Simple Formulation of Concrete Behavior Under Multiaxial Load Histories", *ACI Journal*, **82**, 1985, 213-221.

[Stevens (1987)] Stevens N.J., Uzumeri S.M. and Collins M.P., "Analytical Modelling of Reinforced Concrete Subjected to Monotonic and Reversed Loadings", *Publication No. 87-01*, Department of Civil Engineering, University of Toronto, Toronto, 1987.

[Stevens (1991)] Stevens N.J., Uzumeri S.M., Collins M.P. and Will G.T., "Constitutive Model for Reinforced Concrete Finite Element Analysis", *ACI Structural Journal*, **88**, 1991, 49-59.

[Suaris (1983)] Suaris W. and Shah S.P., "Properties of Concrete Subject to Impact", *Journal of Structural Engineering*, **109**, 1983, 1727-1741.

[Suaris (1987)] Suaris W. and Fernando V., "Ultrasonic Pulse Attenuation as a Measure of Damage Growth during Cyclic Loading of Concrete", *ACI Materials Journal*, **84**, 1987, 185-193.

[Suidan (1973)] Suidan M. and Schnobrich W.C., "Finite Element Analysis of Reinforced Concrete", *Journal of the Structural Division*, ASCE, **99**, 1973, 2109-2122.

[Su (1988)] Su E.C.M. and Hsu T.T.C., "Biaxial Compression Fatigue and Discontinuity of Concrete", *ACI Materials Journal*, **85**, 1988, 178-188.

[Tassios (1984)] Tassios T.P. and Koroneos E.G., "Local Bond-Slip Relationships by Means of the Moiré Method" *ACI Journal*, **81**, 1984, 27-34.

[Tassios (1987)] Tassios T.P. and Vintzeleou E.N., "Concrete-to-Concrete Friction", *Journal of Structural Engineering*, **113**, 1987, 832-849.

[Tasuji (1978)] Tasuji M.E., Slate F.O. and Nilson A.H., "Stress-Strain Response and Fracture of Concrete in Biaxial Loading", *ACI Journal*, **75**, 1978, 306-312.

[Tepfers (1979)] Tepfers R. and Kutti T., "Fatigue Strength of Plain, Ordinary, and Lightweight Concrete", *ACI Journal*, **76**, 1979, 635-653.

[Thomas (1974)] Thomas G.R., "Nonlinear Finite Element Analysis of Thin Shells", *Doctoral Dissertation*, Department of Structural Engineering, Cornell University, Ithaca, 1974.

[Thomas (1975)] Thomas G.R. and Gallagher R.H., "A Triangular Thin Shell Finite Element: Linear Analysis", *NASA Contractor Report 2483*, Washington D.C., 1975.

[Timoshenko (1943)] Timoshenko S., *Theory of Elasticity*, McGraw-Hill, New York, 1934.

[Trost (1967)] Trost H., "Auswirkungen des Superpositionsprinzips auf Kriech- und Relaxationsprobleme bei Beton- und Spannbeton", *Beton- und Spannbetonbau*, **62**, 1967, 230-238 and 261-269.

[Trost (1991)] Trost H. and Paschmann H., "Frühe Kriechverformungen des Betons", *Deutscher Ausschuß für Stahlbeton*, Heft 420, Beuth Verlag, Berlin, 1991, 111-121.

[Truesdell (1955)] Truesdell C., "Hypo-Elasticity", *Journal of Rational Mechanics Analysis*, **4**, 1955, 83-133.

[VGB-Rules (1980)] Bautechnik bei Kühltürmen - Bautechnische Richtlinien, *VGB - Rules for Design and Construction*, VGB Kraftwerkstechnik GmbH, Essen, 1980.

[Van Mier (1984)] Van Mier J.G.M., "Complete Stress-Strain Behavior and Damaging Status of Concrete under Multiaxial Conditions", *Proceedings of the International Conference on Concrete under Multiaxial Conditions*, RILEM - CEB - CNRS, Presses de'l Université Paul Sabatier, Toulouse, 1984, 75-85.

[Van Greunen (1983)] Van Greunen J. and Scordelis A.C., "Nonlinear Analysis of Prestressed Concrete Slabs", *Journal of Structural Engineering*, **109**, 1983, 1742-1760.

[Van Mier (1987)] Van Mier J.G.M., "Examples of Non-Linear Analysis of Reinforced Concrete Structures with DIANA", Heron, **32**, No.3, Delft, 1987.

[Van Zyl (1978)] Van Zyl S.F., "Analysis of Curved Segmentally Erected Prestressed Concrete Box Girder Bridges", *Report No. UC SESM 78-2*, University of California at Berkeley, Berkeley, 1978.

[Vecchio (1982)] Vecchio F.J. and Collins M.P., "The Response of Reinforced Concrete to In-Plane Shear and Normal Stresse", *Publication No. 82-03*, Department of Civil Engineering, University of Toronto, Toronto, 1982.

[Vecchio (1986)] Vecchio F.J. and Collins M.P., "The Modified Compression-Field Theory for Reinforced Concrete Elements Subjected to Shear", *ACI Journal*, **83**, 1986, 219-231.

[Vecchio (1988)] Vecchio F.J. and Collins M.P., "Predicting the Response of Reinforced Concrete Beams Subjected to Shear Using Modified Compression Field Theory", *ACI Structural Journal*, **85**, 1988, 258-268.

[Vidosa (1991)] Vidosa F.G., Kotsovos M.D. and Pavlovic M.N., "Nonlinear Finite-Element Analysis of Concrete Structures: Performance of a Fully Three-Dimensional Brittle Model", *Computers & Structures*, **40**, 1991, 1287-1306.

[Vintzeleou (1987)] Vintzeleou E.N. and Tassios T.P., "Behavior of Dowels under Cyclic Deformations", *ACI Structural Journal*, **84**, 1987, 18-30.

[Walraven (1981)] Walraven J.C., "Fundamental Analysis of Aggregate Interlock", *Journal of the Structural Division*, ASCE, **107**, 1981, 2245-2270.

[Walter (1988)] Walter H., "Finite Elemente Berechnungen von Flächentragwerken aus Stahl- und Spannbeton unter Berücksichtigung von Langzeitverformungen und Zustand II", *Doctoral Dissertation*, Technical University of Vienna, Vienna, 1988.

[Wang (1978)] Wang P.T., Shah S.P. and Naaman A.E., "Stress-Strain Curves of Normal and Lightweight Concrete in Compression", *ACI Journal*, **75**, 1978, 603-611.

[Wang (1987)] Wang C.-Z., Guo Z.-H. and Zhang X.-Q., "Experimental Investigation of Biaxial and Triaxial Compressive Concrete Strength", *ACI Materials Journal*, **86**, 1987, 92-100.

[Wang (1990)] Wang Q.B., van der Vorm P.L.J. and Blaauwendraad J., "Failure of Reinforced Concrete Panels - How Accurate the Models Must Be", in Bićanić N. and Mang H.A. (eds.), *Proceedings of the Second International Conference on Computer Aided Analysis and Design of Concrete Structures*, Pineridge Press, Swansea, 1990, 153-163.

[Welscher (1993)] Welscher S., "Implementierung und Anwendung eines elasto-plastischen Werkstoffmodells für Beton", *Master's Thesis*, Institute for Strength of Materials, Technical University of Vienna, Vienna, 1993.

[Willam (1975)] Willam K.J. and Warnke E.P., "Constitutive Model for the Triaxial Behavior of Concrete", *Proceedings of the International Association for Bridge and Structural Engineering*, **19**, 1975.

[Winkler (1984)] Winkler H., "Fundamental Investigations on the Influence of Test Equipment on Multiaxial test Results of Concrete", *Proceedings of the International Conference on Concrete under Multiaxial Conditions*, RILEM - CEB - CNRS, Presses de'l Université Paul Sabatier, Toulouse, 1984, 9-19.

[Wittmann (1977)] Wittmann F.H., "Grundlagen eines Modells zur Beschreibung charakteristischer Eigenschaften des Betons", *Deutscher Ausschuß für Stahlbeton*, Heft 290, W. Ernst u. Sohn, Berlin, 1977.

[Wittmann (1985)] Wittmann F.H., "Deformation of Concrete at Variable Moisture Content", in Bažant Z.P. (ed.), *Mechanics of Geomaterials*, John Wiley & Sons, 1985, 425-459.

[Yamaguchi (1991)] Yamaguchi E. and Chen W.F., "Microcrack Propagation Study of Concrete under Compression", *Journal of Engineering Mechanics*, **117**, 1991, 653-673.

[Yang (1990)] Yang H.T.Y., Saigal S. and Liaw D.G., "Advances of Thin Shell Finite Elements and Some Applications - Version I", *Computers & Structures*, **35**, 1990, 481-504.

[Yin (1989)] Yin W.S., Su E.C.M., Mansur M.A. and Hsu T.T.C., "Biaxial Tests of Plain and Fiber Concrete", *ACI Materials Journal*, **86**, 1989, 236-243.

[Yoshikawa (1989)] Yoshikawa H., Wu Z. and Tanabe T., "Analytical Model for Shear Slip of Cracked Concrete", *Journal of Structural Engineering*, **115**, 1989, 771-788.

[Yuan (1989)] Yuan K.Y. and Liang C.C., "Nonlinear Analysis of an Axisymmetric Shell Using Three Noded Degenerated Isoparametric Shell Elements", *Computers & Structures*, **32**, 1989, 1225-1239.

[Zhen-hai (1987)] Zhen-hai, G. and Xiu-qin Z., "Investigation of Complete Stress-Deformation Curves for Concrete in Tension", *ACI Materials Journal*, **84**, 1987, 278-285.

[Ziegeldorf (1983)] Ziegeldorf S., "Phenomenological Aspects of the Fracture of Concrete", in Wittmann F.H. (ed.), *Fracture Mechanics of Concrete*, Elsevier, Amsterdam, 1983, 31-41.

[Zienkiewicz (1989,1991)] Zienkiewicz O.C. and Taylor R.L., *The Finite Element Method*, 4th ed., Vol. 1: Basic Formulations and Linear Problems", McGraw-Hill, London, 1989; Vol. 2: Solid and Fluid Mechanics, Dynamics and Non-linearity", McGraw-Hill, London, 1991.

Index

accuracy
    accuracy of the integration, 160
    first-order accuracy, 158, 202
    second-order accuracy, 158, 202
aggregates, 2
aggregate interlock, 38, 134, 173, 178, 261, 263, 270, 293
    models for consideration of aggregate interlock, 134
    interface elements for consideration of aggregate interlock, 259
aging of concrete, 19, 128
algorithms
    closest point projection algorithm, 160
    for consideration of relaxation of the prestressing steel, 208
    for the computation of creep strains, 202, 204
    stability of algorithms, 160
    return mapping algorithms, 160, 166, 196
anchorage slip, 253
angle of similarity, 16, 52, 73
arc-length methods, 154
associated
    associated flow rule, 90, 167
    associated hardening law, 93, 162
    associated plasticity, 161
assumed strain methods, 223
axisymmetric
    axisymmetric analysis, 287
    axisymmetric finite elements, 234, 280
back stress, 87
beam
    deep beam, 235
    slender beam, 235
biaxial
    biaxial behavior of concrete, 10
    biaxial compression, 10
    biaxial cyclic loading, 18
    biaxial elastic-plastic constitutive model, 96

biaxial strength, 12
biaxial tension, 12, 43, 45
equivalent-uniaxial models for biaxial stress states, 55
invariant-based biaxial constitutive model, 71
ultimate strength curve for biaxial stress states, 50, 58, 170
bond, 32, 136
    bond slip, 36, 136, 263
    bond strength, 36
    frictional bond stress, 137
    models for consideration of bond, 136
    interface elements for consideration of bond, 259
bounding curve, 97
bounding surface, 87
C_0-interelement continuity, 218, 235, 237
C_1-interelement continuity, 228, 232, 250
characteristic length, 184
coaxiality of principal axes, 193
coefficient of sliding friction, 253
compliance function, 118, 122
compression
    biaxial compression, 10
    confined compression test, 17
    tension-compression, 13
    triaxial compression, 14
    uniaxial compression of concrete, 3
compressive
    compressive behavior of concrete, 3
    compressive meridian, 15, 53
    compressive strength, 30
    confined compressive loading, 102
    uniaxial compressive strains, 3
concrete
    high-strength concrete, 3, 23
    low-strength concrete, 3, 23
    moist-cured concrete, 115, 121
    normal-strength concrete, 3, 23
    prestressed concrete, 243
    reinforced concrete, 32

steam-cured concrete, 115, 121
configuration
   current configuration, 140, 217
   initial configuration, 140, 217
consistency condition, 83
consistency parameter, 82, 92, 165, 169
constitutive models
   see material models
convergence
   convergence criterion, 143
   order of convergence, 145
   quadratic rate of convergence, 145, 165
   rate of convergence, 145, 177
convexity of the yield function, 91, 161
coordinates
   convective coordinates, 227
   curvilinear coordinates, 221, 224
   surface coordinates, 247, 301
cracking of concrete, 1, 45, 53, 171
   crack opening, 180, 185, 190
   crack band, 181
   crack closure, 179, 186, 190
   crack dilatancy, 38
   crack distribution, 276, 282, 288, 290, 298, 306, 320
   crack-induced orthotropy, 193
   crack models, 171, 172, 174, 175, 176, 181, 191, 194, 258, 270, 278, 284, 293
   crack plane, 177
   crack propagation, 179
   crack stiffness matrix, 135
   crack strain, 182
   crack width, 173, 182
   cracked integration point, 172, 187
creep of concrete, 23, 117, 251, 254
   algorithms for the computation of creep strains, 202
   approximation of the creep compliance by a Dirichlet series, 125
   basic creep, 123
   biaxial creep tests, 25
   correction factors for creep laws, 121
   creep coefficient, 122, 207
   creep compliance, 119, 201
   creep recovery, 24
   creep strain, 24, 113, 200, 202
   delayed elastic creep strain, 122
   discussion of creep laws, 130
   double-power law for creep of concrete, 123
   drying creep, 123
   extension of creep laws to multiaxial stress states, 127
   initial creep strain, 122
   integral-type creep laws, 112, 117, 121, 122, 123
   irreversible creep strain, 122
   logarithmic double-power law for creep of concrete, 125
   long-time creep, 23
   material models for creep of concrete, 117
   rate-type creep laws, 112, 126
   short-time creep, 23
   simultaneous creep and shrinkage, 25
   triaxial creep tests, 25
   triple-power law for creep of concrete, 125
   uniaxial creep tests, 23
criterion
   convergence criterion, 143
   Drucker-Prager criterion, 51
   failure criterion, 49, 98, 107
   loading/unloading criterion, 65, 74, 83, 94, 105
   maximum tensile strength criterion, 51, 181, 194
   Mohr-Coulomb criterion, 51
   Rankine criterion, 174, 194
   strength-based fracture criterion, 179
   Tresca yield criterion, 51
   ultimate strength criterion, 51
   von Mises yield criterion, 51, 96
crushing of concrete, 53, 170
curvature
   changes of curvature, 227
   curvature of a tendon, 246
   curvature tensor, 225

Index

curvatures, 226
curve parameter, 246
cyclic loading of concrete, 9, 18, 30
deep-beam, 235
deformation-controlled experiments, 49
deformation gradient, 142
deviatoric
    deviatoric loading, 67, 77, 108
    deviatoric planes, 15
    deviatoric stress, 52
    deviatoric strain, 72
dilatancy factor, 103
directional derivative, 140
director vector, 216, 218
Dirichlet series, 125
discretization, 146
displacement
    displacement continuity, 174
    displacement-controlled solution scheme, 153
    displacement discontinuity, 173
    displacement vector, 140, 211
    incremental displacements, 143, 149
    normal displacement, 229
    relative nodal displacements, 241
    small displacements, 139, 228
    tangential crack displacement, 185
    tangential displacement components, 229
    total incremental displacements, 149
    total incremental nodal displacements, 149, 156
    total nodal displacements, 149, 156, 211
    virtual displacement, 139
    virtual nodal displacements, 147
double nodes, 173
double-power law for creep of concrete, 123
dowel action, 2, 40, 173, 261
Drucker's inequality, 89
Drucker's stability postulate, 89, 110
Drucker-Prager criterion, 51
effective modulus methods, 206

effective thickness, 115
elastic
    elastic compliance, 119
    elastic loading/unloading, 83
    elastic material moduli, 80
    elastic predictor, 161
    elastic region, 83
    elastic tangent material moduli, 156, 158, 165
    elastic trial stress, 161
    linear-elastic fracture models, 54, 270, 284
elasticity
    asymptotic modulus of elasticity, 123
    Cauchy elasticity, 48
    dynamic modulus of elasticity, 124
    modulus of elasticity, 28, 56
    theory of elasticity, 174
elastic-plastic
    consistent elastic-plastic material tangent moduli, 166, 167
    elastic-plastic constitutive models, 96, 100, 160, 270, 278
    elastic-plastic material tangent moduli, 93, 166, 167
embedded
    embedded elements, 241, 280
    embedded representation of the reinforcement 210, 241, 259
    embedded representation of the tendons, 245, 294
endochronic theory of plasticity, 49, 111
energy norm, 152
equilibrium conditions, 140, 147, 253
    incremental form of the equations of equilibrium, 151, 254
equivalent
    equivalent length, 187
    equivalent plastic strain, 85, 132
    equivalent-uniaxial models 55, 62
    fictitious equivalent-uniaxial strains, 62
Euler integration, 158, 162, 202
Eulerian formulation, 140

experiments
- biaxial experiments, 10, 18, 25
- confined compression test, 17
- deformation-controlled experiments, 49
- direct shear test, 39
- experiments on reinforced concrete specimens, 41
- load-controlled experiments, 49
- triaxial experiments, 14, 26, 66
- uniaxial experiments, 3, 5, 9, 19

finite elements, 209
- axisymmetric elements, 234, 280
- beam elements, 235, 251, 285
- brick elements, 213
- cable elements, 239
- degenerate solid elements, 214, 231, 235, 249
- discrete Kirchhoff elements, 223
- embedded elements, 241, 247, 249, 251, 280
- hybrid elements, 173
- interface elements, 240, 258, 262
- isoparametric elements, 211
- Lagrangian elements, 217
- linkage elements, 173
- plane strain elements, 234
- plane stress elements 232
- plate elements, 231
- quadrilateral elements, 146
- rebar elements, 279
- serendipity elements, 211, 216, 233, 234
- shell elements, 214, 215, 223, 228, 230, 247, 285, 294
- three-dimensional elements, 211, 270
- truss elements, 239, 270

flow potential, 81

flow rule, 80, 85, 97, 102
- associated flow rule, 90, 167
- nonassociated flow rule, 90
- Koiter's generalized flow rule, 90, 169

form invariance, 70, 193

formulation
- Eulerian formulation, 140
- Langrangian formulation, 140
- total Lagrangian formulation, 145, 227
- updated Lagrangian formulation, 145, 284

fracture
- fracture mechanics, 173, 179
- fracture strain, 182
- linear-elastic fracture models, 54, 270, 284
- nonlinear-elastic fracture models, 55, 293
- specific fracture energy, 173, 179, 181, 183, 195
- strength-based fracture criterion, 179

frictional bond stress, 137

frictional force, 253

fundamental form
- first fundamental form of a curved surface, 225
- second fundamental form of a curved surface, 226

Gaussian quadrature
- see numerical integration

geometric linearity, 145, 147, 149, 228

geometric nonlinearity, 139, 228, 307

hardening, 81, 83, 89
- associated hardening law, 93, 162
- combined hardening, 87
- hardening law, 97, 103, 162, 166
- hardening moduli, 85, 87
- hardening parameter, 85
- hardening tangent moduli, 92
- isotropic hardening, 85, 131
- kinematic hardening, 87, 131
- non-associated hardening law, 168

humidity, 21, 113

hydration, 2, 19, 23

hydrostatic
- hydrostatic axis, 53
- hydrostatic loading, 14, 67, 76
- hydrostatic stress, 52

hyperelasticity, 47

hypoelasticity, 47

Index

hypoelastic constitutive models, 62, 72, 158, 270
incremental
   incremental displacements, 143, 149
   incremental form of the equations of equilibrium, 151
   incremental-iterative procedure, 142
   incremental strains, 149
   incremental stresses, 144, 149
   total incremental displacements, 149
   total incremental nodal displacements, 149, 156
induced anisotropic material behavior, 87
initial
   initial configuration, 140, 217
   initial creep strain, 122
   initial displacement matrix, 148
   initial strains, 201
   initial stress matrix, 148
   initial tangent modulus, 3, 55
   initial yield curve, 96
   initial yield surface, 100
integration
   see numerical integration
interface
   concrete-steel interface, 241, 263
   explicit representation of the interface behavior, 259
   implicit representation of the interface behavior, 263
   interface behavior, 2, 32, 134, 258
   interface elements, 240, 258, 262
   models for consideration of the interface behavior, 134, 258
internal damage parameter, 194
internal variable, 48, 80, 151, 165
invariant-based material models, 71, 72, 79, 270
invariants
   invariants of the deviatoric stress tensor, 52
   invariants of the strain tensor, 66
   invariants of the stress tensor, 52
isotropic
   isotropic hardening, 85, 131
   isotropic linear-elastic material, 55
   isotropic nonlinear-elastic material, 56
   isotropic softening, 197
iteration
   iteration step, 163
   iterative solution, 145, 165
   modified Newton iteration, 152
   Newton iteration, 142, 143, 148, 162
   quasi-Newton iteration, 152
Jacobian determinant, 142, 213
Jacobian matrix, 142, 213
Kelvin unit, 126
kinematic
   kinematic hardening, 87, 131
   kinematic hardening modulus, 87
   kinematic softening, 197
kinematic hypothesis
   Bernoulli hypothesis, 235
   Kirchhoff hypothesis, 222, 227, 231
   Reissner-Mindlin hypothesis, 216, 231
Koiter's generalized flow rule, 90, 169
Koiter's nonlinear shell theory, 293
Kuhn-Tucker-conditions, 83
Lagrangian
   Lagrangian elements, 217
   Lagrangian strain tensor, 141
   Lagrangian formulation, 140
layer concept, 230, 232, 242
limit point, 153
linear approximation, 143, 144
linearization, 142, 144, 148, 164
linkage elements, 173
load
   load-balancing concept, 243
   load-controlled experiments, 49
   load-controlled solution scheme, 153
   load history, 200
   load increment, 142
   load parameter, 152
   load step, 167
   load vectors, 149
   ultimate load analysis, 267, 268, 276, 283, 291, 299, 307

loading
    loading/unloading conditions, 92, 95, 161
    loading/unloading criterion, 65, 74, 83, 94, 105
    neutral loading, 83
    nonproportional loading, 68, 78, 109, 111
    plastic loading, 82, 83, 92, 106, 151, 162
    shear loading, 45
    deviatoric loading, 67, 77, 108
local base vectors, 218, 224, 235
long-term behavior of concrete, 113
material compliance matrix, 188
material
    isotropic material, 55
    orthotropic material, 59, 62, 70, 174
    linear-elastic material, 54, 55
    elastic-plastic material, 93, 96, 100, 160, 166, 167, 270, 278
    nonlinear-elastic material, 55, 56, 131, 157, 176
    perfectly-plastic material, 80, 82, 83
material failure, 170
    failure criteria, 49, 98, 107
    failure modes, 14
    mixed-type of failure, 53
    modelling of post-failure material behavior, 170
material models
    crack-band model, 181
    discrete crack models, 172, 258
    elastic-plastic constitutive models, 96, 100, 160, 270, 278
    equivalent-uniaxial models 55, 62
    smeared crack models, 171, 174, 176, 181, 191, 194, 270, 278, 284, 293
    hypoelastic models, 62, 158, 270
    hypoplastic model, 278
    implementation of material models, 197
    invariant-based models, 71, 72, 79, 270
    linear-elastic fracture models, 54, 270, 284
    material models for prestressing steel, 131
    material models for reinforcing steel, 131
    material models for the time-dependent behavior of concrete, 112
    material models based on the theory of plasticity, 79
    material models for consideration of the interface behavior, 134, 136, 258
    nonlinear-elastic constitutive models, 55, 157, 176
    tension stiffening models, 263, 265, 293
    user-defined material model, 197
material moduli
    asymptotic modulus of elasticity, 123
    consistent elastic-plastic material tangent moduli, 166, 167
    dynamic modulus of elasticity, 124
    elastic material moduli, 80
    elastic tangent material moduli, 156, 158, 165
    elastic-plastic material tangent moduli, 93, 166, 167
    hardening moduli, 85, 87
    hardening tangent moduli, 92
    initial tangent modulus, 3, 55
    kinematic hardening modulus, 87
    material tangent moduli, 144
    modified shear modulus, 134, 177, 189, 263
    modulus of elasticity, 28, 56
    plastic modulus, 81, 104
    secant moduli, 47, 176
    shear slip modulus, 185
    softening modulus, 180, 184
    tangent bulk modulus, 73
    tangent shear modulus, 74, 178
    tension moduli, 47
material nonlinearity, 46, 139
matrix
    crack stiffness matrix, 135

Index   361

    element stiffness matrix, 148, 213
    element tangent stiffness matrix, 149
    global tangent stiffness matrix, 149
    initial displacement matrix, 148
    initial stress matrix, 148
    Jacobian matrix, 142, 213
    large displacement matrix, 148
    material compliance matrix, 188
    secant stiffness matrix, 176
    tangent compliance matrix, 62
    tangent stiffness matrix, 62, 153
Maxwell unit, 126
mean stress, 30
membrane locking, 222
membrane strains, 227
mesh-dependent results, 179, 181
metric tensor, 225
nodal forces
    internal nodal forces, 147, 149
    external nodal forces, 147, 149
nonlinear
    isotropic nonlinear-elastic material, 56
    nonlinear-elastic constitutive models, 131, 155, 157, 176
    nonlinear-elastic fracture models, 55, 293
    nonlinear finite element analysis, 151
    nonlinear orthotropic material, 59
    nonlinear uniaxial constitutive model, 55
normal vector, 224
    pseudo-normal vector, 216
notation, 321
numerical integration, 157, 213
    accuracy of the integration, 160
    backward Euler integration, 158, 162
    explicit Euler integration, 158, 162, 202
    forward Euler integration, 158
    generalized midpoint rule, 158, 203
    generalized trapezoidal rule, 158
    implicit Euler integration, 158, 162, 202
    integration point

        see sampling point
    reduced integration, 222, 232
    Runge-Kutta method, 159
    sampling point, 157, 172, 174
    selective integration, 222, 232
    sub-incrementation method, 159
numerical stability, 160, 162
objectivity, 172
octahedral
    octahedral engineering shear strain, 66
    octahedral normal strain, 66
    octahedral normal stress, 66
    octahedral shear stress, 66
plastic
    elastic-plastic constitutive models, 79, 96, 100, 160, 270, 278
    consistent elastic-plastic material tangent moduli, 166, 167
    elastic-plastic material tangent moduli, 93, 166, 167
    equivalent plastic strain, 85, 132
    perfectly-plastic material, 80, 82, 83
    plastic corrector, 161
    plastic dissipation, 92
    plastic fracturing theory, 49, 110
    plastic loading, 82, 83, 92, 106, 151, 162
    plastic modulus, 81, 104
    plastic strains, 151, 165
    plastic work 85, 100, 195
plasticity, 49, 79, 156, 160, 174
    associated plasticity, 161
    computational plasticity, 165
    continuum-based plasticity theory, 79, 165
    endochronic theory of plasticity, 49, 111
    fundamental assumptions of the theory of plasticity, 79
    non-smooth multisurface plasticity, 169
    plasticity-based crack models, 174, 194
    rate-independent plasticity, 79, 82, 160
    strain-space plasticity, 48, 93

plates
    moderately thick plates, 231
    thin plates, 231
Poisson effect, 56
post-failure material behavior, 170
post-peak material behavior, 170
post-tensioned member, 251
postulate of maximum plastic dissipation, 92, 110, 168
prestressed
    prestressed concrete, 243
    prestressed concrete containment, 307
    prestressed concrete panel, 246, 290
    prestressed concrete shell, 246, 290, 299, 307
    prestressed concrete slab, 246, 290, 291
prestressing steel, 29
    algorithm for consideration of relaxation of the prestressing steel, 208
    material models for prestressing steel, 131
    relaxation of prestressing steel, 31
    time-dependent material model for prestressing steel, 133
    time-independent material models for prestressing steel, 131
prestress loss, 252
prestressing forces, 251
pre-tensioned member, 251
principal axes of orthotropy, 59, 60, 70
principal trial stresses, 196
principle of superposition, 118
principle of virtual displacements, 139, 142, 148, 243, 261
proportional limit, 29
rate
    constitutive rate equations, 93, 98, 105, 156, 158, 160, 312
    quadratic rate of convergence, 145, 165
    rate of convergence, 145, 177
    rate-dependent material, 79
    rate-independent plasticity, 79, 82, 160
    rate-type creep laws, 112, 126
    strain rate, 93

stress rate, 93
reinforced concrete, 32
    reinforced concrete cantilever, 268
    reinforced concrete cooling tower, 283
    reinforced concrete cylinder, 276
    reinforced concrete panel, 282
    reinforced concrete shell, 282
    reinforced concrete slab, 282
    reinforced concrete specimen, 41
reinforcement
    discrete representation of the reinforcement, 209, 238
    distributed representation of the reinforcement, 210, 231, 242, 259
    embedded representation of the reinforcement 210, 241, 259
reinforcing steel, 27, 131, 238
    as-rolled steel, 27
    cold-drawn reinforcing steel, 28
    effect of low and high temperatures on reinforcing steel, 31
    material models for reinforcing steel, 131
    representation of the reinforcing steel, 238
    time-independent material models for reinforcing steel, 131
relaxation
    relaxation of the prestressing steel, 31, 133, 208, 251, 254
    relaxation coefficient, 206
    relaxation function, 120, 206
    relaxation times, 125
Rendulic plane, 14, 66, 75
residual
    residual stiffness, 170, 173
    residual strength, 170
    residual vector, 150, 152
    norm of the residual vector, 152
return mapping algorithm, 160, 166, 196
rotation, 216
    finite rotations, 219
    moderately large rotations, 139, 219, 228

Index   363

  nodal rotations, 220
  small rotations, 218, 228, 231, 237
secant
  secant moduli, 47, 176
  secant stiffness matrix, 176
shape functions, 147, 211, 216, 236
shear
  shear correction factor, 221, 237
  shear forces, 2
  shear loading, 45
  shear locking, 221, 232
  shear retention factor, 134, 273
  shear slip modulus, 185
shells
  imperfection sensitive shell, 230
  prestressed concrete shell, 246, 290, 299, 307
  reinforced concrete shell, 282
  thin shells, 215, 222, 227, 230
  moderately-thick shells, 215, 222, 230
shell theory, 223
shrinkage of concrete, 21, 251, 254
  correction factors for shrinkage laws, 115
  discussion of shrinkage laws, 130
  drying shrinkage, 21
  extension of shrinkage laws to multiaxial stress states, 127
  load-induced shrinkage, 25
  material models for shrinkage of concrete, 114
  restrained shrinkage, 21
  shrinkage laws, 114, 115, 116
  shrinkage strain, 113, 201
  simultaneous creep and shrinkage, 25
  stress-induced shrinkage strains, 114
  temperature-induced shrinkage strains, 114
  thermal shrinkage, 21
  unrestrained shrinkage of concrete, 21
Smith-diagram, 31
softening, 81, 83, 184
  exponential softening law, 195
  isotropic softening law, 197

  kinematic softening law, 197
  linear softening law, 195
  softening modulus, 180, 184
  tension softening, 4, 270, 275, 278, 285
stability
  Drucker's stability postulate, 89, 110
  numerical stability, 160, 162
  stability postulate in the strain space, 94
stiffness
  crack stiffness matrix, 135
  element stiffness matrix, 148, 213
  element tangent stiffness matrix, 149
  global tangent stiffness matrix, 149
  residual stiffness, 170, 173
  secant stiffness matrix, 176
  tangent stiffness matrix, 62, 153
strain
  additive decomposition of the strains, 156, 199
  average tensile strain, 264
  Cauchy strain tensor, 140
  crack strain, 182
  creep strain, 24, 113, 200, 202
  delayed elastic creep strain, 122
  equivalent plastic strain, 85, 132
  fictitious equivalent-uniaxial strains, 62
  fracture strain, 182
  Green's strain tensor, 141
  incremental strains, 149
  inelastic strain, 80
  initial creep strain, 122
  initial strains, 201
  instantaneous strain, 113, 129
  irreversible creep strain, 122
  Lagrangian strain tensor, 141
  linearized strain tensor, 140
  membrane strains, 227
  octahedral engineering shear strain, 66
  octahedral normal strain, 66
  parasitic transverse shear strains, 222
  plastic strains, 151, 165

shrinkage strain, 113, 201
strain rate, 93
stress producing strain, 120
stress-dependent strain, 118, 201
stress-independent strain, 201
stress-induced shrinkage strains, 114
stress-induced strain, 114, 129
temperature-induced shrinkage strains, 114
temperature-induced strain, 113, 200
total strains, 149, 156
transverse shear strains, 221, 232, 235
ultimate strain, 3, 28, 50
uniaxial compressive strains, 3
virtual Cauchy strains, 141
virtual Lagrangian strain tensor, 141, 147
viscoplastic strains, 80
volumetric strain, 3
viscoelasticity, 47
strength
   biaxial strength of concrete, 12
   bond strength, 36
   compressive strength of concrete, 3
   fatigue strength of concrete, 10, 18
   residual strength of concrete, 170
   tensile strength of concrete, 12, 29, 172
stress
   average tensile stress, 264
   back stress, 87
   Cauchy stress tensor, 140
   deviatoric stress, 52
   elastic trial stress, 161
   frictional bond stress, 137
   hydrostatic stress, 52
   incremental stresses, 144, 149
   mean stress, 30
   octahedral normal stress, 66
   octahedral shear stress, 66
   peak stress, 50
   Piola-Kirchhoff stress tensor, 141
   principal trial stresses, 196
   stress rate, 93
   stress update, 156
   total stresses, 149, 156
   trial stress, 84, 169
   yield stress, 27
surface
   bounding surface, 87
   ultimate strength surface, 51, 73, 75, 100, 170
   loading surface, 75, 87, 100, 101
   non-smooth yield surface, 92, 169
   surface coordinates, 247, 301
   surface curve, 246
   surface tractions, 140, 142
   yield surface, 80, 100, 161
tangent
   consistent elastic-plastic material tangent moduli, 166, 167
   elastic tangent material moduli, 156, 158, 165
   elastic-plastic material tangent moduli, 93, 166, 167
   element tangent stiffness matrix, 149
   global tangent stiffness matrix, 149
   hardening tangent moduli, 92
   initial tangent modulus, 3, 55
   material tangent moduli, 144
   tangent bulk modulus, 73
   tangent compliance matrix, 62
   tangent shear modulus, 74, 178
   tangent stiffness matrix, 62, 153
tangential crack displacement, 185
tangential displacement components, 229
temperature
   effect of low and high temperatures on reinforcing steel, 31
   temperature-induced shrinkage strains, 114
   temperature-induced strain, 113, 200
tendon, 243
   tendon geometry, 246
   bonded tendons, 253
   curvature of the tendon, 246
   geometric modelling of tendons, 245, 247, 249, 251

Index 363

nodal rotations, 220
small rotations, 218, 228, 231, 237
secant
   secant moduli, 47, 176
   secant stiffness matrix, 176
shape functions, 147, 211, 216, 236
shear
   shear correction factor, 221, 237
   shear forces, 2
   shear loading, 45
   shear locking, 221, 232
   shear retention factor, 134, 273
   shear slip modulus, 185
shells
   imperfection sensitive shell, 230
   prestressed concrete shell, 246, 290, 299, 307
   reinforced concrete shell, 282
   thin shells, 215, 222, 227, 230
   moderately-thick shells, 215, 222, 230
shell theory, 223
shrinkage of concrete, 21, 251, 254
   correction factors for shrinkage laws, 115
   discussion of shrinkage laws, 130
   drying shrinkage, 21
   extension of shrinkage laws to multiaxial stress states, 127
   load-induced shrinkage, 25
   material models for shrinkage of concrete, 114
   restrained shrinkage, 21
   shrinkage laws, 114, 115, 116
   shrinkage strain, 113, 201
   simultaneous creep and shrinkage, 25
   stress-induced shrinkage strains, 114
   temperature-induced shrinkage strains, 114
   thermal shrinkage, 21
   unrestrained shrinkage of concrete, 21
Smith-diagram, 31
softening, 81, 83, 184
   exponential softening law, 195
   isotropic softening law, 197

kinematic softening law, 197
linear softening law, 195
softening modulus, 180, 184
tension softening, 4, 270, 275, 278, 285
stability
   Drucker's stability postulate, 89, 110
   numerical stability, 160, 162
   stability postulate in the strain space, 94
stiffness
   crack stiffness matrix, 135
   element stiffness matrix, 148, 213
   element tangent stiffness matrix, 149
   global tangent stiffness matrix, 149
   residual stiffness, 170, 173
   secant stiffness matrix, 176
   tangent stiffness matrix, 62, 153
strain
   additive decomposition of the strains, 156, 199
   average tensile strain, 264
   Cauchy strain tensor, 140
   crack strain, 182
   creep strain, 24, 113, 200, 202
   delayed elastic creep strain, 122
   equivalent plastic strain, 85, 132
   fictitious equivalent-uniaxial strains, 62
   fracture strain, 182
   Green's strain tensor, 141
   incremental strains, 149
   inelastic strain, 80
   initial creep strain, 122
   initial strains, 201
   instantaneous strain, 113, 129
   irreversible creep strain, 122
   Lagrangian strain tensor, 141
   linearized strain tensor, 140
   membrane strains, 227
   octahedral engineering shear strain, 66
   octahedral normal strain, 66
   parasitic transverse shear strains, 222
   plastic strains, 151, 165

shrinkage strain, 113, 201
strain rate, 93
stress producing strain, 120
stress-dependent strain, 118, 201
stress-independent strain, 201
stress-induced shrinkage strains, 114
stress-induced strain, 114, 129
temperature-induced shrinkage strains, 114
temperature-induced strain, 113, 200
total strains, 149, 156
transverse shear strains, 221, 232, 235
ultimate strain, 3, 28, 50
uniaxial compressive strains, 3
virtual Cauchy strains, 141
virtual Lagrangian strain tensor, 141, 147
viscoplastic strains, 80
volumetric strain, 3
viscoelasticity, 47
strength
  biaxial strength of concrete, 12
  bond strength, 36
  compressive strength of concrete, 3
  fatigue strength of concrete, 10, 18
  residual strength of concrete, 170
  tensile strength of concrete, 12, 29, 172
stress
  average tensile stress, 264
  back stress, 87
  Cauchy stress tensor, 140
  deviatoric stress, 52
  elastic trial stress, 161
  frictional bond stress, 137
  hydrostatic stress, 52
  incremental stresses, 144, 149
  mean stress, 30
  octahedral normal stress, 66
  octahedral shear stress, 66
  peak stress, 50
  Piola-Kirchhoff stress tensor, 141
  principal trial stresses, 196
  stress rate, 93
  stress update, 156
  total stresses, 149, 156
  trial stress, 84, 169
  yield stress, 27
surface
  bounding surface, 87
  ultimate strength surface, 51, 73, 75, 100, 170
  loading surface, 75, 87, 100, 101
  non-smooth yield surface, 92, 169
  surface coordinates, 247, 301
  surface curve, 246
  surface tractions, 140, 142
  yield surface, 80, 100, 161
tangent
  consistent elastic-plastic material tangent moduli, 166, 167
  elastic tangent material moduli, 156, 158, 165
  elastic-plastic material tangent moduli, 93, 166, 167
  element tangent stiffness matrix, 149
  global tangent stiffness matrix, 149
  hardening tangent moduli, 92
  initial tangent modulus, 3, 55
  material tangent moduli, 144
  tangent bulk modulus, 73
  tangent compliance matrix, 62
  tangent shear modulus, 74, 178
  tangent stiffness matrix, 62, 153
tangential crack displacement, 185
tangential displacement components, 229
temperature
  effect of low and high temperatures on reinforcing steel, 31
  temperature-induced shrinkage strains, 114
  temperature-induced strain, 113, 200
tendon, 243
  tendon geometry, 246
  bonded tendons, 253
  curvature of the tendon, 246
  geometric modelling of tendons, 245, 247, 249, 251

tendon layout, 245
  unbonded tendons, 256
  wobble of the tendon, 253
tendon forces, 251, 252
  changes of the tendon forces, 254
  update of the tendon forces, 255
tension
  average tensile strain, 264
  average tensile stress, 264
  biaxial tension, 12, 43, 45
  tension moduli, 47
  tension softening, 270, 275, 278, 285
  tension-compression, 13
  uniaxial tensile strength of concrete, 12, 29, 172
  maximum tensile stress criterion, 51, 181, 194
  tensile behavior of concrete, 5
  tensile meridian, 15, 53
tension stiffening, 34, 263
  concrete-related tension stiffening models, 263, 265
  steel-related tension stiffening models, 263, 265
tensor
  Cauchy strain tensor, 140
  Cauchy stress tensor, 140
  curvature tensor, 225
  fourth-order tensor, 61
  Green strain tensor, 141
  Lagrangian strain tensor, 141
  linearized strain tensor, 140
  metric tensor, 225
  Piola-Kirchhoff stress tensor, 141
  second-order tensor, 61
  virtual Lagrangian strain tensor, 141, 147
testing device, 10
three-dimensional
  three-dimensional hypoelastic constitutive model, 72
  three-dimensional finite elements, 211, 270
  three-dimensional reinforced concrete structures, 268
time-dependent
  time-dependent analysis, 314
  time-dependent behavior of concrete, 2, 19, 112, 199
  time-dependent behavior of prestressing steel, 31
  time-dependent material model for prestressing steel, 133
  time-dependent prestress loss, 251
time-independent
  time-independent material models for concrete, 54
  time-independent material models for prestressing steel, 131
  time-independent material models for reinforcing steel, 131
time step, 167
Timoshenko beam theory, 235
tolerance, 151, 160, 165
total
  total Lagrangian formulation, 145, 227
  total incremental displacements, 149
  total incremental nodal displacements, 149, 156
  total nodal displacements, 149, 156, 211
  total strains, 149, 156
  total stress-strain relation, 58, 157
  total stresses, 149, 156
treshold angle, 192
triaxial
  triaxial tests, 14, 66
  equivalent-uniaxial models for triaxial stress states, 62
  triaxial behavior of concrete, 14
  triaxial elastic-plastic constitutive model, 100, 270, 278
  triaxial hypoplastic constitutive model, 278
  ultimate strength surface for triaxial stress states, 51, 75, 100, 170
twist, 226

ultimate
    ultimate strength surface, 51, 73, 75, 100, 170
    ultimate load analysis, 267, 268, 276, 283, 291, 299, 307
    ultimate load curve, 96
    ultimate strain, 3, 28, 50
    ultimate strength, 3
    ultimate strength criteria, 51
    ultimate strength curve, 50, 58, 170
uniaxial
    fictitious equivalent-uniaxial strains, 62
    nonlinear uniaxial constitutive model, 55
    uniaxial behavior of concrete, 3
    uniaxial compressive strength of concrete, 3
    uniaxial tensile strength of concrete, 12, 29, 172
variation, 139, 144
variational principle, 139
virtual
    virtual Cauchy strains, 141
    virtual Lagrangian strain tensor, 141, 147
    virtual displacement, 139
    virtual nodal displacements, 147
    virtual work, 144, 244
viscoelasticity, 47
volume
    volume compaction, 76
    volume expansion, 76
    volume forces, 140, 142
wobble, 253
yield
    yield function, 161
    yield stress, 27
    yield surface, 80, 161
zero-energy modes, 223